New Frontiers in Regional Science: Asian Perspectives

Volume 55

This series is a constellation of works by scholars in the field of regional science and in related disciplines specifically focusing on dynamism in Asia.

Asia is the most dynamic part of the world. Japan, Korea, Taiwan, and Singapore experienced rapid and miracle economic growth in the 1970s. Malaysia, Indonesia, and Thailand followed in the 1980s. China, India, and Vietnam are now rising countries in Asia and are even leading the world economy. Due to their rapid economic development and growth, Asian countries continue to face a variety of urgent issues including regional and institutional unbalanced growth, environmental problems, poverty amidst prosperity, an ageing society, the collapse of the bubble economy, and deflation, among others.

Asian countries are diversified as they have their own cultural, historical, and geographical as well as political conditions. Due to this fact, scholars specializing in regional science as an inter- and multi-discipline have taken leading roles in providing mitigating policy proposals based on robust interdisciplinary analysis of multifaceted regional issues and subjects in Asia. This series not only will present unique research results from Asia that are unfamiliar in other parts of the world because of language barriers, but also will publish advanced research results from those regions that have focused on regional and urban issues in Asia from different perspectives.

The series aims to expand the frontiers of regional science through diffusion of intrinsically developed and advanced modern regional science methodologies in Asia and other areas of the world. Readers will be inspired to realize that regional and urban issues in the world are so vast that their established methodologies still have space for development and refinement, and to understand the importance of the interdisciplinary and multidisciplinary approach that is inherent in regional science for analyzing and resolving urgent regional and urban issues in Asia.

Topics under consideration in this series include the theory of social cost and benefit analysis and criteria of public investments, socio-economic vulnerability against disasters, food security and policy, agro-food systems in China, industrial clustering in Asia, comprehensive management of water environment and resources in a river basin, the international trade bloc and food security, migration and labor market in Asia, land policy and local property tax, Information and Communication Technology planning, consumer "shop-around" movements, and regeneration of downtowns, among others.

Researchers who are interested in publishing their books in this Series should obtain a proposal form from Yoshiro Higano (Editor in Chief, higano@jsrsai.jp) and return the completed form to him.

More information about this series at https://link.springer.com/bookseries/13039

Subrata Saha • Mukunda Mishra • Anil Bhuimali
Editors

Economic and Societal Transformation in Pandemic-Trapped India

Emerging Challenges and Resilient Policy Prescriptions

Springer

Editors
Subrata Saha
Department of Economics
Raiganj University
Uttar Dinajpur, West Bengal, India

Mukunda Mishra
Department of Geography
Dr. Meghnad Saha College
Uttar Dinajpur, West Bengal, India

Anil Bhuimali
Raiganj University
Uttar Dinajpur, West Bengal, India

ISSN 2199-5974 ISSN 2199-5982 (electronic)
New Frontiers in Regional Science: Asian Perspectives
ISBN 978-981-16-5754-2 ISBN 978-981-16-5755-9 (eBook)
https://doi.org/10.1007/978-981-16-5755-9

This Springer imprint is published by the registered company Springer Nature Singapore Pte Ltd.
The registered company address is: 152 Beach Road, #21-01/04 Gateway East, Singapore 189721, Singapore

Foreword

It is a great privilege to welcome this most timely and pertinent volume, edited by a team of social scientists who seek to enhance understanding of the necessary economic and societal transformations in pandemic-trapped India, within the context of a globalized world. It is important to think about the current challenges and any potential remedies in these massified proportions of disaster presently witnessed. It is a sign of our times that in the wider context of the Anthropocene, international perspectives and state-based recommendations are assuming new prominence in desperate efforts to control the increasingly disastrous manifestations of climate change- and migration-induced global pandemics.

In all of this, India is not unique in this context, because virtually the whole world has been trapped by COVID-19 and its dramatic implications. While this pandemic continues to be devastating in terms of physical and mental health as well as socio-economic consequences for many people and entire nations, and partly precarious also in terms of political fallouts, it is certainly not the first time that humanity has been badly hit in this way. Natural disasters and epidemics have always been part of humanity's experience. We should not forget this when we label current disasters 'unprecedented' or try to suggest new remedies. Learning from past experiences—and failures—remains a very important challenge.

It will not be widely known that fear of infection, natural disasters like flash floods and resulting death and destruction predate modern state structures, let alone efforts at international regulation. As part of the global human experience of living as part of nature, such evidence is recorded since ancient times. Then, as manifestly still now, focus on local scenarios and determined individual action, first of all by affected persons and households, remain crucial, especially regarding epidemics and their spreading. In our times, something like disaster risk reduction, at all levels, is a pertinent concept that requires further consideration.

Modern social scientists are unfamiliar with the fact that even in the most ancient written sources of India's traditional literature, such as the Rigveda and Atharvaveda, there is evidence of abundant concerns about pollution control and management of infectious diseases. While I certainly do not suggest that prayer and

Preface

Humankind is witnessing an unprecedented global pandemic outbreak of COVID-19, which has been devastating in its consequence, breaking all statistics of illness and fatality hitherto known to humankind. The practitioners of Medical Science are working at the frontline of the fight against this pandemic. However, beyond this scenario of an acute health hazard, the pandemic has carried with it the other threats for humanity associated with the economy, society, culture, psychology and politics. The long period of lockdown all over the globe has resulted in environmental reinvigoration and diminishing of the pollution level. On the other hand, the lockdown has affected the world economy to initiate another global economic recession. The fallout of the latter is of grim significance to the entire population of the world. Amidst these multifaceted dimensions of the pandemic, this is high time for global solidarity to save humankind on this only planet in the universe where life exists.

Human society, its ambient environment, the process of socio-economic development, and politics and power—all drives to set the world order. All these parameters are intimately and integrally related. The interconnections of these driving forces have a significant bearing on life, space and time. In parallel, the interrelationship between all these drivers is dynamic, and they are changed drastically with time and space.

India, on its 74th Independence Day, unfortunately, counts the highest numbers of confirmed cases of COVID-19 infection per day and the largest numbers of total cases in Asia. The first COVID-19 case in India was reported on 30 January 2020 in Thrissur of Kerala, and two other cases were reported by 3 February 2020. The initial spreading of the diseases at the multiple clusters at different corners of the country is clearly evidenced by the first 50 COVID-19 cases, which were reported in 41 days and were spread in 12 states and 18 cities or districts; and, interestingly, 23 cases out of this first 50 confirmed cases had a travel history to Italy. However, the country started experiencing a steady growth in the count of COVID-19 infection since mid-April. The number of confirmed cases of COVID-19 infection takes only 15 days to reach from 10^2 to 10^3; another 16 days to 10^4; another 35 days to

10^5 and then take only 68 days to reach 10^6. On 15 August 2020, the total infection caseload in the world is reported by the WHO as '69,239' to make the total confirmed cases '3,044,940'.

With an increasing number of COVID-19 infections, the government has locked down transport services, closed all public and private offices and factories and restricted mobilization. A recent report of the Ministry of Human Resource Development, Government of India, states that there is a job loss of 40 million people in the country, and most of them have occurred in the unorganized sectors. Economists are predicting that the Indian economy may face a recession affecting the unorganized sector and semi-skilled jobholders losing their employment. Moreover, the possibility for a large volume of the population living just above the poverty line to get included below the poverty line (BPL) will cause undue pressure on the national economy due to the financial involvement in the economic reform for the newly added population below the poverty line. Besides, this is the time of eroding trust within and between countries. National leadership is under pressure from growing societal unrest and economic confrontations between major powers.

The labour sector is unorganized, and informal labour sector is probably the worst impacted as they are not provided jobs due to lockdown. As most of the labour sectors are associated with construction companies and daily wage earners, they are to face hardship. Travel restrictions and quarantines affecting hundreds of millions of people have left Indian factories short of labour, putting havoc on the production system. In India, the quarterly GDP growth has been consistently falling since Q4 of FY18. The World Bank had predicted a 3.2% contraction in India's economy during the current fiscal year. In its latest Global Economic Prospects report, the World Bank had assessed that the global economy would be expected to contract by 5.2% due to the COVID-19 pandemic. Experts asserted that it would be the deepest recession in the global economy since the Second World War. However, invalidating all the predictions, the quarterly GDP of India in its first quarter of the financial year 2020–2021 took a nosedive with a negative 23.9%.

The development parameters are evidently linked with one another. While the joblessness and income insecurity looms large, the education system in India witnessed the highest degree of the digital education divide as the pandemic has transformed the teaching–learning system into an online mode. The poor and the marginalized section remained utterly disadvantaged from the newer system. On the one hand, the failure of the public education system to reach online education to the poor and the higher cost involved for the online learning system on the online education system have made the 'new normal' of the education sector inaccessible to the population in the lower-income strata.

The volume attempts to touch upon a large spectrum of issues that may be considered imperative and critical in the sphere of the emerging challenges and resilient policy prescriptions on the economic and societal transformation during the pandemic-trapped period in India from the global perspective. There are 24 chapters that are categorized into six broad parts.

Part I is general in structure. It is devoted to sketching the trajectory of the global pandemic, which deals with several impacts of pandemic outbreaks on the global economy. Chapter 1 depicts the emerging issues of public policies and self-action while India has been in risky surfing through the consequent two deadly waves of pandemic waves. Chapter 2 intends to discuss the facts on sustainable green resilience and globalization or de-globalization in a Post-COVID World and concludes that various associated elements bring forth a different paradigm for the future decisions of the communities. Chapter 3 attempts to analyse the income insecurity, GDP and the future of human development during the COVID-19 period and demonstrates that the COVID-19 pandemic has many negative impacts on life expectancy in India. Chapter 4 describes the impacts of COVID-19 and the corresponding responses on a comparative basis across SAARC countries in the light of the global experiences and provides interesting and significant views. Finally, Chap. 5 contains a relevant study on COVID-19 pandemic and social–spatial inequality in the case of a selected metropolitan city in Brazil. With some imperative results, including a territorial inequality impact, the study documents some policy recommendations.

Part II of the book presents how the same health hazard brings differential shocks to different communities, gender or income groups, which is very relevant from the policy point of view. Chapter 6 attempts to explore the impact of the second wave of pandemic upon the women of the Darjeeling Himalayan region in the state of West Bengal in India. The study has resulted in numerous directions, including that the lockdown has severely impacted the lives of female students. Chapter 7 contains thorough discussions on how India's pre-existing digital education divide is exacerbated amid the pandemic, particularly excluding the marginalized communities. Chapter 8 analyses human resource management during the pandemic period, where the emerging challenges for the public sector employees in India are highlighted.

Part III is entitled 'New Challenges for the Indian Economy', which addresses various economic consequences of the COVID-19 pandemic in India. Chapter 9 empirically analyses the impact of the pandemic outbreak concerning the future of poverty and inequality scenario from the perspective of India. In regard to the issues on unemployment and economic development, the study provides significant outcomes. Chapter 10 concentrates on issues of healthcare expenditure and economic development dynamics in India during the COVID-19 pandemic through an extensive time series econometric analysis. The study views that the impact of public health expenditure on economic growth is relatively small, whereas the economic growth on domestic private healthcare expenditure is relatively strong in India. Chapter 11 aims to investigate the linkages between the agriculture sector, informal economy, and inequality over the pandemic period. It observes the deep-rooted linkages between the agriculture sector, inequality and employment in India. Chapter 12 discusses the impact of the COVID-19 pandemic on the informal labour market in India, and in this context, the study documents how the pandemic and economic slowdowns are linked and that influence informal and migrant workers.

Part IV is concerned with how the effects of pandemic outbreaks influence employment, labour, policies and governance. Chapter 13 raises the allied issues on emerging challenges and opportunities to governance over the new pandemic era, and as the conclusion, the study also presents the perception to the changing socio-economic-political dimension during that period. Chapter 14 demonstrates the employment dynamics and labour mobility amidst the COVID-19 pandemic in India; the author makes a critical appraisal of ILO recommendations. With this objective, the study analyses the dynamics of the vulnerability of domestic migrant workforces in respect of their mobility, gender and health security in India. The last chapter of this part (Chap. 15) brings forth the debates between the politics and policies in the pandemic stuck India, out of which the author does a search for the avenue of the largest democratic system to revive its citizens from the siege of the dreadful health hazard.

Part V is composed of five chapters that discuss sectoral impacts and responses to the COVID-19 crisis in India. Chapter 16 presents a unique study on the migratory beekeepers in West Bengal, and, through the field-based investigation, the authors present how the beekeeping livelihood comes at stake due to the lockdown and travel restrictions imposed nationwide. The study views that migration is part of the beekeeping profession, and it spreads awareness to all human beings for the eco-friendly and agriculture-support activities that the beekeeping industry renders. Chapter 17 tries to analyse the struggle of the apiculture sector during the COVID-19 pandemic in West Bengal, India, through the demand and supply mechanism, and based on that, a few certain feasible midterm planning and policies are prescribed to smoothly run the apiculture sector, gunning for higher production in the 'new normal phase'. Chapter 18 empirically investigates the effectiveness of public expenditure and crop production export in the Indian economy. The results of time series econometric analyses imply that the agricultural sector's emergence as the only area to experience positive growth can be attributed to an increase in agricultural production as a result of a favourable monsoon and targeted government expenditure in this pandemic. Chapter 19 is devoted to discussing the impact of COVID-19 on Indian Micro, Small and Medium Enterprises (MSMEs), and in this connection, the study mainly highlights the issues on employment, efficiency, identification problem in both manufacturing and service sectors. Chapter 20 focuses on the tourism sector in India under the seige of COVID-19, and using the backward and forward linkages methods, it mainly analyses the significant challenges and opportunities in the travel and tourism sectors.

Part VI of the book deals with the silver lining in the COVID-19 cloud of the downturn. Chapter 21 attempts to build up a model and analyse the invasion of the pandemic and the government in the case of the Indian economy by using a general equilibrium approach. The major emphasis of the study is on the health sector, the impact of subsidy and human development. Chapter 22 attempts to find the avenues of reviving the Indian economy through successfully managing the rural livelihood options during the pandemic. It is evidenced that agriculture is the only sector having resilience during the crisis, and growth of this sector does not provide income but, through its linkages, help to start other sectors of the economy. Chapter 23 endeavours to identify the key constraints and surviving strategies for the Micro, Small and Medium Enterprises (MSMEs) in India during the COVID-19 pandemic and lockdown. The study analyses the constraint factors concerning financing, market, technology, infrastructure and technology in this connection. The last chapter of this part (Chap. 24) raises the Classical and Keynesian economists' contradiction, and the author depicts it with the phrase in the title 'Where Classical ends, Keynes proceeds'. The author has placed arguments in this line to find out the reviving avenues for India (as well as the developing nations of the global south) from the siege of COVID-19. It aims to identify the changes in macroeconomic variables during COVID-19 to revisit Keynesian prescription to see how the economy breathes life by following Keynesian suggestions.

So, finally, it can be concluded that the essence of the chapters covered by the present book establishes the existence of significant societal as well as economic impacts of COVID-19 on the Indian economy. After that, many policy measures are also prescribed from different viewpoints, which will be effective for academicians, researchers, policymakers and socio-economic planners.

The practitioners of all academic disciplines under the umbrella of social sciences need a common platform to exchange their ideas that may effectively manage the socio-economic after the health emergency is over. This volume accumulates the scholarly contributions from different disciplines that portray the solidarity of academic knowledge to fight against this global challenge.

Since the topic is quite pertinent in these troubled times, we believe that the thrust area would be novel in terms of its research dimension, and the investigations would be academically beneficial and contribute to growing consciousness about the various dimensions of the pandemic beyond its impact on human health. Furthermore, it is paramount in these times when the entire emphasis has been on the medical sciences and health emergencies to come out with a solution to this menace and remember that the pandemic's social-economic-political impact has been highly formidable. Therefore, there must be ongoing discussions and exchanges of ideas regarding this facet of the crisis to which this volume is a humble submission to worldwide readers.

Uttar Dinajpur, West Bengal, India — Subrata Saha
Uttar Dinajpur, West Bengal, India — Mukunda Mishra
Uttar Dinajpur, West Bengal, India — Anil Bhuimali
15 July 2021

Acknowledgements

Sincere collective efforts have made this book possible amidst this first pandemic of this century. We are sincerely thankful to all of our friends, colleagues, students and others concerned that this venture would not have been completed without whose active support. They all have given us the mental support and inspiration to write.

We express our most profound sense of indebtedness to Professor Werner Menski, *Emeritus Professor of South Asian Laws, School of Law, SOAS, London, UK*. We are thankful to him for his kind advice and valuable time writing the 'Foreword' for this volume.

We are thankful to Professor *Yoshiro Higano of the University of Tsukuba, Japan,* for allowing this space for our volume in the Springer Series of *New Frontiers in Regional Science: Asian Perspectives,* in which he has been disseminating his role as the Editor-in-Chief with profound support to us whenever needed.

Valuable shreds of advice of Professor Bipul Malkar of *Jadavpur University*; Professor R.B. Singh, *Former Professor of Delhi School of Economics, University of Delhi*; Professor Soumeyndra Kishore Dutta and Dr. Abhijit Pakira of the *University of Burdwan*; Professor Anil Kr Bera of the *University of Illinois*; Professor Sankar Kumar Bhaumik of the *Central University of South Bihar*; Professor Abhijit Dutta of *Sikkim Central University*; Professor Jitendra Sahoo of the *University of Gour Banga*; Professor M. Thangaraj of the *University of Madras*; Professor Govinda Choudhury and Professor Anjan Chakraborti of the *University of North Bengal*; Dr. Kumarjit Mandal of the *University of Calcutta*; Dr. Madhabendra Sinha of *Raiganj University*; Professor Kajal Lahiri of the *State University of New York*; Dr. Priya Brata Dutta of *Visva Bharati*; Professor Partha Pratim Sengupta of *National Institute of Technology, Durgapur* remain unparalleled in substantially upholding the quality of the content within its folds.

The role of the contributing authors is precious in an edited volume. We convey our sincere thanks to all the contributors for offering us the opportunity to include their valuable works in this volume. Their prompt response and active cooperation have made this volume successful on time.

8 Human Resource Management during the Pandemic Period: Emerging Challenges for the Public Sector Employees in India 173
Ahana Sen

Part III New Challenges for the Indian Economy

9 Pandemic Outbreak and the Future of Poverty and Inequality Scenario: Indian Perspective 189
Abhijit Dutta, Partha Mukhopadhyay, Madhabendra Sinha, and Anjan Ray Chaudhury

10 Healthcare Expenditure and Economic Development Dynamics in India: Experiences from COVID-19 Pandemic 203
Subrata Saha

11 Mapping Linkages between the Agriculture Sector, Informal Economy, and Inequality amid Pandemic 227
Pooja Sharma and Anjan Chakrabarti

12 Impact of COVID-19 Pandemic on Informal Labour Market in India 243
Anil Kumar Biswas

Part IV COVID-19, Governance and Policies

13 New Pandemic in India: Emerging Challenges to Governance and Responses to Overcome 257
Jitendra Sahoo

14 Employment Dynamics and Labor Mobility amidst COVID-19 Pandemic in India: A Critical Appraisal of ILO Recommendations 277
Siddhartha Sankar Manna

15 India's Tryst with the Second Wave of COVID-19: Politics and Policies at the Crossroads 303
Madhuri Sukhija

Part V Sectoral Impact and Responses

16 Beekeeping Livelihood at Stake Amidst the Pandemic Outbreaks: A Study on the Migratory Beekeepers in West Bengal 319
Anil Bhuimali, Sanghamitra Purkait, and Manish Baidya

17 Struggle of Apiculture Sector in West Bengal, India During COVID-19 Pandemic: Analyzing the Demand and Supply Sides 347
Sanghamitra Purkait and Anindya Basu

18 The Importance of Public Expenditure on Crop Production and Exports in India: An Econometric Analysis in the Pandemic Trapped Era . . . 365
Labonie Mukhopadhyay

19 A Study on the Impact of COVID-19 on Indian MSMEs . . . 379
Rajib Lahiri

20 Indian Tourism Sector Under Siege of the COVID-19 Pandemic . . . 397
Arunava Kumar Choudhury and Subrata Saha

21 Invasion of the Pandemic in Indian Economy and the Government: A General Equilibrium Approach . . . 415
Nilendu Chatterjee and Bappaditya Koley

Part VI Silver Lining in the Cloud of Uncertainty

22 Rural Livelihood Options During the Pandemic in India: Finding Avenues for Revival . . . 433
Govinda Choudhury and Debjani Choudhury

23 COVID-19 and Lockdown: Key Constraints and Surviving Strategies for the Micro, Small, and Medium Enterprises (MSMEs) . . . 455
Arindam Metia

24 Where Classical Ends, Keynes Proceeds: Arguments Under the Perspective of Reviving India from the Impact of COVID-19 . . . 469
Indrani Basu

About the Editors and Contributors

About the Editors

Subrata Saha was awarded a Ph.D. in Economics from the University of North Bengal. Dr. Saha specializes in issues of fiscal policies and effects of structural changes on fiscal relationships in Southeast Asian Countries and India. The area of research interest is Public Finance, Health and Education, and Time Series Econometrics. He is a member of the editorial board of UGC CARE listed journal *Ensemble* and a life member of the Indian Econometric Society and Indian Economic Association. Dr. Saha is presently working as an Associate Professor of Economics at Raiganj University, West Bengal, India. He has supervised four M. Phil. and one Ph.D. and presently supervising four Ph.D. scholars at Raiganj University. He has contributed several research papers to national and international journals and edited research volumes. Besides, he has authored two books to his credit. IRDP, India, has awarded him with the prestigious Sarvepalli Radhakrishnan Lifetime Achievement National award in 2018 in recognition of his contribution to his research activities, and in 2021, the World Education Congress has awarded him West Bengal Education Leadership Award 2021 towards promoting educational excellence.

Mukunda Mishra is an Assistant Professor, Department of Geography, and designated Vice Principal of Dr. Meghnad Saha College in West Bengal, India. The college is affiliated with the University of Gour Banga. Dr. Mishra completed his postgraduate studies in Geography and Environmental Management at Vidyasagar University (receiving top rank in both the B.Sc. and M.Sc. panels of merit) and held a Ph.D. in Geography from the same university. He was selected for the prestigious National Merit Scholarship by the Ministry of Human Resource Development, Government of India. His research chiefly focuses on analysing unequal human development, regional planning, and multi-criteria predictive model building. He has more than 12 years of hands-on experience in dealing with development issues at the ground level in various districts of eastern India. Dr. Mishra has in his credit to publish one monograph and three edited research volumes so far from the house of Springer Nature. He has more than 30 research articles and book chapters published in the journals and books of international repute. Besides serving as the reviewer of several reputed journals published by Springer Nature and Elsevier Inc., he is continuing as the Managing and Publishing Editor of *Ensemble*, a UGC-CARE (India) enlisted journal of repute, since its inception to date.

Anil Bhuimali , *M.A. (Economics), Ph.D. (Economics), D.S.A, D.Litt.*, is presently the Vice-Chancellor of Raiganj University in West Bengal, India. Prior to this, he was designated as the Professor and Head of the Department of Economics, North Bengal University. Alongside, Prof. Bhuimali had also served as the Coordinator in the Centre for Differently-Abled Persons of North Bengal University; Coordinator of SAP-DRS-II and SAP-DRS-III in the Department of Economics of the same University. He has a long 34 years of teaching experience, of which 15 years as a full professor. He has, to his credit, more than 30 years of research experience in the area of Microeconomics, Gender Economics, International Economics, Rural Economics, Dalit Studies, Gandhian Economics, Development Economics, IT Policies and Management. Prof. Bhuimali supervised 20 M.Phil. dissertations, 35 Ph.D. theses and one post-doctoral thesis. He has published more than two hundred research articles and book chapters in reputed journals and 50 books. Prof. Bhuimali is a Member of the Research Board of Advisors, the American Biographical Institute, North Carolina, USA, and a Visiting Fellow at Kannur University, Kerala. He is the recipient of several prestigious awards like the *Banga Ratna Award* by the Government of West Bengal in 2016; *Life Time Achievement Award* by the Confederation of Indian Universities (CIU) in 2016; *Paul Samuelson Royal Economist Award* by South Asia Management Association in 2016; *Ambedkar Social Service Award* by the Indian Academic Researchers' Association (IARA) in 2016; The *Outstanding University Administrator and Academic Excellence Award* 2017 by the Society for Research Development, Kuala Lumpur and many more.

Contributors

Manish Baidya Department of Commerce, Shyampur Siddheswari Mahavidyalaya, Howrah, West Bengal, India

Anindya Basu Department of Geography, Diamond Harbour Women's University, Diamond Harbour, South 24 Parganas, West Bengal, India

Indrani Basu Berhampore College, Berhampore, Murshidabad, West Bengal, India

Anil Bhuimali Raiganj University, Raiganj, West Bengal, India

Anil Kumar Biswas Department of Economics, P. D. Women's College, Jalpaiguri, West Bengal, India

Anjan Chakrabarti UGC-Human Resource Development Centre, The University of North Bengal, Siliguri, West Bengal, India

Nilendu Chatterjee Department of Economics, Bankim Sardar College, South 24 Parganas, West Bengal, India

Anjan Ray Chaudhury Department of Economics, Durgapur Government College, Durgapur, West Bengal, India

Bishal Chhetri Department of Geography, Southfield College, Darjeeling, West Bengal, India

Arunava Kumar Choudhury Department of Economics, Sripat Singh College, Jiaganj, Murshidabad, West Bengal, India

Debjani Choudhury B.S.F. Senior Secondary Residential School, Kadamtala (Siliguri), West Bengal, India

Govinda Choudhury Department of Economics, University of North Bengal, Siliguri, West Bengal, India

André Santos da Rocha Laboratory of Economic and Political Geography (LAGEP), Federal Rural University of Rio de Janeiro, Rio de Janeiro, Brazil

Partha Das Department of Geography, A. B. N. Seal College, Cooch Behar, West Bengal, India

Heitor Soares de Farias Integrated Laboratory of Applied Physical Geography (LIGA), Federal Rural University of Rio de Janeiro, Rio de Janeiro, Brazil

Andrews José de Lucena Integrated Laboratory of Applied Physical Geography (LIGA), Federal Rural University of Rio de Janeiro, Rio de Janeiro, Brazil

Leandro Dias de Oliveira Laboratory of Economic and Political Geography (LAGEP), Federal Rural University of Rio de Janeiro, Rio de Janeiro, Brazil

Gustavo Mota de Sousa Integrated Laboratory of Applied Physical Geography (LIGA), Federal Rural University of Rio de Janeiro, Rio de Janeiro, Brazil

Hasan Dinçer The School of Business, İstanbul Medipol University, İstanbul, Turkey

Abhijit Dutta Department of Commerce, School of Professional Studies, Sikkim (Central) University, Gangtok, Sikkim, India

Pablo Ibanez Laboratory of Economic and Political Geography (LAGEP), Federal Rural University of Rio de Janeiro, Rio de Janeiro, Brazil

Hakan Kalkavan The School of Business, İstanbul Medipol University, İstanbul, Turkey

Hüsne Karakuş The School of Business, İstanbul Medipol University, İstanbul, Turkey

Bappaditya Koley Department of Geography, Bankim Sardar College, South 24 Parganas, West Bengal, India

Rajib Lahiri Department of Commerce, Derozio Memorial College, Kolkata, West Bengal, India

Kabita Lepcha Department of Geography, University of Gour Banga, Malda, West Bengal, India

Siddhartha Sankar Manna Department of Political Science, University of Gour Banga, Malda, West Bengal, India

Arindam Metia Department of Management, Raiganj University, Raiganj, West Bengal, India

Mukunda Mishra Department of Geography, Dr. Meghnad Saha College, Uttar Dinajpur, West Bengal, India

Labonie Mukhopadhyay Awadh Dental College and Hospital, Jamshedpur, Jharkhand, India

Partha Mukhopadhyay Business Excellence Department, Durgapur Steel Plant, Steel Authority of India Limited, Durgapur, West Bengal, India

Sanghamitra Purkait Department of Geography, Diamond Harbour Women's University, Diamond Harbour, South 24 Parganas, West Bengal, India

Arabinda Roy Department of Geography, Dr. Meghnad Saha College, Uttar Dinajpur, West Bengal, India

Subrata Saha Department of Economics, Raiganj University, Raiganj, West Bengal, India

Jitendra Sahoo Department of Political Science, University of Gour Banga, Malda, West Bengal, India

Tanmoy Sarkar Deparment of Geography, Gazole Mahavidyalaya, Malda, West Bengal, India

Ahana Sen Department of Management, Raiganj University, Raiganj, West Bengal, India

Pooja Sharma Daulat Ram College, The University of Delhi, Delhi, India

Madhabendra Sinha Department of Economics and Politics, Visva-Bharati University, Santiniketan, West Bengal, India

Madhuri Sukhija Department of Political Science, Mata Sundri College For Women, Delhi University, Delhi, India

José G. Vargas-Hernández Posgraduate and Research Department, Instituto Tecnológico Mario Molina, Unidad Zapopan, Zapopan, Jalisco, Mexico

Serhat Yüksel The School of Business, İstanbul Medipol University, İstanbul, Turkey

Abbreviations

ABA	Atmanirbhar Bharat Abhiyan
ADB	Asian Development Bank
ADF	Augmented Dickey–Fuller (test)
AISHE	All India Survey on Higher Education
ALA	American Library Association
ATF	Aviation Turbine Fuel
AUC	Area under curve
BMRC	Bangladesh Medical Research Council
BRICS	Brazil, Russia, India, China and South Africa
CFR	Case fatality rates
CMIE	Centre for Monitoring Indian Economy
COVID	Coronavirus disease
CRIW	Cointegrating Regression Durbin-Watson (test)
eNAM	Electronic National Agriculture Market (India)
EPF	Employees provident fund (India)
FAITH	Federation of Associations in Indian Tourism and Hospitality
FAO	Food and Agriculture Organization
FDI	Foreign Direct Investment
FMOLS	Fully modified ordinary least squares
FSC	Federal Supreme Court (of Brazil)
FTA	Foreign tourist arrival
GAVI	Global Alliance for Vaccines and Immunisation
GDP	Gross domestic product
GER	Gross enrolment ratio
GISAID	Global Initiative on Sharing All Influenza Data
GTA	Gorkha Territorial Administration (in India)
GVA	Gross value added
HDI	Human Development Index
HII	Health Infrastructure Index
IBEF	India Brand Equity Foundation

ICAR	Indian Council of Agricultural Research
ICMR	Indian Council of Medical Research
ICT	Information and Communication Technology
ILO	International Labour Organization
IMF	International Monetary Fund
IRF	Impulse response function
LFPR	Labour force participation rate
MARJ	Metropolitan Area of Rio de Janeiro
MDA	Market Development Assistance (India)
MDG	Millennium Development Goals
MERS	Middle East respiratory syndrome
MGREGA	Mahatma Gandhi Rural Employment Guarantee Act (India)
MOOC	Massive Open Online Course (India)
MPI	Multidimensional Poverty Index
MSME	Micro, Small and Medium Enterprise
MSMED	Micro, Small and Medium Enterprises Development (Act, India)
NDMA	National Disaster Management Act (India)
NFHS	National Family and Health Services (India)
NITI	National Institution for Transforming India
NIV	National Institute of Virology
NSIC	National Small Industries Corporation (India)
NSO	National Statistical Office (India)
NSS	National Sample Survey (India)
NUEPA	National Institute of Educational Planning and Administration (India)
OECD	Organisation for Economic Co-operation and Development
OLS	Ordinary least square
PHE	Public health expenditure
PMGKY	Pradhan Mantri Garib Kalyan Yojana (India)
PMJDY	Pradhan Mantri Jan Dhan Yojana (India)
RBI	Reserve Bank of India
RNFE	Rural Non-farm Economy
ROC	Receiver operating characteristic (curve)
RTPCR	Reverse transcription polymerase chain reaction (test)
SAARC	South Asian Association for Regional Co-operation
SAATHI	System for Assessment, Awareness & Training for Hospitality Industry (India)
SARS	Severe acute respiratory syndrome
SC	Scheduled castes (India)
SDG	Sustainable Development Goal
SEM	Structural Equation Model
SHG	Self-help Groups (India)
SSI	Small-scale industry
ST	Scheduled tribes (India)
TSAI	Tourism Satellite Account for India

U-DISE	Unified District Information System for Education (India)
UHS	Unified Health System (of Brazil)
UMPCE	Uniform Monthly Per Capita Consumption Expenditure
UNDP	United Nations Development Programme
UNESCO	United Nations Educational, Scientific and Cultural Organization
VECM	Vector error correction model
VER	Vector autoregression
WHO	World Health Organization
WTO	World Trade Organization

List of Figures

Fig. 1.1 State-wise daily COVID-19 confirmed cases from 14th March–31st October 2020 in India. (Source: Prepared by the authors with the help of Open Government Data (OGD) Platform, Government of India.) (Visit for the datasource https://data.gov.in/) 8

Fig. 1.2 Cumulative confirmed deaths due to COVID-19 (deaths per million). (Source: Map prepared by Our World in Data, Oxford Martin School based on the Data published by COVID-19 Data Repository by the Center for Systems Science and Engineering (CSSE) at Johns Hopkins University for 22nd January 2020–17th July 2021) 10

Fig. 1.3 A comparison between the COVID-19 first wave in 2020 and the lethal second wave in 2021. (Source: Prepared by the authors from the Open Government Data (OGD) Platform of the Government of India) 11

Fig. 1.4 Monthly New COVID-19 cases in rural and urban India, May 2020–April 2021. (Source: Drawn by the authors based on Mahapatra et al. 2021, Down To Earth, 16–31 May 2021; Census of India 2011) 12

Fig. 1.5 Monthly COVID-19 related death in rural and Urban India, May 2020–April 2021. (Source: Drawn by the authors based on Mahapatra et al. 2021, Down To Earth, 16–31 May 2021; Census of India 2011) 12

Fig. 1.6 The continuous fluctuation in Confirmed vs Recovery rate of the COVID-19 in India. (Source: Drawn by the authors based on the data from the Open Government Data (OGD) Platform, Government of India) 13

Fig. 1.7 The COVID-19 cases trend in India, 14th March 2020–10th July 2021. (Source: Drawn by the authors based on the data from the Open Government Data (OGD) Platform, Government of India) 14

Fig. 1.8 State-wise COVID-19 status in India, 14th March 2020–10th June. (Source: Drawn by the authors based on the data from the Open Government Data (OGD) Platform, Government of India) ... 15
Fig. 1.9 (**a**) Current health expenditure (% of GDP), and (**b**) Out-of-pocket expenditure (% of current health expenditure) in India (Source: Prepared by the authors based on the World Health Organization Global Health Expenditure database.) (See WHO datasets in www.apps.who.int/nha/database) ... 18
Fig. 1.10 The General Assembly Election in the Indian States and the COVID-second wave surge (Drawn by the author based on Open Government Data (OGD) Platform India, Government of India) . 20
Fig. 1.11 (**a**) Number of people vaccinated against COVID-19, as of 16th July 2021 and (**b**) Proportion of Total Population vaccinated against COVID-19 as of 16th July 2021. (Source: Our World in Data, Oxford Martin School, the University of Oxford, Accessed on 17th July, 2021) ... 23
Fig. 1.12 India's Vaccination trend by (**a**) doses and (**b**) age. (Source: COWIN Dashboard, MOFFW, GOI, Accessed on 17th July 2021) ... 24
Fig. 1.13 Statewise Vaccination Status in India as of 27th June 2021, 08:00 IST. (Source: Drawn by the author based on GOI database.) (Database accessed from https://www.mygov.in/COVID-19/) ... 25

Fig. 4.1 Spread of Covid-19 in SAARC countries as on (**a**) 31 March, 2020; (**b**) 1 June, 2020; (**c**) 1 August, 2020 and (**d**) 30 September, 2020. (Source: Prepared by the author from worldometers 2020) ... 73
Fig. 4.2 Progression of (**a**) Total cases and (**b**) 7-Days moving average of daily number of cases. (Source: Prepared by the author from worldometers 2020) ... 74
Fig. 4.3 Occurrence of death due to Covid-19 in SAARC countries as on (**a**) 31 March, 2020; (**b**) 1 June, 2020; (**c**) 1 August, 2020 and (**d**) 30 September, 2020 by September, 2020. (Source: Prepared by the author from worldometers 2020) ... 76
Fig. 4.4 Global practices of combatting the coronavirus pandemic ... 77
Fig. 4.5 Comparison of mean daily positivity rates from April to September (Source: Prepared by the author from worldometers 2020) ... 85
Fig. 4.6 Comparison of COVID Situation among SAARC countries between July and September, 2020 (Source: Computed by author from worldometers 2020) ... 85

Fig. 5.1 Location of the Metropolitan Area of Rio de Janeiro (MARJ) in the context of the state of Rio de Janeiro and Brazil/South America (Source: IBGE/SEA 2018) ... 99

Fig. 5.2 Metropolitan Area of Rio de Janeiro (MARJ) and its subdivisions (Source: IBGE/SEA 2018) 99
Fig. 5.3 Prevalence rate for COVID-19 for the city of Rio de Janeiro and its periphery (Source: IBGE/SEA 2018; IBGE 2020; SES-RJ 2021) 103
Fig. 5.4 COVID-19 Mortality rate for the city of Rio de Janeiro and its periphery (Source: IBGE/SEA 2018; IBGE 2020; SES-RJ 2021) 103
Fig. 5.5 COVID-19 Case fatality rate for the municipality of Rio de Janeiro and its periphery (Source: IBGE/SEA 2018; IBGE 2020; SES-RJ 2021) 104
Fig. 5.6 Number of confirmed COVID-19 cases per month between March 2020 and June 2021 in the city of Rio de Janeiro and MARJ (Source: SES-RJ 2021) 105
Fig. 5.7 Number of deaths by COVID-19 per month between March 2020 and June 2021 in the city of Rio de Janeiro and MARJ (Source: SES-RJ 2021) 106
Fig. 5.8 Case fatality rate per COVID-19 per month between March 2020 and June 2021 in the city of Rio de Janeiro and MARJ (Source: SES-RJ 2021) 106
Fig. 5.9 Planning Areas (AP) of the city of Rio de Janeiro (Source: IBGE/SEA 2018; SMS-PCRJ 2021) 107
Fig. 5.10 Number of confirmed COVID-19 cases for the city of Rio de Janeiro on 25 June 2021 (Source: IBGE/SEA 2018; SMS-PCRJ 2021) 108
Fig. 5.11 COVID-19 deaths for the city of Rio de Janeiro on 25 June 2021 (Source: IBGE/SEA 2018; SMS-PCRJ 2021) 109
Fig. 5.12 Case fatality rate of COVID-19 for the city of Rio de Janeiro on 25 June 2021 (Source: IBGE/SEA 2018; SMS-PCRJ 2021) 110
Fig. 5.13 Total vaccinated against COVID-19 in the city of Rio de Janeiro on 25 June 2021 (Source: IBGE/SEA 2018; SMS-PCRJ 2021) .. 111

Fig. 6.1 Location of the study area 122
Fig. 6.2 (**a**) Age group of the respondents; (**b**) Marital status; (**c**) Family size; and (**d**) Type of house 127
Fig. 6.3 (**a**) Occupation; (**b**) Source of COVID-19 infection; (**c**) Risk factor groups; (**d**) Reasons for going outside 128
Fig. 6.4 (**a**) Perception regarding lockdown; (**b**) Willingness to stay confined; (**c**) Impact of female student; and (**d**) Financial impact 129
Fig. 6.5 (**a**) Service that opened during lockdown; (**b**) Availability of vehicles; (**c**) Place of work from home; (**d**) Correlation matrix ... 131
Fig. 6.6 (**a**) PPE provided at work; (**b**) Utilization of PPE; (**c**) Opinion on PPE; and (**d**) Level of comfort of PPE 132

Fig. 6.7 (**a**) Comorbidity factors; (**b**) Accessing health services; (**c**) Self isolation; and (**d**) COVID-related symptoms 133
Fig. 6.8 Correlation coefficients 134

Fig. 7.1 A cyclical approach of education in emergencies (**Source:** UNESCO 2020a) 141
Fig. 7.2 COVID-19 pandemic affected learners in India (**Source**: Prepared by author based on UNESCO 2020a) 143
Fig. 7.3 Age-sex structure of India (**Source**: Based on the Census of India 2011a datasets) 147
Fig. 7.4 Rural–urban disparity of the households' possession of computers and access to internet facility across the quintile classes of UMPCE in India (**Source:** 75th round of National Sample Survey conducted between July 2017 and June 2018) 151
Fig. 7.5 Income distribution over time in India (**Source:** Compiled by the authors from the World Inequality Database [The World Inequality Database (WID) provide open and convenient access to the most extensive available database on the historical evolution of the world distribution of income and wealth, both within countries and between countries. See https://wid.world/data/]) ... 154
Fig. 7.6 The location of the district of Dakshin Dinajpur where the case study on the Namasudra Community was conducted; the white dots with red circles presents the location of interviews conducted, and the yellow hairlines are the village boundaries (**Source**: Drawn by the authors) 158
Fig. 7.7 A graphical presentation of the procedure of fitting a binormal ROC curve to ordinal confidence rating data by ML estimation (**Source**: Drawn by the author following Metz et al. 1998) 163
Fig. 7.8 Maximum Likelihood Estimation of Binomial ROC Curve for **(a)** Event-1; **(b)** Event-2 and **(c)** Event-3. **N.B.** Red symbols and BLUE line show the Fitted ROC curve, and Gray lines show 95% confidence interval of the fitted ROC curve (**Source**: Prepared by the authors) 164

Fig. 8.1 Maslow's Need Hierarchy (Source: Prepared by the author) 178

Fig. 9.1 Triggering of global poverty (Source: Authors' own compilation) 195

Fig. 10.1 Plots of different health expenditures in India (Data Source: The World Bank, 2018) 210
Fig. 10.2 Trends in current health expenditure (% of GDP) of few SAARC Countries (Data Source: The World Bank Data 2018) 212
Fig. 10.3 Roots of AR Characteristic Polynomial (Source: Computed by the author using EViews 10 software) 218
Fig. 10.4 Variance Decomposition of **(a)** GDP, **(b)** GHE, & **(c)** DPHE (Source: Computed by the author using EViews 10 software) 220

Fig. 10.5 Impulse Response Functions Analysis (Source: Computed by the author using EViews 10 software) ... 221
Fig. 10.6 Plots of **(a)** CUSUM and **(b)** CUSUM-Square tests **(Source:** computed by the author using EViews 10 software)) ... 223

Fig. 11.1 Growth Rates of GVA at Basic Prices from 2017–18 to 2020–21 (Source: Ministry of Statistics and Programme Implementation, Govt. of India) ... 230
Fig. 11.2 Changing Gini Coefficient for India from 1990 to 2018 (Source: World Bank Open Data [See here: https://data.worldbank.org/]) . 233
Fig. 11.3 Unemployment Rate (in %) during April 2019 to January 2021 (Source: Centre for Monitoring Indian Economy Pvt. Ltd., CMIE) ... 234
Fig. 11.4 Changing Percentage of Vulnerable Workers in India from 1990 to 2020 (Source: ILOSTAT Database [See here: https://ilostat.ilo.org/]) ... 235

Fig. 12.1 **Sectoral distribution of India's youth losing their jobs** (Source: Calculated from International Labour Organization and Asian Development Bank (2020) Tackling the COVID-19 youth employment crisis in Asia and the Pacific. https://www.ilo.org/wcmsp5/groups/public. Accessed 11th December 2020) ... 248

Fig. 14.1 Migrant Workforces excluding agricultural workers in India; Refer Table 14.1 for details about "Job Type" (Source: Prepared by the author, based on Khanna 2020) ... 280
Fig. 14.2 Subsector-wise Industrial Employment Trends in India (Source: Developed by author, based on Khanna 2020) ... 282
Fig. 14.3 ILO recommended four pillars to respond during the pandemic (Drawn by the author, based on ILO's recommendation) ... 290
Fig. 14.4 Five Policy Visions (Source: Drawn by author) ... 291
Fig. 14.5 *Inclusive* Framework for Migrant Workers (Source: Drawn by author) ... 295

Fig. 16.1 Migration pattern of the beekeepers of West Bengal in two study zones before the pandemic (Source: Drawn by the authors based on field survey datasets) ... 324
Fig. 16.2 Per head colony-holding capacity of the migratory beekeepers at pre-pandemic months (Source: Drawn by the authors based on field survey datasets) ... 325
Fig. 16.3 Number of migratory beekeepers and their professional experience (Source: Drawn by the authors based on field survey datasets) ... 327
Fig. 16.4 Status of professional experiences of the migratory beekeepers in the perspective of hive-holding capacity (Source: Drawn by the the authors based on field survey datasets) ... 327

Fig. 16.5 Average contraction of colonies (in %) due to the lockdown as per hive-holding Capacity-wise class of beekeepers (Source: Drawn by the authors based on field survey datasets) 339
Fig. 16.6 Average decrease of colonies (in %) due to the lockdown as per experience-wise classes of beekeeper Groups (Source: Drawn by the authors based on field survey datasets) 340
Fig. 16.7 Loss of average migration months due to the lockdown by the different classes of beekeepers according to their past hive-holding capacity (Source: Drawn by the authors based on field survey datasets) 341
Fig. 16.8 Loss of average migration months due to the lockdown by the different classes of beekeepers according to their professional experience (Source: Drawn by the authors based on field survey datasets) 342

Fig. 17.1 Declining bee box holding by the selected beekeepers in West Bengal from January 2020 to September 2020 (Data Source: Authors) 353
Fig. 17.2 Loss of migration months of the beekeepers due to lockdown during COVID-19 (Data Source: Telephonic Survey, 2020)...... 353

Fig. 18.1 Stability Diagnostics (CUSUM Test) (Source: Prepared by the author) 373
Fig. 18.2 CUSUM Square Test (Source: Prepared by the author)........... 374

Fig. 20.1 Foreign tourist arrival in top 5 countries (Source: Prepared by the author) 402
Fig. 20.2 International tourist earnings in US$b (Source: Prepared by the author) 404
Fig. 20.3 India's percentage share in FEE (Source: Prepared by the author) 404
Fig. 20.4 Contribution of tourism in GDP in US$b (Source: Satista.com, 2021(See the Statista website for the details which are available under the URL: https://www.statista.com/statistics/313665/direct-contribution-of-travel-and-tourism-to-gdp-in-india/#:~:text=In%202018%2C%20the%20direct%20contribution%20of%20travel%20and,highest%20tourism%20GDP%20contribution%20in%20Asia-Pacific%20in%202,018.), (See the Statista website for the details which are available under the URL: https://www.statista.com/statistics/313724/total-contribution-of-travel-and-tourism-to-gdp-in-india-by-segment/.)) 405
Fig. 20.5 Direct contribution of travel and tourism in the world in thousands (Source: Statista.com (See the Statista website for the details which are available under the URL: https://www.statista.com/statistics/292490/contribution-of-travel-and-tourism-to-employment-in-selected-countries/.)) 406

Fig. 20.6 Total contribution of tourism in employment generation in India in million (Source: Prepared by the author) 407

Fig. 23.1 Employment generation by MSME as a percentage of overall employment globally. (Source: Country-specific MSME Reports, KPMG data & estimates, 2015) 457

Fig. 23.2 Number of estimated MSMEs in millions. (Source: MSME Annual Report, 2018–2019, Ministry of MSME, GOI) 458

Fig. 24.1 Equilibrium in labor market: $N_s^c(W/P)$ is the labor supply curve under containment and $N_D^C(W/P)$ is the labor demand curve under containment. (Source: Author's perception) 481

Fig. 24.2 (**a**) Determination of equilibrium real wage rate and level of employment in agrarian-based rural sector labor market; (**b**) Determination of equilibrium real wage rate and level of employment in industrial and service-based urban sector labor market. (Source: Author's perception) 485

List of Tables

Table 1.1	A summary of Lockdown and Unlock phases in India in the COVID-19 first wave	6
Table 1.2	Some key parameters across WHO regions	27
Table 3.1	Quarterly GDP rates in India (percent)	59
Table 3.2	The list of the factors	60
Table 3.3	The evaluations of the experts	61
Table 3.4	Pairwise Comparison Matrix	61
Table 3.5	Normalized matrix	61
Table 3.6	Result of the analysis	62
Table 4.1	COVID Scenario of SAARC countries as on 30th September, 2020	72
Table 4.2	Mean, standard deviation and coefficient of variation for daily recovery rates	84
Table 5.1	COVID-19—total cases, total deaths, cases/one million, deaths/one million, and population	96
Table 5.2	Confirmed cases, deaths, prevalence rate, mortality rate and case-fatality rate per COVID-19 on 06/23/2021, by municipalities and group of municipalities in the Metropolitan Area of Rio de Janeiro	102
Table 6.1	Association of variables	128
Table 6.2	Impact category based on composite scores	130
Table 6.3	Association of different variables	133
Table 7.1	Trajectory of shutting down of schools in India in response to the first wave of COVID-19	146
Table 7.2	Rural–urban distribution of educational institutes in India	149
Table 7.3	Network strength of rural and urban households in India by residence and sex	150
Table 7.4	Dropout from primary education in India	152

Table 7.5 Expected outcome from expert opinion 159
Table 7.6 Estimation of binomial ROC Curve for Event-1 166
Table 7.7 Estimation of binomial ROC Curve for Event-2 167
Table 7.8 Estimation of binomial ROC Curve for Event-3 168

Table 8.1 Employees perception towards satisfaction of different needs . 182
Table 8.2 Calculation of χ^2**statistics** 182

Table 9.1 Unemployment Rate (%) in India 190
Table 9.2 Commodity price increases as compared to previous years 191
Table 9.3 GDP Growth Rate in India 193
Table 9.4 Analysis of foreign trade, income, and consumption with respect to GDP 199

Table 10.1 Status of public health Infrastructure in India 211
Table 10.2 Unit root test results 214
Table 10.3 Results of Johansen's cointegration Test 215
Table 10.4 Results of VECM Regression 216
Table 10.5 VEC Granger Causality/Block Exogeneity Wald Test 217
Table 10.6 Variance Decomposition of Gross Domestic Product 218
Table 10.7 Variance Decomposition of Government Healthcare Expenditure 219
Table 10.8 Variance Decomposition of Domestic private healthcare expenditure 219

Table 11.1 Growth rates of in various quarters in 2020–2021 230
Table 11.2 Country-wise inequality, expenditure on health and education as a percentage of GDP 232
Table 11.3 Correlation Matrix 237
Table 11.4 Regression results 237
Table 11.5 Model Summary of the Regression model 238
Table 11.6 ANOVA Result 239

Table 14.1 Migrant Workforces excluding agricultural workers in India (percentage) 280
Table 14.2 Subsector-wise industrial employment trends in India 281
Table 14.3 Sub-sector-wise service 282
Table 14.4 Types of employment in non-farm sections (in million) 283
Table 14.5 The unemployment ratio 285
Table 14.6 Prickle in urban regions 286
Table 14.7 The Unemployment ratio in different states of India 287
Table 14.8 Four sectors with most severe effects of the infection and deteriorating the production and manufactures 289

Table 16.1 Statement relating to hive-holding capacity and professional experience of the beekeepers before the lockdown and effect of the lockdown on their colonies and migration periods 326
Table 16.2 Significant honey flows of West Bengal and its features 330

Table 16.3	Migration Paths of the Migratory Beekeepers and their detention fields	332
Table 16.4	Problems faced by the migrant beekeepers due to the lockdown	333
Table 17.1	Problems faced by the respondent beekeepers and their respective weight	354
Table 17.2	Chi-Square Test to know the association between gender and honey consumption	355
Table 17.3	Chi-Square Test to know the association between dwelling place and honey consumption	355
Table 17.4	Chi-Square Test to know the association between annual family income and honey consumption	356
Table 17.5	Chi-Square Test to know the association between total member in a family and honey consumption	356
Table 17.6	Chi-Square Test to know the association between the percentage of children in a family and honey consumption	357
Table 17.7	Chi-Square Test to know the association between prior knowledge about benefits of apiculture products and honey consumption	358
Table 17.8a	Paired samples descriptive statistics	358
Table 17.8b	Paired samples correlations	358
Table 17.8c	Paired samples t-test	359
Table 17.8d	Paired samples effects sizes	359
Table 18.1	Estimated statistics of Unit Root Tests	373
Table 18.2	Results from Bai–Perron structural break estimation	374
Table 18.3	Estimated statistics of Johansen cointegration	374
Table 18.4	Vector Error Correction Model (VECM)	376
Table 19.1	Share of gross value added (GVA) of MSME in all India GDP	380
Table 19.2	Estimated number of MSMEs (activity wise)	381
Table 19.3	Distribution of enterprises (rural and urban area wise)	381
Table 19.4	Distribution of employment by type of enterprises in rural and urban areas	381
Table 19.5	Definition of small-scale industries (SSI)	385
Table 19.6	Investment limits in MSMEs as per MSMED Act, 2006 (valid up to 30 June 2020)	386
Table 19.7	Revised definition of MSMEs (Valid from 1 July 2020 onwards)	387
Table 20.1	Foreign tourist arrival from different nations (Top 5)	402
Table 20.2	Foreign tourist arrival in India, 2013–2014 to 2019–2020	402
Table 20.3	Share of India in international tourism receipts in the world and the Asia-Pacific	403

Table 20.4 Contribution of tourism in GDP (in percentage) ... 404
Table 20.5 Contribution of tourism in employment ... 406

Table 22.1 Quarterly estimate of expenditure on GDP for the first and second quarter at constant (2011–2012) prices ... 439
Table 22.2 Quarterly estimate of GVA at constant (2011–2012) prices ... 440
Table 22.3 Work provided under MGNREGA in April–November 2020 (in million) ... 442

Table 23.1 Classification of manufacturing enterprises and enterprises rendering services ... 457
Table 23.2 Estimated number of MSMEs (in Rs. lakh) ... 458
Table 23.3 Estimated employment in MSMEs (in lakh) ... 458
Table 23.4 Distribution of enterprises by categories ... 459

Table 24.1 Fatality rate and infection rate of COVID-19 and other epidemics ... 470
Table 24.2 Real gross domestic product (GDP) growth rate of India from 2009 to 2021 ... 473
Table 24.3 The recent trend of age structure of India during 2014–2018 . 475
Table 24.4 Relative position of India ... 477
Table 24.5 Distribution of gross domestic product (GDP) across economic sectors from 2008 to 2018 ... 482
Table 24.6 Distribution of the workforce across economic sectors in India from 2009 to 2019 ... 483
Table 24.7 Share of migrant workers in total workers by major sectors ... 484

Part I
The Global Pandemic

Chapter 1
India Gets Through the Waves of the Pandemic: Public Policies and Self-Care Interventions on Health at the Crossroads

Tanmoy Sarkar and Mukunda Mishra

Abstract India, with its preexisting issues of economic inequalities, exclusion, dropout, malnutrition, and a range of sociopolitical unease, faces the consecutive two deadly waves of the COVID-19 pandemic. While the health hazard in the country broke all its statistics ever, the impact of the crisis extended to every corner of the society, economy, culture, and governance. This chapter aims to explore how the consecutive two waves of the pandemic are somehow different in terms of the causative organism, the knowledge of humankind about the obligate pathogens, and the management approach as well. However, India witnessed critical differences in the rate of disease spread, positivity rate, and fatality. While the death rate had been lower in the second wave, the widespread infection in the second largest populous country within a narrower span put unprecedented pressure on the health system to which the country had possibly no answer. Questions on the public healthcare policies had been raised, which the article noted; however, the authors are concerned with the other end of the balance, which is the individuals' responsible actions. The self-urged actions of the citizens stand on reciprocity with the public policies, which had long been in the papers of the policy researchers; now, it is time to practice.

Keywords COVID-19 · Governance · Health hazard · Disease spread, death rate · Positivity rate · Vaccination

T. Sarkar
Department of Geography, Gazole Mahavidyalaya, Malda, West Bengal, India

M. Mishra (✉)
Dr. Meghnad Saha College, Itahar, Uttar Dinajpur, West Bengal, India

S. Saha et al. (eds.), *Economic and Societal Transformation in Pandemic-Trapped India*, New Frontiers in Regional Science: Asian Perspectives 55,
https://doi.org/10.1007/978-981-16-5755-9_1

1.1 Introduction

Human civilization has been confronted with various infectious diseases since its inception. However, a situation like the COVID-19 pandemic rarely happens in a century. The journey of the SARS-COVID-19 virus from epidemic to pandemic has already been witnessed, and a group of experts already predicts its sustenance in endemic form. An unprecedented situation, pandemic waves are coming, human civilization is facing the biggest challenge to survive! Anthropogenic footprints have come to a halt! This devil disease has been attacking humankind and societies at their core.

No time to condolence, no time to mourn; instead, unfortunately, it is the time (very pathetically) to wait for RTPCR test, wait for the report, wait at the hospital lawn to get a bed, unending wait for oxygen, wait to get the body, wait to get a chance for cremation! This is the real "wait" scenario that looms large across the country during the second wave of the COVID-19 outbreak. And, by attempting the tragic sequence of "wait and to get chance," a person might have no time to shed tears.

1.2 The Mournful Locus from the First to the Second Wave

The first human transmissions of SARS-CoV-2 were found in Wuhan, the capital of Hubei province in China, in December 2019 (Huang et al. 2020). A cluster of pneumonia cases without any death in Wuhan in China was initially reported on social media by the World Health Organization (WHO) on 4th January 2020, and the first disease outbreak news was reported on the next day. However, no one could estimate the worse future that human civilization had to experience. The first COVID-positive case outside China was reported on 13th January in Thailand. The human-to-human transmission was evidenced in Wuhan, but WHO advocated for more investigation to confirm such transmission. Within few days, on 30th January, the situation report by WHO confirmed a total of 7818 cases globally, where 82 cases were found in 18 countries outside China. On the same day, the coronavirus attack was declared a Public Health Emergency of International Concern by WHO. And, it was 11th March 2020 when WHO admitted the rampant spread of the fatal disease and declared the COVID-19 attack as a pandemic.

In India, the SARS-CoV-2 attack was started with the three medical students who tested COVID positive on 30th January 2020 in the state of Kerala (Perappadan 2020; Narasimhan 2020). All the three had returned from Wuhan, China. All the countries in Europe, several states in the USA, and countries worldwide had gone for "lockdown." India had announced the nationwide lockdown with the target to flattening the curve on 24th March 2020. The whole world had been experiencing the lethal impact of the coronavirus. It was sudden! There was a serious lack of information and extremely limited investigation on the novel coronavirus. It was a gigantic blow for the administration, public health, and health infrastructure. Lockdown had been imposed to get some time to be prepared, to flattening the curve by restricting the spread of the highly infectious human-to-human transfer of the SARS-COVID-19 virus. India had experienced a complete nationwide lockdown for 21 days that was announced on the evening of 24th March 2020 and was further extended up to 31st May in several phases (Table 1.1). The relaxation in different economic and service sectors was allowed with different phases of lockdowns. The services were started to resume with phased unlock periods which were announced on 1st June 2020 (Table 1.1).

1.2.1 Turbulence in the Economy

The impact of COVID-19 and the resultant nationwide lockdown was extremely severe on the Indian economy. GDP growth had been decreased and reached its last 20 years lowest (Sarkar and Mondal 2021). In the pandemic-stuck market, the growing panic with rising COVID-19 positive cases had guided the investor to restrict investment in the market, and the impact on Sensex and Nifty was havoc in the early days of lockdown. The large business houses, firms, and their allied sectors had faced an unprecedented hit in the aftermath of the COVID pandemic. The small business houses and firms had faced a cash crunch. Excepting the essential services, all the non-essential economic sectors had stressed by severe fall down due to the pandemic forced long-term lockdown. There was a considerable decline in demand and supply chain. India had gone through an extreme economic crisis in the unprecedented pandemic situation. The nation had experienced a sudden drop in GDP in 2020, and the immediate impact on consumption, investment, and trade was inevitable. The large states and metropolitans those who are largely dependent on industry, trade, and commerce, multinational business houses, and services, were affected worst. The state-wise data on COVID-19 confirmed cases reveals that Maharashtra, Tamil Nadu, Delhi, Andhra Pradesh, and Karnataka were severely affected, followed by Uttar Pradesh, West Bengal, Kerala, and other states and UTs in the first wave of the pandemic. Lockdown for more than 2 months at a stretch, complete restrictions on movement, irregularity in banking transactions, and job insecurity are the factors that limit the cash in hand, which in turn badly affect the demand side in the Indian economy. In this adverse demand-side situation, the others conditions like restricted international exchanges, imposition of quarantine and containment zone, and shortage of labor seriously affect the supply side of the economy. The industrial sector was severely affected due to the lockdown and shortage of labors. Apart from the migrant workers, the gig workers had suffered a lot due to this pandemic and prolonged lockdown situation. Nowadays, e-commerce and online business platforms share an important part in the economy, and a large working force is engaged in the daily service of these online products to the doorstep of the buyers. However, the impact of complete halt and restriction in movement on these working groups had been hardy estimated. The untold story of millions of

Table 1.1 A summary of Lockdown and Unlock phases in India in the COVID-19 first wave

Lockdown phases in 2020			Unlock phases in 2020		
Phase	Date	Important announcements	Phase	Date	Important announcements
Phase 1	March 24–April 14	The first announcement of complete shutdown: All services were suspended	Unlock-1	June 1–30	Shopping malls, hotels, restaurants, and religious places were permitted to open from eighth June. No restrictions on interstate movement. Lockdown restrictions were only in containment zones. No gatherings
			Unlock-2	July 1–31	Most of the regulations, as in Unlock-1: States were allowed to impose suitable restrictions. Limited international travel was permitted. Educational institutions, recreational activities, and metros remained closed
Phase 2	April 15–may 3	Lockdown was extended: Conditional relaxation was announced to be implemented after reviewing the COVID situation on 20th April 2020. Lockdown areas were classified into three zones: Red zone (severe), Orange (moderate), and Green zone (no infection)	Unlock-3	August 1–31	Night curfews were restricted. Inter and intra-state movements were allowed
			Unlock-4	September 1–30	Outside the containment zones, relaxations were extended. Metro rail was announced to be run from seventh September. Religious, entertainment, sports, and political gatherings were permitted, with 100 people gathering
Phase 3	May 4–17	Lockdown extended with additional relaxations. The districts across the country were classified into three zones according to the severity of cases: Red (130 districts), Orange (284), and Green (320) zone	Unlock-5	October 1–31	Cinema halls were announced to open on 15th October with 50% of its seat capacity. Tourism sectors were started to resume in few states (Kerala)
Phase 4	May 18–31	Lockdown was extended to two more weeks. States were permitted to decide on COVID zone	Unlock-6	November 1–30	No big changes were made in this phase. States were scaled down restriction gradually

(continued)

Table 1.1 (continued)

Lockdown phases in 2020			Unlock phases in 2020		
		identification. Local administration was given more power to decide on containment and buffer zone identification			outside the containment zones. Educational institutions were started to reopen partially in few states

Source: Ministry of Home Affairs, Govt. of India (Dataset is available at: https://www.mha.gov.in/notifications/circulars-COVID-19)

jobless workers, the economic hardship, and its impact on their families could not be estimated.

1.2.2 The Reverse Migration

Along with the global countries, India had also witnessed an unprecedented health crisis, had experienced an unparalleled social crisis, and had gone through an unexampled economic setback. The nation had witnessed the long March of migrant workers with burdens on their backs and babies on their shoulders, from large cities to their native places (Kumar 2020). Many had to walk for thousands of kilometers without food, and some journeys had ended on-road, reraising the issues of equity of share, right to live, and to the essence of sustainable development, once more. The dream of a trillion economy must be questioned if it is unable to protect common people and if, in every possible situation, the poor fellows continue to be disfranchised (Kumar 2020).

Implementing the nationwide lockdown decision within few hours of the announcement on 24th March 2020 had forced the migrant laborers to get stuck at their then places, but most of them had no adequate food, money, and shelter for the early 21 days. The lack of government concern and improper planning due to insufficient data on migratory laborers in our country was criticized by the opposition. Most importantly, the large numbers of workers had to stay in insufficient space without maintaining the social distancing. Moreover, when the government had started to run "Shramik Special" trains to rescuing the migrant workers from their workplaces to native villages, thousands of laborers had traveled where maintaining social distancing was impossible. So, in many cases, the spread of coronavirus disease had got momentum.

1.2.3 Digital Divide

The educational institutions have been locked for several months, all the important secondary and higher secondary board examinations had been postponed. The

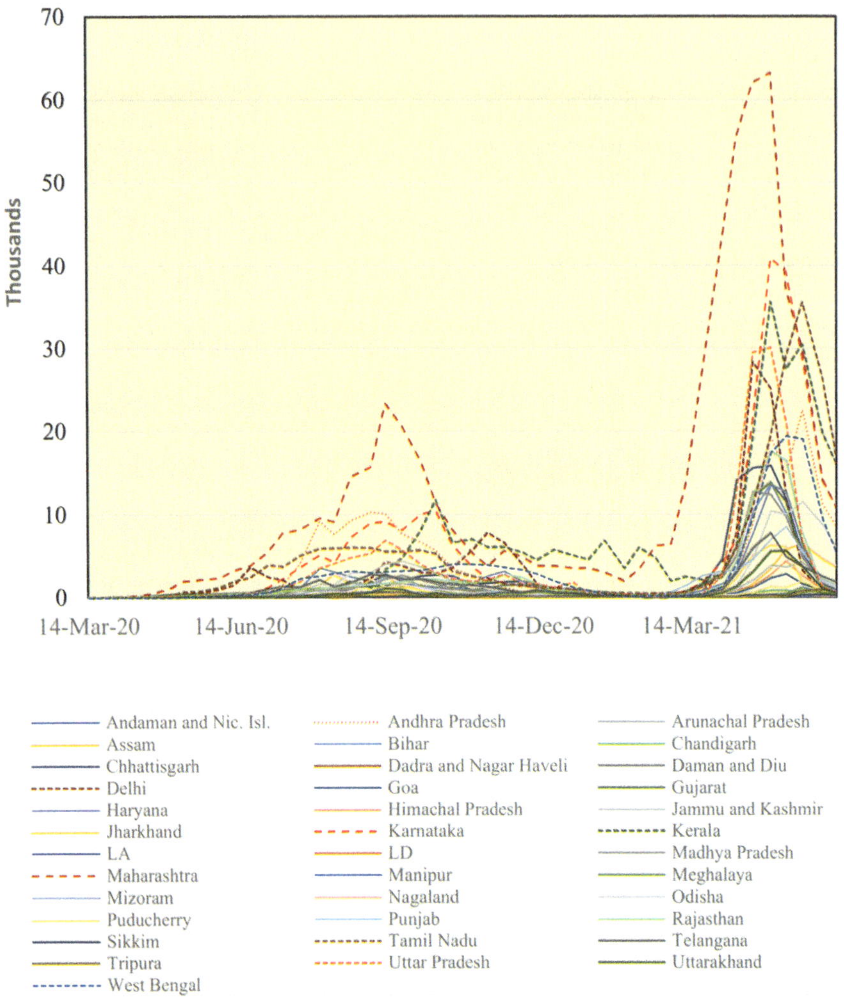

Fig. 1.1 State-wise daily COVID-19 confirmed cases from 14th March–31st October 2020 in India. (Source: Prepared by the authors with the help of Open Government Data (OGD) Platform, Government of India.) (Visit for the datasource https://data.gov.in/)

national-level various entrance tests had been postponed. The education system had been passing through an unprecedented crisis. Online education had emerged as a support practice, but it made a digital divide among the haves and have nots. The "school hood" is somehow had been stolen from the childhood of the children. The teachers and experts are predicting the possibility of huge numbers of drops out. The long-term repercussions of such an educational crisis could not be estimated in this volatile situation. Experts are concerned about the possibly large number of school dropouts and girl child marriages because of the out-of-school situation (Fig. 1.1).

1.2.4 Social Distancing or Social Crisis!

Lockdown had been implemented to restrict the rampant spread of the coronavirus disease. The social distancing had raised an issue of untouchability. In the earlier days of pandemic stuck in the nation, every person with any symptoms of COVID-19 was more concerned about the issue of untouchability and segregation in society instead of getting an earlier doctor and medical support. So, a large section of people was reluctant to be tested if there was any earlier symptom of this devil disease. A certain number of the COVID-positive patients had tried to hide the COVID-positive status because they thought they and their families could be segregated from others. A substantial psychological burden was there for the earlier COVID-positive patients.

1.3 The Calm Before the Storm

The coronavirus attack and its rampant spread were a great challenge for a country with more than 1.3 billion population. The spread of this fatal disease was gradually rising since March 2020 (Fig. 1.7). On 14th March 2020, the total confirmed cases were registered as 81 in India, and it had reached its peak (of the first wave) on 16th September 2020 with 97,860 confirmed COVID-positives cases. The entire month of September 2020, the nation had been confronted with more than 80,000 cases daily. Though the recovery rate was also improved since September and from October 2020, the recovery rate was constantly surplus from confirmed cases. The morbidity rate due to coronavirus was remarkably less in India in comparison to the other European countries and the USA (Fig. 1.2). The lower death rate had been taken as an achievement for the government to fight against coronavirus.

Moreover, the lower death rate somehow had injected overconfidence among a certain section of the population. India had been trying to get back its normalcy, and the situation started to get better since the end of October 2020. But time and life change very quickly. A large section of the population had started to celebrate as the pandemic might have gone forever. The COVID pandemic protocols had been ignored. Large gatherings were started to resume. Usage of the N95 mask had been started to be replaced by the fashionable masks, and a large portion started to believe in a COVID-free world.

The Standing Committee on Health and Family Welfare of the Indian Parliament has predicted a second wave of the COVID pandemic in their report, which was presented to the Chairman, Rajya Sabha, on 21st November and forwarded to the Speaker of Lok Sabha on 25th November 2020. However, the action by the government hardly showed any sign that would confirm that the report has been taken seriously.

Honorable Prime Minister of India, Mr. Narendra Modi, addressed the World Economic Forum's Davos Dialogue and declared that despite the early doomsday

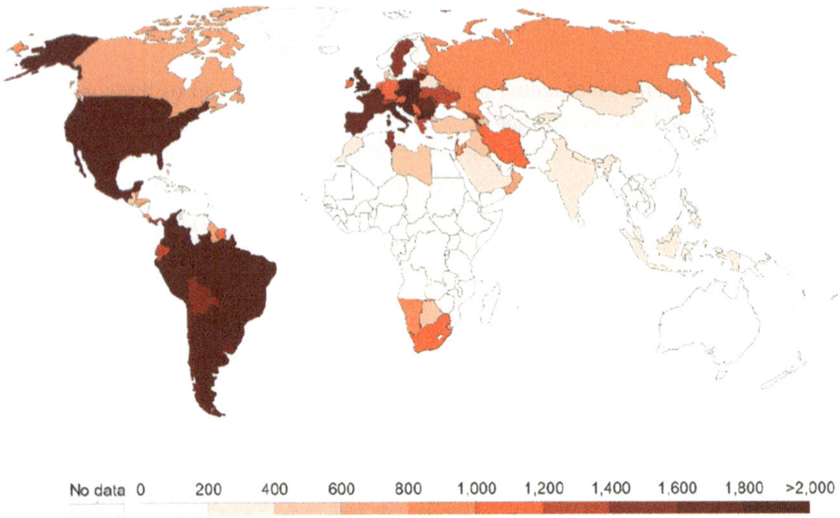

Fig. 1.2 Cumulative confirmed deaths due to COVID-19 (deaths per million). (Source: Map prepared by Our World in Data, Oxford Martin School based on the Data published by COVID-19 Data Repository by the Center for Systems Science and Engineering (CSSE) at Johns Hopkins University for 22nd January 2020–17th July 2021)

predictions on the COVID tsunami, India not only defeated the fatal COVID pandemic but also executed the global responsibility by stretching her helping hand toward 150 countries (India Today, 28th January 2021). He informed the global forum that the nation had been successfully running the world's largest vaccination program by vaccinating 2.3 million health workers in the last 12 days. Prime Minister lauded the nation's fight against the coronavirus and India's pace forward with a pledge to become self-reliant. Political parties have started to organize political rallies with thousands and thousands to show their popularity for the assembly elections in five states. Millions were gathered to take the "Shahi Snan" (the holy bath) in Kumbh Mela. Finally, India has been attacked by the fatal COVID-second wave. The whole country was praying for oxygen; there was an acute scarcity of beds in hospitals, the country has witnessed thousands of burning bodies on the coast of the Holy River Ganga. The honorable PM admitted while addressing the nation on 20th April 2020 that the country has been fighting a big battle against COVID, and the second wave has come like a storm. The entire nation was shocked to experience the lethal impact of the second wave of COVID-19.

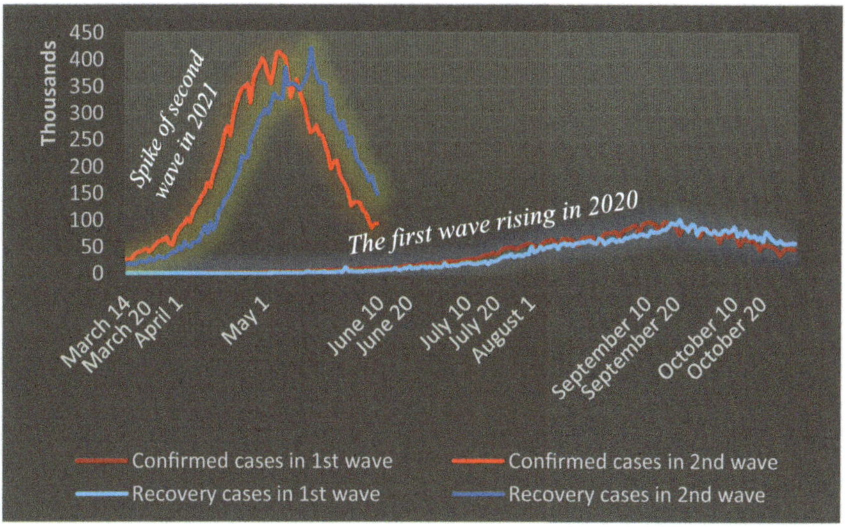

Fig. 1.3 A comparison between the COVID-19 first wave in 2020 and the lethal second wave in 2021. (Source: Prepared by the authors from the Open Government Data (OGD) Platform of the Government of India)

1.4 The Fatal Second Wave

The world has witnessed through the media that Delhi, the capital city of India, gasped for oxygen in the whole month of April (Mahapatra et al. 2021). The extremely high infectivity of the mutant virus has led to the abnormal high spike within few days (Fig. 1.3). From September 2020, the data shows a remarkable fall in COVID-positive cases and January–February (2021) that pretends everything is normal. It was actually like an unexpected recession of water levels before the tsunami strikes. In the entire country, the total COVID-positive cases of 1.1 million in February have risen to more than 6.9 million in March 2021, crosses the nine million mark in June. Between 15th April and 26th May 2021, the entire nation, its administration, and the medical infrastructure look helpless against the most formidable challenge of the fatal COVID-second wave (Figs. 1.4 and 1.5).

The history of pandemics shows that there might be several waves to hit. The Spanish Flu of 1918 came in three waves, and the second wave was considered as deadlier as the people died within few hours or just after few days of the appearance of symptoms.[1] The countries were already reporting on the increasing COVID-positive cases in the second wave. However, in India, the Central and most State governments, local administrations, and the common people denied the reality desperately. The central government had already declared and started to campaign

[1] See the World Atlas discussion on '10 Pandemics Throughout History', available at https://www.worldatlas.com/articles/10-pandemics-throughout-history.html.

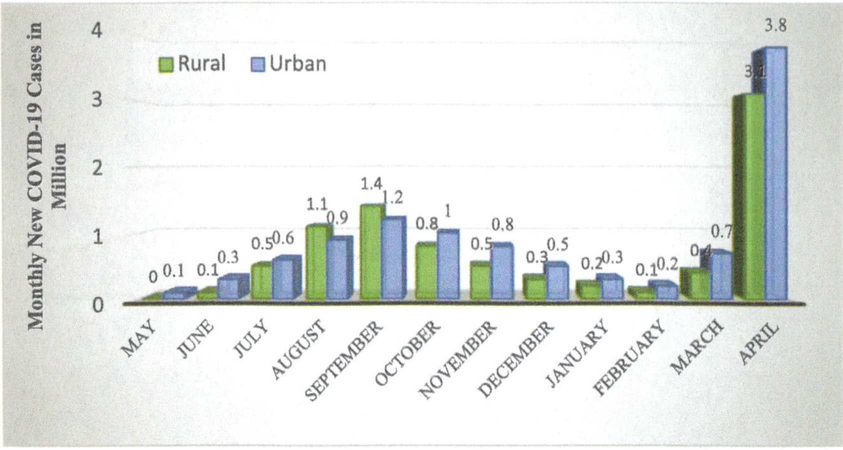

Fig. 1.4 Monthly New COVID-19 cases in rural and urban India, May 2020–April 2021. (Source: Drawn by the authors based on Mahapatra et al. 2021, Down To Earth, 16–31 May 2021; Census of India 2011)

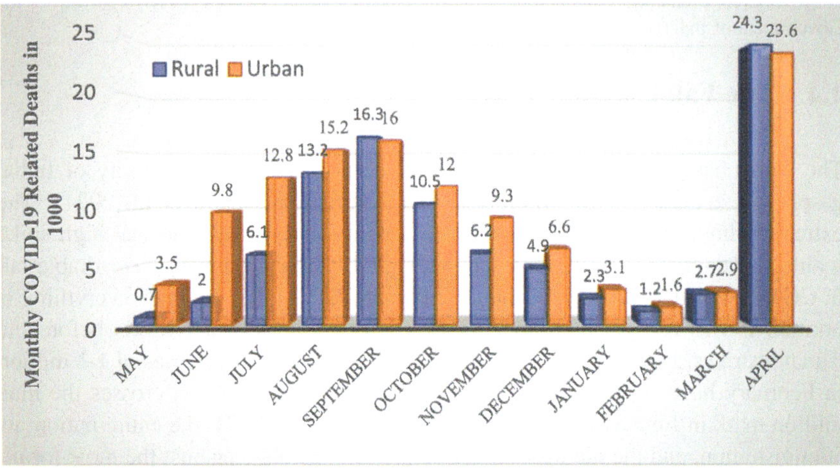

Fig. 1.5 Monthly COVID-19 related death in rural and Urban India, May 2020–April 2021. (Source: Drawn by the authors based on Mahapatra et al. 2021, Down To Earth, 16–31 May 2021; Census of India 2011)

the success story of flattening the COVID curve. Confirming the success story to win over COVID, only 9086 COVID positives were recorded in the entire nation on 15th February 2021. However, the pandemic situation was taking a dramatic change since early March 2021. On 10th March, the total daily cases were recorded as 22,851; on 20th March, it raised to 43,815, nearly doubled in 10 days. The lethal second wave surge crossed the 1,00,000 daily contamination on 4th April, and the spike continued to cross the daily 2,00,000 COVID-positives cases on 15th April. India exceeded

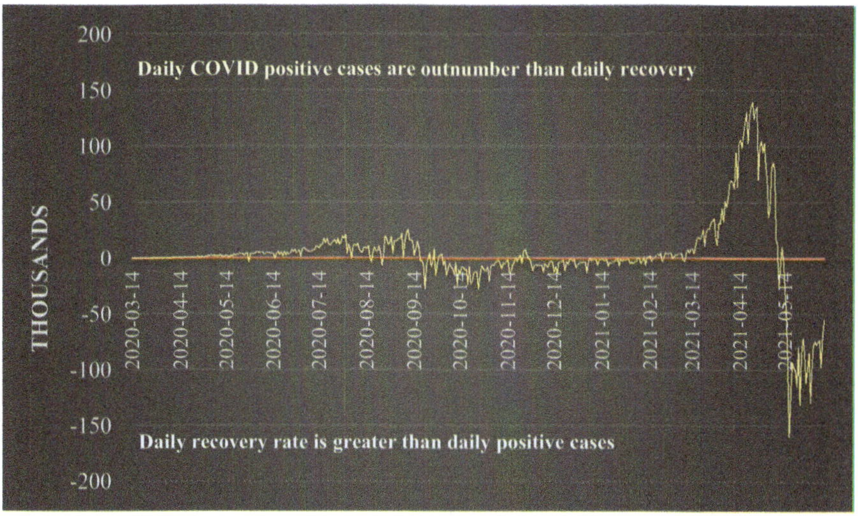

Fig. 1.6 The continuous fluctuation in Confirmed vs Recovery rate of the COVID-19 in India. (Source: Drawn by the authors based on the data from the Open Government Data (OGD) Platform, Government of India)

the limit of one million active cases on 9th April (Fig. 1.6), and by 12th April, the country left Brazil behind to be the second-highest COVID-positive cases in the world. On the last day of April, India was the single country globally to record more than 400,000 COVID-positive cases in a single day (Fig. 1.7). Between 5th May and 8th May, the daily cases in India were constantly recorded with over 400,000. Every state and union territory of India had been affected by the COVID pandemic. However, the pandemic has stuck in some states with greater severity (Fig. 1.8). Maharashtra, Delhi, and Kerala are massively affected, followed by Karnataka, Tamil Nadu, Uttar Pradesh, Andhra Pradesh, and West Bengal, the second wave surge.

There must be several factors to contribute to and accelerate the sudden spike in the COVID-19 s wave. Since the outbreak of coronavirus, there are several mutations to evolve new variants, which wreaked havoc with millions of infections and death tolls across the world. Early in the second wave, it was evidenced as if multiple variants were responsible in different states in India behind the second wave surge. Delhi was the most affected state where B.1.1.7 (Alpha Variable) variant which was initially identified in the United Kingdom, was dominant. A variant B.1.617 was dominant in Maharashtra, whereas B.1.618 was dominant in the state of West Bengal (Vaidyanathan 2021). However, after the initial days, B.1.617 variant wrecked many states and overshadowed other variants. This particular leading variant with havoc potential to spread the disease has had three sub-lineages: B.1.617.1, B.1.617.2, and B.1.617.3.[2,3] However, the first two sub-lineages, i.e.,

---

[2] https://www.who.int/en/activities/tracking-SARS-CoV-2-variants/

[3] https://covariants.org/

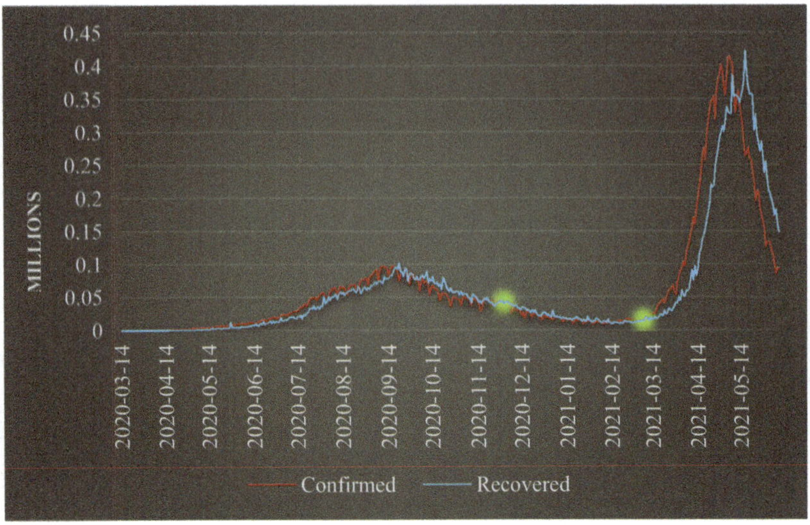

Fig. 1.7 The COVID-19 cases trend in India, 14th March 2020–10th July 2021. (Source: Drawn by the authors based on the data from the Open Government Data (OGD) Platform, Government of India)

B.1.617.1 and B.1.617.2, might have fuelled the brutal second wave in India, viz., these two lineages are responsible for around 70% of the SARS-CoV-2 genomes sampled in India (Kunal et al. 2021) as per the Global Initiative on Sharing All Influenza Data (GISAID) database. Moreover, this B.1.617.2 sub-lineage was first identified in India, and it is labeled by WHO as Delta variable, i.e., a "variants of concern."[4] Kunal et al. (2021) also reported that these variants are not only restricted to the Indian subcontinent, instead they have been transmitted across various countries in the world since late February 2021.

The other factors of brutal COVID-second wave surge include the declaration of premature victory against COVID, extreme lack of preparations in administration, health and safety precautions according to COVID protocols being ignored in social gatherings, festivals, gatherings in thousands and lakhs for religious pilgrimage, and festivals, local and state elections in several states in the country where politicians have addressed rallies, meetings with thousands of populations. The pressure of economic slowdown due to long-run lockdown in 2020 put pressure on the Union and state government to lift the COVID restrictions as early as possible to Kick off the run of the economy. However, this quick response does badly contribute to the second wave of pandemic situations. The infectivity rate is severely high. India gasped for oxygen. In every large city, patients with severe symptoms thronged into hospitals desperately need oxygen.

[4]See WHO website: https://www.who.int/en/activities/tracking-SARS-CoV-2-variants/.

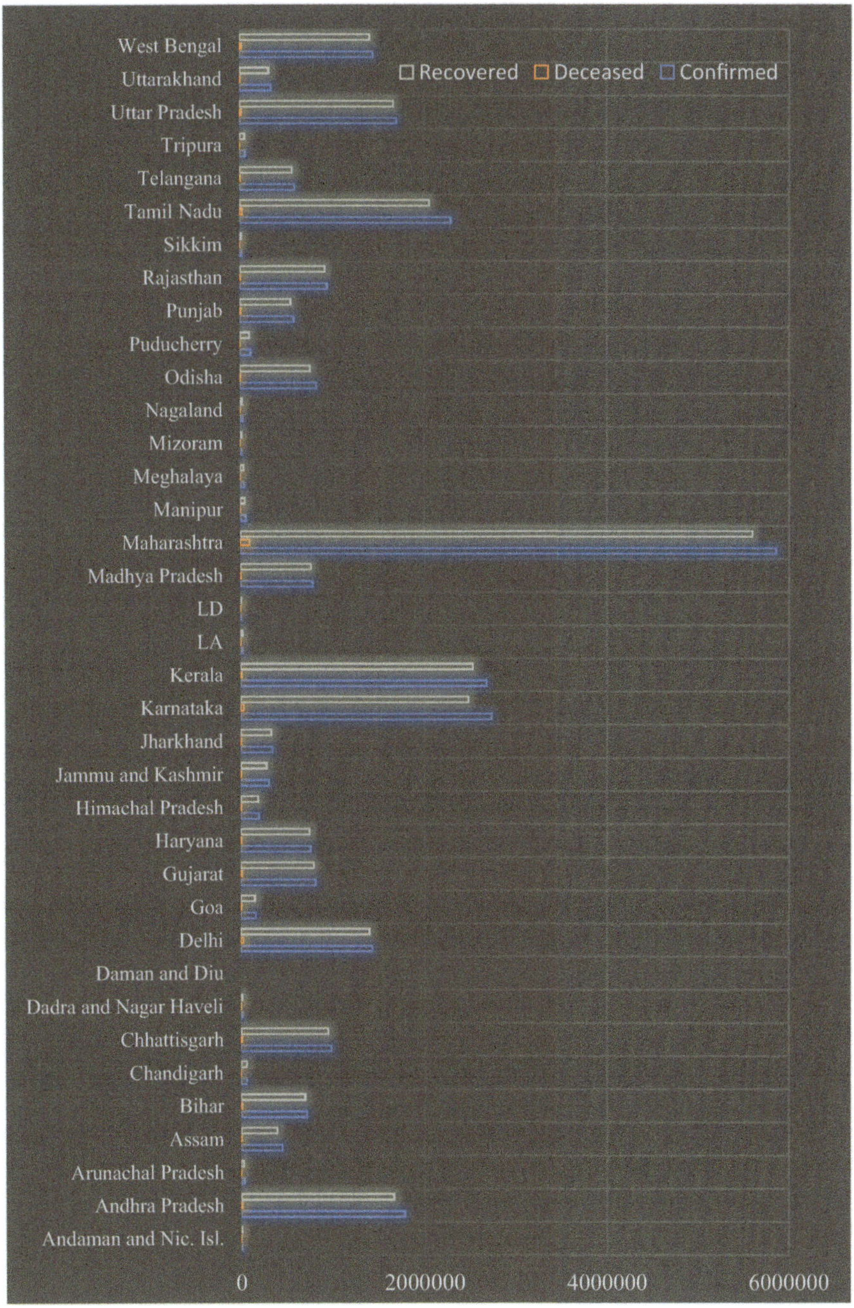

Fig. 1.8 State-wise COVID-19 status in India, 14th March 2020–10th June. (Source: Drawn by the authors based on the data from the Open Government Data (OGD) Platform, Government of India)

1.4.1 At the Doorstep of Rural India

The most striking concern of the COVID-second wave is the spread of this brutal virus in rural India. The relatively poor health infrastructure, lack of hospital beds with ventilation, shilly-shally attitude to get tested, staying at home after being affected, and reluctance to go to the hospital is the attitudinal factors that elevated the risk in rural India (Fig. 1.4 and 1.5).

1.5 Discourses of Neo-liberalism and Governance: Does the Changing Equations between State, Markets and Civil Society Take on Pandemic Management

All manners of exercising control and authority in allocating resources are cumulatively termed as governance (World Bank 1994). So how and to what extent the citizens access resources are closely are associated with the processes and mechanisms of governance. In the Indian context, and also globally, the widening gap between "governance" and "government" has been a great concern as the authoritative role of many actors other than the state itself come at the forefront of decision making in the allocation of resources (Mathur 2009), as Narain et al. (2014) speculates that "the locus of policymaking has moved from the state to other actors: markets and civil society have created greater space for themselves." The growing dilution of the State authority by a more significant influence of other actors has been becoming prominent at both the local and global levels (Kumar and Narain 2014). The global dialogues on the climate change event and its regional action narratives may be cited out of ample examples.

The way the discourses of neo-liberalism and good governance have had changed the relationships between state, markets, and civil society, poorer become more vulnerable. This affects both the state and the poor citizens of the states as Urs and Whittel (2009) speculates that while the discourse of neoliberalism had been founded on the narrative of the inefficient state, the neo-liberal paradigm attracted criticisms on account of its tendency to exclude the poor from the provision of service delivery. The pandemic has exposed it to the daylight for income (e.g., the mass joblessness of the informal and unorganized labor forces), for education (e.g., the ever widened digital education divides), and health (e.g., the fatalities due to severe oxygen crisis during the second wave). It would affect the progress of India toward achieving the Sustainable Development Goals.

1.5.1 Struggling Public Healthcare System and the Poor

India has had been struggling in the public health care system in the form of a lack of infrastructure and a severe shortage of human resources. The health care system is the most critical component of COVID-19 pandemic preparedness. Indian health care is provided in three tiers system-primary, secondary and tertiary health care sector (Danigond 2021). Rural India is provided the health care service through Primary Health Center (PHC), subcenter, and Community Health Center (CHC). The secondary health care system is provided through the district hospitals, and the regional and/or central super-specialty hospitals are considered as the tertiary health care sector (Danigond 2021). With a 1.4 billion population in India, providing proper and adequate health service is a massive task. The doctor-population ratio (1:1456) is much less than what is recommended by WHO, i.e., 1:1000 (Pilla 2020). The issues of shortage of efficient and trained human resources, the inadequacy of infrastructure, high out-of-pocket expenditure (Fig. 1.9b), huge patient load have made the situation challenging for the doctors and administration to fight against coronavirus. Government expenditure on health care seems to be flat for a decade. The inadequate spending in the health sector is much more evident when it is compared in the proportion of GDP (Fig. 1.9a).

A large proportion of the rural and urban populations has to depend on the private sector for health and medical urgency. Moreover, in this time of health emergency, the catastrophic rise in medical expenses has made an economic crisis to the middle, lower-middle, and poor income group of people.

The routine health care system and child immunization program have been affected badly. Due to the closure of immunization centers in the lockdown and the risk of infection, child immunization was disrupted. The current pandemic has shown the lack of preparedness at the workplaces, in public and private establishments to cater to infection prevention control guidelines because of improper architecture and design. Moreover, the role of public health engineering at the time of emergency, imposed by pandemic, war, or by any disaster, must be considered in policy framing.

In the twenty-first century real-time, data-driven evidence-based decision-making is important to fight against any hazard and disaster. Union and State Govt. must focus on strengthening the health management information system. Laboratory and pathological networks must be extended with the basic medical testing facilities and adequate infrastructure in the peripheral primary health centers and subcenters in the rural areas (Garg et al. 2020). An integrated health information and communication system must be developed to manage, store and analyze the public health-related data. The real-time data on public health centers, the public and private hospitals, availability of beds, and other infrastructure and helpline numbers should be available to the common people throughout the year. Public healthcare laws and regulations must ensure the active participation of private hospitals and the corporate health care sector in disease control and provide health services at a reasonable cost under government surveillance. The pandemic shows the necessity of govt.

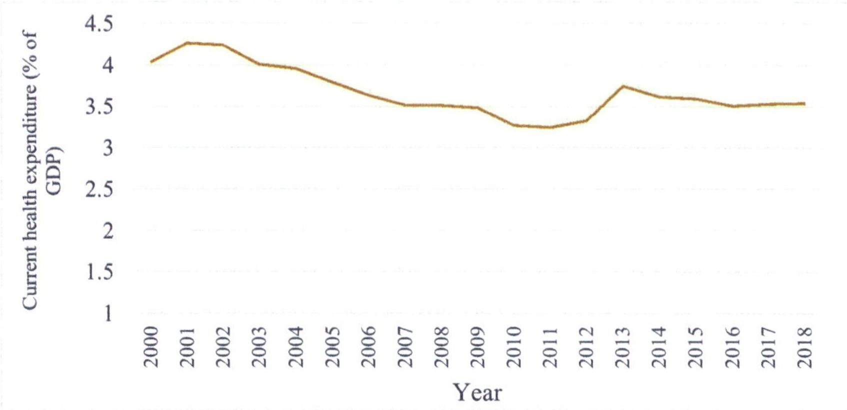

(a)

(b)

Fig. 1.9 (**a**) Current health expenditure (% of GDP), and (**b**) Out-of-pocket expenditure (% of current health expenditure) in India (Source: Prepared by the authors based on the World Health Organization Global Health Expenditure database.) (See WHO datasets in www.apps.who.int/nha/database)

Surveillance in the market provides essential medicines and health equipment (e.g., need of oximeter in the pandemic situation) at a reasonable cost in any health crisis condition. The National Health Policy 2017 has proposed a unique Public Health Management Cadre development concept in all states (National Health Policy 2017; Garg et al. 2020). Moreover, such public health team must have a multidisciplinary character to respond to a health crisis. Health care financing and adequate budgetary allocation in the public health sector are the real-time demand.

1.5.2 Government and Governance

The pandemic effect has made some unprecedented changes in the normal cycle of life and provided the states to execute an unusual power to put people under various degrees of containment, including a prolonged nationwide lockdown. The nation has experienced a lockdown declared on 25th March 2020 and also witnessed the phase-wise lifting of the lockdown since 1st June 2020. The government had taken substantial steps to shut down every human movement to flat the COVID curve. Educational institutions have been closed for a year, the economy has been severely affected, millions of people lost their jobs, parliamentary sessions were truncated, and the COVID curve has been showing some positive sparks with its remarkable reduction in COVID cases since January 2021. However, unfortunately, the governance comes to a halt when it was the time of General assembly Elections in few states. The government has suddenly been a very aspirant for the "game of power." The Election Commission of India (ECI) declared assembly elections on 26th March 2020 for the states of Kerala, Tamil Nadu, Assam, West Bengal, and Puducherry.

To reduce the possibility of the spread of COVID-19 at the time of the election, the Centers for Disease Control and Prevention (CDC) recommends some specific guidelines to conduct the elections. Three guiding principles are followed to restrict the risk of coronavirus spread:

- A wide variety of voting options are recommended to restrict a gathering.
- A longer vote casting period is introduced (7 a.m. to 6:30 p.m.) with more phases.
- Initiatives to reduce the number of voters by adopting all feasible options to reduce the risk.

There are several recommendations for election officials, workers, and poll agents to behave maintain the COVID protocol. However, all those advice were strictly followed in papers, not in practice. The health experts have unequivocally agreed that the election campaign and related gathering accelerated the COVID pandemic second wave surge in several states. WHO and the Honorable Indian Judiciary have also deliberated the same view.

On 26th March, the day the Chief Election Commissioner of India addressed the nation regarding the Assembly election schedule, the COVID-positive cases in the five states of Assam, Kerala, Puducherry (union territory), Tamil Nadu, and West Bengal was recorded as 37, 1825, 96, 1971, and 646, respectively. However, on 2nd May 2021, the day the result of the poll was declared, the COVID-positive cases in the five states of Assam, Kerala, Puducherry (union territory), Tamil Nadu, and West Bengal have recorded 2385, 31,959, 1360, 20,768, and 17,515 respectively (Fig. 1.10).

The ECI announced the general election schedule of the Legislative Assemblies in four states and one union territory-Kerala, Tamilnadu, West Bengal, Assam, and Puducherry. In Kerala, Tamilnadu, and Puducherry, assembly elections were held in single-phase, while in Assam, it was in three phases, and the maximum eight phases were declared for West Bengal. All the political parties had made their campaign

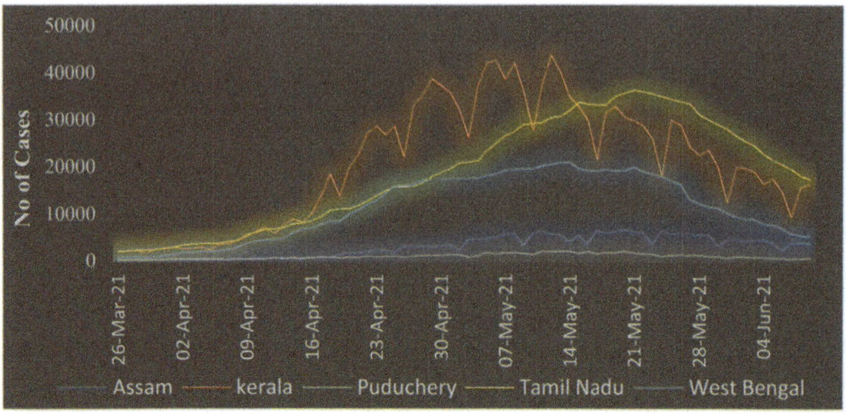

Fig. 1.10 The General Assembly Election in the Indian States and the COVID-second wave surge (Drawn by the author based on Open Government Data (OGD) Platform India, Government of India)

with thousands of gathering. Political leaders addressed the massive rallies of thousands to lakhs of people, where maintaining the COVID regulations and safety protocols is a reverie. The division bench of Calcutta High Court, on 22nd April, criticized the role of ECI and remarked that despite having the full power to execute the COVID norms, the ECI did not take any responsibility to control the massive polling rallies by political parties (Rajaram 2021). On 26th April 2021, Madras Highcourt slammed the ECI for not controlling the political parties from violating the COVID norms during the campaign rallies and blamed the ECI for raging the COVID-second wave surge in Tamilnadu and Puducherry (Chandrababu 2021; Deka 2021; The Hindu 2021). The paramount importance of public health was unfortunately disregarded. Politics and political power were materialized at the cost of common people's life.

It is reported that the Delta variant (B1.617.2) was the primary cause behind the fatal second COVID-second wave surge in India (Thakur 2021). The second wave surge was also very fatal in the states of Maharashtra, Uttar Pradesh, Delhi, and other provinces, where there were no assembly elections. However, that does not justify the inappropriate behavior of the administration and political fraternity during the crisis amid COVID mayhem. When the nation has already experienced the unprecedented impact of the COVID-19 pandemic during the first wave, and when the world was still struggling under the pandemic, possibly, no logic can be tabled in the name of constitutional exigency that can justify the pro-active roles of politics during the crisis (Deka 2021).

Millions of Hindu devotees gathered in the northern Himalayan foothill city of Haridwar in March and April 2021 to participate in the largest religious pilgrimage on Earth, the *Purna Kumbh Mela*, which typically comes about every 12 years. While the nation grappled with the COVID pandemic second wave, the gathering of millions might have escalated the contamination risk as predicted by the health experts. Though the traditional start date of the holy bath, called Makar Sankranti,[5] was in January 2020, due to COVID measures, the time was scaled back to April 2021. Authorities have taken some important measures includes reducing the pilgrimage from traditional three and half months to only 1 month. However, the religious sentiments and devotions of devotees disregarded the COVID protocols to many extents. Hundreds and thousands of devotees have been gathering in Haridwar since mid-January to flock in the Holy river Ganga in Makar Sankranti. The possibility of the spread of coronavirus was extremely high as they have shared quarters or tents for hours, shared common public facilities, enjoyed meals together (Yeung et al. 2021).

The fatality of the COVID-second wave seen in the nation is not just because of the rapid spread of the coronavirus, rather the crunch in the life-saving oxygen across the country, more significantly in Delhi, Maharashtra, Uttar Pradesh along with other states have made the situation worse. Medical scientists and doctors reported that in the second wave, the coronavirus strain seems to be invading the human lungs very quickly, raising the oxygen demand of the contaminated persons within just a few days since getting infected (Palicha 2021). Moreover, the sudden rise in oxygen demand challenges the public health care system, health infrastructure, government policy, plan, and response. Before the COVID pandemic stuck, there was no crisis for oxygen in most of the hospitals across India, as it was commonly available and inexpensive. However, the situation was changed extraordinarily on April–May 2021 with the second COVID wave surge. Oxygen in its pure state is majorly used in industrial uses (75–80%), and 20–25% of its production is used for medical purposes in India (Seshadri 2021). The exponential surge of COVID cases and quick breathing problems had drastically been draining medical oxygen supplies. Within just a few weeks in April 2021, the total health care system seems to be swamped by the rising cases. Several numbers of deaths were reported in the hospitals in large cities and towns across the country because of the severe shortage of oxygen. However, a lack of official database management, undercounting, and a tendency of official denial challenge the actual situation to be perceived. A report prepared by Priya et al. (2021) found that at least 629 people died because of oxygen shortage in 110 hospitals across the nation in the period of 6th April–19th May 2021. Attack of the new COVID strain at a devastating pace, lack of precautions, under-equipped health care system led to happen the situation while lacking interstate cooperation, the political fight between the Union and State Governments, exaggerated oxygen demand by the state in few cases, and finally, the black marketing contributed the oxygen crisis that led to several deaths in India.

The Central Government has responded to fight the situation by working on a comprehensive six-month capacity-building plan to upgrade the health sector to

[5]While most festivals are set by the lunar cycle of the lunisolar Hindu calendar, *Makar Sankranti* is one of the few ancient Indian festivals that is observed according to solar cycles. Being a festival that celebrates the solar cycle, it almost always falls on the same Gregorian date every year i.e., 14th January.

resilient enough in this health crisis period. It has planned to adopt robust steps in public health, including beds in hospitals, availability of medical oxygen, and the basic infrastructure to fight against any illness (Singh 2021). The oxygen production has been scaled up to 9446 MT per day in August 2021, while it was only 5700 MT per day in May 2021 (Prasad 2021). Govt. has taken steps to scale down the industrial use of liquid oxygen, and all the private and public manufacturers were asked to surge the production of liquid oxygen. Besides the manufacturing and production, the oxygen was supplied to the states by "Oxygen Express" run by Indian Railways, and also the Indian Airforce has also taken the initiative to airlift the oxygen containers. However, if such initiatives could have been taken earlier, maintaining coordination with the states, maybe the nation would witness fewer burning bodies on the crematoriums.

1.6 Vaccination Drive: India Is a Forerunner amongst the Developing Nations

India started the world's largest vaccination drive on 16th January 2021, at a total of 3006 session sites across the country. The Govt. of India initially started the administration of COVID-19 vaccination with the approval of the Oxford–AstraZeneca vaccine, manufactured by the Serum Institute of India with the trade name Covishield and the country made Covaxin, developed and marketed by the Bharat Biotech. Now, other internationally reputed Sputnik V and Moderna vaccines with others have been approved. Within 6 months, India has administered a total of 398,577,881 doses that includes the first dose of 31,74,87,620 and the second dose of 8,10,90,261. Moreover, currently, more than four million people are vaccinated daily. India announced its first phase of the vaccination program with a target of 30 million people, including healthcare workers and frontline workers, i.e., police personals, paramilitary forces, disaster management volunteers, and sanitation workers at free of cost (Bhuyan 2021). By 1st March 2021, the target was limited to only 14 million vaccination because of the primary concern on the safety and side effects of the vaccination (Allana 2021). The second phase was declared to vaccinate the persons aged above 60 years, persons between the ages 45–60 with comorbidities. The health care and frontline workers not covered in the first phase were also eligible to get their term in the second phase. From 1st April 2021, the eligibility to get vaccination was extended to all persons aged above 45 years. The country has already started to observe the lethal impacts of the COVID-19 second wave surge. The third phase with liberalized vaccination policy was announced on 19th April to extend the eligibility to all the residents above 18 years from 1st May. In this third phase, Central Govt. has decided to take responsibility for the distribution of 50% of the vaccines produced by the Central Drugs Laboratory.

The policy was to roll out those doses of vaccines from government-run clinics and be provided free of cost to persons 45 years and above and to the frontline

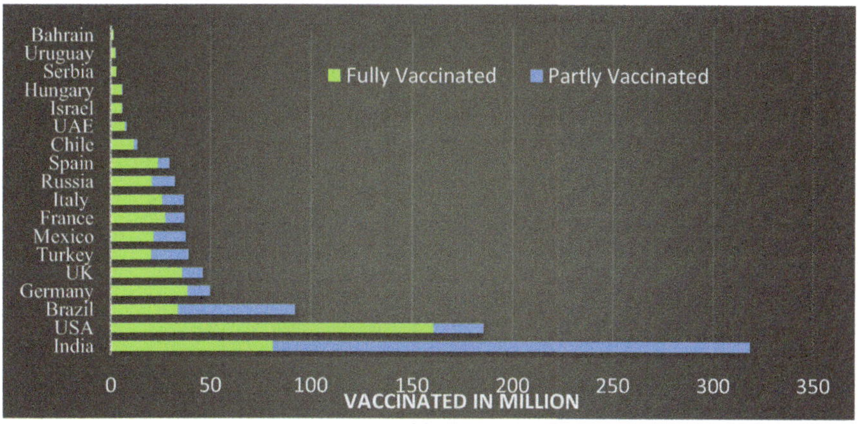

(a)

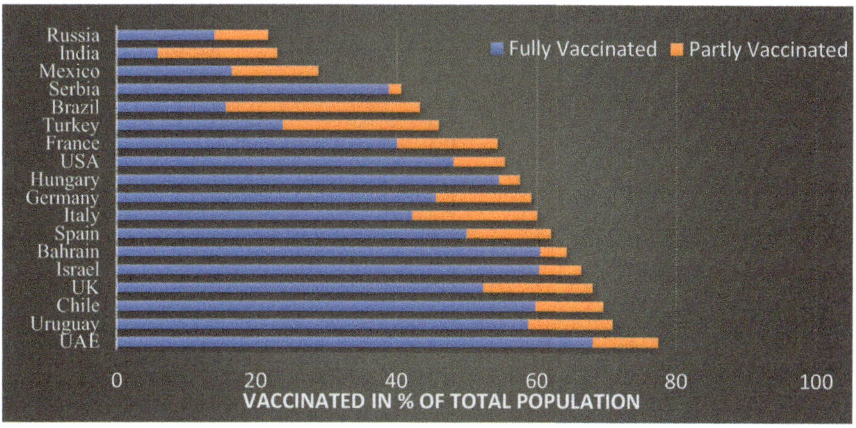

(b)

Fig. 1.11 (**a**) Number of people vaccinated against COVID-19, as of 16th July 2021 and (**b**) Proportion of Total Population vaccinated against COVID-19 as of 16th July 2021. (Source: Our World in Data, Oxford Martin School, the University of Oxford, Accessed on 17th July, 2021)

workers. Despite four million daily doses of COVID vaccination, only 23% of the Indian population has got at least one dose, and around 6% of the population was fully vaccinated, and the proportion of vaccination picture seems to be blurred in comparison to other countries across the world (Fig. 1.11b). However, considering the absolute number of residents vaccinated, India is the leading one (Figs. 1.11a and 1.12). For the Indian states, the same picture is found (Fig. 1.13), as the states with small populations show the better result in percent, though the larger states have administered a larger number of doses. The statewise vaccination status (Fig. 1.13) is quite heterogeneous because of the availability of vaccines, state-specific plan of distribution, and infrastructure. It is an enormous challenge for a country with more than 1.4 billion population to be vaccinated within 6 months, but the vaccination

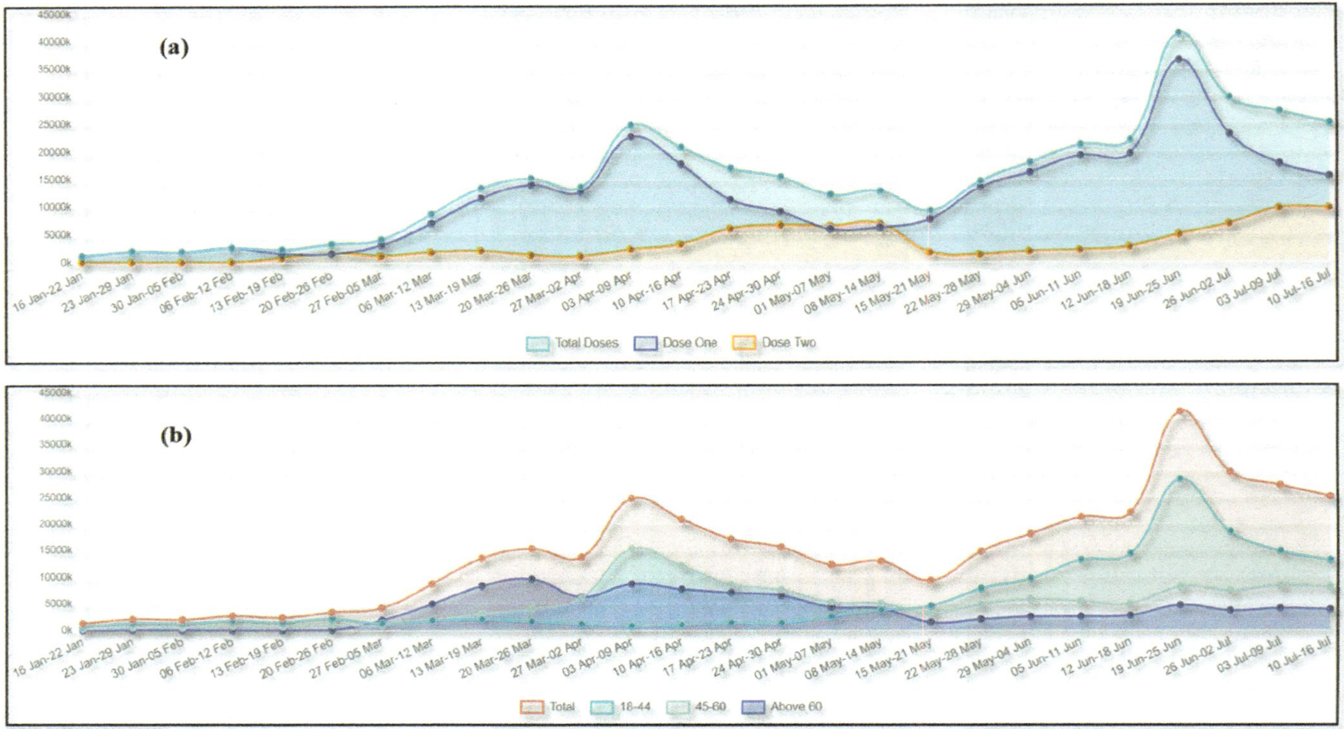

Fig. 1.12 India's Vaccination trend by (**a**) doses and (**b**) age. (Source: COWIN Dashboard, MOFFW, GOI, Accessed on 17th July 2021)

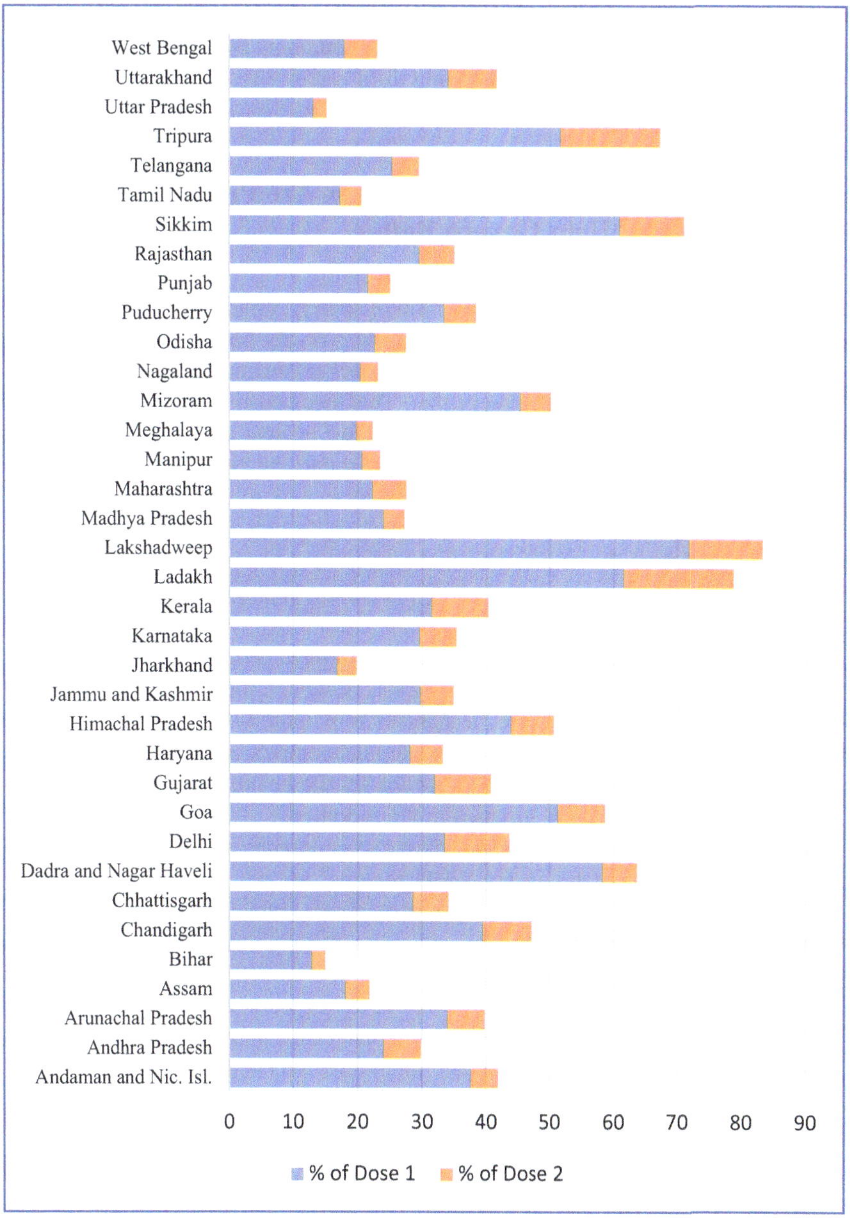

Fig. 1.13 Statewise Vaccination Status in India as of 27th June 2021, 08:00 IST. (Source: Drawn by the author based on GOI database.) (Database accessed from https://www.mygov.in/COVID-19/)

trends, both in respect of doses and by age, shows ray of hope (Fig. 1.12). Scientists and experts advised that rapid vaccination and maintaining the COVID protocols are crucial ways to fight against the upcoming third wave.

1.7 Self Imposed Prevention Measures Are Precious: Self-Care to Help Govern Better

It is often said that while culture is the mind of society, civilization forms the body. Man has planned since the time immemorable with its best effort to control the conditions of life and civilization is the whole mechanism and a kind of organization which has empowered man to achieve that. This mechanism of the human being to control the condition of life is the collective effort against adversities. Society has long been following the ethos of assemblage around the people in distress—both physically and mentally. It works as the support system to the individual or group getting through a stressful ambiance. However, this pandemic-stuck world is critically different where the imposition of the lockdown has been in the forefront to break the chain of contagion by ensuring people are kept socially distant—a topsy turvy to the usual social norms.

The self-care interventions in the health system were getting popular and achieving recognition in a slow but consistent mode. However, the pandemic has made a huge shift in adopting the self-care mechanism as reciprocity to help the state govern better to its citizen. It is significant for developing countries, the global south nations, and the sub-Saharan poverty patch where the fragile health system has repeatedly surrendered to communicable diseases in the recent past.

The health care system has inequalities across geographical locations. The grouping of WHO member states in six WHO regions[6] are not merely a geographical clustering of countries; instead, the regions represent crucial public health differentials. Table 1.2 exhibits the differences across the regions. While the Domestic Government Health Expenditure (DGGHE) is PPP int$1986.9 in Europe, Africa crawling at PPP int$138.9. The Probability of Dying between 15 and 60 years per 1000 population is negatively correlated with DGGHE, making government health expenditure a crucial factor. Though the South East Asia and Latin America (which is different from North America in terms of health service though both belong to the WHO AMRO region) have a better situation than Africa, still the population pressure is the challenging factor where the quality of the health service needs to compromise while reaching to such a vast population. The states that need to expend

[6]WHO Member States are grouped into six WHO regions—African Region (AFRO), Region of the Americas (AMRO), South-East Asia Region (SEARO), European Region (EURO), Eastern Mediterranean Region (EMRO), and the Western Pacific Region (WPRO). See WHO website: https://www.who.int/healthinfo/global_burden_disease/definition_regions/en/.

Table 1.2 Some key parameters across WHO regions

WHO Region	Domestic general government health expenditure (DGGHE) per capita in PPP int$, in 2017	Domestic general government health expenditure as a percentage of gross domestic product (GDP), in 2017	Probability of dying between 15 and 60 years per 1000 population, in 2016		Mean Population density of the region (Person/Sq. Km.)
			Male	Female	
EURO	1986.9	4.9	155	70	110
WPRO	835.3	4.2	104	69	109
AMRO	816.8	4.0	162	89	82
EMRO	956.1	2.6	172	125	112
SEARO	253.3	2.1	206	134	239
AFRO	138.3	1.9	306	248	79

Data source: WHO (2016, 2017), World Bank (2018)

a lot to pull a large section of the population out of poverty are constrained by spending a considerable share of their GDP as people's health expenditure.

Reaching the basic health services to the wider population involves both financial support and a sufficient number of healthcare workers, and during a health emergency, the requirement increases manifold. While WHO estimates 3.6 billion, nearly half the world population, lacks access to essential health services and 100 million had plunged into poverty because of out-of-pocket health care, the world is going to face a shortage of 12.9 million healthcare workers by 2035.[7] At this juncture, the self-care interventions for health have appeared as the essential mechanism for a country like India where the contagion like COVID-19 outbreaks brings forth immeasurable plight to the uncounted with most of the fatalities arises due to the manifold surplus pressure to limited healthcare resources.

With this backdrop, WHO launches its first conceptual framework for the self-care interventions for health with detailing its goal as:

> The conceptual framework recognizes that in addition to the traditional self-care practices that societies have passed on through generations, people are accessing new information, products, and interventions through stores, pharmacies, and the internet. [...] This means access to the following: justice; a strong, accountable, people centered healthcare system; integrated and accessible services of good quality; protection from violence, coercion, and discrimination; social inclusion and acceptance; and knowledge and information, appropriately tailored to different needs.[8]

Self-care is the individual's ability to promote and maintain health, prevent disease, cope with illness, and managing the disability with or without the support of a healthcare provider. People have long been practicing self-care, which is now shifted

[7] See the report in The Hindu, dated 30th June 2019 https://www.thehindu.com/news/national/who-launches-its-first-guidelines-on-self-care-interventions-for-health/article28234153.ece.

[8] See WHO website https://www.who.int/reproductivehealth/self-care-interventions/conceptual-framework/en/.

in the way health care is perceived, understood, and accessed. Traditional self-care practices are now made more effective through the addition of medicines, diagnostics, where technologies play a vital role. However, while the self-care interventions show the silverlining for the lower and lower-middle economy, the WHO has only released the advisory for self-care interventions for sexual and reproductive health and rights (SRHR) for the people to follow amid the COVID-19 keeping a larger area unexplored till the mid-2021.

Why should self-care be an effective mechanism in India? Let the emergencies like the COVID-19 outbreak aside for a while. As per the provisional population totals of Census of India 2011, the total population of India was 1210.2 million. Of this, the rural population stands at 833.1 million. The usual rate of illness arising out of this vast rural population from rural India has no sufficient infrastructure to address. Now add with it the low-income and densely populated urban slopes. It is a fact that 17% of urban India lives in slums and over a third of India's slum population lives in its 46 million-plus cities.[9]

The rural health sector has been undergoing some progress since independence, reflected through the improving health status of the population when health indicators are analyzed with respect to time. Still, there exist high disparities between states, regions, socioeconomic classes, and gender, and also, the hard reality is that though achievements in health sectors appear significant. Yet, the survival rates in India are comparable to the challenging health scenario of sub-Saharan Africa. Interestingly, most rural deaths are due to infections and communicable, parasitic and respiratory diseases in the list of preventable diseases. However, infectious diseases dominate the morbidity pattern in rural India. The COVID-19, when started primarily as the urban health hazard it did not take time to transform to a nationwide outbreak on this preexisting grazing ground of the infectious disease.

However, it is often said that for infectious disease, "prevention" is a better option than "cure," and preventing the disease is also the cheaper option that suits well with the economy with multiple challenges. The success stories of the WASH program of the Center for Disease Control (CDC) are worthy of being cited here. It saves millions of lives and reduces illness by improving global access to healthy and safe water, adequate sanitation, and improved hygiene. The WASH program has proved itself to work on "long-term prevention and control measures for improving health, reducing poverty, and improving socioeconomic development as well as responding to global emergencies and outbreaks of life-threatening illnesses."[10]

While the control of COVID-19 contagion is grounded on the successful breaking of the infection chain by maintaining social distances, prevention depends on easy

---

[9] See the Tomes of India report on 22 March 2013: https://timesofindia.indiatimes.com/india/17-of-urban-india-lives-in-slums-census/articleshow/19118219.cms.

[10] See CDCs WASH program factsheet at https://www.cdc.gov/healthywater/pdf/global/programs/global-wash-overview-fact-sheet.pdf.

steps such as self-action, e.g., wearing face masks, washing hands frequently, and maintaining basic hygiene standards. In an extended version of the self-care interventions in health, the effort of making the communities and localities self-sufficient in basic health monitoring (e.g., body weight, temperature, blood pressure, oxygen level), pathological sample collection, vaccine administering will reduce the pressure on the public health centers and professional health workers in one hand and generate alternative income sources to the unemployed youth. The successful functioning of the *Sahaj Tathya Mitra Kendra Scheme*[11] in the state of West Bengal in collaboration with private partners may be a role model. The arrangement of training and initial financial support could do the wonder in the self-care health intervention model in India and other low-incoming nations.

1.8 Conclusion

The COVID pandemic has challenged human civilization unprecedentedly, and some extraordinary responses need to vanquish this dark period. Every crisis leads to some opportunities. India is trying to be resilient enough to fight against pandemics, epidemics, or any crisis period with the vision of "Atmanirvaar Bharat" (Self-resilient India). The inadequacies in the public health service in all three sectors were prominent and needed to escalate the infrastructure. Specialized human resources, superspecialists doctors with the corresponding upgrade of medical equipment in Govt. hospitals are needed. This pandemic paved the way for new strides in technology, mainly in healthcare services through contact tracing, rapid test kits, and the use of portable instruments for a quick health check-up. The pandemic situation has explored the scope of tele-consultancy, online doctor consultation maintaining the Govt. approved guidelines through various Mobile Apps that lead to a new avenue toward medical communication. Besides, a systematic referral system should be implemented in every state to reduce the burden on the apex city hospitals. Digital technology, especially Artificial Intelligence and Machine Learning (AI&ML), has ample scope to revolutionize real-time data management, sharing, and managing such a crisis. The outbreak of COVID-19 has affected society and the economy through several channels. The supply chain in the Indian economy has been disrupted severely with the pandemic surge at the doorstep of the rural hinterland.

The political scenario in respect of the Center–state relationship must be redefined considering the aspirants of the Indian Constitution. The states have to be self-sustaining in terms of resources, infrastructure, and territorial policymaking, and

[11]The scheme was launched in 2011 by the West Bengal State Rural Development Agency (WBSRDA), under the Panchayats and Rural Development Department of the Government of West Bengal. It promotes the state's e-governance initiatives to help the rural population pay electric bills, download mark sheets, apply for recruitment or filing e-tender, and provide employment to the educated youth of the rural areas.

meaningful coordination between the Center and States should be retained to respect the federal structure. Economic federalism might be a possible way ahead to establish economic development following the principle of equity and sustainability across the country in the post-pandemic world (Khosla 2020). The colonial practice of a Center-centric economy of production, transport, and communication has to be changed to take a more distributive direction toward the states and rural hinterland.

The rural and natural resources exploited in the service of the metropolitans are not brought back to their origin equivalently as by-products. It is at the center of the rising inequality gradient from urban to rural, and the stainability is under question.

The state's policies and its citizens' self-urged actions need to act in reciprocity to achieve the development in its truest sense. However, interventions of other actors to the public policies and governance might appear fatal for the poorer, which the pandemic has brought to the light of the day.

References

Allana A (2021) Opinion—How do you vaccinate 1.3 billion people? The New York Times. ISSN 0362-4331. Retrieved from https://www.nytimes.com/2021/03/15/opinion/india-COVID-vaccine.html

Bhuyan A (2021) A month in, here's how india has fared on COVID vaccination. www.indiaspend.com. Retrieved from https://www.indiaspend.com/COVID-19/a-month-in-heres-how-india-has-fared-on-COVID-vaccination-727088

Census of India (2011) Office of the Registrar General & Census Commissioner, India. Ministry of Home Affairs, Government of India

Chandrababu D (2021) Madras high court blames Election Commission for surge in COVID cases. Retrieved from https://www.msn.com/en-in/news/newsindia/madras-high-court-blames-election-commission-for-surge-in-COVID-cases/ar-BB1g58uP

Danigond A (2021) 5 reasons why India's healthcare system is struggling. Retrieved from https://www.thehindubusinessline.com/profi le/author/Ashvini-Danigond-140863/

Deka K (2021) Is the Election Commission responsible for the second wave of COVID cases? Retrieved from https://www.msn.com/en-in/news/other/is-the-election-commission-responsible-for-the-second-wave-of-COVID-cases/ar-BB1gbSqV

Garg S, Bhatnagar N, Singh MM, Borle A, Raina SK, Kumar R et al (2020) Strengthening public healthcare systems in India; learning lessons in COVID-19 pandemic. J Family Med Prim Care 9:5853–5857. https://doi.org/10.4103/jfmpc.jfmpc_1187_20

Huang C, Wang Y, Li X, Ren L, Zhao J, Hu Y, Zhang L, Fan G, Xu J, Gu X et al (2020) Clinical features of patients infected with 2019 novel coronavirus in Wuhan, China. Lancet 395(10223): 497–506. https://doi.org/10.1016/S0140-6736(20)30183-5

Khosla R (2020) In a post-COVID-19 world, the only way ahead for India is economic federalism. The Wire. Retrieved from https://thewire.in/political-economy/COVID-19-economic-federalism

Kumar R (2020) Migrant in my own country: the long March of migrant workers in India during the COVID-19 pandemic 2020-failure of postcolonial governments to decolonize Bihar and rebuild Indian civilization after 1947. J Family Med Prim Care 9(10):5087–5091. https://doi.org/10.4103/jfmpc.jfmpc_2045_20

Kumar A, Narain V (2014) Public policy and governance in India. Vision 18(4):257–260. https://doi.org/10.1177/0972262914555815
Kunal SDM, Aditi MD, Gupta KMD, Ish PDM (2021) COVID-19 variants in India: potential role in second wave and impact on vaccination. Heart Lung. https://doi.org/10.1016/j.hrtlng.2021.05.00
Mahapatra R, Kaur B, Varshney V, Kapil S, Mishra V et al (2021) Let's fix blame: What led to the all-engulfing second wave? How dis we miss the early signals? Why are we still unprepared? Down to Earth, p 34
Mathur K (2009) From government to governance: a brief survey of the Indian experience. National Book Trust, New Delhi
Narain V, Goodrich CG, Chourey J, Prakash A (eds) (2014) Globalization of water governance in South Asia. Routledge Publications, New Delhi
Narasimhan TE (2020) India's first coronavirus case: Kerala student in Wuhan tested positive. Business Standard India. Retrieved from https://www.business-standard.com/article/current-affairs/india-s-first-coronavirus-case-kerala-student-in-wuhan-tested-positive-120013001782_1.html
Palicha D (2021) On SO2S duty. Down to Earth. www.downtoearth.org.in
Perappadan BS (2020) India's first coronavirus infection confirmed in Kerala. The Hindu. ISSN 0971-751X. Retrieved from https://www.thehindu.com/news/national/indias-first-coronavirus-infection-confirmed-in-kerala/article30691004.ece
Pilla V. (2020) Economic Survey 2020: Expenditure on healthcare continues to be flat. Retrieved from https://www.moneycontrol.com/news/economy/policy/economic-survey-2020-expenditure-on-healthcare-continues-to-be-flat-4888481.html
Prasad RS (2021) To fight COVID, we must all come together. Retrieved from https://indianexpress.com/article/opinion/columns/india-second-wave-COVID-pandemic-ravi-shankar-prasad-bjp-govt-opposition-politics-7323636/
Priya A, Dash S, Ranaware K (2021) The truth about oxygen deaths during COVID second wave. Retrieved from https://indianexpress.com/article/opinion/columns/COVID-second-wave-oxygen-death-coronavirus-cases-7392219/
Rajaram P. (2021) Election Commission has totally failed to implement COVID guidelines, says Calcutta High Court. Retrieved from https://www.msn.com/en-in/news/other/election-commission-has-totally-failed-to-implement-COVID-guidelines-says-calcutta-high-court/ar-BB1fWOKE
Sarkar T, Mondal J (2021) Boon for the environment and bane for the economy: emerging debate in pandemic stuck India. In: Mishra M, Singh RB (eds) COVID-19 pandemic trajectory in the developing world. Springer, Singapore, pp 73–100
Seshadri N (2021) Understanding India's oxygen crisis. Retrieved from https://citizenmatters.in/understanding-indias-oxygen-crisis-COVID-second-wave-crisis-24854
Singh R (2021) Centre sets 6-month capacity building goal to fight COVID. Retrieved from https://www.prokerala.com/news/articles/a1177850.html
Thakur A (2021) All about delta variant, the most dangerous form of COVID that caused second wave in India. Retrieved from https://www.india.com/health/all-about-delta-variant-the-most-dangerous-form-of-COVID-that-caused-second-wave-in-india-4724464/#:~:text=As%20per%20data%2C%20the%20Delta%20variant%20was%20the,Services%20From%20Tomorrow%2C%20Domestic%20Flights%20From%20July%201
The Hindu (2021) Madras HC blames ECI for raging second wave of COVID-19 in T.N., Puducherry. Retrieved from https://www.thehindu.com/news/national/tamil-nadu/madras-hc-blames-eci-for-raging-second-wave-of-COVID-19-in-tn-puducherry/article34413670.ece
Urs K, Whittel R (2009) Resisting reform ? Water profits and democracy. SAGE, New Delhi
Vaidyanathan G (2021, May 11) Coronavirus variants are spreading in India – what scientists know so far. Nature 593:321–322. https://doi.org/10.1038/d41586-021-01274-7

WHO (2016) The global health observatory. World Health Organization. https://www.who.int/data/gho/data/indicators. Accessed 16 Aug 2020

WHO (2017) The global health observatory. World Health Organization. https://www.who.int/data/gho/data/indicators. Accessed 25 July 2020

World Bank (1994) Governance: the world bank's experience. The World Bank, Washington, DC

Yeung J, Mitra E, Suri M (2021) Mass religious festival goes ahead in India, despite COVID fears as country enters second wave. Retrieved from https://cnnphilippines.com/world/2021/4/1/india-mass-religious-festival-COVID.html

Tanmoy Sarkar is an assistant professor in the Department of Geography of Gazole Mahavidyalaya (College) in Malda, West Bengal, India. The college is affiliated to the University of Gour Banga. Mr. Sarkar completed his postgraduate studies in Geography and Environment Management with a specialization in Remote Sensing & GIS at Vidyasagar University. His research chiefly focuses on the application of geospatial tools and techniques and multi-criteria predictive models in addressing physical and human processes on the spatial framework. He has been teaching geographical science at the undergraduate level for more than 10 years. He has working experience as an investigator in different research projects sponsored by DST and ICSSR. Mr. Sarkar has published three research papers in reputed journals, published by Springer Nature.

Mukunda Mishra is an assistant professor in the Department of Geography and designated vice-principal of Dr. Meghnad Saha College in West Bengal, India. The college is affiliated with the University of Gour Banga. Dr. Mishra completed his postgraduate studies in geography and environmental management at Vidyasagar University and holds a Ph.D. in geography from the same university. He was selected for the National Merit Scholarship by the Ministry of Human Resource Development, Government of India. His research chiefly focuses on analyzing unequal human development and creating multi-criteria predictive models. He has more than 10 years of hands-on experience in dealing with development issues at the ground level in various districts of eastern India. Dr. Mishra has, in his credit to publish one monograph and three edited research volumes so far from the house of Springer Nature, and he has more than 25 research articles and book chapters published in the journals and books of international repute. Besides serving as the reviewer of several reputed journals published by Springer-Nature and Elsevier Inc., he is continuing as the Managing and Publishing Editor of Ensemble, a UGC-CARE (India) enlisted journal of repute, since its inception to date.

Chapter 2
Sustainable Green Resilience and Globalization or De-Globalization in a Post-COVID World

José G. Vargas-Hernández

Abstract Humanity is facing a series of important challenges-global warming and the pandemic being two of the most important. Consequently, sustainability and resilience have become key elements in providing a better response to the global crisis and in maintaining an equilibrium between ecology, economics, and various social domains. The analysis departs from the assumption that the economic globalization-deglobalization processes respond to more complex dynamic forces created by the economic, financial, and the most recent sanitary crisis that blocks the continuity of the economic globalization, while biodiversity is able to provide renewal and reorganization capacities for changes in social-ecosystems. All these elements bring forth a different paradigm for the future decisions of communities.

Keywords Dialectical evolution · Environmental sustainability · Post-global · Socio-ecosystem · World economy

2.1 Introduction

Globalization is widely criticized for its adverse effects and loss of valuation; it has fueled nationalist and populist movements reaffirmed by the health and sanitary crisis of the COVID pandemic. The processes of globalization have had devastating effects from the 2008–2009 financial crisis on the jobs of workers in various sectors of the industry. Under the concept of the world economy or economic globalization (Sapir 2011, 2016) the benefits and damages caused by the advances in the international economy are analyzed, questioning the achievements of reciprocity and equivalence between the advances in the development of national economies with diametrical differences. What results from international economic relations are very distant from the economic interrelationships of a globalized economic process.

J. G. Vargas-Hernández (✉)
Posgraduate and Research Department, Instituto Tecnológico Mario Molina, Unidad Zapopan, Zapopan, Jalisco, Mexico

S. Saha et al. (eds.), *Economic and Societal Transformation in Pandemic-Trapped India*, New Frontiers in Regional Science: Asian Perspectives 55,
https://doi.org/10.1007/978-981-16-5755-9_2

Globalization is a very complex phenomenon that has a considerable influence on contemporary societies. The economic dimensions of globalization have evolved concomitantly with dimensions that have independent dynamics of a nature other than economic determinism, such as social, environmental, etc. The components of globalization and international economic integration processes are creating new challenges to national governments due to such responsible causes as the national diversity of jurisdictions for trade liberalization policies, geographical factors, and institutional quality to enforce regulations. International trade is the exchange of goods and services across borders creating global markets.

As a phenomenon, globalization processes indirectly and subjectively affect the sensitivities of human and social activities that, when objectified as a process of improving people's well-being and living conditions, imply concrete and contrasting measurements in terms of access to markets. In relation to economic globalization, the theories of economic growth and development related to the theories of international economic relations and international trade and the theory of social and personal well-being are closely linked to human development (Tugores 2002, p. 233).

Human development has had a profound imprint on nature and co-evolving ecosystems. This has resulted in complex, economic-socioecological challenges for sustainability and future development. Human activity alters the dynamics of ecosystems with its important impact on the atmosphere, climate, land surface, forest, sea, and waters. Cities have been portrayed as predominantly monumental static and architectural structures of ever-evolving and increasing ecological complexity. Disturbances change the resilient capacity of nature to supply ecosystem services, they can degrade socioecological systems and lead to social and economic vulnerability.

The survival of humankind is also dependent on healthy and resilient social-ecological systems and sustainable environments. Meanwhile, human well-being, economic growth, and social development are dependent on the interrelationships between and within regions and environmental sustainability (Arrow et al. 1995; Folke et al. 1998). Uncertainty, diversity, and variability of social-ecosystems are all factors that contribute to their diminishing capacity to cope with disturbance and change within functional groups in the adaptive capacity of ecosystems (Folke et al. 2002; Jackson et al. 2001; Scheffer et al. 2001).

The resilience approach provides a conceptual and theoretical framework for interdisciplinary collaboration with ecological economics, sustainable development, and governance (Lambin 2005). When a few privileged people control the natural resources and the ecosystem, the capacity of adaptive development is reduced, the resilience diminishes, and there is more disequilibrium.

This chapter analyzes the implications of globalization and the de-globalization process due to the most recent economic, financial, and health crisis, leading to more regional and local alternatives to give continuity to the world economy. The analysis intends to show that the current context is complex and uncertain due to the conflicts between the economic and political powers, sieges, and commercial reprisals aggravated by the sanitary crisis which are leading the economic globalization processes in the opposite direction away from the international cooperation. From these points, the analysis move to consider the arguments in favor and against economic globalization to understand the challenges faced by the continuing economic and global financial integration that put it at a crucial turning point.

Finally, this analysis offers a discussion related to all the issues treated in this chapter and tries to identify some of the main points for a continuing debate on the future development of the world economy.

2.2 A Dialectical Evolution of the World Economy

Some authors contend that globalization begun about 60,000 year ago when the human history started as inherent phenomenon. Since then and through human history, civilizations have experienced migration, exploration expeditions, military conquests, exchanging trade and developed commercial routes. The penultimate stage of globalization during the period 1820–1913 presented as relevant elements (Williamson 2013) in commerce and poverty, when and how the backwardness of the Third World began; he relates that the industrial revolution produced an increase in the demand for raw materials for industrial production that provoke the opening of new international markets generate benefits in favor of the most industrialized countries. The colonial expression of the time favored the emerging technologies of communication and transportation but with less bellicosity than current globalization (Findlay and O'Rourke 2007).

The collapse of the economic and financial integration processes during the Great Depression in the 1930s, there were tariff wars between countries with competitive devaluations that culminated in World War II (James 2001). The Great Economic Depression of the 1930s brought down the economic and global integration that culminated in World War II. Van Bergeijk and Brakman (2010) assess overemphasizing the international trade in the similar impact that the Great Depression had despite the different causes of the breakdowns, but he still demonstrates the globalization process was the result of the international trade flows, by the uncertainties of economic and social risks caused by the crisis.

Developing countries have the risks of unfavorable internal and external conditions that constraint the macroeconomic policies aimed to achieve equilibrium for economic globalization. However, this development was confined to very few countries while the nonmarket economies remained isolated and less developed economies choose the import substitution path of development.

International finance market has achieved globalization, becoming the most relevant element serving the needs of international investment and trade activities and the most rapidly developing aspect of economic globalization since the 1970s. Far-reaching financial globalization changes in geographical organization and activities since the 1970s. In the 1970s, total trade flows exceeded the historical maximum GDP of the first wave in 1913.

As we know it today, economic globalization erupted in the 1980s with the promotion of movements and mechanisms for the liberalization of international trade and finance, such as the elimination or reduction of barriers and tariffs, etc. Analyzes of data from economic globalization processes confirm that privatization and liberalization are intensifying with the new free trade agreements and the international financial system that is more globalized and immune to national regulations.

The wave of globalization beginning in the 1980s was questioned by its stance and long-term incidence (Straw and Glennie 2012). Economic globalization has not been exempt from complex contradictions of reality. Globalization was possible due to the collapse of the soviet system bringing delocalization, deindustrialization, unemployment, precariousness, and debt leading to a financial crisis. The processes of economic globalization were driven at a time of power and unipolar euphoria with the announcement of "the end of history" made by Francis Fukuyama after the decline of the USSR, in the interest of postponing the dominance of North America.

Since the 1990s, economic interdependence has intensified globalization among countries while causing a debate on the potential benefits in terms of economic growth to the involved countries. The last global trade liberalization agreement was the Uruguay Round, which ended in 1993, which created the World Trade Organization (WTO). During the Mexican financial crisis of 1995, the IMF (International Monetary Fund) limited the space of the Mexican government to impose macroeconomic policies to facilitate the borrowing needed.

Private equity consolidated after the dot com bubble burst in 2000/2001 and it continues to increase until the outbreak of the financial crisis of 2008–2009. However, with the rapid rise of China and Russia, a multipolar or plural-polar world is given way, marking the decline of unipolar globalization. The long-term planning of China has taken advantage of globalization, dealing with the challenges of supplying labor and reducing production costs at the global scale. The construction of an alternative new globalization 2.0 relies more on infrastructure investments based on trade and financial ties.

During the economic and financial crises of 2008 and 2009, many small businesses also went into financial crisis and finally closed. The recent economic and financial crisis of 2008–2009 and the pandemic COVID-19 are the two major causes, among others, that have led international trade to undergo the brink of world economic collapse. Many of these leverages are activated, especially during periods of economic crisis. In this regard, conclusive evidence follows from the reaction of highly developed countries (Japan, USA, Germany, France, UK, etc.) to the adverse effects of globalization.

Keynesian anti-cyclical fiscal and monetary policy was effective in overcoming the Great Depression crisis and the international financial crisis of 2008–2009. However, these types of policies create addictions that end in fiscal deficits and greater indebtedness. Global trade growth in the 10 years before the 2008–2009 financial crisis was about 6%, and in the years after 2009 it was about 2.5%.

The economic globalization process has the tendency to have been slowing down since 2004 (Bello 2004) and increasing later after 2008 as the consequence of the economic and financial crisis with a slight decrease of the pace after 2004. The economic crisis led on serious consequences due to the slowdown of the globalization processes experienced no only by developed economies but also by developing economies and their reintegration processes has been stagnating. In the midst of the economic and financial crisis of 2008–2009, many small businesses closed their operations.

The slowdown in economic activities in the most developed countries is motivated by reductions in exports and imports relative to GDP. Countries that increased their exports and in attracting foreign investment to sustain their economy now face great challenges outside of a model of globalization and free market and trade integration.

It was reported in 2010 that some world globalized cities have declined significantly in relative financial services connectivity as the result (Hanssens et al. 2010, p. 11; Derudder et al. 2011, p. 4). Contemporary globalization processes show different traits from previous globalizing trends such as the rate of change in the economy, the speed, quantity, scale and scope of commercial exchanges, giving rise to a more interdependent system (Ritzer 2011). However, the risks of globalization have increased with events such as the attack and demolition of the twin towers, the outbreak of the dot-com crisis, the tsunami that destroyed the Fukushima nuclear plant in Japan and the financial collapse of Lehman Brothers. The impact of the crisis on globalization carries enormous risks that imply profound changes at the individual, organizational, government and national levels.

Traditional manufacturing and electronic industries have moved to China due to low labor costs and environmental regulations, but as soon as China started to increase them business industries are moving to other places where they can benefit from labor costs and environmental regulations. The winners from globalization processes will be the losers from globalization. In 1916, China achieved reduced rates of economic growth due to lower levels of international trade. To sustain and even revive economic and commercial activities, China implements economic policies. Global investment flows are also slowing down due to tighter regulations. In 2017, global foreign direct investment fell by 16% and in 2018 by 27%. Globalization was at the center of the world economy for a period of four decades in which China became the world's factory.

Economic globalization showed signs of exhaustion in 2019 with the contraction of trade flows, for the first time since 2009. Trade wars, the weakening of multilateralism, the paralysis of trade agreements and treaties are expressions that show the exhaustion of the economic globalization model. International trade flows grew by 1.2% in 2019, the lowest growth rate in the last decade due to certain deglobalizing forces beyond trade wars. Economic globalization contracted in 2019 and is exacerbated in 2020 by the health emergency that can cause a prolonged disengagement of commercial and financial activities from the world economy driven by businesses, governments, and households in developed countries. However, the health crisis has immediate and profound effects on the processes of economic globalization that affect all economies.

The processes of economic globalization are in difficulties, showing a contraction in international trade flows in 2019, which are aggravated by the responses that companies have given to the health emergency. The outbreak of the coronavirus pandemic has immediate negative effects with a considerable impact on global trade, investment and financial flows.

The health crisis generated by the pandemic has confirmed that dependence on the provision of resources, goods and services from remote locations has direct effects on global trade. With the impact of the pandemic, the effects on governments, companies and households are devastating in terms of debt, unemployment, loss of income due to slow economic activities and consumption.

The coronavirus crisis plays an important role in deepening the transition of globalization processes from one more phase of international cooperation to a more conflictive one, which began after the economic and financial crisis of 2008–2009 and in which it is perceived that the risks outweigh the benefits. However, it is difficult to sustain the arguments that the health crisis has changed the logic of economic globalization processes. What has become clear is that the pandemic crisis has accelerated the trends that were set in motion since the financial economic crisis of 2008–2009.

Unlike the crisis of 2008–2009, the current health crisis has abruptly and profoundly paralyzed a large part of the markets at all levels. An important indicator to measure the impacts of these two crises is to compare unemployment, which has only taken 6 weeks in the current crisis to reach the same level as in 65 weeks in the 2008–2009 crisis. There are sufficient elements to support with evidence that the global economic scenario is one of a greater collapse in the economic growth of the more developed economies than that of the previous economic and financial crisis of 2008–2009. Among these evidences are the fall in production and demand in international markets in all economies without the possibility of meeting the fall in domestic demand in national markets.

The global economic and financial crisis of 2008–2009 and the health crisis of 2020 deepened the negative effects of globalization, the increase in economic inequality, the dislocation of jobs, the systematic destruction of socio-ecosystems and the loss of biodiversity, uncontrollable and unsustainable consumerism, etc. Trade wars and the coronavirus crisis put the continuity of globalization processes and the international trade system in checkmate. Trade tensions reflect the influence of China's economic management approach with a completely different set of values than the more developed economies, one of the factors that has given rise to trade wars. The health crisis of the pandemic has collapsed global trade and that must be faced with supranational institutions that are weak and without leadership. Human movements, tourism, and immigration have been disrupted.

Cross-border lending networks driven by large global banks declined during financial crises, with fewer lending banks and fewer interrelationships. The big global banks occupy the center of the network. The big global banks have reduced their operations by allowing the official banks of China to enter. In other words, with the global financial crisis, cross-border financing networks have shrunk but density has increased with greater risk diversification. Economic and financial crises have a strong impact on global banks that spreads through the networks, affecting loans to developing countries (Conesa et al. 2020).

The current health crisis has had a greater impact on the world economy because it was accompanied by a drop in the prices of commodities and oil, unlike the crisis of 2008–2009. In this situation of the depression of the commodities market, in the pre-crisis there were difficulties in the financial conditions of the less developed economies, such as Latin America, unlike the crisis of 2008–2009, these countries had low levels of external public debt while some were already running surpluses.

The current context of the international economy shows powers in relations of conflict rather than international cooperation, sieges and commercial reprisals that complicate and make the development of globalization processes complex. The international coordination of governments resolved the international financial crisis 2008–2009. International coordination is essential to establish strategic agreements to determine global demand, production and income. However, at present this coordination has been very scarce due to the polarization of interests involved in trade wars and the spraying of efforts demanded by the solution to the pandemic crisis.

2.3 Argument in Favor of Economic Globalization

Globalization means good opportunities but also responsibilities. Economic globalization processes have benefits, opportunities, challenges and problems related to the internationalization of economic activities. Economic globalization has brought challenges and opportunities than have changed the global scenario. However, the consolidation process has had disappointing results in prosperity. The benefits of globalization policies must continue extending openness and economic integration while alleviating their negative side effects. The integration of financial and goods and services markets has evolved faster than the movements of services and labor.

However, some analysts argue that the benefits of globalization have reached everyone equally and because it has deepened dependence on certain raw materials and inputs. The increased dependence of the local economy on other foreign economies also increases the risks of losing control of supplies and supplies of inputs.

The increasing acceleration of economic, trade and investment exchanges has fostered a strong global development and economic growth and created economic wealth uniquely distributed. Economic globalization has direct effects on people's lives by allowing direct access as producers and consumers to goods and services in international markets where local economies contribute what they have comparative and competitive advantages. The global consumer increasingly prefers less standardized and more personalized satisfiers that fit their specific needs, wants and fears.

During this period of ascent of globalization processes, there are advances that cannot be haggled as, for example, the increase in global GDP that has multiplied by

62, global life expectancy has increased from 52 to 72 years. Globalization processes are redefined after three decades of economic growth while promoting economic, commercial and financial exchanges without borders in a more open society.

The trade liberalization policy mix allows taking advantage of globalization (Gurgul and Lach 2014). Some empirical studies confirm that national economies with higher level of trade liberalization have increased the living standards (World Bank 2016; Frankel and Romer 1999; Sala-i-Martin 2002, 2006; Dollar and Kraay 2001a, b). Other studies analyze the trade effects though regressions on the GDP per capita growth, although Rodriguez and Rodrik (2000) argue that trade liberalization is measured by the proportion between GDP and foreign trade. There has not been found direct correlation between trade liberalization and economic growth as a factor to reduce global inequalities and between each country.

Empirical studies have found association between financial liberalization and macroeconomic instability, but in the other hand, have not identified a direct relationship between economic and financial liberalization and sustained economic growth in developing countries (Stallings and Studart 2006; Prasad et al. 2003; OECD 2013; Tamarauntari and Diseye 2013; Yahya et al. 2015). The so called "Tobin tax" is a proposal made by Tobin (1978) to tax any financial transaction.

Technological innovation has fostered trade liberalization supported by a decrease in tariffs and taxes which in the long run are associated with facilitating innovation, increasing productivity and positive results in economic growth. Institutional innovations developed the market economies advantages which resulted in efficiency and legitimacy of markets. The capacity for innovation is the essential characteristic that motivates and energizes the world economy and markets. The capacity for innovation is the essential characteristic that motivates and energizes the world economy and markets. Globalization promised to boost prosperity by enlarging and making more efficient the markets.

Economic globalization has contributed to have more influx of information around the world. Speedy mass communication and dissemination of information through digital means is growing very fast. Local and national cultures are intermingling shared and people are exposed and learning from other cultures. Governments and corporations concerned with ecological and environmental sustainability are talking more trying to sort out the responsibilities between each other.

2.4 Arguments Against the Economic Globalization

The liberal economy and the emerging globalization after the Second World War are currently being seriously questioned, even by the countries that promoted it the most. Although globalization processes promote modernization and economic growth in the winning countries, they also generate inequalities and socioeconomic inequalities in the losing peoples that give rise to social tensions (Molano 2007, p. 12).

The arguments against economic globalization sustain that countries and people who remain outside of this elitist economic process, should consider the elements of the open flows of trade and foreign investments processes, to identify the way to take advantage of any potential benefit for national economies from their insertion from economic globalization. The growing insecurity in investments and international trade is negatively affected by the lack of effective control mechanisms in strategic sectors.

Economic and financial globalization have effects on all economic activities rising concerns on issues of ecology and sustainable development, use of renewable resources and energy, limitation of gas emissions that accelerate climate change and corporate social responsibility, among others. Climate change is a sign that the wealth of nations is changing in the twenty-first century from a base of unlimited growth marked by the beginning of the end of an economic expansion that is not compatible with the limits set by the finite resources of the planet, to a different mechanism connected with mitigation and reduction. The UN Climate Action Summit 2019 (COP25) does not refer to growth but to the mitigation and reduction of CO_2 emissions.

Economic globalization processes of production, distribution and consumption has negative effects on the depletion of natural resources, loss of biodiversity, destruction of ecosystems and sustainable environmental development. Production exceeds effective demand. Transportation and logistics are responsible of environmental problems such as gas emissions, air pollution, plastic pollution and are the main cause of global warming.

The current processes of unbridled globalization are criticized because the effects have not been as expected because it only benefits a part of the world population and provides opportunities for certain business sectors, but it generates the impoverishment of small and medium-sized local businesses, produces a growing wage reduction accompanied by job insecurity, which is intentionally implemented as a strategy to reduce production costs.

The effects of trade liberalization on economic growth are confusing in some countries that undertake internal reforms, besides the difficulties to determine causality (Rodriguez and Rodrik 2000; Rodrik 2007; Steinberg 2005, 2007; Dollar and Kraay 2001a). However, the benefits of trade liberalization are considered ambiguous by the new endogenous growth theory, depending of the comparative advantage provided by the economic resources (Rodrik 2007, p. 219).

Trade and labor market reforms favoring trade liberalization are related with an increase in income inequality (Brohman 1996). It is not clear if income inequality is caused by trade liberalization. Trade competition has hit the lowest paid workers. Globalization by itself does not guarantee the well-being of all. The dynamics of economic globalization have strengthened socioeconomic inequalities and hierarchies of authority that victimize workers and local cultures subject to their localities. The loss of national identity, due to globalization processes is argued in favor of a homogenization of global cultural identity. For example, minority local languages tend to disappear.

In those nations where there are no fiscal compensation mechanisms, the benefits of globalization go hand in hand with an increase in the levels of inequality. The lack of fiscal mechanisms to compensate for income inequality makes it more pronounced. Therefore, fiscal policy must be oriented to be redistributive of income that benefits those with fewer resources to increase their consumption and demand and therefore tends to reduce poverty and economic inequality, as well as increase economic growth (Piketty 2014).

Despite that the orthodox theory argues that free trade promotes increases of income, Berry (1998) argues that in Latin America the income inequality increases when the trade liberalized increases. Free trade is questioned for its compatibility with the principle of equal opportunities. These situations lead to high income disparities, declining wages and higher economic and social inequalities and exclusions being the more unskilled workers the most affected for lack of opportunities and mistreatment. The race-down trend in reducing wages and salaries as a strategy to promote economic growth only reduced private consumption which is required by investments and therefore diminishes the global market (Stockhammer 2012).

The market as the best mechanism to allocate resources that generates economic growth has resulted in non-sustainability. The markets have a very complicated evolution due to the health crisis that has serious implications in logistics and supply problems, work management, erratic impact on demand, oversupply and reduction in consumption, changes in the behavior of consumers who prefer non-perishable products to perishable ones, management at points of sale due to decreed closures, fear or lack of supply of merchandise.

Inequality is increasing and become deeper, damaging an already unfair society. Global inequality has declined in some countries due to the globalization benefits in economic growth (Sala-i-Martin 2006). Economic globalization has favored the strongest economic sectors and the rapid growth of the so called BRICS and the inequality between countries seems to have decreased. However, other studies have found the increase of inequality within each country and changes in income distribution may not be related to the development of international trade (Bhalla 2002; Dollar and Kraay 2001a; Wolf 2005; Majeed 2016).

Longstanding incentive systems and structures of economic globalization have some perverse economic, social and environmental effects. The entire global economic structure is reeling from the health crisis of the pandemic. Economic globalization processes have opportunities, challenges and problems related to the internationalization of economic activities. When the globalization processes blew up, they have caused extensive collateral damage leading to the inevitable backlash.

World economy has integrated countries in very unevenly and different degrees due to the power of international structures and reliance in market forces resulting in tensions, inequalities and socially constructed gendered impacts on components of institutions and organizations of economic globalization. Globalization has created more social and economic inequalities, benefiting the richer who gets more rich while creating more poverty.

Since 2008, developed countries are experimenting dysfunctionalities of international financial markets. The processes of globalization are causing the increased volatility of the financial and currency markets, the increase in unemployment and the precariousness of employment, the impoverishment of the middle classes, the mediocre economic growth of emerging economies. Emerging and less developed economies are the most vulnerable to the impact of the global economic crisis because they already had problems of low economic growth, and a high level of dependence on commodities. Emerging economies that had increased their levels of trade since 1990 have been affected by a decrease due to the commercial and financial crisis of 2008.

Most of the trade agreements have cost many jobs, dysfunctionalities and trade deficits for some countries. The free trade agreements promoted by the globalization processes aimed to eliminate free trade barriers, but some countries have imposed more import and export tariffs, valued added taxes, subsidies, zoogenic and phyto sanitary barriers, etc. Multinational and transnational companies exploit the tax heavens to avoid payment of taxes.

Globalism destroys the sovereignty and borders of nation states. National and local governments have diminished their sovereignty and the institutional efficiency to enforce regulations, establish redistributive policies to improve the social wellness and manage the financial crisis. Globalization is criticized and questioned for its inability to prevent or at least mitigate financial economic crises.

Economic globalization process provides more development opportunities but also is posing enormous risks by expanding the gap between developed and the few developing countries benefited, marking the differences in income per capita, the value of foreign trade, etc. Emerging economies are driven by exports to developed countries of natural resources, raw materials and consumer goods showing signs of exhaustion of the model.

Despite that trade liberalization may have positive effects on economic growth, the effects on income distribution are not clear (Ezcurra and Rodríguez-Pose 2013). Some developed economies have trade imbalances with negative effects on less developed countries. Developed countries lost jobs by the transferences to lower labor costs countries. Workers have little leverage to face pay-cut demands from their employers in developed countries creating a culture of fear.

Despite that economic globalization processes have contributed to increase the standard of living, has also negative effects on emerging local economies and workers. The middle class in developed economies that used to have economic strength are dissatisfied from the economic globalization processes which have represented a threat to the living standards. Competition has not always driven prices down because countries can manipulate currency exchanges to get a competitive advantage based on price over other nations. Although globalization has increased the living standards of a large part of the world's population, it has also generated large gaps in economic inequality and a rapid deterioration of biodiversity and socio-ecosystems.

Globalization processes have increased levels of job insecurity due to a considerable reduction in wages as one of the strategies to reduce costs and increase competitiveness at the expense of workers. While consumers benefit from great variety of goods and services, better quality and lower prices, communities that are dependent on jobs outsourced elsewhere have difficulties to sustain competing with lower cost labor markets always under the threat of corporations in a race to the bottom. The imported low price goods and services do not match the decline of wages and family wage jobs.

Economic globalization processes have led to exploitation and degradation of labor, such as the case of child working with the lowest wage, health and safety standards in inhuman and insane working conditions, increase of human trafficking for slave labor. Labor skills are marketed around the world which most of the times cause conflicts in local existing labor markets with the wages that are pressured downward. Labor restrictions that prevent the hiring of the best global talent limit the competitiveness of companies and the national competitiveness of the countries that impose them (McGrew 1990).

Transnational and multinational companies invest in installing manufacturing plants in less developed countries providing employment, while most of the times these jobs are precarious. Multinational and transnational corporations take advantage of their situation in weak and less developed countries where they commit social inequalities and injustices creating unfair working conditions and lower living standards, mismanage and exploitation of natural resources, environmental sustainability and ecological damages. The existing WTO rules cannot have control over nonmarket measures to favor specific trading partners. The international negotiation mechanism has been incapable of mediating international trading disputes.

Any disturbance of a strong and resilient socioeconomic and ecological system can have the potential to create new opportunities for innovation and advancement. However, when the system is weak, any slight disturbance can be disastrous (Adger 2006).

2.5 Sustainability and Green Economic and Social-Ecosystem Resilience

Resilience is defined as the capacity of socio-ecological systems to self-reorganize after any disturbance. Resilience is a concept and model framework used to operationalize normative sustainability (Childers et al. 2014). It refers to the ability of an ecosystem to respond productively to significant disruptive change and adapt to external variables that threaten its existence. At the same time, resilience is a system's ability to adapt to any kind of disturbance and self-reorganize while undergoing transformation, and retaining its initial forms, roles, identity, and feedback characteristics (Walker et al. 2004).

Resilience is also a dynamic process that forms symbiotic relationships within and between the social-ecosystem and its environment. Social-ecosystem resilience is found in the continuous cycle of adaptation and transformation all while maintaining the system's integrity and viability. It is the reaction of the socio-ecological system toward disruption and destruction, and its capacity to recover and develop in a state of uncertainty, discontinuity, and emergency.

Social-ecosystem resilience happens when self-organization and learning meet adaptation and persistence.

Resilience is concerned with the management of sustainable interactions between human-developed systems and natural ecosystems. Social and economic resilience linked to social change is the global community capacity to cope with extrinsic disturbances to their social infrastructure, such as political turbulence, socioeconomic reforms, and environmental variability (Adger 2000; Anderies et al. 2004). Resilience can interfere and be in conflict with other beneficial social objectives, such as economic efficiency, due to the costs of redundancy. In other words, economic efficiency reduces resilience. The resilience of social-ecosystems requires more adaptability to stress while maintaining stability in the face of extrinsic disturbances and to find a solution to the conflict between stability and resilience for sustainable development in terms of complex system cycles.

Apart from representing the measures taken by a social-ecosystem to self-organize and cope with disturbances while still maintaining its inertia, attraction and capacities for learning and adaptation (Carpenter et al. 2001), resilience can also be an approach for cogitating and critiquing social–ecological systems, with policy implications for sustainable development (Folke et al. 2002).

Social-ecosystem resilience is an essential factor to cope with uncertain and complex systems for sustainable natural resources and ecosystem services (Gunderson and Holling 2002). Any disequilibrium between sustainability and development change leads to the collapse of ecosystems. In addition, it is extremely difficult to transform a resilient ecosystem into a more congenial one (Scheffer et al. 2001; Gunderson and Holling 2002; Walker et al. 2004). In other words, it is possible to prevent degradation by promoting the system's congeniality with nature.

A sustainable and resilient social-ecosystem does not largely depend on human input or activity due to the fact that anthropogenic interferences can interrupt the provision of inputs. Instead, it is more reliant on the ways of nature. The social-ecosystem is currently threatened by anthropogenic activities that find themselves at the brink of collapse, and this requires prompt action. In the context of social-ecological systems, management and flexible collaboration are crucial; they help develop policy frameworks as a basis to build adaptive capacity. Nature should be strengthened to stimulate development through the interaction and interdependence with humankind to enhance resilience in social-ecosystems.

The concept of green resilience may be seen from a narrow interpretation to broader one of the socioeconomic and ecological contexts. An ecological unit is the functioning of components and their relationships and interactions with each other, forming a complex and dynamic whole. Sustainable and green resilience of ecosystems can be measured at event junctures, where naturally occurring regenerative forces interact with the energy released into disturbances. Green resilience is an institutional capacity to cope and deal with stress and conflicts arising from climate change, unforeseen contingencies, unsustainable development to live and other emerging environmental issues within the ecosphere.

Green resilience sources intertwine with complex adaptive systems of dynamic changes in the ecosystem. The dynamic adaptive capacity of the ecosystem is provided by the connection between resilience, development change and sustainability (Smit and Wandel 2006). The key to coping with change in social systems is in anticipating and combating disturbances, in adaptive development and in integrating resilience in the interactions with ecological sustainable development. Adaptive development in the social-ecosystems has the ability to carry out the ecological assessment of actual events and taking corrective action.

Sustainable development pretends to reduce social and ecological damage on both a local and a global scale through the operation of fairness and social inclusion, protection of the environment, and economic efficiency. Green resilience-building in complex, uncertain and unpredictable urban ecosystems is supported by structured scenarios and active adaptive management for sustainable development. Green resilience management enhances sustainable development in changing complex environments where the future is uncertain and unpredictable (Walker et al. 2004; Adger et al. 2005). The strategies in any system of urban green resilience focus on green infrastructure and aim to transform and adapt various resources to face future challenges such as climate change and food insecurity.

The ability of a social-ecosystem to sustain itself relies on its ability to adapt to the environmental changes that often occur in multiple-equilibrium systems and human-dominated environments (Folke et al. 1996; Norberg et al. 2001; Luck et al. 2003).

2.6 Discussion

The global economic system has been in chronic crisis for more than half a century. If the world economic history is analyzed, the processes of economic globalization have not been linear. In fact, the dialectical developments of increasing economic, financial, trade, people and ideas toward a more integrated world economy is followed by the opposite movements in the direction of deglobalziation. The global economic system has been in chronic crisis for more than half a century. The dynamics of globalization processes have been interrupted by decisions that put into question the world economic system that facilitates exchanges through commercial, financial, people and information flows.

The processes of economic, commercial and financial globalization have contributed to many nations increasing their wealth with the opening of their markets to international trade, while many nations have been ruined, mainly those that remained closed. History provides evidence that those nations that resist signing free trade agreements under equitable conditions, block their development in some way.

Globalization requires sustaining itself on values, cultures, ideas, ways and trends that promote the uniformity of tastes and traditions so that they are shared by local communities in such a way as to develop the existence of an international community. Culture, values, traditions, etc., have been changed by the economic globalization. The integration of globalization has to be spontaneous and not interference by right or obligation that emerge from the evolution and sociocultural development of peoples. What has been erroneously called uniglobalization refers to the assimilation of cultures, values, customs, traditions, habits, etc. that are replicated in a single direction and with commercial and economic effect.

Economic globalization is becoming deindustrialization of some developed countries due to the outsourcing practices in both manufacturing blue and white collar jobs. National economic policies and strategies are designed and implemented to make economic globalization processes more equitable. Economic and financial globalization may be more transformational transition oriented toward more ecological and sustainable world if the environmental concerns are treated as a coordinated response of all agents and actors for a global synergy.

The transition from globalization to the post-global phase is identified with a greater awareness and responsibility for socio-ecosystems, biodiversity and the quality of life of people and their population-spatial burden, taking the economic side into the background. Some of these concerns that require coordinated efforts are the reduction of gas emissions and waste, transition to renewable energies, etc. The health, sanitary and economic crisis have given rise to rethinking the entire business ecosystem. Global competition must play the same rules and regulations (Lascurain and Villafuerte 2016). Greater global scalability of production in global companies can increase competitiveness, but reduce competition and innovation capacity.

International institutions and national governments are redefining their roles as instruments to find better integration processes into the global economy. Globalization processes must redesign institutional structures or create new institutions to build integration based on the socioeconomic and ecosystem needs of nations and communities of all peoples. It is necessary to establish agreements on binding international mechanisms to address economic and financial crises. Also, it is necessary to implement expedited procedures in customs for the import and export of goods that are critical to its sustainability, while promoting and enhancing international agreements and treaties.

Environmental sustainability is one of the great challenges. The recovery from the economic crisis caused by the pandemic and followed by a global recession is an opportunity to reset the world economy based on more sustainable development. Economic globalization processes are leading to the worldwide spreads of human diseases. The initiative to change economic relations in the processes of globalization that originate ecological imbalances, due to an orientation toward greater care for the planet's ecosystems, implies recognizing the finite limits of natural resources and ecosystems, a greater respect for cycles vital of nature and the environment, decarbonize productive activities to preserve the biodiversity of natural resources.

Another challenge of globalization is to achieve a better equitable distribution of global benefits in such a way that it closes the gap of economic and social inequalities and allows people from emerging countries to get out of poverty. Emerging countries require a financial architecture with support mechanisms to regain liquidity and access to international financial markets. Countries that require solvency need to restructure their debt with the support of multilateral institutions and facilities that give more support to their debt markets.

The impact of the processes of economic globalization on the economic growth and social development of the countries explains the increase in inequality. An entirely free global market seems to be an unattainable goal. From 1960 to 1998, the period of rapid growth in international trade and investment economic and social inequalities have worsened. Economic organization is socially digested when it does not involve a destructive acceleration of growth. Multilateral trade relations are under review to guarantee greater autonomy for global economies. Economic globalization processes are not working to benefit the majority of people around the world.

Globalization processes create multiple forms of cooperation. The interests of less developing countries need the development of a new international economic order changing the conditions of the current economic globalization model as the base of an old international economic order that cannot contribute to develop a fairer and reasonable new international economic order to benefit developing countries. The national economy needs the complementation of the international economy to achieve self-sufficiency in agricultural and industrial production. The interests of developing countries must have to be respected and enlarged in such a way that their development may result in larger share of growth.

For this reason, developing nations need to participate and negotiate actively on the economic globalization process in order to achieve a significant share on development provided the big interests that developed economies have to take advantage of the benefits. The benefits obtained from the new international order of the world economy should be made available for the improvement of development to all the participating countries, not only a few ones. However, in the long run the economic interests of developed countries would be affected if developing countries do not benefit.

The debate on globalization points to the implementation of a more inclusive model in such a way that the losers of the more advanced process receive compensation through the implementation of active labor and income protection policies for the unemployed. It is required to analyze the situation, identify the risks inherent to globalization to establish regulation and control measures. The defrontierization initiative as a proposal for the reconfiguration of the current phase of globalization to allow the free cross-border transit of people, promoting diversity, would be a more solidary and fraternal world, but unfortunately it is utopian. Globalization has fostered the attachment to the material and consumerism over the exercise of ethical values and social solidarity. Achieving an economic recovery to achieve previous levels of income and employment requires reactivation instruments.

Pressures arising from climate change are related to some extent, to the increasing scarcity of land for urban use and other natural resources, such as water and natural nutrients. In many places, population is continuing to rise and becoming an important concern for the policy agenda.

The phenomenon of green resilience in sustainable development may lead to some policy recommendations to improve the interrelationships between the economic efficiency and the social-ecosystems and biosphere, to develop flexible and innovative relationships of collaboration, and to achieve sustainability and its operationalization in the context of economic and socioecological resilience. Other policy recommendations should focus on the development of indicators to measure any change in the level of sustainable green resilience and to signal and monitor uncertainties of social-ecosystem variables and to manage diversity.

References

Adger WN (2000) Social and ecological resilience: are they related? Prog Hum Geogr 24:347–364

Adger WN (2006) Vulnerability. Glob Environ Chang 16(3):268–281

Adger WN, Hughes TP, Folke C, Carpenter SR, Rockstrom J (2005) Social– ecological resilience to coastal disasters. Science 309:1036–1039

Anderies JM, Janssen MA, Ostrom E (2004) A framework to analyze the robustness of social-ecological systems from an institutional perspective. Ecol Soc 9(1):18. http://www.ecologyandsociety.org/vol9/iss1/art18/

Arrow K, Bolin B, Costanza R, Dasgupta P, Folke C, Holling CS, Jansson BO, Levin S, Mäler KG, Perrings C, Pimentel D (1995) Economic growth, carrying capacity and the environment. Science 268:520–521

Bello W (2004) Deglobalization: ideas for a new world economy. Zed Books. ISBN 9781842773055

Berry A (1998) The impact of globalization and the information in Latin America. In: Bhalla AS (ed) Globalization, growth and marginalization. Macmillan Press, London

Bhalla S (2002) Imagine there's no country: poverty, inequality, and growth in the ear of globalization. Institute for International Economics, Washington, DC

Brohman J (1996) Popular development, rethinking the theory and practice of development. Backwell Publishers

Carpenter SR, Walker B, Anderies JM, Abel N (2001) From metaphor to measurement: resilience of what to what? Ecosystems 4:765–781

Childers DL, Pickett STA, Grove JM, Ogden L, Whitmer A (2014) Advancing urban sustainability theory and action: challenges and opportunities. Landsc Urban Plan

Conesa M, Lotti G, Powell A (2020) Resilience and fragility in global banking. IDB Working Paper Series IDB-WP-1133

Derudder B, Hoyler M, Taylor PJ (2011) Goodbye Reykjavik: international banking centres and the global financial crisis. Area 43(2):173–182

Dollar D, Kraay A (2001a) Growth is good for the poor, policy research working paper no 2587, World Bank. Abril, Washington, DC

Dollar D, Kraay A (2001b) Comercio exterior, crecimiento y pobreza, Finanzas y Desarrollo. Revista trimestral del FMI 38(3)

Ezcurra R, Rodríguez-Pose A (2013) Does economic globalization affect regional inequality? A cross-country analysis. World Dev 52:92–103

Findlay R, O'Rourke K (2007) Power and plenty: trade, war and the world economy in the second millennium. Princeton University Press

Folke C, Holling CS, Perrings C (1996) Biological diversity, ecosystems and the human scale. Ecol Appl 6:1018–1024

Folke C, Pritchard L, Berkes F, Colding J, Svedin U (1998) The problem of fit between ecosystems and institutions. International Human Dimensions Programme (IHDP). IHDP Working Paper No 2. www.uni-bonn.de/IHDP/public.htm

Folke C, Carpenter SR, Elmqvist T et al (2002) Resilience and sustainable development: building adaptive capacity in a world of transformations. Ambio 31:437–440

Frankel J, Romer D (1999) Does trade cause growth? Am Econ Rev 89(3):379–399

Gunderson LH, Holling CS (eds) (2002) Panarchy: understanding transformations in human and systems. Island Press, Washington, DC

Gurgul H, Lach L (2014) Globalization and economic growth: evidence from two decades of transition in CEE. Econ Model 36:99–107
Hanssens H, Derudder B, Taylor PJ, Hoyler M, Ni P, Huang J, Yang X, Witlox F (2010) The changing geography of globalized service provision, 2000–2008. The Service Industries Journal 31:2293–2307
Jackson JBC, Kirby MX, Berger WH, Bjorndal KA, Botsford LW, Bourque BJ, Bradbury RH, Cooke R, Erlandson J, Estes JA, Hughes TP, Kidwell S, Lange CB, Lenihan HS, Pandolfi JM, Peterson CH, Steneck RS, Tegner MJ, Warner RR (2001) Historical overfishing and the recent collapse of coastal ecosystems. Science 293:629–638
James H (2001) The end of globalization: lessons from the great depression. Harvard University Press
Lambin EF (2005) Conditions for sustainability of human-environment systems: information, motivation, and capacity. Glob Environ Chang 15:177–180
Lascurain M, Villafuerte LF (2016) Primera globalización económica y las raíces de la inequidad social en México. Ensayos de Economía 26(48):67–90
Luck GW, Daily GC, Ehrlich PR (2003) Population diversity and ecosystem services. Trends Ecol Evol 18:331–336
Majeed M (2016) Economic growth, inequality and trade in developing countries. Int J Develop 15(3)
McGrew A (1990) A global society. In: Hall S, Held D, McGrew A (eds) Modernity and its futures. Polity Press, Cambridge
Molano G (2007) El interregionalismo y sus límites. Estudios Internacionales. Revista del Instituto de Estudios Internacionales de la Universidad de Chile, N° 158, septiembre-diciembre, p 12
Norberg J, Swaney DP, Dushoff J et al (2001) Phenotypic diversityand ecosystem functioning in changing environments: a theoretical framework. Proc Natl Acad Sci U S A 98:11376–11381
OECD (2013) Multidimensional review of Myanmar, vol 1. Initial assessment
Piketty T (2014) El capital en el siglo XXI. Fondo de Cultura Económica, México
Prasad ES, Rogoff K, Wei S-J, Ayan Kose M (2003) Effects of financial globalization on developing countries: some empirical evidence. IMF, Washington, DC
Ritzer G (2011) Globalization: the essentials. Wiley-Blackwell
Rodriguez F, Rodrik D (2000) Trade policy and economic growth: a skeptics guide to the cross-national evidence. NBER Working Paper N° 7081
Rodrik D (2007) One economics many recipes, globalization, institutions and economic growth. Princeton University Press
Sala-i-Martin X (2002) The Distributing rise of global income inequality, NBER, Working Paper No 8904
Sala-i-Martin X (2006) Globalización y reducción de la pobreza. FAES
Sapir J (2011) La Démondialisation. Éditions du Seuil, París. Disponible en web: http://ekladata.com/qCMpnHCs71H2aM95aBc5H3ZgCBI.pdf
Sapir J (2016) Jacques Sapir: Donald Trump, president of the demondialisation?. In: Le Figaro, November 10, 2016, available on the web. http://www.lefigaro.fr/vox/monde/2016/11/10/31002-20161110ARTFIG00233-jacques-sapir-donald-trump-president-de-la-demondialisation.php
Scheffer M, Carpenter SR, Foley JA, Folke C, Walker BH (2001) Catastrophic shifts in ecosystems. Nature 413:591–596
Smit B, Wandel J (2006) Adaptation, adaptive capacity and vulnerability. Glob Environ Chang 16(3):282–292
Stallings B, Studart R (2006) Financiamiento para el desarrollo: América Latina desde una perspectiva comparada, Libros de la CEPAL No 90, Santiago de Chile
Steinberg F (2005) Cooperación y Conflicto en el Sistema Comercial Multilateral: La Organización Mundial de Comercio como Institución de Gobernanza Económica Global, Tesis Doctoral presentada en el Departamento de Análisis Económico: Teoría Económica e Historia

Económica de la Facultada de Ciencias Económicas y Empresariales de la Universidad Autónoma de Madrid, España
Steinberg F (2007) Cooperación y Conflicto Cooperación Internacional en la era de la Globalización. Ediciones Akal, Madrid
Stockhammer E (2012) Financialization, income distribution and the crisis. Investigación Económica LXXI(279)
Straw W, Glennie A (2012) The third wave of globalization. In: IPPR review on the future of globalization. Institute for Public Policy Research, London
Tamarauntari MK, Diseye BT (2013) Macroeconomic uncertainty and foreign portfolio investment: evidence from Nigeria. IISTE 3(12):229–236
Tobin J (1978) A proposal for international monetary reform. East Econ J 4(3–4):153–159
Tugores J (2002) Economía internacional, globalización e integración regional. McGraw-Hill, Madrid
Van Bergeijk PAG, Brakman S (2010) The gravity model in international trade: advances and applications. Cambridge University Press
Walker BH, Holling CS, Carpenter SR, Kinzig AP (2004) Resilience, adaptability and transformability in social–ecological systems. Ecol Soc 9(2):5. http://www.ecologyandsociety.org/vol9/iss2/art5/
Williamson JG (2013) Trade and poverty: when the third world fell behind, Vol. 1, 2nd edn, 0262518598. MIT Press Books
Wolf M (2005) Why globalization works. Yale Nota Bene
World Bank (2016) World bank data. Consulted in September 2016. http://data.worldbank.org/?display=default
Yahya W et al (2015) Macroeconomic factors and foreign portfolio investment volatility: a case of South Asian countries. Fut Bus J 1(1–2):65–74

José G. Vargas-Hernández is a professor at Instituto Tecnológico Mario Molina, Unidad Zapopan, Mexico. Besides, he serves as the member of the National System of Researchers (Level I) in his country and he is the visiting scholar at Laurentian University, Sudbury, Ontario, Canada; University of California-Berkeley and Carleton University, Ottawa, Canada. He also served as a professor of the Doctorate in Administrative and Organizational Sciences, LaSalle University, Autonomous University of Durango and as an Academic Advisor of the Doctorate in Administration of the Autonomous University of Aguascalientes. His research area is organizational economics and he is associated with the research project entitled 'Conacyt and BMBF (Germany) Business bio-economics in urban areas' at CUCEA-UDG and Kaiserslautern University, Germany and RE-CITY EUROPEAN UNION research project on 'Reviving shrinking cities - innovative paths and perspectives toward livability for shrinking cities in Europe'. There are more than 1000 research articles, authored by him, which is published in international journals, edited research volumes and conference proceedings of repute and he has authored 23 books so far. He also served as the editor of several journals of academic significance.

Chapter 3
Income Insecurity, GDP, and the Future of Human Development: An Analysis for COVID-19 Period

Hasan Dinçer, Hakan Kalkavan, Serhat Yüksel, and Hüsne Karakuş

Abstract This study will focus on the social and economic impacts of the COVID-19 pandemic on India. In this context, firstly, the impact of the COVID-19 outbreak in the country from the start date to the present will be examined. Next, the measures that the state has already taken to manage this epidemic will be discussed. In the last part, the importance for the country to effectively manage this epidemic will be emphasized. In this context, an analysis will be carried out using the Entropy method. The findings indicate that life expectancy is the most crucial criterion for India. Similarly, GDP growth is another crucial issue in this regard. However, adult literacy rate and health expenditure have the lowest weights. It is evident that the COVID-19 pandemic has many negative impacts on life expectancy in India. This situation explains that Indian people have become very anxious about their futures since lots of people died in the country due to the COVID-19 pandemic. In this period, the governments have two different alternatives, giving importance to reducing the pandemic or taking actions to increase trade volume. These two choices create a paradox for governments as they affect each other negatively. The analysis results of this study demonstrate that the Indian government should mainly focus on the necessary ways to reduce pandemics.

Keywords Economic growth · Development · Entropy · Pandemics · Labor force

3.1 Introduction

COVID-19, also known as coronavirus, is one of the acute respiratory diseases. The treatment of the disease has not been found yet, and it is transmitted by the respiratory. This disease, which is transmitted from animals to other animal species, has also started to be transmitted from humans to humans by mutation. This virus

H. Dinçer · H. Kalkavan · S. Yüksel (✉) · H. Karakuş
The School of Business, İstanbul Medipol University, İstanbul, Turkey
e-mail: serhatyuksel@medipol.edu.tr

S. Saha et al. (eds.), *Economic and Societal Transformation in Pandemic-Trapped India*, New Frontiers in Regional Science: Asian Perspectives 55,
https://doi.org/10.1007/978-981-16-5755-9_3

was first seen in Wuhan, the capital of China's Hubei province (Nove 2020). It has spread to many countries due to the insufficient measures taken over time. In this context, some measures have been taken by countries to reduce the number of cases. Many businesses were physically closed, and work from home was started, schools were vacationed, public places such as places of worship, theaters, and cinemas were closed, and air transport was blocked. The epidemic has tried to be brought under control by introducing lockdowns (Gasana and Shehab 2020).

The measures were taken against coronavirus, which impresses the whole world, negatively affects countries socially and economically. Social distance should be reduced in order to prevent an increase in cases. For this, restrictions are placed in areas where social distance is close. This situation affects countries both socially and economically negatively. Epidemic causes panic among society. With the introduction of curfews, this panic is exacerbated. In this process, people tend to spend money to meet their essential needs, worrying about starvation. This situation decreases the general consumption expenditures. Foreign investors and tourists are leaving the country with concerns about viruses (Fernandes 2020). This means that foreign currencies flee the country. There is an increase in exchange rates due to the foreign currency leaving the country. This situation adversely affects countries with foreign currency debt (Luo and Tsang 2020). However, the government is focusing on healthcare spending to reduce the effects of the virus. All these situations negatively affect the gross domestic product of the countries. In this process, income insecurity has been experienced. In other words, people's incomes are beginning to be insufficient and irregular. This situation also causes the emergence of psychological problems.

On the other hand, the measures introduced in countries increase the unemployment rates and thus increase income insecurity (Lancet 2020). The coronavirus restricts educational activities, shortens human lives, and negatively affects the income opportunity that provides access to resources. These problems lower the human development index. Since the human development index makes completely human-oriented calculations and focuses on life expectancy, literacy rates, and income generation. In this context, the negativities experienced negatively affect the human development process (Anand and Sen 2000).

The coronavirus affects more than 100 countries. One of these countries is India. India has a population of approximately 1.3 billion. In this respect, it is the second-most populous country in the world. The first case was seen on 30 January 2020. However, the first deaths began to occur on 12 March. With the emergence of the cases and deaths, the government has taken some measures. A curfew has been declared, and many companies have switched to work from home.

Moreover, public spaces were either restricted or closed (Vijai and Suryalakshmi 2020). Those measures are taken affect countries negatively, both economically and socially. The gross domestic product of the country has decreased by 23.9%, and the country has gone toward economic shrinkage. Besides, unemployment rates rose to 23%. Social deterioration has been experienced, and people who are locked in their homes have experienced mental and psychological breakdowns. After all, it is observed that healthcare workers especially suffer from mental problems depending on the risk of contracting the disease (Zhu et al. 2020). Income insecurity arises due to a decrease in income and an increase in unemployment rates. Human is a social entity, so the social restrictions imposed negatively affect the human development process. In short, the number of people who lost their lives at an early age due to coronavirus is increasing, people who cannot benefit from educational activities lack information, and people cannot fully meet their needs due to volatility in their income. Hence, there are problems in the human development process in India. These problems threaten India's economic and social development (Mehta and Jha 2020). Therefore, measures to reduce the spread of the virus are extremely important. Considering all these issues, the purpose of this study is to conduct a critical analysis of the social and economic impacts of the COVID-19 outbreak. For this purpose, India was studied by using the Entropy method. It is thought that the scope and method in question will strengthen the literature.

3.2 The Theoretical Background

3.2.1 General Information About the Concepts

Income, which is one of the important factors determining the living conditions, satisfies the individual when it is regular and sufficient. Because desires and needs are met with regular and sufficient income. In this context, individuals should earn regular income in order to survive. However, individuals cannot always earn regular and sufficient income. Some factors that cause income to decrease, disappear and expenditures to increase unexpectedly lead to irregular income. Income insecurity emerges depending on the irregularity and insufficiency of income (Wilthagen and Tros 2004). Income insecurity is defined as the risk of loss of income due to inadequate and regular income. Occupational accidents, birth, diseases, unemployment, permanent incapacity are among the risks that lead to a decrease in income. The person who has an occupational accident may lose his ability to work due to the damage. This situation may cause him to lose his income. However, the disease causes the person to be unable to continue their work. In this process, the income of the person decreases due to the job loss. Income insecurity occurs mostly in the labor market (Vulkan 2012; Western et al. 2016). Developments that adversely affect the labor market increase income insecurity. For example, developments that require the reduction of the workforce cause individuals to lose their jobs. The income of the individual, who loses his job, decreases and his income starts to be insufficient. The well-being of the individual, who cannot meet his needs and desires, falls. Income security is possible with the regular provision of a sufficient income. For this, actors such as public institutions, various insurance companies, and financial markets take part (Mishra and Chatterjee 2018). Income insecurity, especially in family environments, is important because family members collectively meet their desires and needs. Therefore, the loss of sufficient income of one of the family members and their irregularity make life in the family difficult. Therefore, measures should be taken to reduce domestic income insecurity (Knijn 2002). On the other hand, the risk of loss of income for those who continue their job is also an indicator of income insecurity. Individuals who work overtime, have the same income, and are at risk of losing their income become alienated from their jobs and experience serious mental problems (Standing 1999; Hellgren and Sverke 2003).

Gross domestic product (GDP) is one of the important criteria used in evaluating the economic performance of countries because GDP shows the market value of final goods and services produced in a certain period. In this context, final goods are considered in the GDP calculation (Reddy 2012). In other words, market values of ready-made products purchased by consumers are included in the calculation. Intermediate goods used in the production of ready-made products are not considered in the calculation. However, only the production of final goods and services in a specific period is taken into consideration in the GDP calculation. Old period production is not included in the calculation. Therefore, it is called 'gross' domestic product (Xianchun 2004). The sum of consumption, investment, and public expenditures for final goods and services gives the GDP value. If the total of these expenditures has increased compared to the previous period, the country grows economically. However, net exports are added to the expenditures in the calculation. In this way, the foreign trade balance in the country is also determined. In this context, while evaluating the economic performance of the countries, it is necessary to pay attention to the consumption expenditures of the households, the investment expenditures made by the companies on the vehicles such as machinery and equipment, and the public expenditures made by the government in social and human areas (Keidel 2001). Certain production factors are used in the production of final goods and services. The sum of the wages, interest, rent, and profits obtained by these production factors also shows the GDP value. However, GDP can be determined by adding depreciation to indirect taxes and adding national income to the value obtained by deducting subsidies from the total. Considering these methods, it is seen that the production, expenditure, and income method is used in GDP calculation. These methods provide information about the country's economy. Information obtained from the country's economy reflects the state of living conditions (Landefeld et al. 2008). The information obtained for the national economy is important for investors because an economically growing country leaves a positive image in the eyes of investors. Investors make direct or indirect investments in the country. Foreign exchange and various resources help countries' economic development (Jackson and Marks 1999).

The economic growth of countries is an indicator of increased welfare. Because people, governments, and institutions spend more on consumption, investment, and the public. However, the market value of the final goods and services produced is high. It is thought that countries that experienced economic growth in the 1950s developed because individuals with high per capita income spend more, and their welfare level increases (Keidel 2001). However, economic growth and development are different concepts from each other. This difference is detected in the 1970s. It is seen that economic growth has increased in less developed countries, but the poverty and unemployment levels have also increased. This point shows that economic growth and development are not the same concepts (Griffin and Knight 1992). Development means improvement in all living conditions. Therefore, not only economic issues but also human-oriented issues affect the level of development. In this context, the concept of human development is beginning to gain importance. Human development refers to the level of development as a result of people using every opportunity they have. However, it also means expanding the possibilities given to people (Ranis et al. 2006; Hicks 1997). Considering the importance of this definition, the United Nations Development Program (UNDP) has been publishing a Human Development Report since 1990. The results of the Human Development Index calculations are given in this report. This index takes into account three human-focused issues. The first of these issues is a long, healthy life. Its measurement is made with the average life span (Dinçer and Yüksel 2019). Another important issue is obtaining information and education level. Literacy rate, primary school, secondary school, and university records are taken into account in its measurement. Another issue taken into consideration in the measurement is the standard of living that should be. In this regard, it focuses on access to resources necessary for strong living conditions. In its measurement, per capita income and purchasing power are taken into account (Noorbakhsh 1998). Thanks to these calculations, the economic and human-oriented issues are determined. The development level of countries emerges as a result of this index (Anand and Sen 2000).

3.2.2 Social and Economic Effects of COVID-19

Coronavirus, also known as COVID-19, is a viral infection that is transmitted by the respiratory, involves the lungs, and manifests itself with respiratory symptoms. The virus, which started to show itself in December 2019, was first seen in Wuhan, China's Hubei province (Akkurt et al. 2020; Nove 2020). This type of virus, which was previously only seen in animals, has been mutated and started to be seen in humans. Studies conducted on the treatment not responding show that the virus in question is a new type of virus. On 07 January 2020, the virus was diagnosed and determined to be a coronavirus (Eurosurveillance Editorial Team 2020; Stoecklin et al. 2020). The coronavirus spreads rapidly around the world and causes sudden deaths. Therefore, the World Health Organization (WHO) declared a pandemic on 11 March 2020 and stated that countries should take measures against this virus. The virus that emerged in China also manifested itself in Iran and later in countries such as Thailand, South Korea, Japan, and Italy. Coronavirus is a virus that affects more than 100 countries today and shakes the balance of the world. Therefore, certain measures are taken to minimize the effects of the virus (Cheng et al. 2020).

COVID-19, which affects the whole world, causes countries to experience many social and economic problems. Some measures are taken in countries to prevent the spread of the virus in question. With the onset of the virus, the production and service activities of companies and the education and training activities in schools were prevented from taking place in the physical environment (Ahani and Nilashi

2020). However, places of worship were closed, many sports competitions were postponed, and entry and exit abroad were closed. In this way, it is aimed to minimize the spread of the virus by providing social distance (Yüksel et al. 2020). On the other hand, the measures taken put countries in economic difficulties. Production and service activities of many sectors have come to a standstill. Therefore, companies face the risk of bankruptcy. The foreign trade, transportation, and tourism sector are among the most affected sectors (Ayittey et al. 2020; Siddiquei and Khan 2020). Depending on the effects of these sectors, an increase in poverty and unemployment rates is expected in the long term, especially in developing countries. In this process, it is thought that countries will experience a recession (Kalkavan and Ersin 2019). People have experienced a great panic with the coming of the curfew and are especially focused on food spending. People started to spend more on their essential needs. Accordingly, there is a decrease in consumption expenditures. However, governments are increasing their spending on the health sector (Shi et al. 2019). In this context, the GDP of the countries is negatively affected, and budget deficits are starting to emerge in the countries. When all these are evaluated, countries are faced with serious problems in the macroeconomic field (Luo and Tsang 2020).

COVID-19 also causes people to experience a lot of social anxiety. Social distance is tried to be achieved by closing and restricting places such as theaters and cinemas. However, with the introduction of curfews, social activities are reduced. People are affected by these restrictions socially and psychologically. People who cannot continue their jobs face the risk of loss of income and experience psychological problems. However, they spend too much food on the psychology of fasting (Nicola et al. 2020). Especially healthcare workers stay away from their families and social environments, and the risk of getting sick is high. This situation negatively affects healthcare professionals mentally (Zhang et al. 2020). The society stays away from its social habits and cultures. Therefore, priority should be given to treatment studies developed for coronavirus (Qiu et al. 2020; Alsubaie et al. 2019).

3.2.3 General Information About Indian Economy Under COVID-19

India, one of the South Asian countries, hosts approximately 1.3 billion people from different cultures and languages (Khumbongmayum et al. 2006). In this respect, it is the second-most densely populated country in the world after China. However, with an area of approximately 3 million 300 thousand square meters, India is the seventh country with the largest area in the world. India's main livelihood is the manufacturing and mining industry. However, it is also a country rich in iron reserves. India's industrialization and development process, which is rich in underground resources and iron reserves, is accelerating day by day. A part of the country with a wide geographical area makes their living from agriculture. It ranks first in the world to

Table 3.1 Quarterly GDP rates in India (percent)

Period	Actual (%)	Previous (%)
Q4 (fourth quarter)	4.7	5.1
Q1 (first quarter)	3.1	4.7
Q2 (second quarter)	−23.9	3.1
Q3 (third quarter)	–	−23.9

Source: Ministry of Statistics and Programme Implementatio (2020)

grow products like lentils, chickpeas, tea, and sesame, while it is in second place in the world to produce products like sugar cane, rice, and hemp. In this context, agriculture is an important source of income for India (Ahmad et al. 2020).

The coronavirus, which emerged in December 2019, has infected more than 100 countries. One of these countries is India. On 30 January, the first coronavirus case in the country was detected. Additionally, the first death occurred on 12 March. The number of cases has exceeded four million over time, and the number of those who lost their lives has approached 72 thousand. Therefore, the government of India has declared a curfew as of 24 March (Chakraborty and Maity 2020). The curfew, which was planned to last for 3 weeks, was extended until the end of May due to the increased number of cases. In this process, companies across the country were closed, and the home-working system was introduced. Places of worship, educational institutions, public spaces have been closed, and social and economic activities have come to a standstill (Dev and Sengupta 2020). With the introduction of the curfew, many immigrants, tourists, and students were allowed to return to their countries. Although the measures are taken slow down the increase in cases, they affect the country economically negatively. Significantly the country's gross domestic product is adversely affected (Lokhandwala and Gautam 2020). The effect of coronavirus on GDP is shown in Table 3.1.

Table 3.1 shows the quarterly GDP rates in India. While the country's gross domestic product grew by 3.1% in the first 3 months of 2020, with the introduction of the curfew, the country's gross domestic product shrank by 23.9% in the second quarter. Citizens of countries experiencing economic shrinkage increased their spending for their mandatory needs. In this context, it is seen that there are problems regarding income in India. In India, which has a population of approximately 1.3 billion, the pre-quarantine unemployment rate is calculated as 7.76%. The unemployment rate increased to 23% with the closure of public activity areas and the halt of company activities. Approximately 140 million people lost their jobs during the quarantine period (Center for Monitoring Indian Economy 2020). A curfew was declared in March due to the occurrence of the cases on 30 January. With the introduction of the curfew, it is seen that the unemployment rate has increased to 23%. During this period, many people lost their jobs. However, after May, the curfew started to be applied on certain days. This situation led to a decrease in unemployment rates (Mishra and Rampal 2020). The Indian economy has shrunk due to the coronavirus, and many people have lost their jobs. This situation reveals some social and psychological problems. Income insecurity increases with increasing unemployment rates and decreasing incomes. Educational activities have come to a halt, life span has shortened due to deaths, and people have become limited to access resources due to the decrease in income. Disruptions in living conditions negatively affect the human development index (Lancet 2020).

3.3 The Effect of COVID-19 on Human Development Index

In this part of the study, an evaluation has been conducted to see the effect of COVID-19 on the human development index. For this purpose, India is taken into consideration in the analysis process. Concerning the human development index, six different factors are identified based on the three different dimensions. The details are demonstrated in Table 3.2.

Three different experts evaluate these six factors. These experts have sufficient knowledge about the human development factors for India. Hence, they can easily make a recommendation about these issues. Additionally, Entropy methodology is taken into consideration in the analysis process. This method is mainly used to weigh the criteria. These experts used 5-scale evaluation factors. The evaluations of all experts are indicated in Table 3.3.

After that, a pairwise comparison matrix is created by taking the average values of them. This matrix is detailed in Table 3.4.

Later, all values in this pairwise comparison matrix are divided into the total values of columns. With the help of this situation, the normalized matrix can be generated. This matrix is illustrated in Table 3.5.

In the next step, entropy values are calculated. By considering these values, the weights of the criteria can be computed. Table 3.6 gives information about the analysis results.

Table 3.6 indicates that life expectancy (C1) is the most significant criterion since it has the highest weight (0.2270). Additionally, GDP growth (C6) is another important factor for India. On the other side, the adult literacy rate (C3) and health expenditure (C2) have the lowest weights. This situation indicates that the COVID-19 pandemic has negative impacts on life expectancy. Because many people died due to this pandemic, people become very anxious about their futures. There is a

Table 3.2 The list of the factors

Dimensions	Criteria	Supported literature
Health (D1)	Life expectancy (C1)	Dinçer et al. (2020)
	Health expenditure (C2)	Eti et al. (2020a, b)
Education (D2)	Adult literacy rate (C3)	Colliver et al. (2020)
	The number of students (C4)	Shahzad et al. (2020)
Economy (D3)	Income distribution (C5)	Wang et al. (2020)
	GDP growth (C6)	Eti et al. (2020a, b)

Source: Prepared by the authors

Table 3.3 The evaluations of the experts

Criteria	C1	C2	C3	C4	C5	C6
Evaluation of expert 1						
C1	0	2	4	4	3	3
C2	5	0	3	3	4	4
C3	2	2	0	3	2	2
C4	1	2	2	0	3	3
C5	1	2	3	2	0	4
C6	1	1	3	3	4	0
Evaluation of expert 2						
C1	0	3	3	4	4	3
C2	5	0	2	3	3	4
C3	3	2	0	2	2	2
C4	2	3	2	0	3	3
C5	2	2	2	2	0	4
C6	1	2	3	3	4	0
Evaluation of expert 3						
C1	0	4	3	4	4	3
C2	4	0	2	4	3	4
C3	2	2	0	3	2	2
C4	2	3	3	0	4	3
C5	2	3	3	2	0	4
C6	2	2	3	4	4	0

Source: Tabulated by the authors

Table 3.4 Pairwise Comparison Matrix

Criteria	C1	C2	C3	C4	C5	C6
C1	0.00	3.00	3.33	4.00	3.67	3.00
C2	4.67	0.00	2.33	3.33	3.33	4.00
C3	2.33	2.00	0.00	2.67	2.00	2.00
C4	1.67	2.67	2.33	0.00	3.33	3.00
C5	1.67	2.33	2.67	2.00	0.00	4.00
C6	1.33	1.67	3.00	3.33	4.00	0.00

Source: Computed by the authors

Table 3.5 Normalized matrix

Criteria	C1	C2	C3	C4	C5	C6
C1	0.00	0.26	0.24	0.26	0.22	0.19
C2	0.40	0.00	0.17	0.22	0.20	0.25
C3	0.20	0.17	0.00	0.17	0.12	0.13
C4	0.14	0.23	0.17	0.00	0.20	0.19
C5	0.14	0.20	0.20	0.13	0.00	0.25
C6	0.11	0.14	0.22	0.22	0.24	0.00

Source: Computed by the authors

Table 3.6 Result of the analysis

Criteria	Weights
Life expectancy (C1)	0.2270
Health expenditure (C2)	0.1538
Adult literacy rate (C3)	0.1457
The number of students (C4)	0.1575
Income distribution (C5)	0.1561
GDP growth (C6)	0.1599

Source: Computed by the authors

paradox for the Indian government to take action for this situation. They may prefer to make the rules stricter. In this framework, the trade activity in the country will slow down, and the economy will shrink. On the other side, in order to increase commercial mobility, the rules should be stretched. This situation causes the epidemic to spread rapidly. The government has to choose one of these issues. The analysis results of this study demonstrate that the Indian government should mainly focus on the necessary ways to reduce pandemics.

3.4 Discussion and Conclusion

The COVID-19 outbreak first appeared in Wuhan, China, in January 2020. After that, it quickly spread to other countries around the world, becoming a global threat. Although a long time has passed, there is no drug produced for this virus yet. A large number of people have died in different parts of the world due to this disease. In addition, the health systems of countries have started to experience many problems due to the fact that too many people are sick at the same time. Countries have started to take some measures in order to eliminate the adverse health effects of this epidemic. In this context, curfews and the closure of some workplaces can be shown as examples of this issue.

Due to these issues, the COVID-19 outbreak has also negatively affected the economies of the country. In other words, the general strategy applied to stop the spread of the virus is to keep people at home and not be taken out unless necessary. Due to these measures, trade activity in the countries has come to a standstill. India is a country that has been adversely affected by the COVID-19 outbreak. India has risen to second place in the world after the USA in COVID-19 cases. While the number of cases in the country exceeded four million, the number of people who died due to this epidemic approached 72,000. This situation threatens the economic and social development of India. Considering that India has a developing economy and the country is overcrowded, it would be appropriate for the country to take urgent measures to manage this epidemic effectively. Otherwise, the country will inevitably experience both social and economic problems in the future. This study will focus on the social and economic impacts of the COVID-19 pandemic on India.

The study carefully determined the relative importance of the measures that the country could implement to manage this epidemic effectively. In this context, an analysis has been carried out using the entropy method. At first, the social and economic factors that the country could apply for outbreak management have been listed. After that, it has been determined which of these factors would be more important, applying the entropy method. Following the analysis results, the anti-epidemic strategies could be presented for India. In this way, it will be possible to contribute to the future sustainable economic and social development of the country.

The method, designed and applied here, can determine which criteria would be most effective in combating the epidemic will guide both policymakers and researchers. The analysis results specified here are a guide not only for India but also for other developing countries. Another contribution of the study to literature is in a methodological sense. The entropy method will be considered for the first time in research on the COVID-19 outbreak. This situation is thought to increase the methodological originality of the study.

It is concluded that life expectancy is the most crucial criterion because it has the highest weight. Moreover, GDP growth is another significant issue for India. Nevertheless, adult literacy rate and health expenditure have the lowest importance. It is evident that the COVID-19 pandemic has many negative impacts on life expectancy in India. Because many people died due to this pandemic, people become very anxious about their futures. There are mainly two different alternatives for the Indian government to manage this problem. They may prefer to make the rules stricter so that the adverse effects of the pandemics can be minimized. However, within this context, the trade activity in the country will slow down, and the economy will shrink. On the other side, in order to increase commercial mobility, the rules should be stretched. This situation causes the epidemic to spread rapidly. The government has to choose one of these issues. The analysis results of this study demonstrate that the Indian government should mainly focus on the necessary ways to reduce pandemics.

References

Ahani A, Nilashi M (2020) Coronavirus outbreak and its impacts on global economy: the role of social network sites. J Soft Comput Dec Support Syst 7(2):19–22

Ahmad F, Talukdar NR, Uddin M, Goparaju L (2020) Climate Smart Agriculture, need for 21st century to achieve socioeconomic and climate resilience agriculture in India: a geospatial perspective. Ecol Quest 31(1):1–21

Akkurt YD, Küçükkartallar T, Kandemir B, Cihan FG (2020) Experıences of a Unıversıty Hospıtal during the Covıd-19 pandemic in Turkey. Konuralp Tıp Dergisi 12(2):344–346

Alsubaie S, Temsah MH et al (2019) Middle East Respiratory Syndrome Coronavirus epidemic impact on healthcare workers' risk perceptions, work and personal lives. J Infect Develop Count 13(10):920–926

Anand S, Sen A (2000) The income component of the human development index. J Hum Dev 1 (1):83–106

Ayittey FK, Ayittey MK, Chiwero NB, Kamasah JS, Dzuvor C (2020) Economic impacts of Wuhan 2019-nCoV on China and the world. J Med Virol 92(5):473–475

Centre for Monitoring Indian Economy (2020) Unemployment in India. https://unemploymentinindia.cmie.com/

Chakraborty I, Maity P (2020) COVID-19 outbreak: migration, effects on society, global environment and prevention. Sci Total Environ. https://doi.org/10.1016/j.scitotenv.2020.138882

Cheng VC, Wong SC, Kwan GS, Hui WT, Yuen KY (2020) Disinfection of N95 respirators by ionized hydrogen peroxide during pandemic coronavirus disease 2019 (COVID-19) due to SARS-CoV-2. J Hosp Infect 105(2):358–359

Colliver Y, Arguel A, Parrila R (2020) Formal literacy practices through play: exposure to adult literacy practices increases child-led learning and interest. Int J Early Years Educ:1–19

Dev SM, Sengupta R (2020) Covid-19: impact on the Indian economy. Indira Gandhi Institute of Development Research, Mumbai

Dinçer H, Yüksel S (2019) Identifying the causality relationship between health expenditure and economic growth: an application on E7 countries. J Health Syst Polic 1(1):5–23

Dinçer H, Yüksel S, Gökalp Y, Eti S (2020) SERVQUAL-based evaluation of service quality in Turkish health industry with fuzzy logic. In: Interdisciplinary perspectives on operations management and service evaluation, pp 213–233

Eti S, Gökalp Y, Tosun NO (2020a) Identifying the relationship between health investment and economic development: a cluster analysis for developing, developed, and least developed countries. In: Multidimensional perspectives and global analysis of universal health coverage. IGI Global, pp 1–30

Eti S, Kalkavan H, Dinçer H, Yüksel S (2020b) Predicting the role of Islamic banking on sustainable economic development: an analysis for Turkey with ARIMA model. In: Handbook of research on creating sustainable value in the global economy. IGI Global, pp 146–164

Eurosurveillance Editorial Team (2020) Note from the editors: novel coronavirus (2019-nCoV). Eur Secur 25(3):2001231

Fernandes N (2020) Economic effects of coronavirus outbreak (COVID-19) on the world economy. Available at SSRN 3557504

Gasana J, Shehab M (2020) Coronavirus disease (COVID 19): handling challenges in Kuwait. Science 2(2):40

Griffin K, Knight J (1992) Human development: the case for renewed emphasis. In: The political economyof development and under development

Hellgren J, Sverke M (2003) Does job insecurity lead to impaired Well-being or vice versa? Estimation of cross-lagged effects using latent variable modelling. J Organiz Behav Int J Indust Occupat Organiz Psycho Behav 24(2):215–236

Hicks DA (1997) The inequality-adjusted human development index: a constructive proposal. World Dev 25(8):1283–1298

Jackson T, Marks N (1999) Consumption, sustainable welfare and human needs—with reference to UK expenditure patterns between 1954 and 1993. Ecol Econ 28(3):421–441

Kalkavan H, Ersin I (2019) Determination of factors affecting the South East Asian crisis of 1997 probit-logit panel regression: the South East Asian crisis. In: Handbook of research on global issues in financial communication and investment decision making. IGI Global, pp 148–167

Keidel A (2001) China's GDP expenditure accounts. China Econ Rev 12(4):355–367

Khumbongmayum AD, Khan ML, Tripathi RS (2006) Biodiversity conservation in sacred groves of Manipur, Northeast India: population structure and regeneration status of woody species. In: Human exploitation and biodiversity conservation. Springer, Dordrecht, pp 99–116

Knijn T (2002) Family solidarity–social solidarity; communicating vessels? In: Social values, social policies conference. Tilberg University, pp 29–31

Lancet T (2020) India under COVID-19 lockdown. Lancet (London, England) 395(10233):1315

Landefeld JS, Seskin EP, Fraumeni BM (2008) Taking the pulse of the economy: measuring GDP. J Econ Perspect 22(2):193–216

Lokhandwala S, Gautam P (2020) Indirect impact of COVID-19 on environment: a brief study in Indian context. Environ Res 188:109807
Luo S, Tsang KP (2020) How much of China and world GDP has the coronavirus reduced? Available at SSRN 3543760
Mehta K, Jha SS (2020) COVID-19: a nightmare for the Indian economy. UGC Care J 31(20)
Ministry of Statistics and Programme Implementatio (2020) Ouarterly of GDP in India. http://www.mospi.gov.in/national-account-statistics-releases-2020
Mishra M, Chatterjee S (2018) Application of Analytical Hierarchy Process (AHP) algorithm to income insecurity susceptibility mapping–a study in the district of Purulia, India. Socio Econ Plan Sci 62:56–74
Mishra K, Rampal J (2020) The COVID-19 pandemic and food insecurity: a viewpoint on India. World Dev 135:105068
Nicola M, Alsafi Z, Sohrabi C, Kerwan A, Al-Jabir A, Iosifidis C et al (2020) The socio-economic implications of the coronavirus pandemic (COVID-19): a review. Int J Surg 78:185–193
Noorbakhsh F (1998) A modified human development index. World Dev 26(3):517–528
Nove C (2020) The epidemiological characteristics of an outbreak of 2019 novel coronavirus diseases (COVID-19) in China. Zhonghua liu xing bing xue za zhi=. Zhonghua liuxingbingxue zazhi 41(2):145
Qiu Y, Chen X, Shi W (2020) Impacts of social and economic factors on the transmission of coronavirus disease 2019 (COVID-19) in China. J Popul Econ 1
Ranis G, Stewart F, Samman E (2006) Human development: beyond the human development index. J Hum Dev 7(3):323–358
Reddy DL (2012) Impact of inflation and GDP on stock market returns in India. Int J Adv Res Manag Soc Sci 1(6):120–136
Shahzad A, Hassan R, Aremu AY, Hussain A, Lodhi RN (2020) Effects of COVID-19 in E-learning on higher education institution students: the group comparison between male and female. Qual Quant:1–22
Shi X, Li J, Wang F, Dinçer H, Yüksel S (2019) A hybrid decision-making approach for the service and financial-based measurement of universal health coverage for the E7 economies. Int J Environ Res Public Health 16(18):3295
Siddiquei MI, Khan W (2020) Economic implications of coronavirus. J Public Aff 20(4):e2169
Standing G (1999) Global labour flexibility: seeking distributive justice. Palgrave Macmillan
Stoecklin SB, Rolland P, Silue Y et al (2020) First cases of coronavirus disease 2019 (COVID-19) in France: surveillance, investigations and control measures. Eur Secur 25(6):2000094
Vijai C, Suryalakshmi SM (2020) Awareness, attitudes, and practices related to coronavirus (COVID-19) pandemic among public in Chennai city, India. Cape Comorin, Special Issue, 2
Vulkan P (2012) Labour market insecurity: the effects of job, employment and income insecurity on the mental Well-being of employees. Int J Organiz (9):169–188
Wang S, Ha J, Kalkavan H, Yüksel S, Dinçer H (2020) IT2-based hybrid approach for sustainable economic equality: a case of E7 economies. SAGE Open 10(2):2158244020924434
Western B, Bloome D, Sosnaud B, Tach LM (2016) Trends in income insecurity among US children, 1984–2010. Demography 53(2):419–447
Wilthagen T, Tros F (2004) The concept of 'flexicurity': a new approach to regulating employment and labour markets. Transf Europ Rev Lab Res 10(2):166–186

Xianchun XU (2004) China's gross domestic product estimation. China Econ Rev 15(3):302–322
Yüksel S, Dinçer H, Kalkavan H, Karakuş H, Ubay GG (2020) Determination of the most optimal economic recovery package for Turkey with DEMATEL regarding the damage caused by COVID-19 pandemics. Gaziantep Üniversitesi Sosyal Bilimler Dergisi 19(COVID-19 Special Issue):326–339
Zhang C, Li R, Xia Y, Yuan Y, Dinçer H, Yüksel S (2020) Analysis of environmental activities for developing public health investments and policies: a comparative study with structure equation and interval type 2 fuzzy hybrid models. Int J Environ Res Public Health 17(6):1977
Zhu Y, Chen L, Ji H, Xi M, Fang Y, Li Y (2020) The risk and prevention of novel coronavirus pneumonia infections among inpatients in psychiatric hospitals. Neurosci Bull:1–4

Hasan Dinçer is professor of finance at Istanbul Medipol University, Faculty of Economics and Administrative Sciences, Istanbul-Turkey. Dr. Dincer has BAs in Financial Markets and Investment Management from Marmara University. He received his PhD in Finance and Banking with his thesis entitled "The Effect of Changes on the Competitive Strategies of New Service Development in the Banking Sector". He has work experience in the finance industry as a portfolio specialist and his major academic studies focusing on financial instruments, performance evaluation, and economics. He is the executive editor of the International Journal of Finance and Banking Studies (IJFBS) and the founder member of the Society for the Study of Business and Finance (SSBF). He has about 200 scientific articles and some of them are indexed in SSCI, SCI-Expended and Scopus. In addition to them, he is also editor of many different books published by Springer and IGI Global.

Hakan Kalkavan has completed PhD in Economics from Istanbul Medeniyet University in 2018. Previously, he finished his master's degree in Business Administration, Istanbul Şehir University and his bachelor's degree in International Business Administration and Management from HS Osnabrück University, Germany. Since 2015, he has been working at Istanbul Medipol University, Department of Economics and Finance as an assistant professor. During the 2019–2020 period, he worked as a TUBITAK postdoctoral researcher at Durham University, UK. His main research topics are political Economy, Islamic Economics, Economic equality, Sustainable economy, Religion-Ethics-Economic relations and Business Ethics.

Serhat Yüksel is an associate professor of finance in İstanbul Medipol University. Before this position, he worked as a senior internal auditor for 7 years in Finansbank, Istanbul-Turkey and 1 year in Konya Food and Agriculture University as an assistant professor. Dr. Yüksel has a BS in Business Administration (in English) from Yeditepe University (2006) with full scholarship. He got his master degree from the economics in Boğaziçi University (2008). He also has a PhD in banking from Marmara University (2015). His research interests lie in energy economics, banking, finance, and financial crisis. He has more than 140 scientific articles and some of them are indexed in SSCI, SCI, Scopus and Econlit. Also, he is the editor of some books that will be published by Springer and IGI Global.

Hüsne Karakuş received the degree in business administration from İstanbul Medipol University and the degree from the Banking and Insurance and International Trade and Finance Department, İstanbul Medipol University in 2020. Her research interests include energy finance, solar energy, wave energy, and renewable energy projects. She has some articles and international book chapters regarding these topics.

Chapter 4
Impact of COVID-19 and Responses to It: A Comparative Study of SAARC Countries in Light of Global Experiences

Partha Das

Abstract The first pandemic of the twenty-first century, COVID-19 has caused worldwide devastation. Exceptionality of this coronavirus pandemic is evident from its global spread, coincidence with the age of scientific and technological superiority of humankind, and the concerted manner of never seen before global response it receives. Studying the pandemic scenario in South Asia becomes important as it is home to one-quarter of the global population and one of the poorest regions on earth. No single member of the South Asian Association for Regional Co-operation (SAARC) was prepared to combat the pandemic when the outbreak started there in early 2020. Considerable variations in these countries' physiography, population, society, and economy have resulted in differences in the pattern of spread of the disease, death toll from it, and in their responses to tackle it. Again these developing countries equally suffer from the stresses of unemployment, poverty, social and economic discrimination. So some comparability is there in their COVID experiences. In such context, the present study aims to perform an appraisal of the pandemic scenario in the SAARC region and discuss the progress of the member nations in combatting the disease in comparison to the various measures taken across the world.

Keywords SARS-CoV-2 · Pandemic · South Asia · Severity · Measures · Relative success

4.1 Introduction

By the time the calendar year 2020 gets over, "COVID-19" is undoubtedly going to become the "word of the year." Designated by World Health Organization (WHO) as "Severe Acute Respiratory Syndrome Corona Virus 2" or SARS-CoV-2, this novel

P. Das (✉)
A. B. N. Seal College, Cooch Behar, West Bengal, India

S. Saha et al. (eds.), *Economic and Societal Transformation in Pandemic-Trapped India*, New Frontiers in Regional Science: Asian Perspectives 55,
https://doi.org/10.1007/978-981-16-5755-9_4

coronavirus disease with a zoonotic origin has brought the world into a global pandemic scenario almost after a 100 years and to the first one of the twenty-first century. Though COVID-19 is not unprecedented as a pandemic because the world has experienced many such before in the near and remote past, it definitely has some unique features compared to the earlier ones that require to be studied. Originating from the Chinese city of Wuhan in presumably November 2019, this contagious respiratory illness has spread over 200 nations, has infected more than 46 million people and has killed more than 1.2 million people by the start of November, 2020 (worldometers 2020). South Asian Nations have also not been an exception, with around a quarter of global COVID-19 infection being reported from this region, co-incidentally home to one in four persons in the world (United Nations 2019a, b). Three of the land-locked nations of this region, Bhutan, Nepal, and Afghanistan, are counted among ten higher average elevation countries globally, whereas the island nation of Maldives is the lowest average elevation country in the world. This region accommodates the world's second and fifth most populous countries, India and Pakistan, respectively, along with the world's most densely populated country, Bangladesh. People living here are diverse concerning their ethnicity, culture, religion, language, and society. South Asia is again home to the world's second-highest number of poor people (530 million) just after Sub-Saharan Africa (United Nations Development Programme 2020). The region is also politically volatile as two of the largest nations here, India and Pakistan, have been fighting over the disputed Kashmir region for more than 70 years now, while Afghanistan has been a global theater of warfare for centuries. In such a varied physiographic, economic, and socio-cultural scenario, considerable variation has been observed in the severity of COVID-19 and the countries' responses toward it.

So the present article aims to measure the spatial variation in the spread and impact of the disease in South Asian Region in general and SAARC (South Asian Association for Regional Co-operation) nations in particular. It further discusses the global response to the ongoing pandemic and the progress of SAARC countries in this regard. The paper investigates how the varying sociopolitical and economic scenarios observed among these nations impact their fight against COVID-19. A comparison among these nations is attempted concerning their existing COVID situation and maps the road ahead. Finally, the article goes down the history of epidemics and pandemics observed worldwide, particularly in the twentieth century, to compare their impact with that of SARS-CoV-2.

4.2 Literature Review

COVID-19 might be new in the academic discussion, but extensive works on it are being done by scholars and researchers worldwide. There are certain limitations in studying the disease at present though some noteworthy successes have already been achieved, particularly in the field of medical research. Such studies include a range of topics from serological survey reports (Kshatri et al. 2020) to COVID-19 reinfection (Tillett et al. 2020) to vaccine research (Krause et al. 2020) etc. Statistical studies ranging from the topics of the spread of the pandemic (Mustafa 2020) to forecasting (Yousaf et al. 2020) to the preparedness and responses of individual countries (Tripathi et al. 2020), (Koirala et al. 2020) to the economic impact of the pandemic (Centre for Disease Dynamics, Economics and Policy 2020) have also been extensively performed based on data provided by different national and international agencies. WHO and Health Departments of different countries have published status reports and issued directives to tackle the pandemic from time to time, further aiding academic research. Some comparative analyses among two to four countries have also been attempted, but an extensive regional study with comparable indices is missing in general. A few such attempts have been made for European nations or African and American countries (Balmford et al. 2020), but it is absent for South Asian nations which might be due to considerable differences in their areal extents, population, and sizes of their economies.

There are other constraints in COVID-19 research also. World Health Organization has assumed that the pandemic would stay for two more years[1] which puts any conclusive research findings based on current data at risk of being obsolete. Again WHO apprehends that by September 2020, around 10% of the world population has been infected by the disease, and this number is 20 times more than the officially diagnosed cases (Times of India 2020). So any statistical inference based on currently available data might give false representations. Another limitation in COVID-19 research has been that most of the works so far are framed on data that often miss the social, political, ethical, and philosophical implications of this global pandemic. As Lal (2020) states in his book that "In the response of each country to the pandemic we can see shades of that country's history" (p. xiv).

The present work is a regional study of the pandemic situation in SAARC nations till September 2020. The study is divided into three parts; where the first part deals with pandemic experiences of these nations in terms of severity and responses; the second part is a comparison of the measures taken by individual countries and an evaluation of their respective success; the final part discusses the earlier epidemics and pandemics to find out the exceptionality in the nature of ongoing COVID-19 global pandemic.

4.3 Objectives

The present study aims to delineate the severity of the COVID-19 pandemic and to compare the pandemic scenario among South Asian countries. It discusses the global response to this pandemic and evaluates the measures taken by SAARC nations in

[1] Dr. Tedros Adhanom Ghebreyesus, WHO Chief told reporters on 21 August 2020 that they hoped to finish the pandemic within 2 years. This is available under the URL: https://www.livemint.com/news/world/coronavirus-likely-to-vanish-faster-than-spanish-flu-who-chief-11598075277127.html.

the backdrop of their society, politics, and economy and their relative success in contrast to each other. It goes on to compare the overall COVID scenario in the said countries and their future prospects. Finally the study attempts to identify if the COVID-19 pandemic is exceptional or unprecedented in comparison to earlier epidemics and pandemics in the world starting from the twentieth century onward.

4.4 Materials and Methods

For the present study, eight (8) SAARC member nations, namely Afghanistan, Bangladesh, Bhutan, India, Maldives, Nepal, Pakistan, and Sri Lanka, have been taken as a broader study region. In the absence of ample published materials in a pandemic situation, COVID-19 datasets have been collected from the official websites of the health ministries of different SAARC nations[2,3,4,5,6] and other reliable web platforms like "worldometer"[7] recognized by the American Library Association (ALA),[8] and "COVID-19 INDIA"[9] etc. Though the COVID scenario rapidly changes over time, the present study discusses the COVID situation as experienced between April and September 2020 as any quantitative research requires to be performed in a specific spatial and temporal framework. By April 2020, all SAARC member nations had officially been treating COVID patients in their countries. While by the end of September 2020, all these nations experienced a fall in active SARS-CoV-2 caseload at least once. This might indicate that all SAARC countries were over at least one transmission cycle of the coronavirus outbreak by the end of the month of September.[10] Again after September 2020, the island nation of Maldives stopped publishing its official data on coronavirus testing regularly. After 27 September, 2020, Govt. of Maldives had released test related information on 5 October and then again on 29 November (see footnote 5).

[2]Ministry of Health and Family Welfare, Govt. of India website: https://www.mohfw.gov.in/.

[3]Epidemiology Unit, Ministry of Health, Govt. of Sri Lanka website: http://epid.gov.lk/web/index.php?lang=en.

[4]Ministry of Health, Royal Govt. of Bhutan website: http://www.moh.gov.bt/.

[5]Ministry of Health, Republic of Maldives website: http://health.gov.mv/.

[6]Government of Pakistan Health Advisory web-platform: https://covid.gov.pk/.

[7]See worldometer website: https://www.worldometers.info/coronavirus/.

[8]Reference and User Services Association, A Division of the American Library Association website: https://www.ala.org/rusa/sections/mars/marspubs/marsbestfreewebsites/marsbestref2011.

[9]See COVID19INDIA website: https://www.covid19india.org/.

[10]WHO reported a 7% fall in weekly new cases and a 3% fall in weekly new deaths within 21–27 September, 2020 in its South-East Asia Region comprising six SAARC nations except Afghanistan and Pakistan. This was published on WHO's official website on 28 September 2020 under the headline "Weekly epidemiological update—28 September 2020" which is archived under the URL: https://www.who.int/publications/m/item/weekly-epidemiological-update%2D%2D-28-september-2020.

SAARC countries have considerable differences in their areal extent, population size, and economy. Any comparison based on absolute data values (like total or daily COVID cases, total or daily fatalities due to COVID, number of recoveries, etc.) might turn out to be misleading. So this study uses relative data measures like cases, deaths, and tests per million population, positivity rates (% of people tested positive for COVID infection out of the total number of people tested), case fatality rates (CFR), recovery rates as ideal measures of comparison. Simple and standard statistical techniques like moving average, coefficient of variation, and correlation have been applied to infer the data collected. Cartographic techniques of Line graphs, Scatter diagrams, Choropleth Maps, etc., have been applied using Ms. Excel and QGIS 3.10 software.

4.5 SARS-CoV-2 Scenario Among SAARC Nations

Analysis of a health situation among SAARC countries who differ extensively over the area, total population, size of their economy, society, and culture; is a challenging task putting up a valid question that if this comparison is justified enough. Again, all of these countries are developing nations that deal with similar problems of poverty, unemployment, lack of infrastructure, inadequate health services, social and economic discrimination, etc.; hence the comparative study becomes rational enough. Any comparison based on absolute values might be misleading between countries like India with a 1.3 billion population and Bhutan or Maldives, where less than one million people live. Thus this study is going to be based upon relative measures as stated earlier.

4.5.1 Spread of COVID-19 in SAARC Region

The first confirmed cases of COVID-19 infection, also known as Index Cases, in all SAARC countries have been imported in nature. All these patients were either foreign nationals on a business or leisure tour in these countries or citizens of these countries who returned from abroad. By the end of March, the disease spread in all countries of this region through varied spread rates. While in Sri Lanka and Bhutan, the disease has shown a slow progression, it has been rapid in the rest of the countries. All these countries represented no clear indication of how infection is occurring, indicating a hidden chain of transmission (The Economic Times 2020). By the end of September 2020, India has recorded more than six million cases which is worrying but not shocking from the second-most populous country in the world and because it has conducted more than 75 million tests by then (Table 4.1) which shows it has a low positivity rate. Bhutan is best positioned with the lowest number of cases and lowest positivity rate (Fig. 4.5) while Sri Lanka stays at a comfortable position with the lowest cases per million (Fig. 4.1d). The situation appears alarming

Table 4.1 COVID Scenario of SAARC countries as on 30th September, 2020

Country	Index case	Total COVID cases	Cases/ million	Total tests	Tests/ million	Total deaths	Deaths/ million	Case fatality rate (%)	Recovery rate (%)
India	January 30	6,310,267	4656	75,619,781	55,342	98,708	72	1.56	83.51
Bangladesh	March 08	363,479	2229	1,959,075	12,015	5251	32	1.44	75.79
Pakistan	February 26	312,263	1442	3,548,376	16,385	6728	31	2.15	95.07
Nepal	January 23	77,817	2720	1,033,947	36,141	498	17	0.64	72.51
Afghanistan	February 24	39,268	1032	111,310	2926	1458	38	3.71	83.5
Maldives	March 07	10,291	19,380	149,135	280,857	34	64	0.33	88.5
Sri Lanka	January 27	3380	158	289,588	13,580	13	0.61	0.38	95.56
Bhutan	March 06	281	368	136,947	179,485	0	0	0	77.94

Source: worldometers (2020), United Nations (2019a, b)

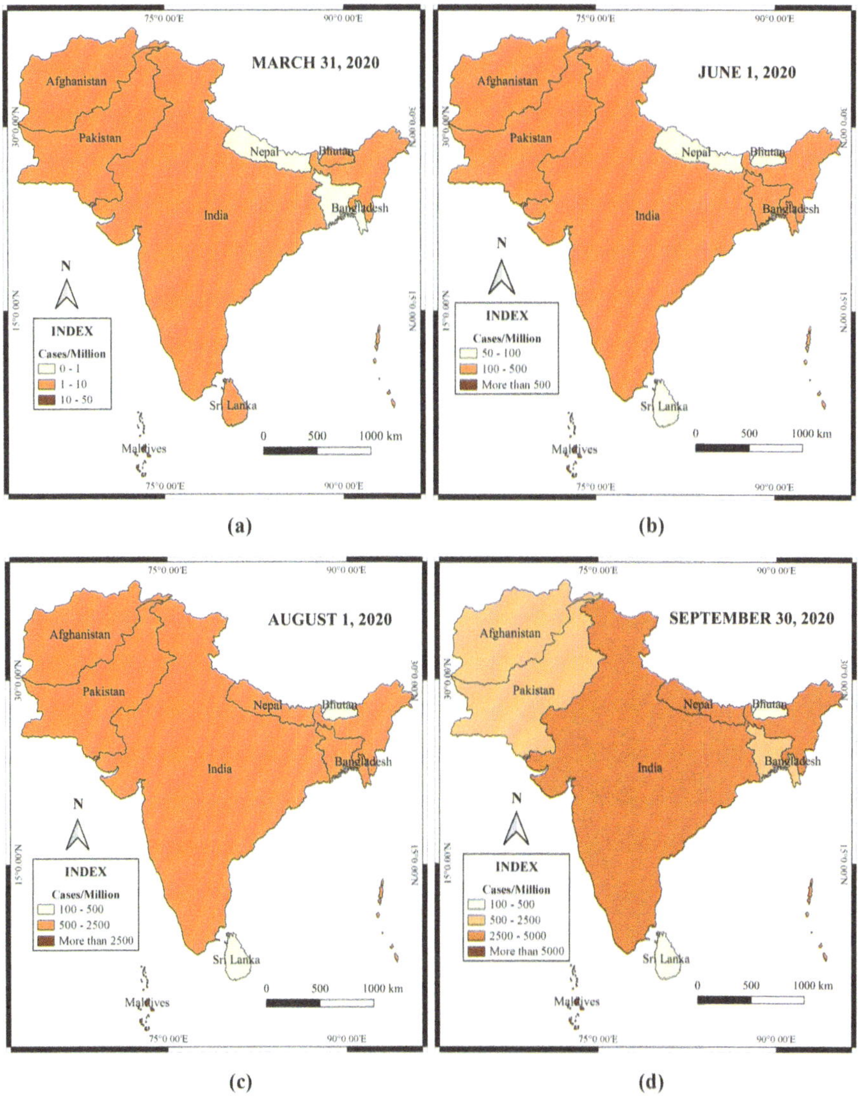

Fig. 4.1 Spread of Covid-19 in SAARC countries as on (**a**) 31 March, 2020; (**b**) 1 June, 2020; (**c**) 1 August, 2020 and (**d**) 30 September, 2020. (Source: Prepared by the author from worldometers 2020)

for the Maldives as this tiny island nation has experienced the rapid spread of coronavirus with around 2000 cases per million population by the end of September (Fig. 4.1). Here it is to be noted that with more tests, more patients are detected. Maldives has conducted more vigorous testing, which explains its high COVID graph.

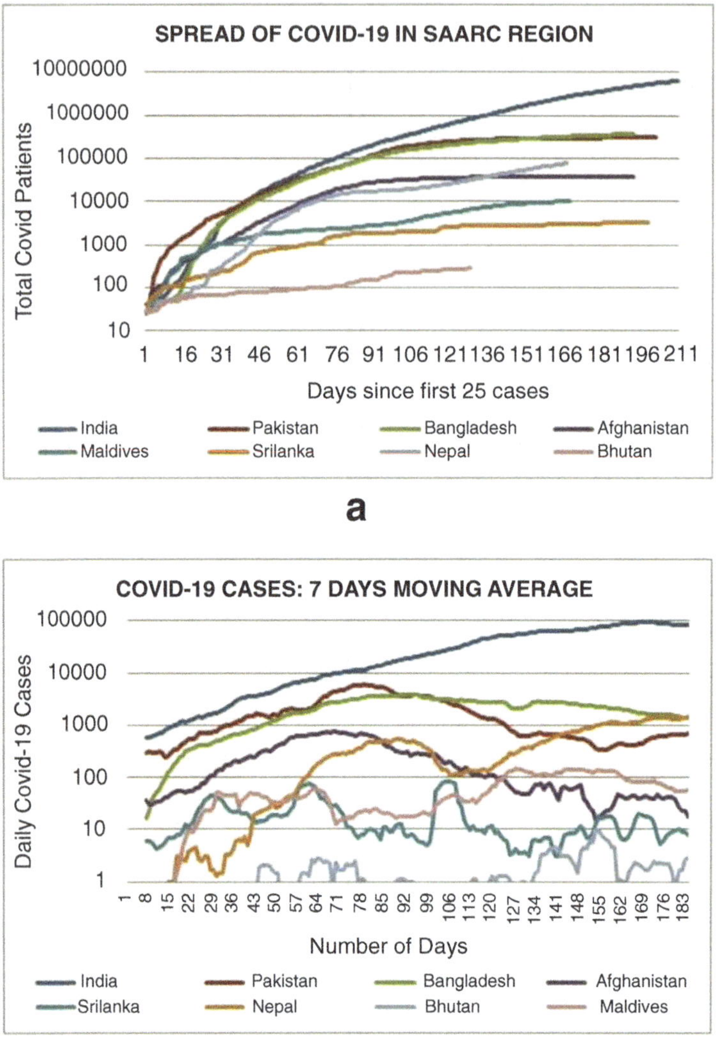

Fig. 4.2 Progression of (**a**) Total cases and (**b**) 7-Days moving average of daily number of cases. (Source: Prepared by the author from worldometers 2020)

Apart from India, all SAARC countries have successfully flattened the curve of COVID-19 spread by September, 2020 (Fig. 4.2). But seven-days moving average curves show a lowering in the number of daily cases in India also by that time which indicates a potential slowing down of the virus spread in the future. The moving average curves (Fig. 4.2) indicate that all the SAARC countries have completed at least one transmission cycle of the disease, if not more. While a slow rise of cases can be seen in Pakistan, Nepal, and Sri Lanka in September, it is stable or lower in others.

4.5.2 Death Toll due to Coronavirus Pandemic

The fatalities due to COVID-19 in the SAARC region are evidently much lower than what has been experienced in other parts of the world. Though India has experienced the third-highest number of absolute deaths in the world unfortunately, its values of CFR (% of patients dying out of the total number of infected people) and Deaths per million (lowest among 20 most COVID-19 affected countries) remain much lower than the world average (see footnote 7). It shows a better survival probability of COVID patients in this country. The country of Bhutan has successfully resisted any death of its citizens so far till early December. Sri Lanka is another success story in this regard where both the CFR and Deaths per million remain very low till September. The performance of Nepal is also commendable, where though death per million is gradually rising, the CFR remains in check resisting any sudden deterioration of the COVID situation in the country. Regarding deaths per million, Afghanistan, Bangladesh, and Pakistan face a similar scenario to each other with smaller values than the global average. However, CFR for Afghanistan and Pakistan remains much higher, indicating comparatively critical survival chances for COVID patients in those countries (Table 4.1). Maldives has successfully checked the CFR, but deaths per million in the country remain higher just next to India in this region (Fig. 4.3).

4.6 Response to the Pandemic, from Global Practices to Experiences of SAARC Nations

Dealing with COVID-19 has been a challenge to all countries in the world. Though it is the first pandemic to hit humankind during their peak of scientific and technological progress, the impact has been very hard, causing loss of lives across international borders irrespective of the country's economic development scenario. By November 2020, almost 1 year since the pandemic outbreak, no definite preventive or curative measure to fight this disease has been discovered or invented. Though scientific and medical communities across the globe are working very hard to find a cure to this pandemic, no success has been achieved so far. The global practices in the fight against COVID-19 are classified as given in Fig. 4.4.

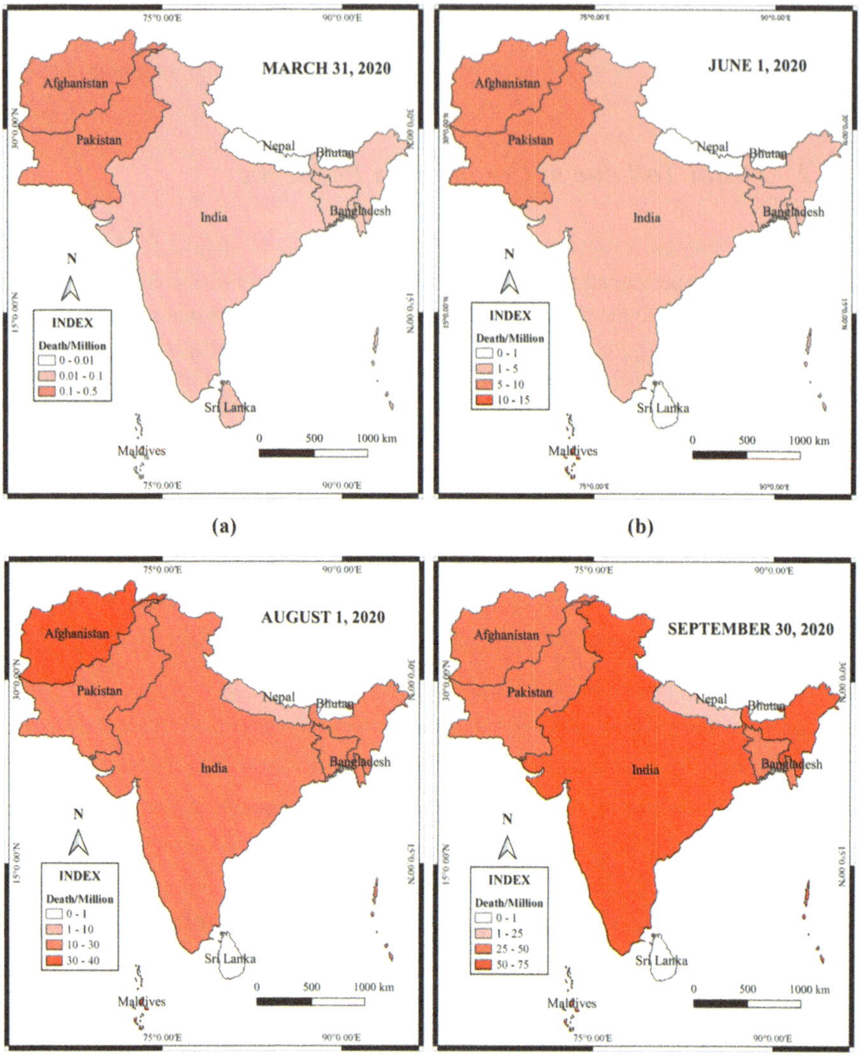

Fig. 4.3 Occurrence of death due to Covid-19 in SAARC countries as on (**a**) 31 March, 2020; (**b**) 1 June, 2020; (**c**) 1 August, 2020 and (**d**) 30 September, 2020 by September, 2020. (Source: Prepared by the author from worldometers 2020)

4.6.1 Preventive Measures Against COVID-19

Vaccination has been the most popular and reliable measure against the spread of any contagious disease. Vaccines are the attenuated or non-virulent and sometimes inactivated viral particles that mimic the virus or part of it they protect against by

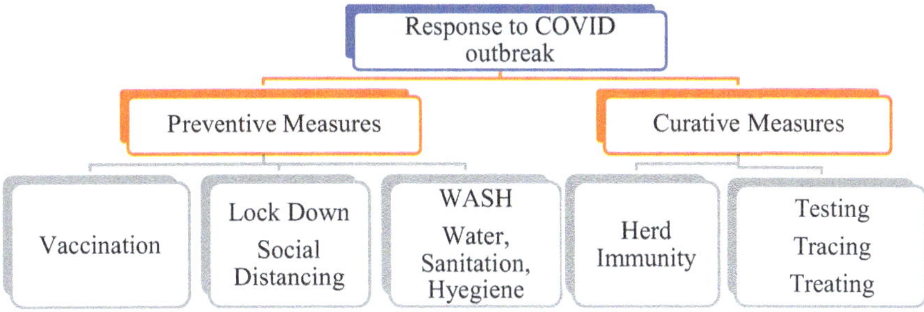

Fig. 4.4 Global practices of combatting the coronavirus pandemic

stimulating the immune system to develop antibodies (Owen et al. 2013). These are supposed to follow the highest safety standards as these are given to healthy people. So before the final stage of approval for general public use, vaccines go through four stages of clinical trials. At the first pre-clinical stage, vaccines are applied to animals to trigger an immune response in them. In the next Phase-1, it is given to a very small group of human volunteers to check the safety and the immune response it provokes. Then in Phase-2, the vaccine is given to a larger number of people to learn about its safety and correct dosage. Next, in Phase-3, it is given to thousands of people to confirm its effectiveness, safety, and side-effects before being finally approved for wide usage.[11] By the end of July, around 150 vaccines were being prepared across the world, out of which twenty-six (26) were at phase-1 human trials, and three (3) were at phase-2 clinical trials (Hindustan Times 2020a, b). This number of probable vaccine candidates increased to 321 by early September. By this time, thirty three (33) vaccines reached the phase-2 and phase-3 clinical trials (Le et al. 2020). Among these, the vaccine jointly developed by Oxford University and AstraZeneca firm (AZD1222) is seen as a front runner as it is found to be safe and efficacious against symptomatic COVID-19 during clinical trials (Voysey et al. 2020) Other promising candidates are two vaccines manufactured by Pfizer and Moderna as they have submitted their clinical trial data to the regulatory authorities of different countries for emergency use (see footnote 11). Apart from these, the Chinese pharmaceutical company, Sinovac was running phase-2 and phase-3 trials of their vaccine by mid-August, and Russia had already approved their vaccine (*Sputnik V*) without completing clinical trials. In India, Bharat Biotech Pharmaceuticals, in assistance with the Indian Council of Medical Research (ICMR) and National Institute of Virology (NIV), were running phase 2 trials of its vaccine (*Covaxin*) in September[12] Another Indian Vaccine, *Covishield,* jointly prepared by

[11] WHO reported in its official website under the headline 'Update on COVID-19 vaccine development-CORONAVIRUS UPDATE 45' published on 21 December 2020 which is available under the URL: https://www.who.int/docs/default-source/coronaviruse/risk-comms-updates/update45-vaccines-developement.pdf?sfvrsn=13098bfc_5.

[12] Indian online news portal. www.livemint.com covered this news on 17 July 2020, which is available under the URL: https://www.livemint.com/news/india/india-s-first-covid-19-vaccine-covaxin-human-trial-starts-well-key-updates-11594977657239.html.

AstraZeneca and Oxford University along with Pune-based Serum Institute of India (SII) has been running phase-3 trials in October 2020. At the same time, Zydus Cadila Biotech was running phase-2 trials of their vaccine (*ZyCOV-D*) at October end[13] . Bangladesh Medical Research Council (BMRC) had granted "ethical approval" for phase-3 human trials of Chinese Sinovac vaccine on healthcare professionals in the month of July (Reuters 2020a, b). Govt. of Bangladesh has signed a Memorandum of Understanding with Serum Institute of India in early November (WION 2020), to buy 30 million doses of its potential coronavirus vaccine (*Covishield*). During the same time, the Government of Pakistan approved phase-3 trials for a vaccine made by Chinese CanSino Biologics (*Ad5-nCoV*) in association with Pakistan's National Institute of Health (NIH) (Reuters 2020a, b). No coronavirus vaccine trial is running in Sri Lanka until September 2020, and the country has joined the Global Alliance for Vaccines and Immunisation (GAVI) to procure vaccines when prepared (XINHUANET 2020). While Bhutan has collaborated with India in the phase-3 trial of a potential Indian vaccine, the governments of Afghanistan, Maldives, and Nepal are still at the stage of forming national steering committees for procuring the SARS-CoV-2 vaccine. In this regard, experts are worried that political and economic pressure for the rapid introduction of a vaccine might produce a weakly effective vaccine (Krause et al. 2020). They believe that only a vaccine with more than 50% efficacy could appreciably reduce the incidence of COVID-19 among vaccinated individuals. England's Chief Medical Officer, Prof. Chris Whitty, apprehended in July that an effective and safe coronavirus vaccine would only arrive in the winter of 2021 (Sky News 2020).

In the absence of any vaccine or confirm drug, all the COVID-affected countries in the world have imposed some form of restrictive lockdown or "cordon sanitaire." These measures include closing down borders, suspending international air travel, limiting intra-national travel, suspending nonessential services (restaurant, school, shop, gym), maintaining self-isolation and social distancing. This measure of lockdown and social distancing is highly effective in breaking the chain of disease transmission (World Economic Forum 2020). It can successfully reduce the "R-Effective" to below 1, which means that an infected individual shall be able to infect less than one person on average. Here it is to be noted that lockdown can only be an initial measure at the beginning of pandemic spread. Imposing indefinite lockdown is an unscientific and futile measure to limit the spread of the pandemic. Non-pharmaceutical interventions such as quarantine to contain and control the spread of the disease are not evidence based and often neither protect the public health nor safeguard civil liberties (Greenberger 2018). Long-term lockdown measures are not sustainable for developing countries as these have various social, psychological and economic impact (Meo et al. 2020). India has observed a four-phased nationwide lockdown between 25 March and 31 May 2020. Though it had

[13]Indian online news portal, www.india.com covered this news 29 November 2020, which is available under the URL: https://www.india.com/news/india/covid-vaccine-update-india-covaxin-covishield-zycov-d-4230743.

slowed down the spread of the disease, it could not deter it; even the deaths from COVID-19 continued rising.

Further, it caused severe damage to people's lives and livelihoods as they lost jobs and failed to get proper healthcare due to limited transportation and local hospital services. These observations remain true for India's neighbors, like all the countries in the SAARC region, having similar economic conditions, health scenario, and employment opportunities. Pakistan has never opted for a strict lockdown, and it's nationwide "Smart Lockdown Policy" operating between 1 April and 9 May, 2020, with the economic sector being open with certain restrictions, was lifted after the month of May. A nationwide lockdown was observed in Bangladesh from 26 Mar to 30 May, 2020 with certain exemptions. Later the government had divided the country into Red, Yellow, and Green zones and extended separate restriction measures for each zone to be followed between 1 July and 3 Augues. Nepal observed countrywide lockdown between 24 March and 21 July 2020 and then went on to observe limited lockdowns at local levels. Sri Lanka had also observed a 52 days long strict lockdown between March and May 2020. On 28 June, the government lifted the lockdown all over the country. Afghanistan had never gone for a total lockdown and opted for regional lockdowns based on the requirement. The Royal Government of Bhutan imposed its first nationwide lockdown late in the month of August, 2020 though restrictions on the arrival of foreign tourists are there in operation since 6 March 2020. The government of Maldives declared a public health emergency on 12 March, which continued till 30 April (World Health Organization 2020). This tiny island nation, dependent on tourism, had no option but to welcome international travelers back since 15 July, 2020.

Safely managed water, sanitation, and hygiene (WASH) services are essential for preventing the spread of the pandemic and protecting human beings (The World Bank 2020). These are cost-effective strategies that can be ideal for developing countries for pandemic preparedness. The practices of using safe water, maintaining self-hygiene, hygiene at home or workplaces, maintaining clean sanitation help in being safe from disease transmission and in the recovery phase. WHO and other national agencies globally circulated advisory on WASH and waste management in the wake of COVID-19. These include frequent and proper hand hygiene, safe management of drinking water and sanitary services, mandatory use of face covers or masks, and maintaining respiratory etiquettes (Ministry of Health and Family Welfare 2020). But success of this WASH strategy too has been put under scrutiny by the pattern of COVID spread in South Asian nations. A study conducted by the Centre for Scientific and Industrial Research (CSIR), India, has shown that as Indians are exposed to pathogens from early in life due to their less desirable hygiene levels, they develop better immunity which reduces their risk of death from coronavirus (Chatterjee et al. 2020)

4.6.2 *Curative Measures for COVID-19*

Herd immunity has been a popular measure of epidemic control. Here a proportion of the population needs to be infected with the disease either naturally or through vaccination (John and Samuel 2000). People who survive become immune from the disease; as more and more people become immune, the chain of virus transmission breaks, and the epidemic settles down (The Royal Society 2020). Herd immunity is achieved when one infected person transmits the virus to less than one person, and thus the value of "R-Effective" goes below 1 (Randolph and Barreiro 2020). Researchers from Princeton University, Johns Hopkins University and the Centre for Disease Dynamics, Economics and Policy, New Delhi, had opined that the high proportion of the young population in India is the key to the lower risk of hospitalization and death (Klein et al. 2020). So the country should opt for achieving herd immunity which would rescue India from the economic burden of lockdown and reduce pressure on existing medical establishments by not treating in isolation those patients who have mild COVID conditions. The same could go true for any SAARC nation as they have higher proportion of youth population (United Nations 2019a, b) and they also appear to have inadequate pandemic preparedness based on their low to average performance in terms of health security (Global Health Security Index 2019). Zaidi et al. (2017) has highlighted the existing disparities in the availability of basic health interventions in South Asian countries between rich and poor, rural and urban. Samaranayake (2020) has shown how poor public health infrastructure coupled with lack of trained health workers have resulted into inequities in health care in these countries. Keeping these in view, herd immunity appears a logical escape route from COVID-19 for any SAARC country. Studies also revealed that people, who have been infected by SARS-CoV-2 and recovered later, develop antibodies that might temporarily protect them from reinfection (Kirkcaldy et al. 2020) Thus even if people are not tested and quarantined, they would develop natural immunity and live. WHO's chief scientist Dr. Soumya Swaminathan has stated that around 60% population needs to be infected by coronavirus to reach the stage of natural herd immunity, which requires much longer time and more waves of infection (Swaminathan 2020). At the same time, some other researchers have put doubt on the possibility of achieving a herd immunity as new strains of coronavirus keep emerging and acquired immunity cannot provide long-term protection (Nature 2021). . Govt. of India has never accepted herd immunity as a reliable measure for epidemic control.[14] United Kingdom had adopted it initially, and they dropped it later due to controversy (Clemente-Suárez et al. 2020). Pozzer et al. (2020) has shown that air pollution is an important cofactor increasing the risk of mortality from COVID-19. Hence in India and other SAARC countries with poor health

[14] India's digital media platform The Print covered the news under the headline 'India far from herd immunity, can only come at very high human cost, says top govt official' on 30 July 2020 which is available under the URL: https://theprint.in/health/india-far-from-herd-immunity-can-only-come-at-very-high-human-cost-says-top-govt-official/471464/.

infrastructures and sanitary conditions, severe air pollution, the prevalence of other diseases, this strategy of herd immunity might lead to more hospitalization and death, straining a country's healthcare system. Again a Lancet study has revealed that exposure to COVID-19 for once and getting cured later does not guarantee immunity. The patient can again be infected with more severe symptoms (Tillett et al. 2020).

The triple "T" method or the combined method of Testing, Tracing, and Treating has been the most popular and reliable measure followed by the world to combat coronavirus. It was first applied in South Korea (Lexlife India 2020). The method involves fast identification of coronavirus patients in a population by a rapid and large number of testing. Then the people who made close contact with confirmed COVID patients are tracked and kept in isolation. Finally, people infected with the disease are treated while in institutional or in-home quarantine. The advantages of this method are—early detection and containment of the virus, better treatment for patients. While the challenges are—huge resource employment for the purpose of testing, tracing, and treating; exposure of medical personnel and other frontline workers (e.g., police, sanitation workers) to the virus; risk of community spread. Primarily there are two types of COVID-19 tests—Diagnostic Tests (*Rapid Antigen Test*, *RT-PCR Test*) looking for active coronavirus infection in human mucus or saliva and Antibody Tests (*Serology Test*) looking for development of antibodies in human blood against coronavirus (Health 2020). The biggest challenge for SAARC countries initially was the absence of adequate testing facilities for novel coronavirus, which had delayed deployment of this triple "T" method, and many suspected cases went undetected. At the beginning of June, 877 government and private laboratories in India were conducting tests; then, such facilities increased to more than 1600 by August end (Indian Council of Medical Research 2020). In Pakistan, 107 testing facilities with an average daily capacity of around 46,730 tests were performing by mid-June (National Institute of Health, Islamabad 2020). While in Bangladesh number of testing facilities increased from 49 to 82 between May and July (Huq and Biswas 2020). Only eight testing facilities were running in Afghanistan in the month of May, which later increased to 32 public and private laboratories in October with a maximum daily capacity of 5500 tests[15] With respect to testing, Sri Lanka looked prepared better from the beginning, with 17 facilities running from March (Ministry of Health and Indigenous Medical Services 2020), but as cases started rising in September, the country could not increase the number of laboratories at the required pace. While Nepal had only one testing laboratory in March, they increased the testing facilities to 10 by June, then to 22 by July, and then to 77 by the end of November, 2020.[16] While the Maldives was operating three testing

[15]See excerpts from WHO's Regional COVID-19 mission to Afghanistan, available under the URL: http://www.emro.who.int/afg/afghanistan-news/regional-covid-19-mission-to-afghanistan-concludes.html.

[16]WHO reported in its Nepal country office website under the headline 'Situation Update #33 – Coronavirus Disease 2019 (COVID-19)' published on 2 December 2020 which is available under the URL: https://www.who.int/docs/default-source/nepal-documents/novel-coronavirus/who-nepal-sitrep/-33_who_nepal_sitrep_covid-19.pdf?sfvrsn=9e666b77_7.

laboratories by August, Bhutan was operating five testing facilities by September. Tracing potential coronavirus patients from among the close contacts of a person diagnosed with the disease has been an issue for almost all SAARC countries. In the absence of proper health consciousness and due to the spread of misinformation, people have been observed hiding their health issues to avoid COVID testing. Both WHO and US Centre for Disease Control and Prevention have stated that it takes 5–14 days between exposure to the virus and showing its symptoms. It is a well-known fact that most COVID patients do not even show symptoms of the disease. In this scenario people, hiding from the test risk multiplication of the disease. Govt. of India launched a mobile app for corona tracking to combat this situation, known as "Aarogya Setu." Other nations in the SAARC region launched similar apps too, for example "COVID-19-GOV-PK" in Pakistan, "Corona Tracer BD" in Bangladesh, "COVIDShield" in Sri Lanka, "Druk Trace" launched in Bhutan and also operating in the Maldives. In Nepal and Afghanistan, no such nationwide mobile applications for tracing the virus were launched. Concerning treatment, the common method applied worldwide has been home or institutional quarantines of varying duration. In India and most other SAARC nations, a 14-day quarantine period has been recommended by the governments for both symptomatic and asymptomatic patients. It is only Bhutan that has a mandatory quarantine period of 21 days which might be one of the reasons for the country's great performance against COVID-19. Till now, no drug or medical procedure has been totally successful in curing the virus, but different countries are using different drugs and methods with varying success. While *Dexamethasone* steroid is said to reduce death from COVID-19 successfully, Remdesivir resists the virus from taking severe form and results in quick recovery for patients (The Conversation 2020a, b). Hydroxychloroquine drug has been used in India to treat SARS-CoV-2 patients along with *Convalescent Plasma Therapy* in limited cases, but it was discouraged later (ICMR 2020). This same therapy was being practiced in Pakistan without any state advisory. Bangladesh produced Remdesivir in the country and provided it free to COVID patients admitted in state-run hospitals (Government of the People's Republic of Bangladesh 2020). Ministry of Health and Family Welfare, Govt. of India, reported around 15,378 dedicated COVID treatment facilities providing medical care to patients by the end of the year 2020.[17] Whereas in Pakistan, 35 designated tertiary hospitals were treating coronavirus patients at the beginning of August.[18] Bangladesh, in June, reported having only 1267 ventilators what was grossly inadequate as per WHO protocol of 15–20% COVID patients admitted to hospitals requiring ventilator

[17]Press Information Bureau, India's state-run media agency published this news on behalf of Ministry of Health and Family Welfare under the headline 'Ministry of Health and Family Welfare 2020 ACHIEVEMENTS' on 30 December 2020 which is archived under the URL: https://pib.gov.in/PressReleasePage.aspx?PRID=1684546.

[18]Govt. of Pakistan published this document entitled 'List COVID-19 Designated Tertiary Care Hospitals' in their COVID-19 Health Advisory Platform which is available under the URL: https://covid.gov.pk/facilities/List%20of%20COVID-19%20Designated%20Tertiary%20Care%20Hospitals%20Pakistan.pdf.

support (Swarajya 2020). Treating the patients has been one of the biggest challenges in Afghanistan, with only 3230 hospital beds and 300 ventilators all over the country by April end, which increased to 500 by the end of July (The Conversation 2020a, b). Difficulty in accessing remote regions and lack of female access to healthcare in a deeply conservative society are other challenges in the country (Financial Express 2020). In February, there were no isolation wards or beds in any health facilities in Nepal, and all healthcare staff across the country lacked any protective gear against the virus (The Kathmandu Post 2020). Presently 32 hospitals are treating COVID patients in the country.[19] In the Maldives, five flu clinics were treating patients at the end of July. By the month of May, only two hospitals were treating COVID patients in Bhutan. Presently 36 designated COVID hospitals are running in Sri Lanka (Health Promotion Bureau 2020). This triple "T" method of fight against the pandemic has also been criticized as it increases pressure on existing medical establishments, denies or limits the treatment of patients suffering from other diseases, risks the lives of medical personnel and frontline workers by increasing viral loads upon them.

4.7 Assessing the Relative Success of SAARC Nations in Fight Against the Pandemic

All the COVID-affected countries in the world, as well as SAARC nations, are fighting against the pandemic with more or less success. Comparing the countries based on total cases, deaths, or on their preparedness by means of medical infrastructure and capacity might be misleading as the countries are acting as per the severity of the disease within their boundaries and their economic strength. Now comparing the impact of the measures taken by individual nations could help assess the pandemic situation, analyzing their relative success and assuming the path ahead. For this purpose, two measures of comparison have been attempted.

The mean and standard deviations of daily recovery rates (RR) of individual countries are calculated to determine the Coefficient of Variation (CoV). The CoV values for daily recovery rates explain how consistent has been the recovery scenario for individual SAARC nations between April and September 2020 (Table 4.2).

Table 4.2 indicates that the best recovery scenario has been observed in Sri Lanka, where the mean daily recovery rate is found to be highest till the month of September, with the lowest CoV value indicating the most consistent recovery of patients. Though the mean value of the daily recovery rate is high for Maldives and Pakistan, the values for CoV are also exceptionally high, indicating their failure to

[19] This article entitled 'Designated hospitals to treat COVID-19 Cases in Nepal' was published by the IOM (International Organization for Migration) Nepal Mission and it is available in their official website under the URL: https://nepal.iom.int/sites/nepal/files/COVID-19/Designated_COVID-19_Hospitals_ENG.pdf.

Table 4.2 Mean, standard deviation and coefficient of variation for daily recovery rates

Countries	Mean of RR	Standard deviation of RR	COV of RR
Afghanistan	38.70036	31.57878	81.60
Bangladesh	38.2045	23.3586	61.14
Bhutan	51.43113	28.74566	55.89
India	47.9443	25.4879	53.16
Maldives	52.53087	30.4596	57.98
Nepal	44.36267	28.38556	63.98
Pakistan	51.68214	34.07943	65.94
Sri Lanka	61.67901	30.97651	50.22

Source: Computed by author

stay consistent with recovery opportunities for their patients. Bhutan and India have succeeded in achieving a moderately good mean daily recovery rate by the end of September and smaller COV values, indicating that the recovery scenario has been consistently good during this period of study. Mean daily recovery values for both Nepal and Bangladesh have been low, with greater values of COV between April and September. This indicates not such good recovery opportunities for COVID patients in these countries, which are in sync with the statement of a public health expert in Bangladesh who claimed that a curative approach to deal with COVID would be suicidal in the country as the health system was not prepared enough (BBC News 2020). In terms of recovery between the months of April and September, the worst-performing nation has been Afghanistan, where the value for mean daily recovery is lowest among SAARC nations while COV for the same has been very high, indicating a highly inconsistent recovery scenario in the country. This is further evident because there was only one physician and 9.4 skilled health professionals serving 10,000 Afghan citizens in the month of July (The Conversation 2020a, b).

The second measure of comparison is a more extensive one involving three indices of tests per million, positivity rates, and fatality rates; then correlation values of tests per million with the rest two are put into scatter diagrams. While tests per million in any country increase over time, both positivity and fatality rates fluctuate based on the country's response to the disease situation. Vigorous testing and isolating the infected persons might help in reducing the positivity rate by checking the spread of the virus, whereas better treatment ensures the survival of COVID patients and thus reduces the fatality rate. However, if the positivity rate does not come down with the increasing number of tests, it shows the country has failed in tracing and isolating the infected people. Similarly, if the correlation value between tests per million and the fatality rate for a country remains positive, it shows the poor medical scenario of that nation. Figure 4.5 shows that mean daily positivity rates are coming down over the months in all SAARC nations except Nepal and Maldives. The mean daily positivity rate is alarmingly high for Afghanistan, which shows its failure in containing the virus. This finding is further validated by the result of the Afghan Health Ministry conducted serological survey in July–August, 2020 (Islamic

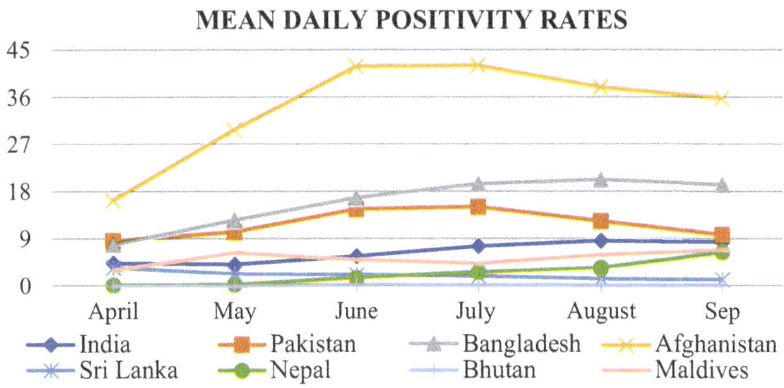

Fig. 4.5 Comparison of mean daily positivity rates from April to September (Source: Prepared by the author from worldometers 2020)

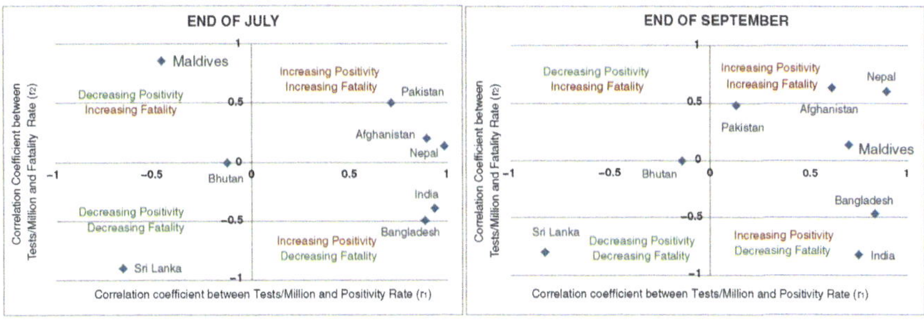

Fig. 4.6 Comparison of COVID Situation among SAARC countries between July and September, 2020 (Source: Computed by author from worldometers 2020)

Republic of Afghanistan 2020). It is estimated that around 31.5% of Afghan nationals have contracted the virus by that time (NDTV 2020).

Figure 4.6 shows a bi-monthly change in the COVID situation from July to September 2020 among SAARC nations with the help of scatter diagrams. Here correlation coefficient (r_1) values between Tests per Million and Positivity Rates are put on the horizontal x-axis while correlation coefficients (r_2) between Tests per Million and Fatality Rates are put on the vertical y-axis. As both r_1 and r_2 contain positive and negative values, four segments of "Increasing Positivity-Increasing Fatality," "Increasing Positivity-Decreasing Fatality," "Decreasing Positivity-Increasing Fatality," and "Decreasing Positivity-Decreasing Fatality" are found. While the first one is the worst place to be at, the last one is the best. With a minimum number of infections and zero death to date, Bhutan is found to be reducing its positivity rate over time. Likewise, Sri Lanka, with a limited number of cases and deaths, has been able to reduce both its positivity and fatality between July and September. Thus these two countries appear as the best performers among SAARC nations. India and Bangladesh have been able to save their COVID patients by providing them better health care, but they have failed to check the spread of infection, which is evident from the fact that positivity is increasing while fatality is decreasing in these two countries. Maldives has always struggled to check fatality from COVID, but its initial success to check the spread of infection is wasted as between July and September, its positivity rate also increased. Pakistan, Afghanistan, and Nepal turn out to be the worst performers in the fight against the pandemic as both the positivity and fatality rates in these countries kept on increasing with the number of tests per million between July and September 2020.

4.8 Discussions

Since antiquity, different disease outbreaks have troubled human beings in different parts of the world. Such outbreaks have been classified as Endemic, Epidemic, and Pandemic, respectively, based on their geographical expansion and their supposed threat to human lives. The Dictionary of Epidemiology defines an endemic disease as "the constant presence of a disease or infectious agent within a given geographic area or population group; may also refer to the usual prevalence of a given disease within such an area or group." At the same time, an epidemic has been defined as "the occurrence in a community or region of cases of an illness, specific health-related behavior, or other health-related events clearly in excess of normal expectancy" (Porta 2008). An endemic disease stays all times in a particular population, environment, or region in a contained manner. The same disease turns into an epidemic when there is a sudden outbreak over a short period, affecting many people and spreading through one or several communities. A Pandemic is a global outbreak of any disease spreading across countries and continents, affecting and killing more people (WebMD 2020). From prehistoric times to the twenty-first century, different epidemics and pandemics have taken a severe toll on human lives.[20] Antonine Plague (165–180), Plague of Justinian (541–542), Cocoliztli epidemic (1545–1548), Great Plague of London (1665–1666), West African Ebola (2014–2016), Zika Virus epidemic (2015–present day) are few examples of epidemics which killed millions of people in recorded history. On the other hand, the Black Death (1346–1353) spreading over Europe and Asia, the Flu pandemic (1889–1890) spreading across the globe, Influenza pandemic or the Spanish Flu (1918–1920) killing almost 100 million people worldwide, Asian Flu (1957–1958) killing a million people, HIV-AIDS (1981–present day) killing 35 million people worldwide, H1N1 Swine Flu (2009–2010) killing half a million people are examples of pandemics (see footnote 20).

[20] Science news website LIVE SCIENCE covered the news entitled '20 of the worst epidemics and pandemics in history' prepared by Owen Jarus and it published on 21 March 2020 which is available under the URL: https://www.livescience.com/worst-epidemics-and-pandemics-in-history.html.

In this scenario, though the SARS-CoV-2 pandemic (2019–present day) cannot be termed unprecedented both in terms of its spread and severity, it is definitely "without precedent in the experience of any living person" (Lal 2020). Similar to diseases like AIDS (Acquired Immuno Deficiency Syndrome), SARS, MERS (Middle East Respiratory Syndrome) or H1N1 Swine Flu; SARS-CoV-2 is a zoonotic disease caused by viruses that have infected human beings coming from other animals. Common COVID-19 symptoms include fever, dry cough, and tiredness, among other symptoms like sore throat, diarrhea, loss of taste or smell, or both. Patients in serious conditions might experience difficulty in breathing, chest pain etc. (CDC 2020). The primary measures are taken in response to the spread of COVID-19; (i.e., observance of social distancing, closure of non-essential services) have been similar to the earlier epidemics and pandemics. Researchers from Imperial College London have termed it as "the most serious public health crisis in generations" (Imperial College London 2020). The exceptionality of COVID-19 lies in the facts that it is the first global pandemic of the twenty-first century occurring after almost a 100 years gap, it has coincided with a global climate change crisis, and it is the first pandemic in history that has caused a worldwide pattern of state intervention in the form of lockdowns leading to a complete closure of economic activities all over the globe (Lal 2020).

Now COVID-19 has been a global pandemic in the truest sense as it has affected more or less every single country in the world, including those in the SAARC region. It has been observed that India is the worst-hit country in terms of total infections and deaths within the region of study. However, if India's size and demography are taken into consideration, it is found to be better positioned than its neighbors like Afghanistan and Pakistan. Tiny countries of Bhutan and Sri Lanka have outperformed their larger neighbors in the fight against the pandemic in the SAARC region. Though it is too early to make any conclusive remark on the COVID situation, from the pattern of the disease, it can be asserted that along with Bhutan and Sri Lanka, the performance of India and Bangladesh has been commendable with respect to their demographic challenges. For the rest of the SAARC countries, the situation is more worrying.

4.9 Conclusion

The Impacts of COVID-19 are not limited only to the number of people dying of the disease or the knee-jerk responses of the governments to build up health sectors in their respective countries. The pandemic has left its positive impact upon the environment by reducing pollution levels; it has taught people to be prepared for emergency situations, it has led up to rapid upliftment of medical infrastructure. Again it has caused an economic crisis when the poorest and most vulnerable section of the society is hurt the most. It is said to have increased domestic violence, crime against women and children, and social discrimination too. Another important aspect is the volatile nature of coronavirus, which puts any prediction at risk. The month of October has observed the second wave of infection in European nations and also in

SAARC countries like Nepal, Pakistan, and Sri Lanka. Sri Lanka particularly has been walloped by this second wave of infection and has failed to repeat its success to tackle the COVID situation like earlier. When the article is being written, the second wave is yet to hit the rest of the SAARC nations as their respective governments have reiterated an apprehension of the same. Ministry of Home Affairs, Govt. of India has issued guidelines for "Surveillance, Containment and Caution" for the month of December with an aim to consolidate the substantial gains that have been achieved against the spread of COVID-19 in the country.[21] Vaccines turn out to be the most promising solution to this disease, and the whole world has been eagerly waiting for an effective one to come for months now. By the end of November, Pfizer, Moderna, and AstraZeneca—the frontrunners in vaccine research as mentioned earlier—have claimed high efficacy for their products. While their products are soon to reach the global market, phase-3 trials for India-made *Covaxin* has started in the first week of December. Now it is to be seen how well the vaccines are distributed among countries without any social or economic discrimination. While the rich can afford doses of highly priced vaccines, the poor and marginal people of society who are most vulnerable to this pandemic should be vaccinated as early as possible. How different nations on earth will acquire and distribute the vaccines among their citizens in the coming months; could be an exciting topic for future research. However, caution needs to be kept as vaccines will probably reduce but not completely eliminate the chance of contracting COVID-19 disease, and thus people will have to keep wearing masks and follow social distancing for years to come (Hindustan Times 2020a, b). Another reason to worry is the outcomes of various serological survey reports conducted across the globe, which tell that much more people have been infected by the virus than has been diagnosed. Again some patients are found to be "super-spreaders" who more readily transmit the disease than other COVID active patients (Laxminarayan et al. 2020). The silver lining of the aforesaid events could be the fact that with more people being infected and not reported, the actual CFR turns out to be much smaller than what is being calculated now. So more people are growing immunity and surviving the pandemic, and the possibility of reaching herd immunity is getting brighter in the coming months. So a detailed study of the pandemic has many facets and call for the collaboration of academicians from different fields. The social and economic impact of this pandemic and its measures are yet to be realized and demand future research. The present article focuses only on the direct impact of the disease on human health and responses to protect the same, particularly in the domain of SAARC countries. While the COVID situation remains varying from one country to another and changes over time, there has been considerable progress in fighting against the pandemic worldwide. Dr. T A Ghebreyesus, the Director General of WHO, has declared in the first week of December that the world "can begin to dream about the end of the pandemic" in the wake of promising

[21] Indian digital media outlet The Wire published this news on 25 November 2020, which is available under the URL: https://thewire.in/government/covid-19-unlock-lockdown-mha-guidelines.

vaccine research (Hindustan Times 2020a, b). However, it would be wise to conclude with the precautionary message playing over the phones in India while making a call, "COVID-19 is not over yet; we need to observe maximum precaution."

References

Balmford B, Annan JD, Hargreaves JC et al (2020) Cross-country comparisons of covid-19: policy, politics and the price of life. Environ Resour Econ 76:525–551. https://doi.org/10.1007/s10640-020-00466-5

BBC News (2020) bbc.com. http://bbc.com/news/world-asia-53054785. Accessed 30 Nov 2020

CDC (2020) Centre for disease control and prevention. https://www.cdc.gov/coronavirus/2019-ncov/symptoms-testing/symptoms.html. Accessed 31 Oct 2020

Centre for Disease Dynamics, Economics and Policy (2020) COVID-19 in India: potential impact of the lockdown and other longer-term policies. New Delhi, Washington DC

Chatterjee B, Karandikar RL, Mande SC et al (2020) The mortality due to COVID-19 in different nations is associated with the demographic character of nations and the prevalence of autoimmunity. medRxiv. https://doi.org/10.1101/2020.07.31.20165696

Clemente-Suárez VJ et al (2020) Dynamics of population immunity due to the herd effect in the COVID-19 pandemic. Vaccine 8(2):236. https://doi.org/10.3390/vaccines8020236

Financial Express (2020) financialexpress.com. http://financialexpress.com/world-news/afghanistan-faces-catastrophe-as-covid-19-cases-grow-afghan-red-crescent-society/2024140/. Accessed 30 Nov 2020

Global Health Security Index (2019) 2019 global health security index: building collective action and accountability. John Hopkins Centre for Health Security. https://www.ghsindex.org/wp-content/uploads/2020/04/2019-Global-Health-Security-Index.pdf

Government of the People's Republic of Bangladesh, Ministry of Health and Family Welfare (2020) Bangladesh preparedness and response plan for COVID-19

Greenberger M (2018) better prepare than react: reordering public health priorities 100 years after the Spanish flu epidemic. Am J Public Health 108(11):1465–1468. https://doi.org/10.2105/AJPH.2018.304682

Health (2020) health.com. https://www.health.com/condition/infectious-diseases/coronavirus/covid-19-test-types. Accessed 30 Nov 2020

Health Promotion Bureau, Government of Sri Lanka (2020) hpb.gov.lk. https://hpb.health.gov.lk/en/covid-19. Accessed 30 Nov 2020

Hindustan Times (2020a) hindustantimes.com. http://www.hindustantimes.com/india-news/mask-social-distancing-will-be-crucial-even-after-covid-19-vaccine-scientist/story-RzEKADAQU4wNR4H10iWgUL.html. Accessed 31 Oct 2020

Hindustan Times (2020b) hindustantimes.com. https://www.hindustantimes.com/world-news/world-can-start-dreaming-of-pandemic-s-end-un-health-chief/story-dZ6Da6VFNAIN30uxcLmKZP.html. Accessed 05 Dec 2020

Huq S, Biswas RK (2020) COVID-19 in Bangladesh: data deficiency to delayed decision. J Glob Health 10(1). https://doi.org/10.7189/jogh.10.010342

Imperial College London (2020) imperial.ac.uk. http://www.imperial.ac.uk. Accessed 31 Oct 2020

Indian Council of Medical Research, Ministry of Health and Family Welfare, Government of India (2020) Evidence based advisory to address inappropriate use of convalescent plasma in COVID-19 patients. https://www.icmr.gov.in/pdf/covid/techdoc/ICMR_ADVISORY_Convalescent_plasma_17112020_v1.pdf. Accessed 1 July 2020

Islamic Republic of Afghanistan, Ministry of Public Health (2020) Prevalence of COVID-19 and its related deaths in Afghanistan: a nationwide, population-based seroepidemiological study

John TJ, Samuel R (2000) Herd immunity and herd effect: new insights and definitions. Eur J Epidemiol 16:601–606. https://doi.org/10.1023/A:1007626510002
Kirkcaldy RD, King BA, Brooks JT (2020) COVID-19 and postinfection immunity: limited evidence, many remaining questions. JAMA 323(22):2245–2246. https://doi.org/10.1001/jama.2020.7869
Klein E, Lin G et al (2020) COVID-19 for India updates. https://cddep.org/wp-content/uploads/2020/03/covid19.indiasim.March23-2-4.pdf. Accessed 1 July 2021
Koirala J, Acharya S, Neupane M et al (2020) Government preparedness and response for 2020 pandemic disaster in nepal: a case study of COVID-19. Dig J SSRN Electr J. https://doi.org/10.2139/ssrn.3564214
Krause P, Fleming TR, Longini I et al (2020) COVID-19 vaccine trials should seek worthwhile efficacy. Lancet 396(10253):741–743. https://doi.org/10.1016/s0140-6736(20)31821-3
Kshatri JS, Bhattarcharya D, Kanungo S et al (2020) Findings from serological surveys (in August 2020) to assess the exposure of adult population to SARS CoV-2 infection in three cities of Odisha, India. medRxiv. https://doi.org/10.1101/2020.10.11.20210807
Lal V (2020) The fury of COVID-19: the politics, histories, and unrequited love of the coronavirus. Pan Macmillan India, New Delhi
Laxminarayan R, Wahl B, Dudala SR et al (2020) Epidemiology and transmission dynamics of COVID-19 in two Indian states. Science 370(6517):691–697. https://doi.org/10.1126/science.abd7672
Le TT, Cramer JP, Chen R, Mayhew S (2020) Evolution of the COVID-19 vaccine development landscape. Nat Rev Drug Discov 19:667–668. https://doi.org/10.1038/d41573-020-00151-8
Lexlife India (2020), lexlife.in. https://lexlife.in/2020/04/20/dealing-with-covid-19-the-triple-t-method/. Accessed 15 Nov 2020
Meo SA, Abukhalaf AA, Alomar AA et al (2020) Impact of lockdown on COVID-19 prevalence and mortality during 2020 pandemic: observational analysis of 27 countries. Eur J Med Res 25 (56). https://doi.org/10.1186/s40001-020-00456-9
Ministry of Health and Family Welfare (2020) Guidelines on preventive measures to contain spread of COVID-19 in workplace settings. Directorate General of Health Services (EMR Division), Ministry of Health & Family Welfare. Government of India
Ministry of Health and Indigenous Medical Services (2020) SUWASIRIPAYA. Government of Sri Lanka
Mustafa N (2020) Coronavirus Disease (COVID-19) Research and Statistics. Int J Syst Dyn Applicat 10(3):1–20. https://doi.org/10.4018/IJSDA.20210701.ao1
National Institute of Health (2020) COVID-19 laboratory capacity. Ministry of National Health Services, Regulation and Coordination, Government of Pakistan, Islamabad
Nature (2021) Why herd immunity for COVID is probably imposible. Nature 591:520–522. https://doi.org/10.1038/d41586-021-00728-2
NDTV (2020) ndtv.com. http://ndtv.com/world-news/nearly-10-million-infected-with-coronavirus-in-afghanistan-officials-2274898. Accessed 28 Nov 2020
Owen JA, Punt J, Stranford SA, Jones PP (2013) Kuby immunology, 7th edn. W H Freeman and Company, New York
Porta M (ed) (2008) A dictionary of epidemiology, 5th edn. Oxford University Press, New York
Pozzer A, Dominici F, Haines A et al (2020) Regional and global contributions of air pollution to risk of death from COVID-19. Cardiovasc Res 116(14):2247–2253. https://doi.org/10.1093/cvr/cvaa288
Randolph HE, Barreiro LB (2020) Herd Immunity: understanding COVID-19. Immunity 52 (5):737–741. https://doi.org/10.1016/j.immuni.2020.04.012
Reuters (2020a) reuters.com. https://www.reuters.com/article/us-health-coronavirus-bangladesh-idUSKCN24L0KO. Accessed 2 July 2021
Reuters (2020b) reuters.com. https://www.reuters.com/article/us-health-coronavirus-vaccine-pakistan-idUKKBN2601CG. Accessed 2 July 2021

Samaranayake N (2020) Covid-19 and competition for influence in South Asia. https://www.nbr.org/publication/covid-19-and-competition-for-influence-in-south-asia/

Sky News (2020) news.sky.com. http://news.sky.com/story/coronavirus-foolish-to-assume-covid-19-vaccine-will-be-here-for-winter-whitty-12054756. Accessed 30 Nov 2020

Swaminathan S (2020) Hindustan times. http://www.hindustantimes.com/world-news/no-covid-19-herd-immunity-yet-says-who-chief-scientist-soumya-swaminathan/story-ho7p9cZpArmLMIDbpNolvo.html. Accessed 30 Nov 2020

Swarajya (2020) swarajyamarg.com. http://swarajyamarg.com/world/bangladesh-slipping-into-grave-crisis-fast-emerging-as-one-of-worst-covid-19-hotspots. Accessed 30 Nov 2020

The Conversation (2020a) theconversation.com. http://theconversation.com/remdesivir-study-finally-published-an-expert-in-critical-care-medicine-gives-us-his-verdict-147862?utm_medium=ampwhatsapp&utm_source=whatsapp. Accessed 30 Nov 2020

The Conversation (2020b) theconversation.com. http://theconversation.com/afghanistan-covid-19-crisis-has-been-fuelled-by-armed-conflict-141924. Accessed 30 Nov 2020

The Economic Times (2020) economictimes.com. http://m.economictimes.com/news/international/world-news/covid-is-spreading-in-unexplained-ways-dimming-containment-hope/articleshow/76992507.cms. Accessed 31 Oct 2020

The Kathmandu Post (2020) kathmandupost.com. http://kathmandupost.com/health/2020/02/12/nepal-s-hospitals-have-no-beds-to-treat-coronavirus-patients-doctors-say. Accessed 31 Oct 2020

The Royal Society (2020) Herd immunity in the epidemiology and control of COVID-19. https://royalsociety.org/-/media/policy/projects/set-c/set-c-herd-immunity.pdf. Accessed 1 July 2021

The World Bank (2020), worldbank.org. https://www.worldbank.org/en/topic/water/brief/wash-water-sanitation-hygiene-and-covid-19. Accessed 29 Nov 2020

Tillett RL, Sevinsky JR, Hartley PD et al (2020) Genomic evidence for reinfection with SARS-CoV-2: a case study. Lancet Infect Dis. https://doi.org/10.1016/s1473-3099(20)30764-7

Times of India (2020) indiatimes.com. http://timesofindia.indiatimes.com/world/europe/who-10-of-worlds-people-may-have-been-infected-with-virus/articleshow/78493850.cms Accessed 28 Nov 2020

Tripathi R, Alqahtani SS, Albarraq AA et al (2020) Awareness and preparedness of COVID-19 outbreak among healthcare workers and other residents of South-West Saudi Arabia: a cross-sectional survey. Front Public Health 8:1–13. https://doi.org/10.3389/fpubh.2020.00482

United Nations (2019a) 2019 ESCAP population data sheet. The Economic and Social Commission for Asia and the Pacific. https://www.unescap.org/resources/2019-escap-population-data-sheet

United Nations (2019b) World population prospects—data booklet. Department of Economic and Social Affairs

United Nations Development Programme (2020) Global Multidimensional Poverty Index 2020 charting pathways out of multidimensional poverty: achieving the SDGs. Oxford Poverty and Human Development Initiative

Voysey M, Clemens SAC, Madhi SA, Folegatti PM et al (2020) Safety and efficacy of the ChAdOx1 nCoV-19 vaccine (AZD1222) against SARS-CoV-2: an interim analysis of four randomized controlled trials in Brazil, South Africa, and the UK. Lancet 397(10269):99–111. https://doi.org/10.1016/S0140-6736(20)32661-1

WebMD (2020) webmd.com. https://www.webmd.com/cold-and-flu/what-are-epidemics-pandemics-outbreaks. Accessed 28 Nov 2020

WION (2020) Bangladesh signs deal with Serum Institute of India. https://youtu.be/MQWwpW2iS9s. Accessed 28 Nov 2020

World Economic Forum (2020) weforum.org. https://www.weforum.org/agenda/2020/03/why-lockdowns-work-epidemics-coronavirus-covid19/. Accessed 15 Oct 2020

World Health Organization (2020) who.int. http://who.int/southeastasia/news/feature-stories/detail/who-helps-maldives-build-laboratory-capacity-to-test-covid-19. Accessed 15 Oct 2020

worldometers (2020) worldometer.org. http://worldometers.info/coronavirus/#countries. Accessed 28 Nov 2020
XINHUANET (2020) http://www.xinhuanet.com/english/2020-11/26/c_139545074.html. Accessed 29 Nov 2020
Yousaf M, Zahir S, Riaz M et al (2020) Statistical analysis of forecasting COVID-19 for upcoming month in Pakistan. Chaos Solitons Fractals 138:1–4. https://doi.org/10.1016/j.chaos.2020.109926
Zaidi S, Saligram P, Ahmed S, Sonderp E, Sheik K (2017) Expanding access to health care in South-Asia. BMJ 357(j1645):1–4. https://doi.org/10.1136/bmj.j1645

Partha Das is an assistant professor in geography, associated with West Bengal Education Service, currently posted at ABN Seal College, Cooch Behar in remote north eastern part of India. Apart from teaching, he is involved with different research works, some of which have already been published in national level journals and edited book volumes. His research interest lies in topics of Population and Tribal Studies, Regional Studies, Tourism Geography, Cartographic and Statistical Techniques, Philosophy of Geography and Environmental Studies.

Chapter 5
The COVID-19 Pandemic and Socio-spatial Inequality: A Study from the Metropolitan Area of Rio de Janeiro, Brazil

Pablo Ibanez, Gustavo Mota de Sousa, Andrews José de Lucena, Heitor Soares de Farias, Leandro Dias de Oliveira, and André Santos da Rocha

Abstract The COVID-19 pandemic in Brazil is a catastrophe and stands out negatively around the world. Brazil is a country with the Unified Health System (SUS), which is public, has a good history in mass vaccination campaigns, produces vaccines, but the number of deaths caused by the coronavirus surpassed all the most pessimistic projections and reached more than 500,000 dead by July 2021. The state of Rio de Janeiro and its metropolitan region are part of this process. This chapter aims to analyze the diffusion of the new coronavirus in the Metropolitan Area of Rio de Janeiro (MARJ), emphasizing the city of Rio de Janeiro and its spatiality in what we call "pandemic spaces on the periphery." A territorial inequality of impact is evidenced, which is linked to the conditions of the territory. In the areas analyzed, the absence of support from health structures is evident, as is the effects of the perversity of globalization.

Keywords Unified Health System (UHS) · Mismanagement in Brazilian politics · Vaccination campaigns · Urban periphery · Pandemic spaces on the periphery

5.1 Introduction

It has been more than a year since the WHO's recognition of the global pandemic. The impact of the new coronavirus was confirmed in economic, social, and health crises in most countries on the periphery and semi-periphery of the world. In the

P. Ibanez (✉) · L. D. de Oliveira · A. S. da Rocha
Laboratory of Economic and Political Geography (LAGEP), Federal Rural University of Rio de Janeiro, Seropédica, Brazil

G. M. de Sousa · A. J. de Lucena · H. S. de Farias
Integrated Laboratory of Applied Physical Geography (LIGA), Federal Rural University of Rio de Janeiro, Seropédica, Brazil

S. Saha et al. (eds.), *Economic and Societal Transformation in Pandemic-Trapped India*, New Frontiers in Regional Science: Asian Perspectives 55,
https://doi.org/10.1007/978-981-16-5755-9_5

early stage of the pandemic cycle, there was a conception of "a democratic pandemic"—in the sense that everyone, irrespective of class, would be affected. However, this conception is dismantled due to the qualitative nature of the impact of the disease, which is due to the high number of contaminations and lethality's that occurred in areas where more impoverished people do live.

In the Latin American context, Brazil remains the epicenter of the pandemic. The policies implemented for federal governance out of step with the actions of the municipalities and states of the federation create a favorable scenario for the spread of the virus. In addition, the voices of scientific denial present with great force in social media, but which also occupied other media diffusion spaces, helped to promote the circulation of false news and disinformation for most people. The cases of promotion of the use of chloroquine, ivermectin, and other drugs without proven therapeutic efficacy, the stimuli for not using the face mask, the lack of encouragement to attachment, the measures of social distancing by many political leaders are some reasons that led Brazil to have obtained a second big wave of contamination in 2021. As a result of all this, we have an increase in the number of deaths, the construction of new strains of the virus (such as the Cepa in the city of Manaus), and the deepening crisis in pandemic territories that affect peripheral areas with greater force.

The differences of these impacts on the territory are significantly unequal, so we reinforce, as pointed out in our last text (Lucena et al. 2020), the importance of observing the perspective of Santos and Silveira (2001) on the "use of territory," so that we can assess not only general values about the pandemic but also observe the differences produced in the territories and thus analyze their impacts on the expansion of internal inequalities. Therefore, we understand that observing the case of the metropolitan area of Rio de Janeiro serves as a theoretical-methodological option to assess the repercussions of the advance of the Brazilian pandemic.

In this sense, this text aims to present the characteristics of the diffusion of the new coronavirus in the metropolitan area of Rio de Janeiro, with emphasis on the city of Rio de Janeiro, and to analyze the spatiality of the processes that materialize in what we call "pandemic spaces of the periphery."

In the first part of the text, we will present data regarding the second wave of the pandemic in Brazil, highlighting the actions of leaders and governments, informing political facts that led the country to assume the third position in the number of cases in the world and the highest daily fatality rate per one million inhabitants of the world. In the second stage of the work, we present the singularity of the metropolitan context of Rio de Janeiro, highlighting aspects of the "use of the territory" and how there is an unequal production of the impacts of diseases in the municipalities. In the third stage, we seek to show, from the spatial cut-out of the city of Rio de Janeiro, how much the aspects of the uneven formation of its territory reflect on the spatiality of the contamination, the lethality rate, and how the progress of vaccination is reflected. Finally, in the final considerations, some discussions of this introductory section are taken up in connection with the presentation of data and comments throughout the text, reflecting on the importance of studies that stimulate discussion in the context of economic, social, and geographic factors in the assessment of the framework pandemic.

5.2 Pandemic Panorama in Brazil

Lucena et al. (2020) wrote about the Brazilian situation in the COVID-19 pandemic in September 2020 and pointed out that Brazil, in addition to unfavorable socioeconomic and sanitary conditions, faced another serious problem: the Federal Executive government inability to promote together with states and cities in favor of human health, causing misinformation and many conflicts and tensions. Several studies have shown statistically and analytically that the effects of the lack of federal coordination were disastrous, placing the country in second place in the number of deaths, in addition to demanding responses from subnational governments practically without support from the health ministry of President Jair Bolsonaro's government (Touchton et al. 2021; Lima et al. 2020; Fleury 2021; Cabral et al. 2021; Bahia et al. 2021).

In the case of the purchase of vaccines, for example, the nine governors of the Northeast region created a consortium to speed up obtaining immunizations. As a result, today, two poor states in this region, Bahia and Maranhão, have become a reference both in terms of vaccination and the reduction in cases and deaths. On the other hand, places where the president of Brazil, Jair Bolsonaro, was the most voted during the 2018 elections, had higher rates of contamination by the disease (CABRAL et al. 2021). In other words, there is concrete evidence that federal policy in Brazil—or lack thereof—has played a decisive role in the alarming number of infected people and deaths.

Regarding mortality, at the end of 2020, it was possible to account for 24% more deaths in relation to the period 2015–2019 (Azevedo e Silva et al. 2021), for this year, the statistic should be even worse. From March to December 2020, 194,976 deaths were recorded by COVID-19.[1] From January to July 2021, the total number of deaths was 323,270.[2] There were 128,294 more deaths in a 4-month shorter period. Table 5.1 shows Brazil's position in July 2021. It is a frightening number for a country with a free public health system, has always been a reference in vaccination campaigns for various types of diseases, and has two public institutions of excellence in vaccine production, Butantan, and Fiocruz. It is not by chance that the first doses applied in the national territory came from these places. Even with the

[1] See details at: https://g1.globo.com/bemestar/coronavirus/noticia/2020/12/31/casos-e-mortes-por-coronavirus-no-brasil-em-31-de-dezembro-segundo-consorcio-de-veiculos-de-imprensa.ghtml.

[2] See details at: https://g1.globo.com/bemestar/coronavirus/noticia/2021/06/30/brasil-registra-mais-de-2-mil-mortes-por-covid-em-24-horas-mas-ve-queda-simultanea-nas-medias-moveis-de-casos-e-obitos.ghtml.

Table 5.1 COVID-19—total cases, total deaths, cases/one million, deaths/one million, and population

Country	Total Cases	Total deaths	Cases/1 M pop	Deaths/1 M pop	Population
World	**185,045,443**	**4,003,010**	**23,74**	**513.5**	
USA	34,598,361	621,335	103,911	1866	332,960,810
India	30,619,932	403,310	21,971	289	1,393,679,418
Brazil	18,792,511	525,229	87,781	2453	214,083,318
France	5,786,999	111,197	88,46	1,7	65,419,339
Russia	5,658,672	139,316	38,759	954	145,997,664
Turkey	5,449,464	49,959	63,921	586	85,253,573
UK	4,930,534	128,231	72,247	1879	68,245,821
Argentina	4,552,750	96,521	99,811	2116	45,613,849
Colombia	4,375,861	109,466	85,085	2128	51,429,416
Italy	4,263,797	127,68	70,625	2115	60,371,997

Data Source: Worldometer (https://www.worldometers.info/coronavirus/)

start of vaccination on 18 January 2021, the number of deaths continued to rise, reaching marks above 2000 daily for months.

Although several experts were warning about the dangers of the lack of control and prevention of the disease, the federal government maintained a policy contrary to the WHO recommendations. Moreover, the president himself was in several unmasked agglomerations, including political acts with the presence of a former minister of health.[3] As a result, the politicization of the pandemic became evident. As an example, one can cite the direct disputes between President Bolsonaro and governors not aligned with his speech, which ended up including the Brazilian Federal Supreme Court (FSC). According to the president, the judiciary would not be letting the federal government work by allowing states to promote policies to restrict the movement of people, against the recommendation of the presidency, which was guided by the maintenance of economic activities even in times of peaks of contamination. Wang (2021) analyzed some of these disputes in the supreme court and reached the conclusion that the decisions were: favorable to preserving the autonomy of subnational entities, especially States, to respond to the pandemic; respect the decision-making space of managers and oblige managers to observe the scientific evidence and technical knowledge produced by the scientific community and international bodies, in particular, the World Health Organization (WHO) worked to ensure that recommendations from the world's health authorities were followed.

Politicization was also present in the evaluation, production, and purchase of vaccines. The most notorious cases were in the Northeast, described above, and in the state of São Paulo, which produced the first doses authorized to be applied in the country. The governor of that state, João Doria, had already come into conflict with

[3] https://www.cnnbrasil.com.br/politica/2021/05/23/ex-ministro-pazuello-participa-de-ato-ao-lado-de-bolsonaro-sem-mascara.

the president in one of the few meetings that the federal government held with the presence of all the governors.[4] However, with the arrival of the Coronavac vaccine, in partnership with a Chinese laboratory, this dispute became even more fierce. The president began to publicly criticize the vaccine for being of Chinese origin, which caused diplomatic problems and not having such high efficacy. Again, we experienced unbelievable scenes. During a pandemic, the president of the Republic put the use of the vaccine in check because it was presented by a political disaffection.

Since the beginning of the pandemic, President Jair Bolsonaro has defended the use of a drug without proven efficacy, hydroxychloroquine, to the detriment of the search for agreements for the production and marketing of vaccines. This ended up being one of the reasons for creating a Parliamentary Inquiry Commission (PIC) by the Brazilian Federal Senate, which began in May 2021, and until the time of delivery of this chapter, it had not yet ended.[5] Notwithstanding all the conflicts mentioned so far, the PIC revealed that vaccine maker Pfizer tried to contact the government through dozens of unanswered emails.[6] One of the goals was for Brazil to serve as an example to the world in the application of the vaccine. At least 70 million doses had a destination other than Brazil.

In addition to all these events mentioned, since the beginning of the pandemic, we have had four health ministers. This fact already indicates that driving has been wrong. The longest-serving minister was not a doctor, it was a general who publicly announced that he did not know the Brazilian Unified Health System (UHS).[7] The minister was chosen because of his experience in logistics. According to senior government officials, the expertise would be useful to organize the challenges facing the demands of the pandemic. However, the tragedy in some states precisely due to lack of logistics was even worse. In the state of Amazonas, in January 2021, there was a lack of oxygen, causing the health system to collapse.[8] Aid for supplies even came from Venezuela, which is a declared enemy of the Bolsonaro government.[9] The state of Amazonas ended up becoming one of the most lethal for its inhabitants during this period of the pandemic, especially considering the fact that its population is younger than that of the more developed states. In other words, there was an accentuated mortality in groups younger than 60 years (Azevedo e Silva et al. 2021), contrary to the logic of the lethality of the virus, which is more accentuated in the elderly.

[4] https://g1.globo.com/jornal-nacional/noticia/2020/03/25/doria-e-bolsonaro-discutem-em-reuniao-de-governadores.ghtml.

[5] https://www.cartacapital.com.br/opiniao/os-motivos-para-a-instalacao-da-cpi-da-pandemia/.

[6] https://www.bbc.com/portuguese/brasil-57104347.

[7] https://www.cnnbrasil.com.br/politica/2020/10/07/pazuello-diz-que-antes-de-cargo-no-governo-nao-sabia-o-que-era-o-sus.

[8] https://g1.globo.com/am/amazonas/noticia/2021/02/14/crise-do-oxigenio-um-mes-apos-colapso-em-hospitais-manaus-ainda-depende-de-doacoes-do-insumo.ghtml.

[9] https://g1.globo.com/am/amazonas/noticia/2021/02/14/crise-do-oxigenio-um-mes-apos-colapso-em-hospitais-manaus-ainda-depende-de-doacoes-do-insumo.ghtml.

The panorama of COVID-19 in Brazil, although much more complex, sought to present some of the main issues that made the country the second world record holder for deaths in the pandemic. Even with a public health system recognized worldwide, the mistakes presented here were fundamental for us to reach this mark. The state of Rio de Janeiro, the main objective of this chapter, represents part of these errors and one of the highest lethalities in COVID-19, both in absolute and relative terms (Azevedo e Silva et al. 2021). Below is an in-depth analysis of the metropolitan area of Rio de Janeiro and the city of Rio de Janeiro.

5.3 The COVID-19 Pandemic in the Metropolitan Area of Rio de Janeiro (MARJ)

The Metropolitan Area of Rio de Janeiro (MARJ) is formed by the state capital - the city of Rio de Janeiro, the most populous and politically and economically important—and 21 other municipalities: Nova Iguaçu, Duque de Caxias, São Gonçalo, Belford Roxo, São João de Meriti, Magé, Itaboraí, Mesquita, Nilópolis, Maricá, Queimados, Itaguaí, Japeri, Seropédica, Guapimirim, Paracambi and Tanguá, and the recently added Rio Bonito, Cachoeiras de Macacu and Petrópolis (the last, only in 2018, and which historically remained linked to the Fluminense Mountain Region) (Fig. 5.1). Second in importance in Brazil, behind only the Metropolitan Area of São Paulo with its more than 20 million inhabitants, the Metropolitan Area of Rio de Janeiro (MARJ) aggregates more than 75% of the state's population, according to IBGE (2020) estimates, with more than 13 million residents.

It is possible to point out that the MARJ can be methodologically subdivided into some sub-regions (Fig. 5.2), in addition to the capital itself: (1) Greater Niterói (or Metropolitan East), formed by the municipalities located in the eastern part of Guanabara Bay and with cities heavily influenced by the former state capital, Niterói; in the western part of the Bay, is (2) Baixada Fluminense, the essential and historic immediate periphery of the capital, whose principal cities are Nova Iguaçu and Duque de Caxias, large, populations around one million inhabitants, with consolidated urbanization and powerful, productive activities, plus the set of surrounding cities (Fortes et al. 2020).

In this sense, the Baixada Fluminense itself presents different urban-productive densities: if the urban block formed by the immediate surroundings of Nova Iguaçu and Duque de Caxias presents characteristics such as high population and social problems typical of peripheral urbanization, cities such as Seropédica, Paracambi, Queimados, Itaguaí and Japeri, located in the extreme west of the RMRJ, are still characterized by presenting rural features, as they have recently witnessed an increase in their population numbers, in their still small urban centers and in their own industrial and commercial activities (Lucena et al. 2020); finally, still with little integration due to the physical obstacle that is the road ascent of Serra dos Órgãos, is Petrópolis, much more integrated to the Serrana Fluminense Area and its neighbors

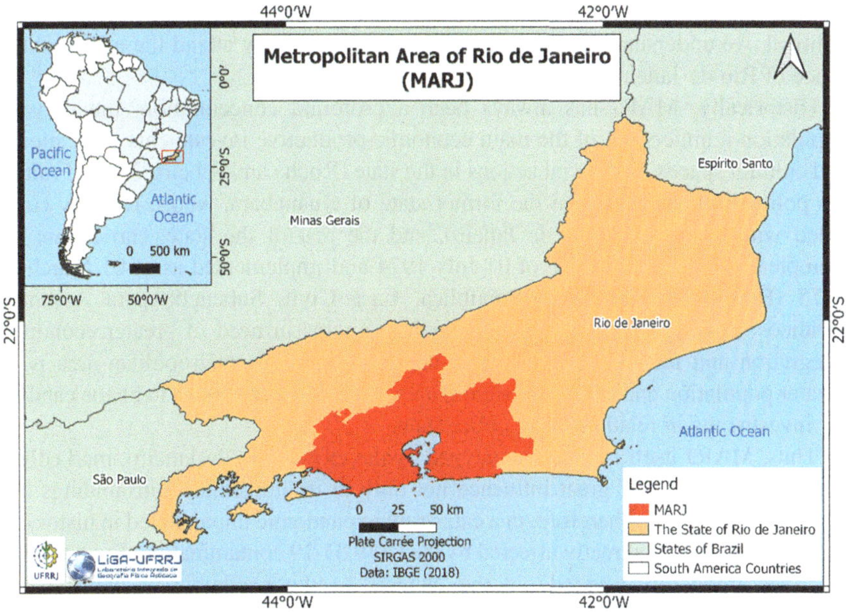

Fig. 5.1 Location of the Metropolitan Area of Rio de Janeiro (MARJ) in the context of the state of Rio de Janeiro and Brazil/South America (Source: IBGE/SEA 2018)

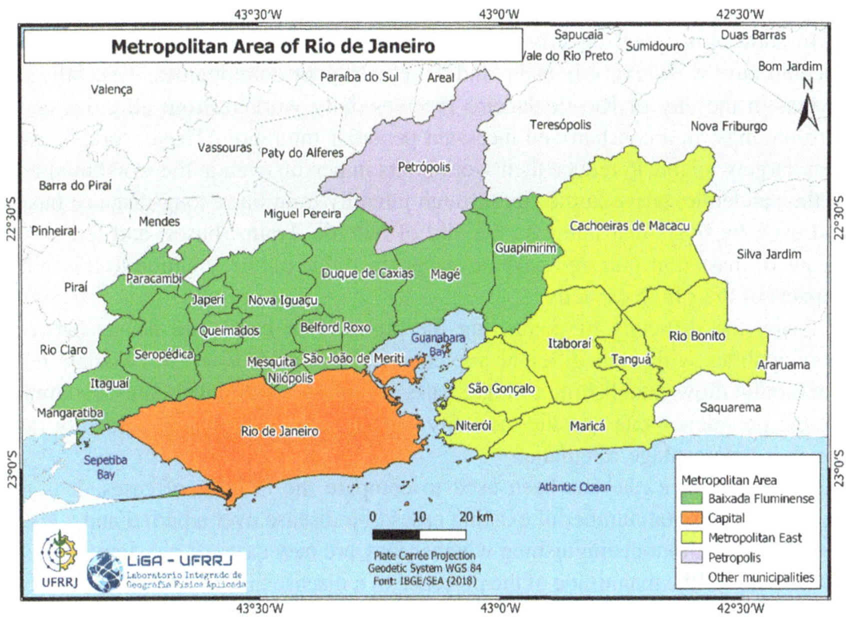

Fig. 5.2 Metropolitan Area of Rio de Janeiro (MARJ) and its subdivisions (Source: IBGE/SEA 2018)

Teresópolis and Nova Friburgo. As metropolitan borders in Brazil are peremptorily political, we understand that Petrópolis does not even truly attend the metropolitan space of Rio de Janeiro.

Historically, MARJ has always been a profound concentration, not only of population numbers but of the main economic-productive investments, educational and cultural spaces, and social actions in the state (Rocha and Ribeiro 2020). Fruit of the political merger between the former state of Guanabara, whose territory coincided with the city of Rio de Janeiro, and the rest of the state, carried out by Complementary Law No. 20, of 01 July 1974 and implemented as of 01 March of 1975 (Brasil, Presidência da República, Casa Civil, Subchefia para Assuntos Jurídicos 1974), the state of Rio de Janeiro remains in need of greater economic integration and income distribution in its domains. If the metropolitan area is of greater population and economic significance, the capital continues to be the catalyst for investments in relation to the entire state.

Thus, MARJ itself is deeply catalyzed by the city of Rio de Janeiro, markedly a national metropolis of great influence not only in its immediate surroundings but throughout the state. Therefore, in a catastrophic pandemic unparalleled in history, it will be the city most directly affected by the COVID-19 contamination. Even though we have frequently pointed out the dangers of the pandemic in the Baixada Fluminense with its social ills, few public health facilities, restricted access to water, poor sewerage system, and precarious conditions of instruction and ability to interpret the dangerous reality inserted (Rocha 2021; Fortes et al. 2020; Rocha et al. 2021), the numbers reinforce how the city of Rio de Janeiro is truly the main scope of the hard impacts of COVID-19.

In addition to great poverty in its own domains, it is a deeply segregated and unequal city, with luxury housing and extremely poor communities, especially the favelas in the city of Rio de Janeiro receives daily workers from all cities of its surroundings, in a constant and incessant pendular migration. These workers, who were largely unable to reduce their work hours in person even in the worst moments of the pandemic, arrive in the city through intercity train lines, long-distance buses, and even by boats that integrate the city of Niterói. Trains, buses, and ferries are means of mass transport and have never received due public attention to lessen the impacts of their high capacity.

Given the difficulty of comparing the numbers of cases and deaths between municipalities with such different populations, it was necessary to calculate rates that would allow establishing a single base, such as 100,000 inhabitants. Examples are the prevalence rate and the mortality rate, in addition to the case fatality rate, which is a percentage calculation.

The prevalence rate has been used to compare the number of cases. It is the measure of the total number of existing cases of a disease over a period and a given population, without distinguishing whether they are new cases or not. Prevalence is an indicator of the magnitude of the presence of a disease in the population (Barbosa et al. 2010). The calculation is: prevalence rate = (total cases/population) $\times$ 100,000, where the result shows the number of cases per 100,000 inhabitants.

The mortality rate has been used to compare the number of deaths, which is the measure of the total number of deaths over a period and in a given population. The mortality rate expresses the intensity with which mortality affects a given population. The calculation is: mortality rate = (total deaths/population) × 100,000, where the result shows the number of deaths per 100,000 inhabitants.

Para os óbitos utilizou-se também a Taxa de Letalidade, que é a medida do número total de óbitos na população que foi acometida pela doença. A Taxa de Letalidade expressa a gravidade da situação, indicando o percentual de pessoas que morreram por essa doença (Barbosa et al. 2010). O cálculo é: Taxa de Letalidade = (Total de casos/Total de óbitos) × 100, onde o resultado traduz o percentual de mortes entre os doentes.

For deaths, the case fatality rate has also been used. It is the measure of the total number of deaths in the population that was affected by the disease. The case fatality rate expresses the seriousness of the situation, indicating the percentage of people who died from this disease (Barbosa et al. 2010). The calculation is: case fatality rate = (total cases/total deaths) × 100, where the result reflects the percentage of deaths among patients.

The results were expressed in tables and maps, and by municipalities and grouped by immediate and extended peripheries, and the capital, demonstrating the behavior of COVID-19.

5.3.1 The Spread of COVID-19 in the Metropolitan Area of Rio de Janeiro: A Look at the Outer Regions

Recent dates show that Rio de Janeiro was the city that had the most deaths resulting from COVID-19 in the entire country. The numbers are impressive since Rio de Janeiro has more than 360,000 cases and 28,000 deaths, but these are absolute numbers that ignore the proportionality of the Prevalence and Mortality Rates that help understand the seriousness of COVID-19 calculated in Table 5.2.

The prevalence rate was higher in municipalities such as first Guapimirim (7679), second Rio de Janeiro (5342), third Magé (4764), and fourth Paracambi (4622), according to Table 5.2. Thus, among the four first municipalities with higher prevalence for COVID-19, three are from the Extended Periphery, showing that this disease is more aggressive in the poorest municipalities, with less infrastructure, from basic sanitation to the health care network (Fig. 5.3).

Regarding the mortality rate, Rio de Janeiro (418) has the highest value, followed by Guapimirim (277), Nilópolis (276), and Itaguaí (254) (Table 5.2). Among the first four municipalities, two are in the extended periphery, more vulnerable, and therefore had a higher prevalence of COVID-19 cases and higher mortality. Even so, the mapping of the mortality rate shows a greater concentration in the capital and in the smaller municipalities bordering it, in addition to the metropolitan borders, in the extreme east and west. On the other hand, the municipalities that make up the Baixada Iguaçuana nucleus had the lowest mortality rates (Fig. 5.4).

Table 5.2 Confirmed cases, deaths, prevalence rate, mortality rate and case-fatality rate per COVID-19 on 06/23/2021, by municipalities and group of municipalities in the Metropolitan Area of Rio de Janeiro

Municipality	Population	Prevalence rate (100 thousand inhab.)	Mortality rate (100 thousand inhab.)	Case fatality rate
Rio de Janeiro	**6,747,815**	**5342**	**418**	**7.83%**
Belford Roxo	513,518	3980	136	3.41%
Duque de Caxias	924,624	2361	157	6.64%
Mesquita	176,569	1884	227	12.02%
Nilópolis	162,693	2157	276	12.79%
Nova Iguaçu	823,302	2198	198	9.03%
São João de Meriti	472,906	2074	221	10.67%
Periferia imediata	**3,073,612**	**2505**	**185**	**7.37%**
Guapimirim	61,388	7679	277	3.61%
Itaguaí	134,819	3796	254	6.68%
Japeri	105,548	983	75	7.61%
Magé	246,433	4764	217	4.55%
Paracambi	52,683	4622	222	4.80%
Queimados	151,335	3770	114	301%
Seropédica	83,092	2683	211	7.85%
Periferia estendida	**835,298**	**3948**	**190**	**4.82%**

Data Source: SES-RJ (2021) and IBGE Cidades (2020)

Regarding the case fatality ratio, the highest value was found in first Nilópolis (12.79%), second Mesquita (12.02%), third São João de Meriti (10.67%) and fourth Nova Iguaçu (9.03%), according to Table 5.2. Among the first four cities, all are from the Immediate Outskirts, completely reversing the logic found in the previous rates, which confirms the suspicion about the low number of deaths registered in the border cities, mainly (Fig. 5.5).

The case fatality rate shows the elements of territorial formation in these peripheries, where the highest percentage is found in the capital and in the immediate periphery. It decreases as it moves toward municipalities on the metropolitan edge (extended periphery). This question expresses aspects of the territory used (Santos and Silveira 2001) in the sense that it can translate the more intense flows contained between municipalities in the Baixada Fluminense, with more consolidated urbanization and bordering the city of Rio de Janeiro, with places of employment, primarily contained in the capital. Thus, in a way, the expression of the geography of COVID-19 is a translation of the geography of the territorial division of labor, its flows, and uses in the metropolis.

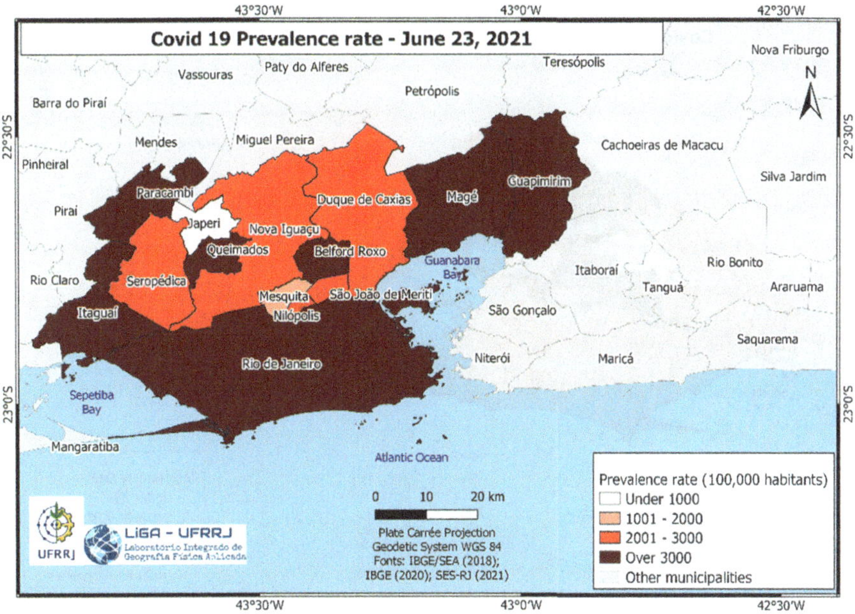

Fig. 5.3 Prevalence rate for COVID-19 for the city of Rio de Janeiro and its periphery (Source: IBGE/SEA 2018; IBGE 2020; SES-RJ 2021)

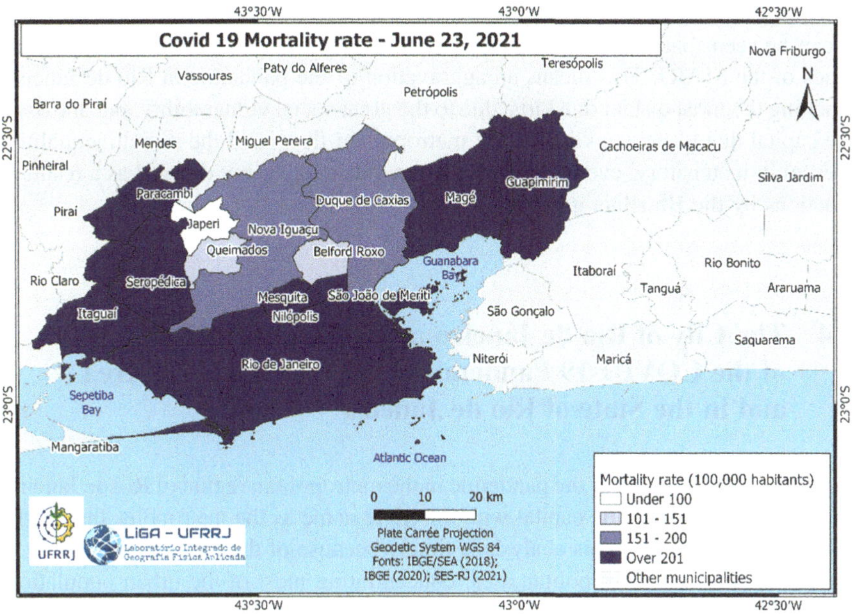

Fig. 5.4 COVID-19 Mortality rate for the city of Rio de Janeiro and its periphery (Source: IBGE/SEA 2018; IBGE 2020; SES-RJ 2021)

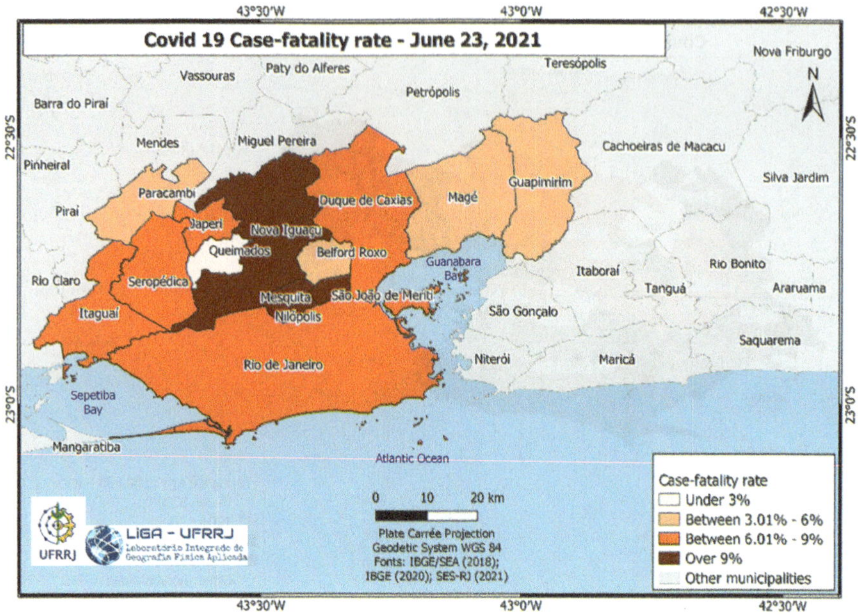

Fig. 5.5 COVID-19 Case fatality rate for the municipality of Rio de Janeiro and its periphery (Source: IBGE/SEA 2018; IBGE 2020; SES-RJ 2021)

Comparing the data for June 2021 with those accumulated up to December 2020, it can be seen that both the number of confirmed cases and deaths doubled in the cities of the MARJ. This means an aggravation of the pandemic in Rio de Janeiro, reaching the most distant outskirts, due to the great social vulnerability, but above all the capital due to the convergence of metropolitan flows and the social inequalities present in its territory, even with the ongoing vaccination, but delayed as a result of inactions by the Brazilian government.

5.4 The City of Rio de Janeiro as the Epicenter of the COVID-19 Pandemic in the Metropolitan Area and in the State of Rio de Janeiro

Continuing the analysis of the pandemic in the metropolitan region of Rio de Janeiro, its nucleus stands out, the capital with the same name as the metropolis, that is, the city of Rio de Janeiro. This analysis is timely because of the function of the city in the entire state and metropolitan area: concentrating most of the urban population, economic, financial, and income activities, the transport network, health services, education, and other specialties, as well as most of the problems and conflicts of social and environmental order.

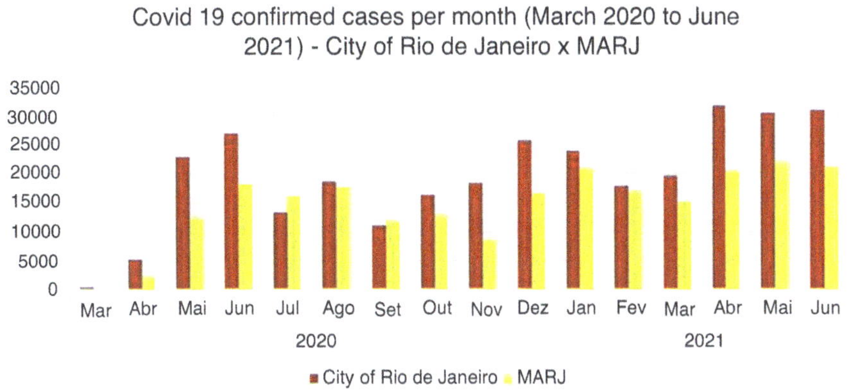

Fig. 5.6 Number of confirmed COVID-19 cases per month between March 2020 and June 2021 in the city of Rio de Janeiro and MARJ (Source: SES-RJ 2021)

Figure 5.6 shows the supremacy of the city of Rio de Janeiro over its metropolitan area in the number of COVID-19 cases since the beginning of the pandemic in Brazil, in March 2020, until the month of June 2021.[10] Figure 5.5 compares the data of the city of Rio de Janeiro with its metropolitan area (MARJ)—for purposes of comparison in this analysis, the city of Rio de Janeiro was excluded from the metropolitan area. In every month, except for July and August 2020 (but in these months by a minimal difference), the city of Rio de Janeiro surpasses the metropolitan area in the number of cases. In May 2020, the difference was above 10,000 cases, and in the months of December/2020 and April, May, and June 2021, the difference is awfully close to 10,000. Although the purpose of this section of the text is not to analyze the number of confirmed cases, it is essential to highlight the large number of cases registered in this phase of the pandemic, that is, after a year of pandemic, the record in the number of cases is very high (above 20,000 in the metropolitan region; above 30,000 in the city of Rio de Janeiro), a cause for much concern.

In the comparison of the city of Rio de Janeiro with the metropolitan region considering the total number of deaths, Fig. 5.7 maintains the pattern found in the previous analysis regarding the total number of confirmed cases, that is, the city of Rio de Janeiro presents a higher quantity in the total of deaths every month. It is impressive that every month the city of Rio de Janeiro is superior to the metropolitan region. In the months of January, April, May, and June 2021, the difference is above 1000 deaths, and in May 2020, the difference approached almost 2000 thousand deaths. Another important fact is the high number of deaths recorded in the last 3 months (April, May, and June) of this year 2021: in the city of Rio de Janeiro, the number was just above 3000, and in the metropolitan region, the number ranged between 1400 and 1900. Only in May 2020, when Brazil was going through its third

[10] June was the last month considered until the closing of this publication.

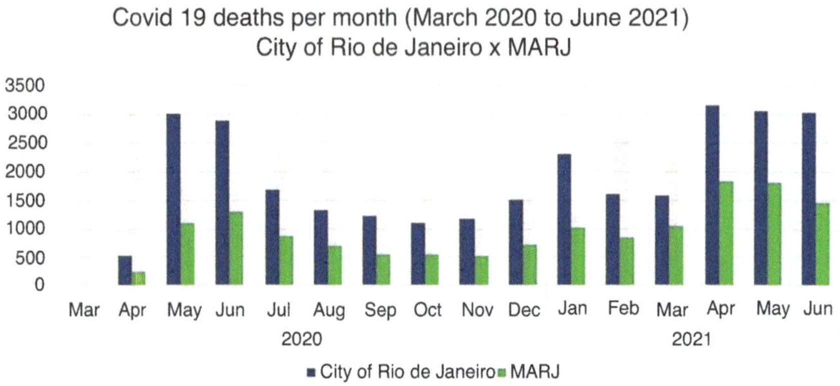

Fig. 5.7 Number of deaths by COVID-19 per month between March 2020 and June 2021 in the city of Rio de Janeiro and MARJ (Source: SES-RJ 2021)

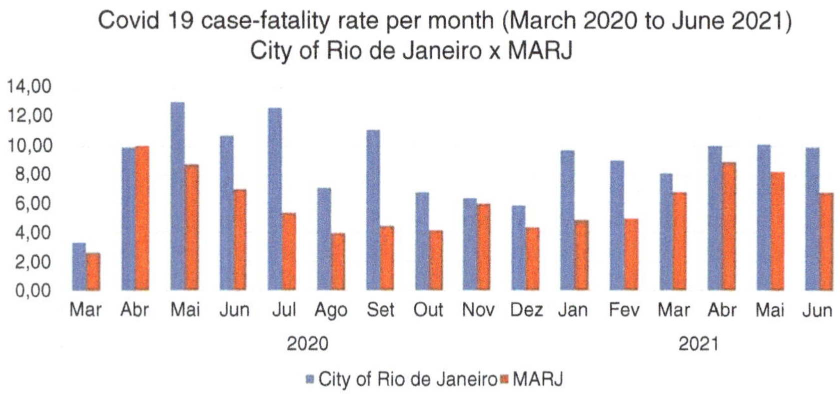

Fig. 5.8 Case fatality rate per COVID-19 per month between March 2020 and June 2021 in the city of Rio de Janeiro and MARJ (Source: SES-RJ 2021)

month of the pandemic, the city of Rio de Janeiro approached 3000 cases (3004 is the exact number), which confirms the seriousness that we are still going through. In the metropolitan region, in June 2020, with 1290 deaths, the quantitative approached the months of April, May, and June 2021, but the records are much higher in these last three, as observed.

When comparing the city of Rio de Janeiro with the metropolitan area considering the case fatality rate, Fig. 5.8 confirms the pattern observed in the two previous analyses, that is, the city of Rio de Janeiro has the highest rates, the only exception being the month of April 2020. Over the entire period, the rates are very high, which is a worrying fact, but especially in 2020, with emphasis on the months of May, June, July, and September in the city of Rio de Janeiro. In the metropolitan area, the highest rates are in the months of April and May, of both years.

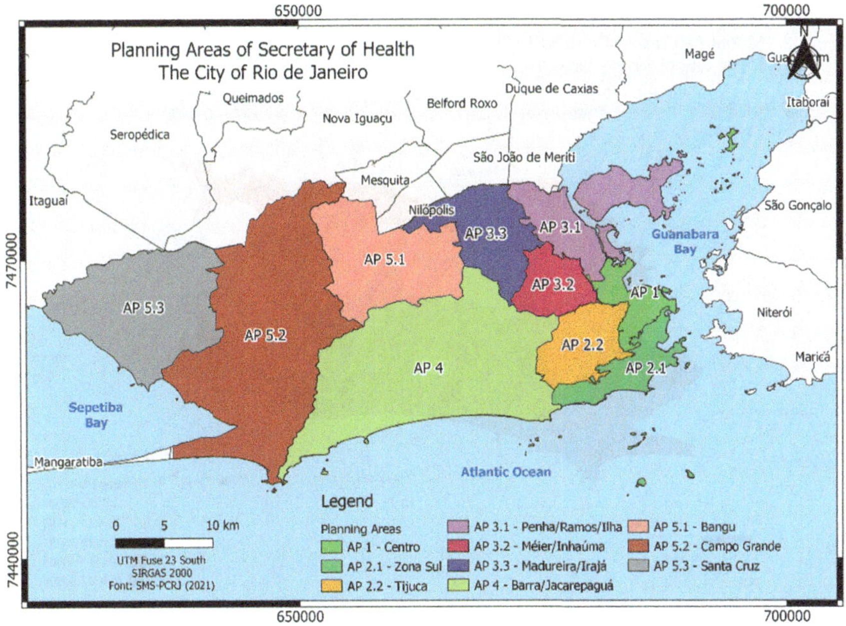

Fig. 5.9 Planning Areas (AP) of the city of Rio de Janeiro (Source: IBGE/SEA 2018; SMS-PCRJ 2021)

The city of Rio de Janeiro, with 163 districts, is geographically divided into the following zones: Center, South, North, and West and is regionalized into five Planning Areas (AP) in order to define better the actions of government policies (Fig. 5.9). The Center, located to the east, and composing the Planning Area 1 (AP-1), is the embryo of the city (456 years since its foundation), where it began and built its physical, social, and cultural structure. Currently, the Center concentrates on financial activities and a good part of commercial and cultural activities, with low residential occupancy. The other areas of the city, South, North, and West are mostly residential and have important service activities (such as health and education) and commercial and cultural activities.

In the South Zone (Planning Area 2.1—AP 2.1), the middle and upper-middle-class population is concentrated, with the best social indicators (per capita income, access to health and education services)[11] in the city, but with many segregated enclaves, like the slums. The North (Planning Areas 2.2 and 3—AP-2.2 and AP-3) and West (Planning Areas 4 and 5—AP-4 and 5) represent the periphery/suburb, with a significant low-income population, mainly in the favelas, but in many neighborhoods, there is an important middle-class population with better social indicators. Special emphasis is given to the West Zone, which occupies almost

[11] Human Development Index (HDI) which considers three pillars: per capita income; education (the average length of study); health (healthy lifetime/longevity).

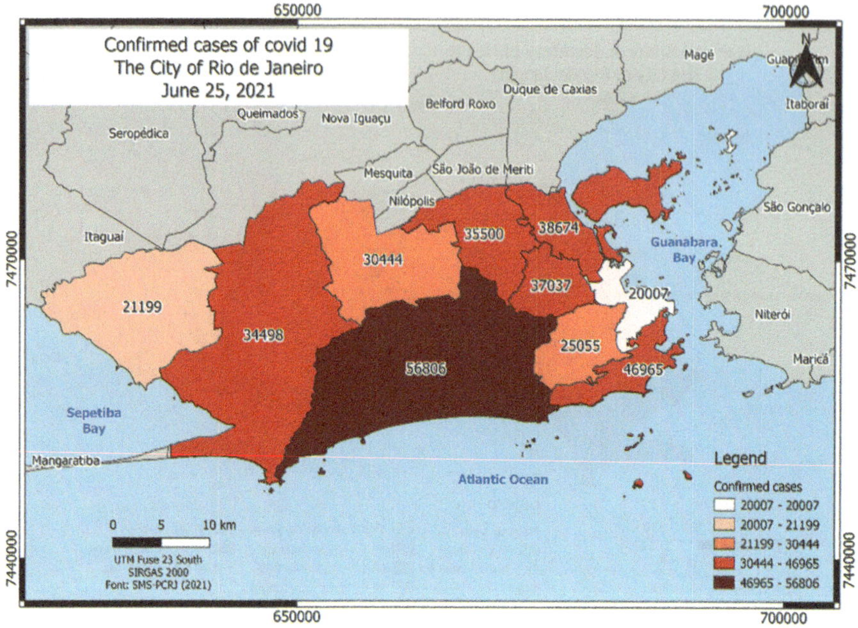

Fig. 5.10 Number of confirmed COVID-19 cases for the city of Rio de Janeiro on 25 June 2021 (Source: IBGE/SEA 2018; SMS-PCRJ 2021)

70% of the municipal territory, and is configured as the city's new expansion area, and is divided into two very distinct areas: Planning Area 5 (AP-5) with traditional neighborhoods of low-income, such as Bangu, Campo Grande and Santa Cruz, and Planning Area 4 (AP-4) with more recent neighborhoods occupied by a middle and upper-middle-class population, such as Jacarepaguá and Barra da Tijuca. Finally, it is important to point out that in all of the city's neighborhoods, regardless of the conditions of social indicators, there are dwellings in favelas, which are striking spaces in the landscape of the city of Rio de Janeiro, which in many cases are confused as neighborhoods.

For this section of analysis of COVID-19 in the city of Rio de Janeiro, a sample of 1 day in the month of June (the last date before sending this publication) will be considered, according to the spatial division by Planning Areas (AP), a criterion also adopted by the Municipal Health Department.

Considering the number of confirmed cases on 25 June 2021 (Fig. 5.10), Planning Area 4, Barra da Tijuca and Jacarepaguá in the West Zone leads with more than 46,000 cases. The neighborhoods in the North Zone (AP-3) and in the South Zone (AP-2.1) follow in sequence with numbers of cases between 30,000 and 46,000. Planning Area 5 is divided into three groups: cases between 20,007 and 21,199 (AP-5.3), cases between 21,199 and 30,444 (AP-5.1), and cases between 30,444 and 46,965 (AP-5.2), the diversity that is largely due to the territorial size, the largest in the city. Finally, in the Center (AP-1), there is the lowest number of cases (below

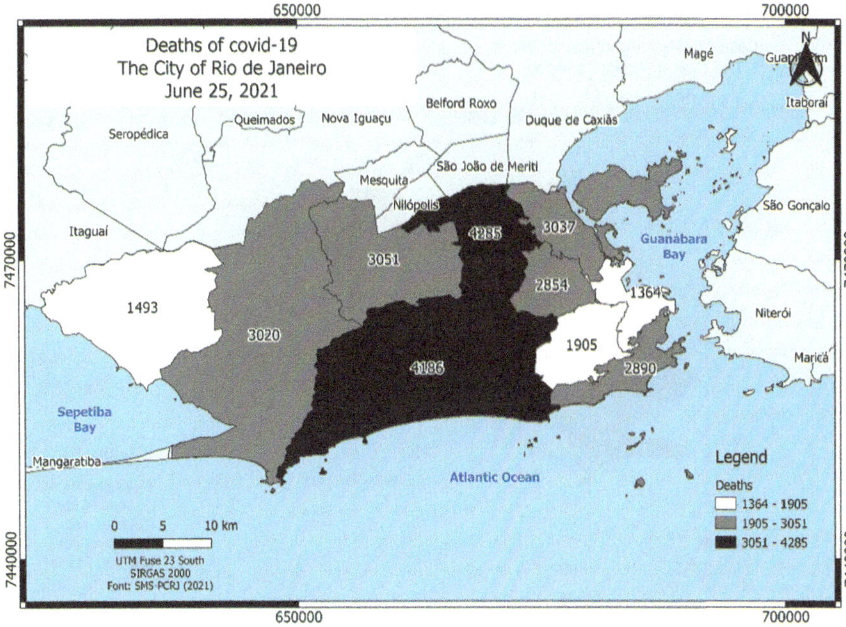

Fig. 5.11 COVID-19 deaths for the city of Rio de Janeiro on 25 June 2021 (Source: IBGE/SEA 2018; SMS-PCRJ 2021)

20,007), which is due to the area's commercial-financial characteristic, sparsely inhabited by residential units, where many commercial buildings are closed by the account of the restrictions generated by the pandemic. A possibility of analysis to understand the high cases in the West Zone (AP-4) and in the South Zone (AP-5.1) may be associated with the diversity of options for existing leisure activities, such as bars, restaurants, squares, and urban parks, beaches, and the most significant tourist spots in the city, which end up attracting a large population, who limit themselves or are distracted in meeting the minimum health restrictions necessary against COVID-19 (social distancing, use of masks and application of alcohol factor 70).

Considering the number of confirmed deaths on 25 June 2021 (Fig. 5.11), this map converges with the previous map (Fig. 5.9) in some areas of the city. Part of the North Zone (AP-3.3) and the West Zone (AP-4) concentrates the highest number of deaths (between 3051 and 4285), being higher in the former. Both already had a high number of confirmed cases of COVID-19, especially AP-4, which led to death. The second group of deaths from the disease spreads across the West Zone (APs 5.1 and 5.2), North Zone (APs 3.1 and 3.2), and South Zone (AP 2.1) with numbers between 1905 and 3051. Finally, the third group of deaths is found in the Center (AP), North Zone (AP-2.2), and West Zone (AP-5.3). In this sample of the geography of deaths, there is no evidence in the numbers that can show a relationship, even if minimal, of deaths by social areas (low or high income), that is, the zoning of deaths does not comply with the income conditions of the groups social. Evidently, studies in this

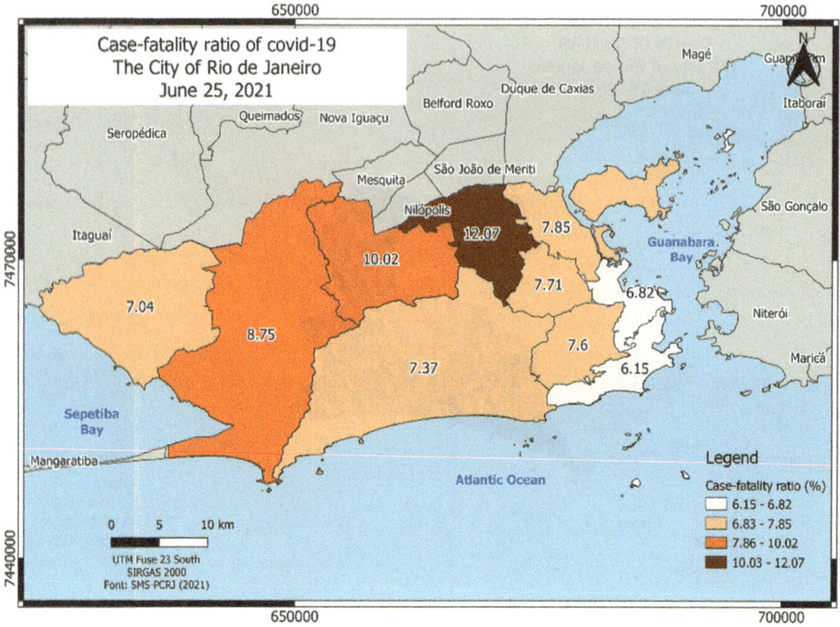

Fig. 5.12 Case fatality rate of COVID-19 for the city of Rio de Janeiro on 25 June 2021 (Source: IBGE/SEA 2018; SMS-PCRJ 2021)

category need to be explored (Cestasi et al. 2021), which may bring greater clarity in the analysis between deaths by COVID-19 and more vulnerable social groups.

Considering the lethality of 25 June 2021 (Fig. 5.12), the AP-3.3, in the North Zone, leads (between 10.02 and 10.07), followed by two areas in the West Zone (AP-5.1 and 5.2), with rates between 7.85 and 10.02. In large part of the city, which includes the West Zone (AP-4 and AP-5.3) and the North Zone (AP 3.1, 3.2 and 2.2), the rate recorded is between 6.82 and 7.85 and, finally, the lower rates are in the South Zone (AP-2.2) and the Center (AP-1). In this geography of lethality, three important considerations: the Center remains isolated from the rest of the city, as in previous maps of confirmed cases and deaths, with very low values, and standing out as a distinct area, a "bubble," in the context of the city; the West Zone (AP-4), unlike previous maps of confirmed cases and deaths, stands out with lower lethality numbers, which may denote the reasonable efficiency of access to health services by this population, even if it is not a rule that residents of this area access the health system close to their residence; AP3.3, in the North Zone, shows the highest concentration of lethality, which is largely due to the high demographic density of the area, the largest in the city, while APs 5.1 and 5.2, in the West Zone, despite the high absolute population, presents much lower demographic density than the North Zone and the South Zone, which should be of concern to health policy authorities.

Considering the vaccination against COVID-19 on 25 June (Fig. 5.13), AP-4 in the West Zone leads the ranking in the city of Rio de Janeiro, with doses between

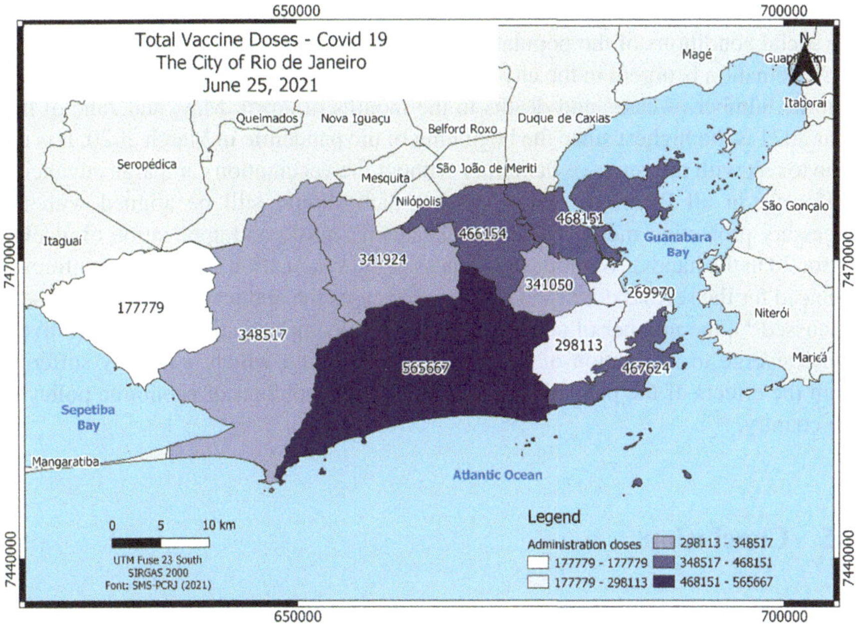

Fig. 5.13 Total vaccinated against COVID-19 in the city of Rio de Janeiro on 25 June 2021 (Source: IBGE/SEA 2018; SMS-PCRJ 2021)

468,151 and 565,667 to be the area with the highest number of cases and deaths from the disease, which indicates a situation of hope. In second place in the ranking are the neighborhoods of the North Zone (AP-3.1 and 3.3) and the South Zone (AP-2.1), with the number of vaccinated between 348,517 and 468,151. The AP-3.3 has stood out as the second area with the highest number of COVID-19 cases and the highest number of deaths and cases fatality ratio, a cause for much concern. Therefore, this second position in the vaccination ranking is satisfactory and needs to advance. In third place in the ranking is another area in the North Zone (AP-3.2) and two in the West Zone (AP-5.1 and 5.2) with numbers between 298,113 and 348,517. In the West Zone, vaccination needs more agility, since which registered the second-highest number of deaths and cases fatality ratio, a very needy region in the city. Likewise, in AP-3.2, it is necessary to speed up vaccination, as it also registered the second-highest number of deaths. In fourth place in the ranking are the Center (AP-1) and AP-2.2 (North Zone), with 177,779 and 298,113 vaccinated, two areas that had the lowest number of deaths, especially the first. However, this condition should not suggest a slow pace of vaccination, especially in AP-2.2, which has an absolute and relatively high population in the city. Last but not least, the extreme west of the West Zone (AP-5.3), has the lowest number of vaccinated (177,779) and is also the lowest in the number of deaths. As discussed in the previous analysis, regarding AP-2.2 and AP-1, even though the number of deaths is low, there is an urgent need to expand vaccination, mainly because it is a very poor area in the city, with low indicators (income, education, and access to health), which can aggravate the social conditions of the population.

Vaccination is urgent in the city of Rio de Janeiro. Figures 5.6, 5.7, and 5.8 show that the number of cases and deaths in the months of April, May, and June of this year 2021 is the highest since the beginning of the pandemic in March 2020. It is not time to rest with measures of flexibility without the vaccination campaign advancing uniformly in all Planning Areas (APs), which should still be aligned with the necessary protection measures (social distancing, mask, and application of alcohol factor 70 in the hands). Without seriousness in the vaccination campaign, reinforcing the need for the second dose (and that in some cases the application of a third dose is discussed[12]), the number of confirmed cases and deaths can still surprise negatively and converge to a situation of social chaos in the city, which is already suffering from the effects of the pandemic and the disastrous conduct of economic policy in the country.

5.5 Conclusion

The geography of the pandemic seen from the territorial scale of the Rio de Janeiro metropolis and the city of Rio de Janeiro allows us to observe that, although there is a global aspect of the new coronavirus in terms of its dissemination, what prevails is a territorial inequality of its impact, which is linked to the conditions of the territory.

Regarding the Brazilian reality, the second wave of the pandemic is still a reality that has not been overcome, despite the slow advance of vaccination, since by 02 June, only 23% of the population had taken at least the first dose, being ranked 72 out of 190 countries and territories according to Our World in Data.[13] This high rate of contamination, which also sustains the second wave, is due to political aspects involving conflicts between federal, municipal, and state powers, disinformation campaigns (fake news) and discouragement of vaccination, and non-adoption of social distancing measures.

The sum of these factors with the differentiating components in the urban-metropolitan territories of Rio de Janeiro confers the constitution of the territories of the "periphery pandemic spaces," as presented in the Introduction, the municipalities of the metropolitan periphery together with peripheral neighborhoods of the city of Rio de January have high disease fatality rates. These areas are designed both by the absence of support from health structures and by the effects of the perversity of globalization, as pointed out by Harvey (2007) on the effects of neoliberalism,

[12]The mayor of the city of Rio de Janeiro, Eduardo Paes, is considering the application of a third dose (or "booster dose") for the elderly, but this is still under discussion (Look: https://www.cnnbrasil.com.br/saude/2021/07/01/paes-planeja-terceira-dose-de-vacina-contra-a-covid-19-para-idosos-no-rj). One of the goals is to protect this age group that is more subject to possible complications from the Indian variant of coronavirus (Delta), already present in Brazil.

[13]https://www.bbc.com/portuguese/brasil-56680167.

since this model, as part of capitalist production, imposes an acute territorial division of labor, circumscribing different uses to territories. The areas heavily affected by prevalence rates, for example, are areas where most workers (both formal and informal) live in the metropolitan area of Rio de Janeiro. Inevitably, these workers are forced to commute daily, crowding into public transport, and further exposing themselves to contamination. An example is the spatiality of the Cases fatality ratio in the city of Rio de Janeiro, which materializes in one of the main axes of urban circulation in the city of Rio de Janeiro.

Likewise, there is an urgent need for more studies that seek to interpret the territorial logics of the pandemic in peripheral spaces, whether in Brazil, India, or other countries in the periphery and semi-periphery of the world. In this sense, it is necessary for public authorities to create mechanisms for the disclosure and transparency of data with greater detail so that we can think about the impacts of these spatialities in the intra-urban sphere in order to allow detailed analyzes that can be carried out to promote a territorial justice, which takes into account social, political, economic and cultural aspects of spaces, which is the use of the territory.

References

Azevedo e Silva G, Jardim BC, Lotufo PA (2021) Mortalidade por COVID-19 padronizada por idade nas capitais das diferentes regiões do Brasil (in Portuguese). Cad Saúde Pública 37(06): 1–9

Bahia B, Chade J, Dedecca CS et al (2021) A tragédia brasileira do coronavírus/covid-19: uma análise do desgoverno federal, 2020–2021. (in Portuguese). https://www.unicamp.br/unicamp/sites/default/files/2021-05/tragedia-brasileira-covid_final.pdf. Accessed 4 July 2021

Barbosa J et al (2010) Módulos de princípios de epidemiologia para o controle de enfermidades. Módulo 1: apresentação e marco conceitual. Organização Pan-Americana da Saúde, Brasília, 48 p

BBC (2021a) CPI da COVID: executivo da Pfizer confirma que governo Bolsonaro ignorou ofertas de 70 milhões de doses de vacinas (in Portuguese). https://www.bbc.com/portuguese/brasil-57104347. Accessed 4 July 2021

BBC (2021b) Afinal, Brasil vacina pouco ou muito? Confira 5 dados do ranking global (in Portuguese). https://www.bbc.com/portuguese/brasil-56680167. Accessed 4 July 2021

Brasil, Presidência da República, Casa Civil, Subchefia para Assuntos Jurídicos (1974) Lei complemen-tar nº 20, de 1º de julho de 1974 (in Portuguese). http://www.planalto.gov.br/ccivil_03/leis/lcp/lcp20.htm. Accessed 3 July 2021

Cabral S, Ito NC, Pongeluppe LS (2021) The disastrous effects of leaders in denial: evidences from COVID-19 crises in Brazil. SSRN. https://papers.ssrn.com/sol3/papers.cfm?abstract_id=3836147 Accessed 4 July 2021

Carta Capital (2021) Os motivos para a instalação da CPI da pandemia (in Portuguese). https://www.cartacapital.com.br/opiniao/os-motivos-para-a-instalacao-da-cpi-da-pandemia/. Accessed 4 July 2021

Cestasi et al (2021) Social vulnerability and COVID-19 incidence in a Brazilian metropolis. Cien Saude Colet 26(3):1023–1033

CNN (2021a) Ex-ministro Pazuello participa de ato ao lado de Bolsonaro sem máscara (in Portuguese). https://www.cnnbrasil.com.br/politica/2021/05/23/ex-ministro-pazuello-participa-de-ato-ao-lado-de-bolsonaro-sem-mascara. Accessed 4 July 2021

CNN (2021b) Pazuello diz que antes de cargo no governo não sabia o que era o SUS (in Portuguese). https://www.cnnbrasil.com.br/politica/2020/10/07/pazuello-diz-que-antes-de-cargo-no-governo-nao-sabia-o-que-era-o-sus. Accessed 4 July 2021

CNN (2021c) Paes planeja terceira dose da vacina contra COVID-19n para idosos no RJ (in Portuguese). https://www.cnnbrasil.com.br/saude/2021/07/01/paes-planeja-terceira-dose-de-vacina-contra-a-covid-19-para-idosos-no-rj. Accessed 4 July 2021

Fleury S (2021) Políticas de isolamento na pandemia: confrontação federativa, disputas discursivas e consequências político-sanitárias (in Portuguese). In Santos, AO, Lopes, LT (orgs) Coleção COVID-19. Principais elementos. https://www.conass.org.br/biblioteca/volume-1-principais-elementos/ Accessed 6 July 2021

Fortes A, Oliveira LD, Sousa GM (2020) A covid-19 na Baixada Fluminense: Colapso e apreensão a partir da periferia metropolitana do Rio de Janeiro (in Portuguese). Espaço e Economia 18. https://doi.org/10.4000/espacoeconomia.13591

Globo (2021a) Brasil registra mais de 2 mil mortes por COVID em 24 horas mas vê queda simultânea das médias móveis de casos e óbitos (in Portuguese). https://g1.globo.com/bemestar/coronavirus/noticia/2021/06/30/brasil-registra-mais-de-2-mil-mortes-porcovid-em-24-horas-mas-ve-queda-simultanea-nas-medias-moveis-de-casos-e-obitos.ghtml. Accessed 4 July 2021

Globo (2021b) Casos e mortes por corona vírus no Brasil em 31 de dezembro segundo consórcio de veículo de imprensa (in Portuguese). https://g1.globo.com/bemestar/coronavirus/noticia/2020/12/31/casos-e-mortes-por-coronavirus-no-brasil-em-31-dedezembro-segundo-consorcio-de-veiculos-de-imprensa.ghtml. Accessed 4 July 2021

Globo (2021c) Crise do oxigênio um mês após colapso Manaus ainda depende de doações do insumo (in Portuguese). https://g1.globo.com/am/amazonas/noticia/2021/02/14/crise-do-oxigenio-um-mes-apos-colapso-em-hospitais-manaus-aindadepende-de-doacoes-do-insumo.ghtml. Accessed 4 July 2021

Globo (2021d) Doria e Bolsonaro discutem em reunião de governadores (in Portuguese). https://g1.globo.com/jornalnacional/noticia/2020/03/25/doria-e-bolsonaro-discutem-em-reuniao-de-governadores.ghtml. Accessed 4 July 2021

Globo (2021e) Paes planeja terceira dose de vacina contra a COVID-19 para idosos no RJ (in Portuguese). https://www.cnnbrasil.com.br/saude/2021/07/01/paes-planeja-terceira-dose-de-vacina-contra-a-covid-19-para-idosos-no-rj

Harvey D (2007) Brief history of neoliberalism. Oxford University Press, New York

IBGE Cidades (2020) Brasil, Rio de Janeiro, Panorama Municipal. [S. l.]: IBGE, 2020. Available at https://cidades.ibge.gov.br/. Accessed 13 April 2020

IBGE/SEA (2018) Continuous Vector Cartographic Base of the State of Rio de January on Scale 1: 25,000, 2018. Available at https://portaldemapas.ibge.gov.br/portal.php#mapa221075. Accessed 20 January 2020

Instituto Brasileiro de Geografia e Estatística [IBGE] (2020) Estimativas de População, Brasil, 1.° de julho de 2020 (in Portuguese). https://www.ibge.gov.br/estatisticas/sociais/populacao/9103-estimativas-de-populacao.html. Accessed 6 July 2021

Lima L, Pereira AMMP, Machado CV (2020) Crise, condicionantes e desafios de coordenação do estado federativo brasileiro no contexto da COVID-19 (in Portuguese). Cad. Saúde Pública 36(7)

Lucena AJ, Oliveira LD, Ibañez P et al (2020) The geography of COVID-19 in Rio de Janeiro, Brazil: conflicts, tensions, and challenges. In: Mishra M, Singh RB (eds) COVID-19 pandemic trajectory in the developing world: exploring the changing environmental and economic milieus in India. Springer, Singapore, pp 51–69

Rocha AS (2021) Globalização, gestão e acesso aos sistemas público e privado de saúde: a Baixada Fluminense no contexto da pandemia (in Portuguese). Espaço e Economia 18. http://journals.openedition.org/espacoeconomia/12672. Accessed 4 July 2021

Rocha ACL, Ribeiro MAC (2020) A expansão da metrópole do Rio de Janeiro e a formação da franja periurbana e perimetropolitana (in Portuguese). Continentes. https://www.revistacontinentes.com.br/index.php/continentes/article/download/250/209/. Accessed 6 July 2021

Rocha AS, Sousa GM, Fortes A et al (2021) A expansão da covid-19 na Baixada Fluminense - RJ: seus caminhos e efeitos sociais na periferia (in Portuguese). GEO UERJ, Rio de Janeiro. https://doi.org/10.12957/geouerj.2021.51431. Accessed 5 July 2021

Santos M, Silveira ML (2001) O Brasil: território e sociedade no início do século XXI (in Portuguese). Record, Rio de Janeiro

SES-RJ (2021) Daily epidemiological bulletin COVID-19, 2021. Available at https://coronavirus.rj.gov.br/boletins/. Accessed 30 April 2021

SMS-PCRJ (2021) Covid 19 Boletim epidemiológico - Centro de Operações de Emergência. Available at https://coronavirus.rio/boletim-epidemiologico/. Accessed 25 June 2021

Touchton M, Marie Knaul F, Arreola-Ornelas H et al (2021) A partisan pandemic: state government public health policies to combat COVID-19 in Brazil. BMJ Glob Health 6(6)

Wang DWL (2021) Atuação do sistema de justiça durante a pandemia de covid-19: Uma análise da jurisprudência do STF (in Portuguese). In: Santos AO, Lopes LT (eds) Coleção COVID-19. Principais elementos

Pablo Ibanez is graduated in geography from the State University of Campinas (2002), master's (2006), and PhD (2012) in human geography from the University of São Paulo. He is currently an adjunct professor at the Federal Rural University of Rio de Janeiro and a researcher linked to the research groups: Welfare in Health, Development and Territory, at the Faculty of Medicine of the University of São Paulo; Research Center on Public Policies and Territory, at Fluminense Federal University; and Group of Studies and Research on Politics and Territory (GEOPPOL), from the Federal University of Rio de Janeiro. He was a visiting researcher at Fudan University, Shanghai. He has experience in the research area, with emphasis on Geopolitics, Political, Economic and Regional Geography, acting mainly on the following themes: science, technology and innovation; regional development, regionalization and decentralization; analysis of public policies, industrial and innovation policies; development and territory.

Gustavo Mota de Sousa is bachelor in geography from Federal Fluminense University—UFF (2004); master's (2009) and PhD (2013) in geography from the Federal University of Rio de Janeiro—UFRJ. He is currently an adjunct professor of the Department of Geography in undergraduate courses in Geography and Geology and of the Postgraduate Course in Geography (PPGGEO) at the Federal Rural University of Rio de Janeiro (UFRRJ). He conducts research at the Integrated Laboratory of Applied Physical Geography (LiGA/UFRRJ), Laboratory of Cartography (GEOCART/ UFRJ) and the Laboratory of Social Dimensions Applied to Physical Activity (LABSAFE/UFRRJ). It has teaching, research, and extension projects with guidance from undergraduate and graduate students through themes on susceptibility to wildfires; geoecological mappings; cartographic representations with studies of tactile cartography, topographic models, Augmented Reality and Digital Fabrication (3D printing and laser router) in addition to geotechnologies applied to participatory mapping.

Andrews José de Lucena is graduated in geography from the Federal University of Rio de Janeiro—UFRJ (2002), master's in geography from the State University of Rio de Janeiro—UERJ (2005), and PhD in atmospheric sciences from UFRJ (2012). Since 2004 he has been teaching in higher education and is currently an Associate Professor in Geography undergraduate and graduate courses in at the Federal Rural University of Rio de Janeiro (UFRRJ). He works in Physical Geography area (Atmospheric Sciences) with an emphasis on Urban Climatology and Atmosphere Remote Sensing. He is interested in environmental changes in the city (ies) and its various associated phenomena, such as the Urban Heat Island (ICU). He participated and integrates projects and works together with researchers from Geography, Meteorology and Engineering, areas of knowledge where he guided and guides undergraduate and graduate students. His main work area is the city and Metropolitan Area of Rio de Janeiro (RMRJ) with a focus on its urban climate. He manages the website www.climatologia.com.br, a portal with information on the urban climate at

RMRJ, especially on the Continental Surface Temperature (TSC), obtained by Remote Sensing, from 1984 to the present. He coordinates the Integrated Laboratory of Applied Physical Geography (LIGA)/UFRRJ.

Heitor Soares de Farias is graduated in geography from the Federal University of Rio de Janeiro—UFRJ (2004), master's in geography from the UFRJ (2007), and PhD in geography from the Federal Fluminense University (2012). Since 2013, he teaches in higher education and is currently an adjunct professor in geography at undergraduate and graduate courses in at the Federal Rural University of Rio de Janeiro (UFRRJ). He works with environmental planning based on geographical climatology, researching the health risk associated with climatic phenomena, such as islands of heat, rains and air pollution, mainly. He coordinates the Integrated Laboratory of Applied Physical Geography (LIGA)/UFRRJ.

Leandro Dias de Oliveira is graduated in geography from the State University of Rio de Janeiro (UERJ), master's in geography from the State University of Rio de Janeiro (UERJ), Doctor of Geography from the State University of Campinas (UNIcamp) and Post-Doctor in Public Policy and Human Training from the University of the State of Rio de Janeiro (UERJ). Associate Professor of the Department of Geography at UFRRJ—Federal Rural University of Rio de Janeiro (DGG-IA), headquarters campus, and professor of the permanent staff of the Graduate Program in Geography (PPGGEO-UFRRJ) and the Graduate Program Interdisciplinary Graduation in Digital Humanities (PPGIHD-UFRRJ). He integrates and coordinates the Laboratory of Economic and Political Geography (LAGEP).

André Santos da Rocha received the graduate degree in geography for Faculty Philosophy, Science, and letters of Duque de Caxias (2005). Have postgraduate (lato sensu) in Territorial Policies for State University of Rio de Janeiro—UERJ (2007), master's in geography for Federal Fluminense University—UFF (2009) and PhD in geography for Federal University of Rio de Janeiro—UFRJ (2014). He is a professor in undergraduate and graduate courses (master) at UFRRJ. He works in human geography, especially with thematic of political and economic geography. Analyses on the Baixada Fluminense and its political and economic dynamics stand out in their studies. He is also dedicated to studies of International Cooperation with a phase in Brazil-Africa-Latin America relations. Coordinates and participates in research projects with other researchers from UFRRJ and colleagues from Latin American Network Space and Economy (REELE). He serves as editor of the Academic Journal Continentes (UFRRJ), and coordinator the Laboratory of Economic and Political Geography (LAGEP).

Part II
Differential Shock

Chapter 6
Women in Pandemic: The Realities of the COVID-19 in the Darjeeling Himalayan Region

Bishal Chhetri and Kabita Lepcha

Abstract The novel coronavirus that was first reported in the year 2019 has been unabated even in 2021. People who are in the old age category and those with comorbidities are more vulnerable to it. The Government of India had imposed a lockdown in 2020 to curb its spread. The initial decline in the number of cases in the country during February 2021 led to relaxations, which saw the mass mobilization of people for religious gatherings, elections, protests, etc. that led to the emergence of the deadlier second wave. Once again, life has been brought to a standstill by the deadly virus, with various state governments imposing lockdowns and partial lockdowns. These preventive measures put forward by the governments seldom consider the gender sensitivities and realities that significantly affect women's lives. The lockdown and pandemic have greatly affected various spheres of life of all the women, whether it may be working, students or homemakers. This chapter tries to explore the impact of the second wave of pandemics upon the women of the Darjeeling Himalayan region under the Gorkha Territorial Administration (GTA) in the West Bengal state of India.

Keywords Social media · Second Wave · Red Zone · Lockdown · PPE · GTA · Comorbidity

6.1 Introduction

The novel coronavirus, or simply COVID-19, is said to have been originated at Wuhan, Hubei Province of China at the flag end of the year 2019 (Chang 2020; Liu et al. 2020), still wreaks havoc worldwide even in mid-2021. COVID-19 is an

B. Chhetri (✉)
Department of Geography, Southfield College, Darjeeling, West Bengal, India

K. Lepcha
Department of Geography, University of Gour Banga, Malda, West Bengal, India

S. Saha et al. (eds.), *Economic and Societal Transformation in Pandemic-Trapped India*, New Frontiers in Regional Science: Asian Perspectives 55,
https://doi.org/10.1007/978-981-16-5755-9_6

infectious disease caused by a newly discovered coronavirus. Most people infected with the COVID-19 virus experience mild to moderate respiratory illness and recover without requiring special treatment. People in old age (say 60 years and above) and those with underlying comorbidity factors like cardiovascular disease, diabetes, chronic respiratory disease, and cancer are more likely to develop severe illness (Kodge 2021). On 12th March 2020, the World Health Organization (WHO) declared the situation arising out of coronavirus-induced infection as the global pandemic (WHO 2020). The first case in India was reported on 30th January 2020, and the prime minister of India called for a nationwide curfew for 1 day termed as, "Janta Curfew" on 22nd March and subsequently imposed a nationwide lockdown on 24th march 2020 (Gupta et al. 2021).

In the Gorkhaland Territorial Administration (GTA)[1] area (Fig. 6.1), the first case of COVID-19 was reported in the town of Kalimpong on 28th March 2020; the patient who had tested positive was a 44-year-old woman with travel history, she later succumbed to the infection on 30th March 2020. Following her death, ten members of the family and a maid were treated for the infection, and Kalimpong town was declared a Red Zone[2]; strict vigilance on the movement of people was kept, which resulted in the GTA area remaining COVID-19 free for nearly two months after the initial infection was tackled. Moreover, throughout the different phases of nationwide lockdown and during the unlock phases, the GTA area had very low COVID-19 cases. In fact, the total active cases in the GTA area on 31st August 2020 was only 478 cases (Giri 2020), which was the lowest in the state. As a result, during the phase of *Unlock4.0*, the Government of West Bengal decided to permit the hotels in the GTA area to function from fifth September 2020 onward with some restrictions to revive the tourism industry, which had been paralyzed for nearly 6 months.

Likewise, during the same phase, India reported the highest number of daily infections of 97,894 persons as of 16th September 2020, which was the highest in the world and after that saw a steady decline in its number of daily infections (Sinha 2021). As the different unlock phases in the nation continued, the number of active cases in India declined by 85% from its peak that was reached on the 28th January 2021 (Sinha 2021). The belligerent hill economy, which heavily depended on tourism, slowly started signs of recovery. Many residents of the Darjeeling Himalayan area working in the hospitality industry in other parts of the country had returned to the hills and started promoting village tourism by establishing homestays. As a result, hotels and homestays started seeing steady footfalls with an 80–90% occupancy rate (PTI 2020). However, even though the homestays and the hotels had started functioning, most of the tourists directly booked the hotels and

[1]GTA is an autonomous self-governing body comprising of Kalimpong, Kurseong and Darjeeling hill areas along with some mouzas of Siliguri terai of Darjeeling and Kalimpong districts. It was established in 2012 to replace the erstwhile DGCH. Its objective is to bring socio-economic, cultural, educational and infrastructural development of the area governed.

[2]Red zone or the hotspots were those districts which had several active cases and high doubling rate of confirmed cases in 2020.

homestays, bypassing the many travel agencies, and started arriving in their own arranged vehicles at their respective destinations. This changing trajectory of the tourism industry adversely affected many people who were either employed in different agencies or were owners of those agencies and local taxi drivers whose sole sustenance depended on tourism.

By mid-February 2021, there were only 8000 daily cases in India (Kumar 2021). This decline in the number of daily infections led to optimism where people neglected COVID protocols like social distancing and wearing masks in public places. The government too relaxed many restrictions such as reopening of public spaces, permission for big religious gatherings like Kumbh Mela, and holding of elections to legislative assemblies and panchayat elections in several states like Assam, Kerala, West Bengal, Tamil Nadu, Puducherry, and Uttar Pradesh and organization of political rallies before the polls. The graph of caseloads also started signs of being rejuvenated. The number of daily infections rose to 23,000 by tenth March, which went up to 44,000 on 20th March, 53,500 on 30th March, 153,000 on tenth April, and by 20th April, it had reached 295,000 cases (Kumar 2021). India, on 25th March, 2021 announced that the Double Mutant variant, now called as Delta variant, was detected from the collected samples and was responsible for the deadly second wave (Pandey and Nazmi 2021). During the first wave, the central government had imposed the nationwide lockdown; however, the lack of coordination between the central government, state governments, and health agencies was evident during the second wave (Kar et al. 2021). The Central Government decided not to impose nationwide lockdown, and on this backdrop, various state governments started imposing their own lockdowns. West Bengal government imposed restrictions from 16th May to 30th May, which was subsequently extended to 15th June and then to first July 2021 (Fig. 6.1).

6.2 Relevance of the Study

Women have exhibited much better resistance and higher survival rates than men against COVID-19 (Chang 2020) even though their chances of being infected are equal to men (The Novel Coronavirus Pneumonia Emergency Response Epidemiology Team 2020). Although women have less mortality rate than men and the chances of being infected remain the same for both sexes, women have been disproportionately impacted by the pandemic. Historically, all pandemics and humanitarian crises have always disproportionately impacted the most vulnerable section of society, including women and girls (Davies and Bennett 2016; Pinchoff et al. 2020). Women all around the world have not only been impacted in their physical health but also in their family, work, and their daily life (Chang 2020). The entrenched gender inequalities present in the society in terms of access to education, employment, and health care often leave women inadequately equipped to effectively protect themselves and their families against infection during an outbreak, and

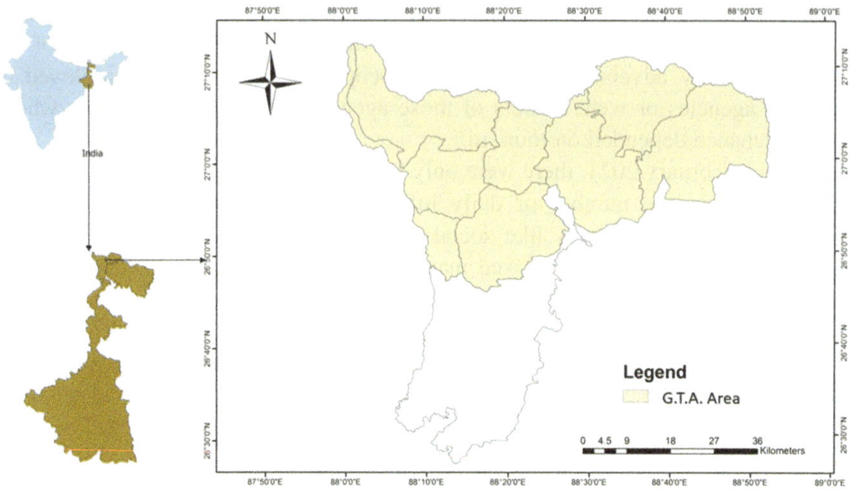

Fig. 6.1 Location of the study area

they are disproportionally vulnerable to secondary adverse effects of prolonged crisis, such as economic insecurity or challenges accessing essential health services (Kapoor et al. 2019; Pinchoff et al. 2020).

Furthermore, the suspension of offices and educational institutions by laying emphasis on work from home and learning from home has increased the hardships for women; women have found more difficulty in balancing their household chores and their official duties or online classes in the case of the students. This increased burden often leads to an increase in stress and may negatively impact their psychological wellbeing. In addition to this, the pandemic has affected the tourism and hospitality industry, which employs many females (Chang 2020). Women who are low-wage earners, female bosses of small businesses, and women working in the informal sector are said to have been hit the hardest (Folayan et al. 2020; Chang 2020).

Since the emergence of the pandemic, most of the literature written on the COVID-19 pandemic beyond medical science regarding the virus does not address women's issues. The available literature is mostly concerned with surveillance and monitoring of the spread of infection at global and regional levels through models and statistical analysis (e.g., Adekunle et al. 2020; Sarkodie and Owusu 2020, etc.). At the same time, some have examined the government responses and policies (e.g., Kar et al. 2021), others have tried to analyze the psychological impact upon youths and the general populace (e.g., Zhong et al. 2020, Varshney et al. 2020, Basu et al. 2020, etc.), and only a few have focused on the issues related to women (Liu et al. 2020; Chang 2020, etc.).

Adekunle et al. (2020) constructed spatial variations of clusters that examined the links between deaths and the number of COVID-19 cases in Africa. Desjardins et al. (2020) applied statistical techniques for detecting space-time clusters and emphasized the importance of focusing surveillance on emerging and active clusters rather than the previous cluster that no longer threaten public health for allocating resources and implementing various mitigation measures in the US. Sarkodie and Owusu (2020) used time-series data and a panel model to examine the relationship between COVID-19 cases and death in China. Gupta et al. (2021) have explained how COVID-19, initially being urban-centric, spread to rural areas due to the reverse migration which had occurred because of the unplanned lockdown imposed by the government and how the functioning of transport networks post-lockdown again led to a rise in infections in the urban areas of India. Some studies like that of Kodge (2021), Biswas et al. (2020) have studied the spread of COVID-19 at a regional level. Kar et al. (2021) have highlighted how the administrative barriers existing between the central government, the state governments, and health agencies have led to severe lapses in putting forward adequate responses during the second wave in India. One of the earliest studies conducted by using social media to assess the people's perspective regarding COVID-19 infection was done in China by Zhong et al. (2020) on first February 2020. Similar projects were run in the UK and US from 23rd February to second March 2020 (Oliver et al. 2020). Varshney et al. (2020) used the snowball sampling method in India to measure the psychological impact of the COVID-19 infection upon the citizens through an online survey called FEEL-COVID. Oliver et al. (2020) also used the snowball sampling method to conduct the COVID-19 impact survey in Spain through social media.

Basu et al. (2020) used similar techniques to carry out a perspective study on youths during the pandemic in India; others like Pinchoff et al. (2020) conducted a rapid telephone survey in Uttar Pradesh and Bihar to assess the gender-specific differences in knowledge, behavior, and health among adolescents and young adults during the pandemic. Liu et al. (2020) investigated the prevalence and predictors of post-traumatic stress symptoms (PTSS) in China. They explored gender differences in PTSS and discovered that the women reported significantly higher levels of PTSS. Chang (2020) tried to understand how the pandemic has impacted women and has highlighted the different difficulties that women face during times of pandemics and diseases.

More research is needed to understand the nature of the gender inequalities that the pandemic has brought about. One needs to understand the gender-specific impact of the pandemic upon the women and has to explore whether the policies and measures taken in place to check the pandemic are gender-sensitive. Have those measures taken into account the gender realities? Have the gender-neutral policies, in fact, adversely impacted women? These questions need to be further explored. This study tries to explore all the important points raised above while trying to understand the pandemic through women's perspective in Darjeeling Himalayas.

This study was conducted in the English language, and through social media platforms, only those respondents who had access to the internet and were familiar with the language could take part in the survey. This has led to the exclusion of the section that does not have internet access and is not familiar with the language. Therefore, this sets the major limitation in the generalization of the findings.

6.3 Methods

6.3.1 Study Design

This study employs a cross-sectional survey design, which intends to explore the impact of the second phase of COVID-19 lockdown upon the women residing in the Darjeeling Himalayas of Gorkhaland Territorial Administration areas (GTA) of West Bengal, India. The study was designed in such a way that the respondents would be derived from across the society, which would include students, home-makers, frontline workers other than health, health workers, government employees, informal workers, and small business owners. The survey questionnaire contains 32 items related to various dimensions like socioeconomic, health, work, etc. The questionnaire was designed after reviewing available literature on the subject along with personal experiences (being a part of the study area) during the 2020 lockdown. Subsequently, a multi-round pilot test was conducted where some questions were added and some modified. The questionnaire was administered in English and was designed to investigate women's experiences during the partial lockdown imposed due to the second wave spread of COVID-19 infection in West Bengal, India.

We have used an online self-selection survey of non-probability sampling technique to derive the information from the different participants through social media platforms like Whatsapp and Facebook. This method is said to have been highly suitable during confinement situations, as presented by the lockdown when the mobility and social contacts are greatly limited (Oliver et al. 2020). Four different nonparametric tests, i.e., Spearman's rank correlation, Chi-Square test, Kruskal–Wallis H-test, and Mann–Whitney U-test, have been applied to analyze the data and Cronbach's alpha test was used to check the reliability of the Likert scale. These tests have been selected as appropriate for the categorical data that the present study deals with.

Spearman's rank correlation has been calculated by using the following formula to determine correlations.

$$r_R = 1 - \frac{6\sum_i d_i^2}{n(n^2-1)} \tag{6.1}$$

where, n is the number of data points of the two variables and d_i is the difference in the ranks of the ith element of each random variable considered. The Spearman correlation coefficient, ρ, can take values from +1 to −1.

Chi-square test has been conducted by using the following formula to determine the test of independence.

$$\chi^2 = \sum \frac{(O - E)^2}{E} \tag{6.2}$$

where, O is the number of observed frequencies, and E is the number of expected frequencies.

Cronbach's Alpha test was used to examine the reliability of the Likert scale and has been calculated by using the following formula.

$$\alpha = \frac{N.\overline{c}}{\overline{v} + (N - 1).\overline{c}} \tag{6.3}$$

where, N is the number of items, $\overline{c}$ is the average covariance between item pairs and $\overline{v}$ is the average variance.

Kruskal–Wallis H-test has been used for comparing two independent samples of different sizes by using the following formula.

$$\mathrm{H} = (N - 1)\frac{\sum_{i=1}^{g} n_i(\overline{r}_i - \overline{r})^2}{\sum_{i=1}^{g}\sum_{j=1}^{n_i}(r_{ij} - \overline{r})} \tag{6.4}$$

where N is the total observations, g is the number of groups, n_i is the number of observations in g_i, r_{ij} is the rank among all the observations of observation j from group i, $\overline{r}_i$ is the average rank of all the observations in the group i and $\overline{r}$ is the average of all the r_{ij}.

Mann–Whitney U-test has been done to see the dependency of the samples by the following formula.

$$U = \sum_{i=1}^{n}\sum_{j=1}^{m} s(x_i, j_i) \tag{6.5}$$

$$\text{with, } S(x, y) = \begin{Bmatrix} 1, \text{if } y < x \\ \frac{1}{2}, \text{if } y = x \\ 0, \text{if } y > x \end{Bmatrix}$$

6.3.2 Data Collection

We deployed the survey from third of June 2021 (8:30 PM) till tenth June 2021 (8: 30 PM) through Whatsapp and Facebook by preparing the questionnaire in Google Form. We employed the snowball sampling technique where we distributed the questionnaire to our immediate contacts, which in turn distributed to their contacts and so on. The snowball sampling technique has already been used successfully by Basu et al. (2020), Oliver et al. (2020), Varshney et al. (2020), and many more scholars during the 2020 COVID-19 lockdowns. The data was collected anonymously without collecting any information that could reveal the identity of the respondent.

6.4 Results and Discussions

Descriptive statistics and nonparametric tests have been conducted for the socioeconomic, workplace-related, health, and nutrition data. Multiple choice questions and questions based on the Likert scale were asked; multiple-choice questions were used to collect various socioeconomic and health-related data, whereas Likert items were used to measure the respondents' perspective on a particular question on a 4 point scale. Altogether, 52 multiple questions were used for collecting the data, and 9 Likert items were employed during the survey. 158 responses were received during the one-week-long online survey.

6.4.1 Demographic Characteristics

The majority of the respondents are single (unmarried) and are less than 40 years of age; around 91.1% of the respondents belonged to this category, while only 1% of the respondents fall within the age group of 60–69 years (Fig. 6.2a, b); this phenomenon can be due to the relatively younger individual's capacity in using the internet and social media being higher than the other age groups. Moreover, if we closely analyze the less than 40 years category, we can find that the majority of the respondents (67.1%) are less than 30 years of age, while only 24.1% belong to the 30–39 years of age bracket. Most of the respondents come from those households where the family size is large, i.e., the number of family members is more than 5 (Fig. 6.2c) and have single free-standing houses, around 7% of the total respondents comprised of the *frontline COVID warriors*; a term referred to denote health and emergency services like police, fire, etc. (Fig. 6.3a) workers.

6.4.2 Impact on Social Behavior

Nineteen percent of the respondents had other comorbidity factors which are considered to pose a high risk of infection. When asked about social contact behavior, 13.9% of the respondents reported having come in contact with the infected persons within the last 2 weeks. The most common contact was either with the infected

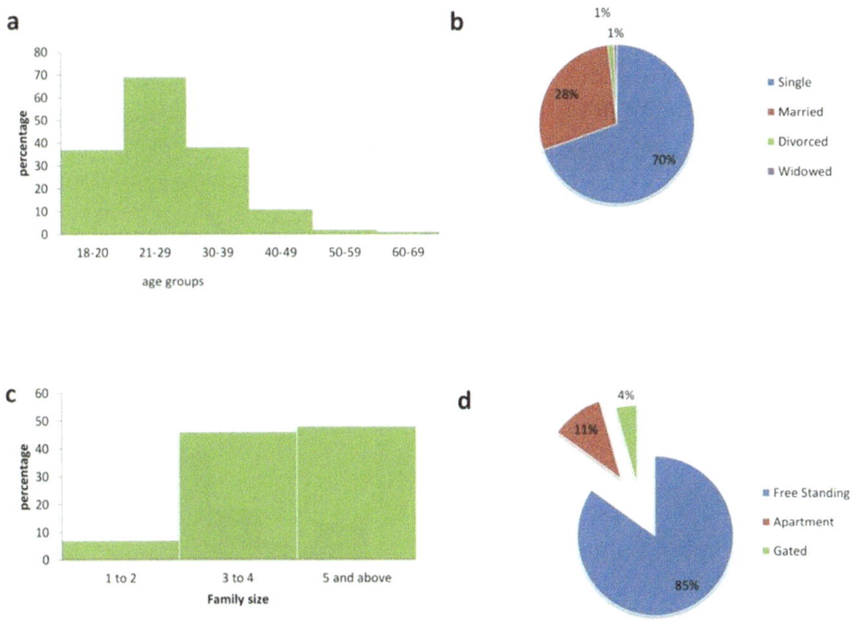

Fig. 6.2 (**a**) Age group of the respondents; (**b**) Marital status; (**c**) Family size; and (**d**) Type of house

family member (19%), with relatives (46%), or with the patients (18%), as in the case of health workers (Fig. 6.3b). When the risk assessment of the women in context to existing comorbidities was analyzed, it was observed that 55% of the women fell under very low-risk factor, 26% fell under low-risk factor, 4% fell under high-risk factor, and 15% in very high-risk factor (Fig. 6.3c). The risk of contracting the COVID-19 virus is higher in those with comorbidities than those without comorbidities (CDC 2020; Sanyaolu et al. 2020). Therefore, it has been said that older people and those with underlying medical conditions are more prone to getting infected by COVID-19 virus. When the social contact behavior and the risk factor of the respondents are correlated, it has been found that both are significantly negatively correlated (Table 6.1). It indicates that the women with higher risk factors are aware and have been following isolation and social distancing norms to minimize their chances of acquiring infection from others, while those with low-risk factors have higher chances of being infected due to social contacts.

6.4.3 Impact Upon Individual's Social and Economic Life

During the lockdown, it was observed that only 34.2% of the respondents stayed home while the majority of the respondents had to go out for various purposes

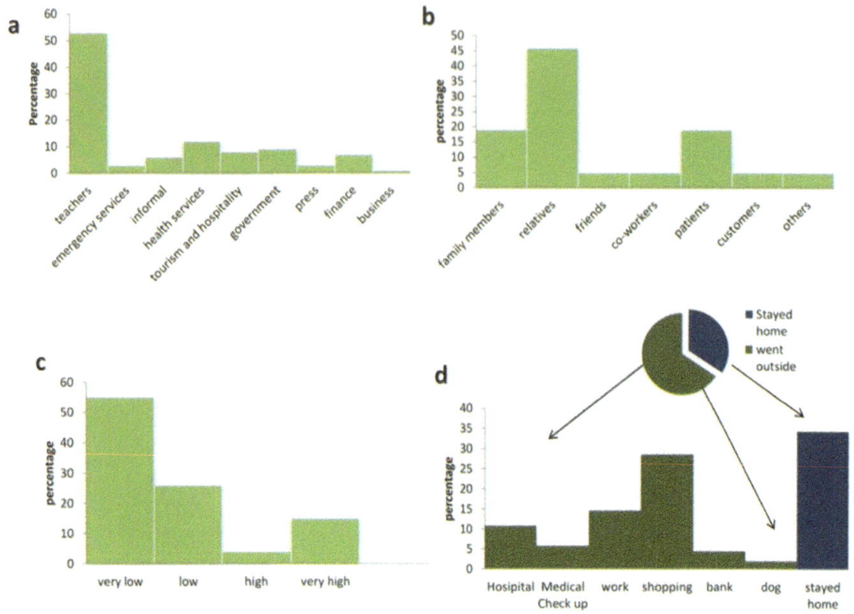

Fig. 6.3 (**a**) Occupation; (**b**) Source of COVID-19 infection; (**c**) Risk factor groups; (**d**) Reasons for going outside

Table 6.1 Association of variables

Sl. No	Variables	Test value	df	P
1	Social contact with risk factor	-0.172^{a} (r_s)		0.031
2	Age with confinement	5.832 (χ^2)	1	0.016
3	Age with opinion about government measures	1.439 (χ^2)	3	0.696
4	Age with duration to remain confined	15.725 (χ^2)	3	0.001
5	Actual confinement to willingness to remain confined for longer duration	9.134 (χ^2)	3	0.028
6	Reliability of the scale	0.739 (α)		

[a]Correlation is significant at the 0.05 level (2-tailed)

(Fig. 6.3d). It was also observed that the respondents mostly had gone out for those tasks which could have been altogether avoided, like shopping and walking the dog; while others had to go out due to those tasks which were unavoidable and demanded urgency like going for hospital visits and medical checkups or for attending work. Since the women are mainly employed as nurses in hospitals and nursing homes, we find a higher incidence of women attending work during the period of lockdown. When the age is compared with the respondents confining themselves at home, it was seen that the age of the respondent affected whether the respondent stayed

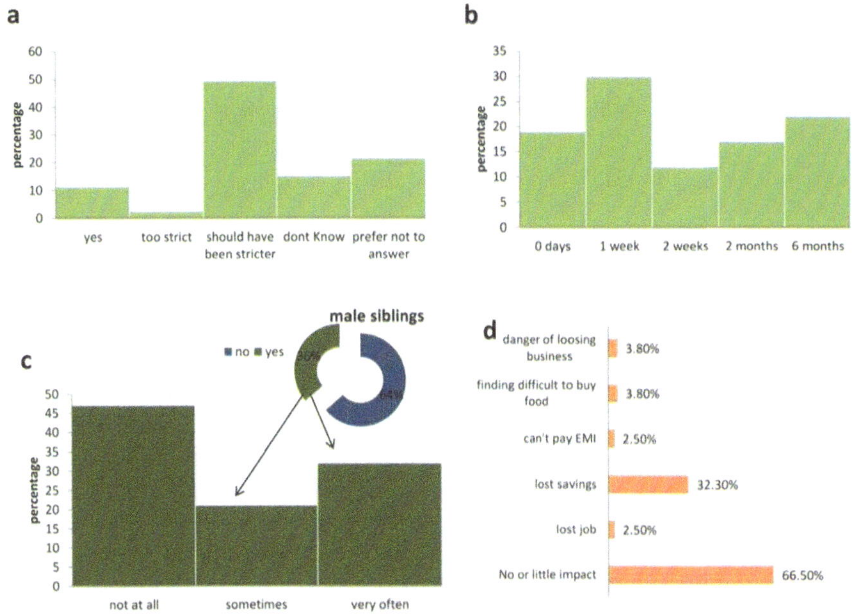

Fig. 6.4 (**a**) Perception regarding lockdown; (**b**) Willingness to stay confined; (**c**) Impact of female student; and (**d**) Financial impact

confined or not. Since the χ^2 value is 5.832 at 1 df. (Degrees of Freedom) with *P*-value = 0.016; the null hypothesis is rejected. It concludes that the home confinement behavior and age group of the respondent are dependent. 49% of the respondents felt that the government should have taken stricter lockdown measures against 11.4% who felt that adequate lockdown measures had been undertaken during the second wave; another 2.5% felt that the lockdown measures were too strict, while 15.2% were not sure and 21.5% preferred not to answer (Fig. 6.4a). When the respondents were asked about their willingness to remain confined, 19% felt that they could stand even for a day, 30% felt they could remain confined for up to 1 week, while 22% felt they could continue for 6 months (Fig. 6.4b). When the opinions by the respondent about measures adopted by the government according to their age category were explored, there was no significant difference of opinion among the different groups as the χ^2 value is 1.439 at 3 df. with *P*-value = 0.696. Hence Null hypothesis is not rejected; whereas, it was also observed that the respondent's willingness to remain confined for a longer duration during the lockdown was depended upon the age (χ^2 value of 15.725 at 3 df. with *P*-value = 0.001) and the actual act of confining oneself indoors greatly influences the one's willingness to remain confined for longer duration (χ^2 value of 9.134 at 3 df. with *P*-value = 0.028) (Table 6.1).

The COVID-19 infection has also affected the lives of female students; while the educational institutions are conducting various online classes, it was observed that

Table 6.2 Impact category based on composite scores

Impact categories	Frequency	Percentage
7–11 (low)	33	20.89
11–15 (medium)	48	30.38
15–19 (high)	48	30.38
19–25 (very high)	29	18.35

nearly 33% of the female students had missed classes in order to perform household chores while their male siblings attended classes (Fig. 6.4c). This shows that the pandemic has greatly affected the learning capacity of the female students by increasing the workload, which may have long time consequences. Women worldwide, especially those working in the informal sector, have been adversely affected by the pandemic. Moreover, the lockdown measures that have severely affected the tourism industry and other service industries will take a long time to recover, which may negatively affect the women engaged in that category (Chang 2020). The majority of the respondents (66.5%) felt that the lockdown had little or no economic impact on them. However, 32.3% of the respondents lost their savings; another 2.5% could not pay the loan EMIs, 3.8% of the respondents have found it challenging to buy food items, 2.5% of the respondents have been severely affected by job loss, and another 3.8% are in danger of losing their business (Fig. 6.4d).

The effect of lockdown induced by the second wave upon the social and economic life of the women was further explored by nine questions based on a four-point Likert scale. These nine Likert items were employed to calculate the composite score. The reliability of the Likert scale was checked by using Cronbach's Alpha. The α-value of 0.707 can be considered acceptable for determining the internal consistency of the composite score. The following questions were asked to the respondents:

(Q1) How likely are you going to tolerate the lockdown?
(Q2) How difficult was it for you to avail sanitary pads during the lockdown?
(Q3) How often were you forced to perform physical relationships against your wish during the lockdown?
(Q4) How difficult was it for you to work from home and look after the children during the lockdown?
(Q5) How has the workload of household chores increased recently during the lockdown?
(Q6) How difficult was it for you to manage the home during the lockdown?
(Q7) How has the increased workload affected you mentally during the lockdown?
(Q8) How often have you had heated discussions with the family members during the lockdown?
(Q9) How has the lockdown affected you financially?

The ranks of each question given by each respondent were summed up to develop a composite score. It was subsequently categorized into four impact categories, i.e., low, medium, high, and very high (Table 6.2). It has been observed that only 20.89% of the women fall under the low impact category, while 18.35% of the women have experienced a very high impact, and 30.38% of the women fall under both the medium and high impact groups.

6.4.4 *Impact on the Working Environment*

With the lockdown in effect and only a few services like health, police, fire, NGOs, etc. functioning, many women of the GTA area employed in these services (Fig. 6.5a) have increasingly found it difficult to report to their duties as most of the vehicles have remained closed (Fig. 6.5b). Many women have reported that as their place of residence was far or very far from the place of work (Fig. 6.5c), it became challenging for them to discharge their duties properly. The ability to regularly attend the offices depends upon the availability of transport and distance from the place of work as they have a statistically significant relationship with each other (Fig. 6.5d). The survey also observed that only 16.5% of the women were provided PPE suits at their workplace (Fig. 6.6a), but surprisingly 64.9% of the women confessed that they never wore PPE even when PPE suit was provided (Fig. 6.6b); this may be due to the fact that the majority of the women felt that the suit was not designed for the women (Fig. 6.6c) and reported it to be very uncomfortable

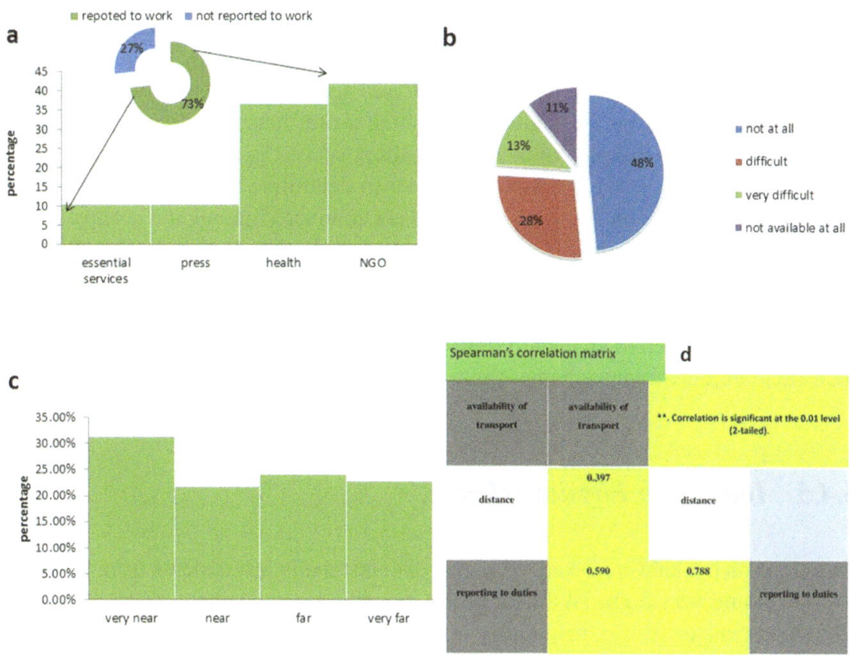

Fig. 6.5 (**a**) Service that opened during lockdown; (**b**) Availability of vehicles; (**c**) Place of work from home; (**d**) Correlation matrix

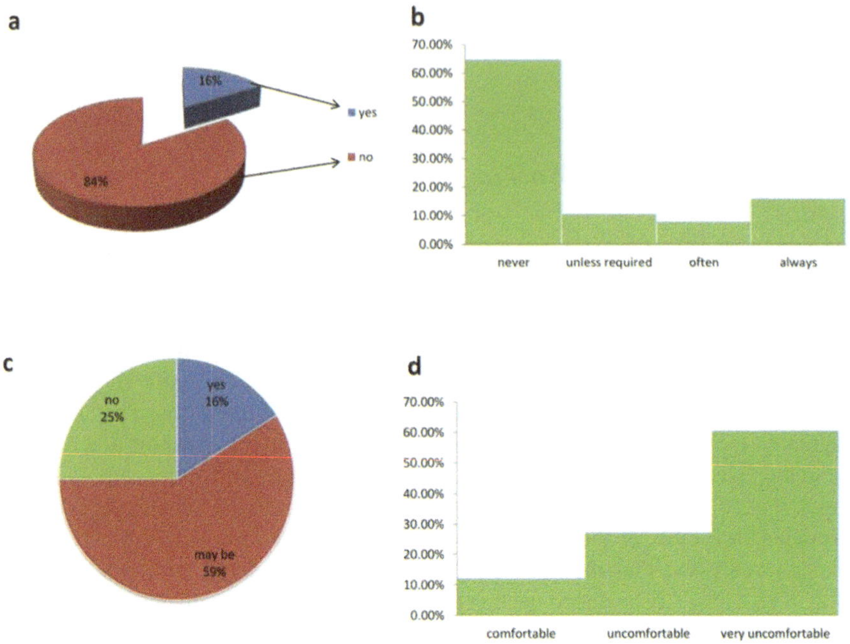

Fig. 6.6 (**a**) PPE provided at work; (**b**) Utilization of PPE; (**c**) Opinion on PPE; and (**d**) Level of comfort of PPE

during the menstruation period (Fig. 6.6d). It was also found that the comfort level of wearing PPEs and how frequently the women wore them significantly differs as determined by the Kruskal–Wallis *H*-Test with χ^2 of 39.136 at 2 degrees of freedom with a *P*-value of <0.001. However, in order to determine the variation within the groups, a series of Mann–Whitney *U*-tests were conducted among the groups. It was observed that there was no significant statistical difference between uncomfortable and very uncomfortable, but the difference between comfortable with uncomfortable and comfortable with very uncomfortable was statistically significant (Table 6.3). Thus it can be inferred that the comfort level of the PPE suit, especially during menstruation, determines whether the women will frequently use it or not.

6.4.5 *Impact on Physical Health*

Questions on health were asked to the respondents; many respondents from the high-risk factor and very high-risk factor group reported having diabetes (42%), suffered from hypertension (34%), respiratory illness (15%), and immune-suppressant disorder (Fig. 6.7a). In addition to this, 34% have admitted to having smoked, which makes them vulnerable to chest infection. All these comorbidity factors make them susceptible to COVID-19 infection. Around 21% of these high-risk factor group and

Table 6.3 Association of different variables

Sl. No	Variables	Test	Results
1	Comfort level of PPE and how frequently it was worn	Kruskal–Wallis test	$\chi^2 = 39.136$, 2 d.f., P-value = <0.001
2	Comfortable and uncomfortable	Mann–Whitney U	$Z = -4.807$, P-value = <0.001
3	Comfortable and very uncomfortable	Mann–Whitney U	$Z = -6.116$, P-value = <0.001
4	Uncomfortable and very uncomfortable	Mann–Whitney U	$Z = -1.753$, P-value = <0.080

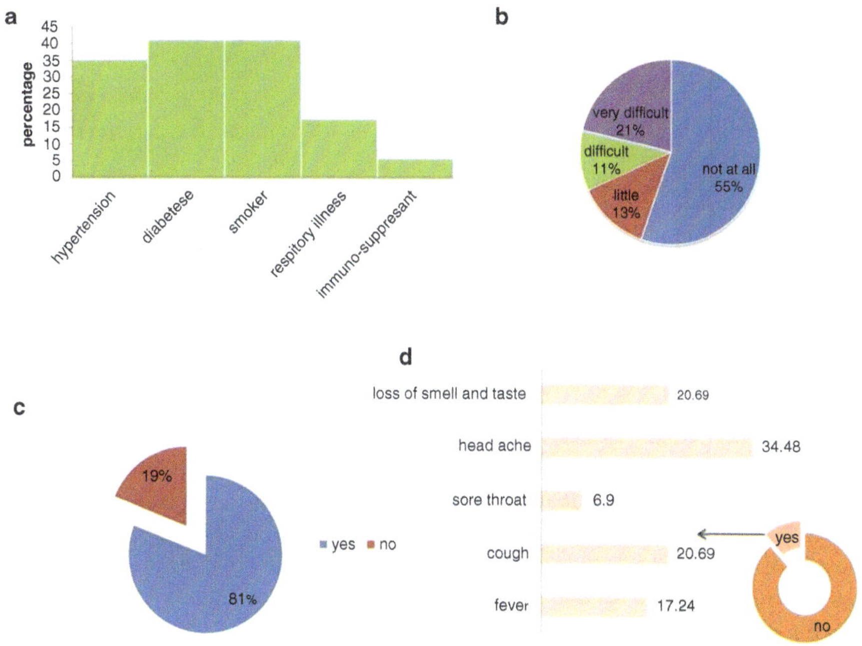

Fig. 6.7 (**a**) Comorbidity factors; (**b**) Accessing health services; (**c**) Self isolation; and (**d**) COVID-related symptoms

very high-risk factor group people found it very difficult to avail health services during the recent lockdown (Fig. 6.7b). When the respondents were asked whether they could adequately isolate themselves if they were diagnosed with COVID-19 infection, 81% of the respondents said they could, while only 19% replied that they would be unable to do so (Fig. 6.7c); the capacity to isolate oneself is not significantly related to any other variables. 87.3% of the respondents replied that they didn't have any COVID-related symptoms against 26.7% who had (Fig. 6.7d), it was

	Tested for Covid 19	Duration of illness	Covid 19 Symptoms	Risk Group	Nutritious meal	Level of information	Vaccine in future	Age	Self isolate	Do you have children
Tested for Covid 19		.012	.298**	.172*	-.076	.112	-.090	.093	-.018	-.004
Duration of illness			.309**	.085	-.012	.044	.078	.173*	-.023	-.155
Covid 19 Symptoms				.078	-.116	.151	-.101	.211**	.050	-.148
Risk Group					-.041	.125	.260**	.189*	-.039	-.129
Nutritious meal						-.080	-.057	-.210**	.103	.000
Level of information							.169	.009	.066	.002
Vaccine in future								.205*	.121	-.044
Age									-.037	-.363**
Self isolate										-.133
Do you have children										

*. Correlation is significant at the 0.05 level (2-tailed).

** Correlation is significant at the 0.01 level (2-tailed).

Fig. 6.8 Correlation coefficients

also observed that those who had identified themselves as having COVID-19 symptoms were sick for a longer duration of time and were more likely to get COVID-19 test done (Fig. 6.8). The people belonging to high and very high-risk groups had a higher tendency to get tested and get vaccinated in the future than the others. It was also observed that the age of the respondents was a significant factor as concerned to the impact of second wave infection of the COVID-19. The people belonging to a higher age group (i.e., >40 years), although most likely to be vaccinated than the lower age group, are in a very high and high-risk group due to the presence of numerous comorbidity factors and their poor choice of diet, they were more likely to exhibit COVID-19 symptoms and be sick for a longer duration of time as compared to younger age groups.

6.5 Conclusion

The study has tried to understand the effect of the pandemic and lockdown upon women of Darjeeling Himalayan region, GTA West Bengal, India. From the above analysis, it has been understood that the pandemic has affected women's lives in various spheres. The lockdown has severely impacted the lives of female students. With educational institutions remaining closed and all classes being conducted online, they have found it difficult to balance studies with household chores; many have reported having missed studies to attend the household works compared to their male siblings. This severely disables the learning capacity of the female students and may cause disadvantages during the exams where their performance may be lower than that of their male counterparts. The female students seem to have been uniquely disadvantaged by the lockdown and pandemic. In context to social contact behavior during the second wave of COVID-19 infection, it seems that the women have become self-aware of their risks, and high-risk females have started following the COVID protocols of social distancing and isolation, yet it has been observed that most of them have come in contact with the family members and relatives. If any of the male members within the family gets infected, then the woman of that family has a higher chance of getting infected. This is especially true in the case of younger females in low-risk groups, as they were more likely to go out or come in contact with infected persons. With lockdown measures in force, the female workforce seems to be significantly affected than the males, as the females are primarily employed in the hospitals and nursing homes as compared to other sectors. The non-availability of vehicles during the lockdown greatly causes inconveniences to the females. Another sphere where women are greatly disadvantaged is wearing protective clothing like PPE suits. As it was evident from the study, most women found it very uncomfortable during the menstruation period, which caused many women not to wear at all, thereby increasing their risks of contracting COVID-19.

Moreover, in those services like health, where one needs to wear a PPE suit for an extended period of time, menstruating women may find it very annoying and cause a higher level of distress. Although most of the respondents in this study reported minimum economic impact, there were some significant economic effects like closing off of the businesses and job losses reported by a few. This needs to be studied further. The pandemic has also significantly impacted women's lives belonging to the 40 years and above age group. Although they seem to exhibit more rational social behavior, the higher prevalence of comorbidity factors and lack of nutritional diet makes them more vulnerable. Lastly, around 48.73% of the women falling under the high to very high impact category are the ones who have rated higher mental stress, forceful physical intercourses, heated discussions, increased workloads, and financial losses. Therefore, the preventive measures adopted by the government like lockdown and distribution of PPE kits to workers in emergency services, although they seem gender-neutral on a surface level, tend to cause significant disadvantages to the women. These measures should adopt some degree of gender sensitivity. The government should set up counseling cells and women helplines where support and counseling can be provided to the women who have face mental stress and physical violence. Broadcasting of messages that create awareness through mobile service providers should be done regarding physical, verbal, and sexual violence so that women may be able to identify it correctly and report it. Female students should be involved in framing the class routines. The routines should be framed by keeping the concerns of the female students rather than the authorities' concerns. The concerned authorities of the health and emergency services that employ a large number of female workforce should keep an inventory of the biological cycle of the female colleagues, and no duties should be assigned to them during that period. Moreover, accommodations should be arranged during the lockdowns from the employer's part for those who stay far away from the workplace. Also, a range of flexibility is to be shown in exempting the breastfeeding women from the duties and responsibilities whatever possible. Gender audit of the various government departments, as well as that of private enterprises, should be routinely carried out to ascertain whether the workplaces are women-friendly or not; if found wanting, then the recommendations mentioned above can be taken in order to make the workplaces more sensitive to women issues and concerns.

References

Adekunle IA, Onanuga AT, Akinola OO, Ogunbanjo OW (2020) Modelling spatial variations of coronavirus disease (COVID-19) in Africa. Sci Total Environ 729:138998. https://doi.org/10.1016/j.scitotenv.2020.138998

Basu A, Roy A, Hazra AK, Pramanick K (2020) Analysis of youth's perspective in India on and during the pandemic of COVID 19. Soc Sci Q. https://doi.org/10.1111/ssqu.12839

Biswas B, Roy R, Roy T, Chowdhury S, Dhara A, Mistry K (2020) Geographical appraisal of COVID-19 in West Bengal. India Geo J. https://doi.org/10.1007/s10708-021-10388-4

CDC (2020) Coronavirus (COVID-19): symptoms of coronavirus. Centers for Disease Control and Prevention. https://www.cdc.gov/coronavirus/2019-ncov/symptoms-testing/symptoms.html. Accessed 20 May 2020

Chang WH (2020) Understanding the COVID 19 pandemic from a gender perspective. Taiwan J Obstet Gynecol 59:801–807. https://doi.org/10.1016/j.tjog.2020.09.004

Davies SE, Bennett B (2016) A gendered human rights analysis of Ebola and Zika: locating gender in global health emergencies. Int Aff 92:1041–1060. https://doi.org/10.1111/1468-2346.12704

Desjardins MR, Hohl A, Delmelle EM (2020) Rapid surveillance of COVID-19 in the United States using a prospective space-time scan statistic: detecting and evaluating emerging clusters. Appl Geogr 118:102202. https://doi.org/10.1016/j.apgeog.2020.102202

Folayan MO, El Tantawi M, Vukovic A, Schroth R, Gaffar B, Al-Batayneh OB et al (2020) Women's economic empowerment, participation in decision-making and exposure to violence as risk indicators for early childhood caries. BMC Oral Health 20(1):54. https://doi.org/10.1186/s12903-020-1045-5

Giri P (2020) Darjeeling logs maximum COVID-19 cases, deaths in North Bengal; experts blame trading hub Siliguri. Hindustan Times. https://www.hindustantimes.com/india-news/darjeeling-kalimpong-open-for-tourists-sikkim-to-follow-suit-in-october/story-hcGm0Atz1DUEVq5DhKFsUO.html

Gupta D, Biswas D, Kabiraj P (2021) COVID-19 outbreak and urban dynamics: regional variations in India. GeoJournal. https://doi.org/10.1007/s10708-021-10394-6

Kapoor M, Agrawal D, Ravi S, Roy A, Subramanian SV, Guleria R (2019) Missing female patients: an observational analysis of sex ratio among outpatients in a referral tertiary care public hospital in India. BMJ Open 9:e026850. https://doi.org/10.1136/bmjopen-2018-026850

Kar SK, Ransing R, Arafat SMY, Menon V (2021) Second wave of COVID-19 pandemic in India: barriers to effective governmental response. EClinical Medicine 36. https://doi.org/10.1016/j.eclinm.2021.100915

Kodge BG (2021) A review of current status of COVID19 cases in Maharastra state of India using GIS: a case study. Spat Inf Res 29:223–229. https://doi.org/10.1007/s41324-020-00349-3

Kumar A (2021) To fight the deadly second wave, a lockdown is the only available strategy. The wire. https://thewire.in/government/to-fight-the-deadly-second-wave-a-lockdown-is-the-only-available-strategy

Liu N, Zhang F, Wei C, Jia Y, Shang Z, Sun L, Wu L, Sun Z, Zhou Y, Wang Y, Liu W (2020) Prevalence and predictors of PTSS during COVID-19 outbreak in China hardest-hit areas: gender differences matter. Psychiatry Res 287. https://doi.org/10.1016/j.psychres.2020.112921

Oliver N, Barber X, Roomp K (2020) The COVID19 impact survey: assessing the pulse of the COVID 19 pandemic in Spain via 24 questions. J Med Internet Res 22(9):e21219. https://doi.org/10.2196/21319

Pandey V, Nazmi S (2021) COVID-19 in India: why second wave is devastating. BBC News. https://www.bbc.com/news/world-asia-india-56811315

Pinchoff J, Santhya K, White C, Rampal S, Acharya R, Ngo TD (2020) Gender specific differences in COVID-19 knowledge, behavior and health effects among adolescents and young adults in Uttar Pradesh and Bihar, India. PLoS One 15(12):e0244053. https://doi.org/10.1371/journal.pone.0244053

PTI (2020) Bengal tour operators left in lurch as travelers opt for direct bookings. Business Standard. https://www.business-standard.com/article/current-affairs/bengal-tour-operators-left-in-lurch-as-travellers-opt-for-direct-bookings-120123000840_1.html

Sanyaolu A, Okorie C, Marinkovic A et al (2020) Comorbidity and its impact on patients with COVID-19. SN Compr Clin Med 2:1069–1076. https://doi.org/10.1007/s42399-020-00363-4

Sarkodie SA, Owusu PA (2020) Investigating the cases of novel coronavirus disease (COVID-19) in China using dynamic statistical techniques. Heliyon 6(4):e03747. https://doi.org/10.1016/j.heliyon.2020.e03747

Sinha A (2021) A year with COVID-19: peak behind India. The Indian Express. https://indianexpress.com/article/explained/a-year-with-covid-peak-behind-india-7167126/

The Novel Coronavirus Pneumonia Emergency Response Epidemiology Team (2020) The epidemiological characteristics of an outbreak of 2019 novel corona virus disease (COVD-19). China CDC Weekly 2:113–112. http://weekly.chinacdc.cn/en/article/doi/10.46234/ccdcw2020.032

Varshney M, Parel JT, Raizada N, Sarin SK (2020) Initial psychological impact of COVID-19 and its correlates in Indian Community: an online (FEEL-COVID) survey. PLoS One 15(5): e0233874. https://doi.org/10.1371/journal.pone.0233874

WHO (2020) Director-general's opening remarks at the mission briefing on COVID-19. https://www.who.int/dg/speeches/detail/who-director-general-s-opening-remarks-at-the-mission-briefing-on-covid-19-12-march-2020. Accessed 16 Mar 2020

Zhong BL, Luo W, Li HM, Zhang QQ, Liu XG, Li WT et al (2020) Knowledge, attitudes, and practices towards COVID-19 among Chinese residents during the rapid rise period of the COVID-19 outbreak: a quick online cross-sectional survey. Int J Biol Sci 16(10):1745–1752. https://doi.org/10.7150/ijbs.45221

Bishal Chhetri is an assistant professor in Department of Geography, Southfield College, Darjeeling. He has been teaching geography since 2009. He completed his BA (Hons) from Darjeeling Government College and MA from University of North Bengal. His area of interest is Developmental studies and has conducted UGC minor research project in this field. He has published number of national and international research articles and has co-edited one book.

Kabita Lepcha is an assistant professor in Department of Geography, University of Gour Banga, Malda. She completed her BA (Hons) from Loreto College, Darjeeling and MA from University of North Bengal with Fluvial Geomorphology as specialization; her field of interest lies in fluvial geomorphology and sociocultural studies. She has been pursuing her PhD from University of Gour Banga, Malda. She has published various research articles in both national and international journals and has published two books.

Chapter 7
Exacerbated Digital Education Divide and the Marginalized: Experiences from India

Arabinda Roy and Mukunda Mishra

Abstract Education links the individual with the socialization process and livelihoods opportunities, it works as the guiding principle of the nation-building process, and it is the beacon light for the world citizens to direct in which direction the world would change for good. Thus, education is treated as a right, and it sets the pathway for attaining the SDGs. This chapter discusses how the COVID-19 pandemic exacerbates the preexisting digital divide in India not only in its geographical perspectives but also vigorously to the underprivileged and marginalized population in India. The study is qualitative and exploratory in design which first, constructs the idea from the published sources, authentic datasets, and reports and then validates them with the field data collected through judgment sampling and expert opinion. The principle of binomial Receiver Operating Characteristic curve (ROC) from ordinal confidence rating data, through the maximum likelihood estimation, grounds the validation. The study depicts that the COVID-19 pandemic magnifies the digital education divide in the country to which the economically marginalized suffer the most. Finally, the study recommends some simple policy directions toward the resolution.

Keywords COVID-19 pandemic · Digital divide · Education · Marginalized sections · Development · Future strategy

7.1 Prologue to the study

The opening of this chapter finds it most relevant to quote what the United Nations Educational, Scientific, and Cultural Organisation (UNESCO 1996), in one of its reports on education, mentioned about education having its four pillars:

A. Roy (✉) · M. Mishra
Department of Geography, Dr. Meghnad Saha College, Uttar Dinajpur, West Bengal, India

S. Saha et al. (eds.), *Economic and Societal Transformation in Pandemic-Trapped India*, New Frontiers in Regional Science: Asian Perspectives 55,
https://doi.org/10.1007/978-981-16-5755-9_7

"Learning to live together

Learning to know

Learning to do

Learning to be."

It opines that an education system should lead to build the capacity to absorb new knowledge (i.e., *to know*), to gainful employment (i.e., *to do*), to the building of a unique all-round personality (i.e., *to be*), and perform the vital function of social integration (i.e., *to live together*). It would make education an instrument of socialization. Through its sociocultural context, education empowers people in such a way that no other process could be its alternative.

Education has become one of the most delicate indicators of life outcomes, a strong predictor of attitudes and well-being, and helps individuals generate awareness, stain efficient living conditions, meet livelihood opportunities, and turn out into moral and ethical human beings. Education is the guiding principle of developing a nation, the tools for climbing up the social ladder, and perspectives about how the world can change for good. Since time immemorable human beings on the earth felt the *importance* of education; the renaissance proposed it to be the *right* of humankind; the modern civilization crowns it as the key to Sustainable Development as evident by its role acknowledged in the SDGs.

The essential conditions for sustainable development are the existence and implementation of human rights. Development is the premier part of the change in human existence, and the existence of human beings and their welfare is possible only by protecting human rights. However, the process of development requires a broad swath of wisdom, a barrier-free learning ecosystem, and the active participation of populations even at the furthest corner of the globe. Sen (1999) speculates *freedom* as the primary objective and the principal meaning of development. Achieve freedom requires the human world to be free from any kinds of discrimination that are guaranteed by rights granted. Finetuned with the same, Article 1 of the Declaration of Right to Development states:

> "The right to development is an inalienable human right by virtue of which every human person and all peoples are entitled to participate in, contribute to, and enjoy economic, social, cultural and political development, in which all human rights and fundamental freedoms can be fully realized." (Declaration on the Right to Development 1986)

Development is necessary to achieve a prosperous world, and education is the cornerstone of development in its truest sense. Therefore, education is the pathway to achieve social change and provisioning a better world. The 17 Sustainable Development Goals (SDGs) recognize that ending poverty and other deprivations must be concordance with strategies that improve health and education for the long run, with discouraging inequality and spurring economic growth might be other essentials (United Nations 2015). Education as a human right has made its place in Article 26 of the UDHR, 1948, which states that:

> "Everyone has the right to education. Education shall be free, at least in the elementary and fundamental stages. Elementary education shall be compulsory. Technical and professional

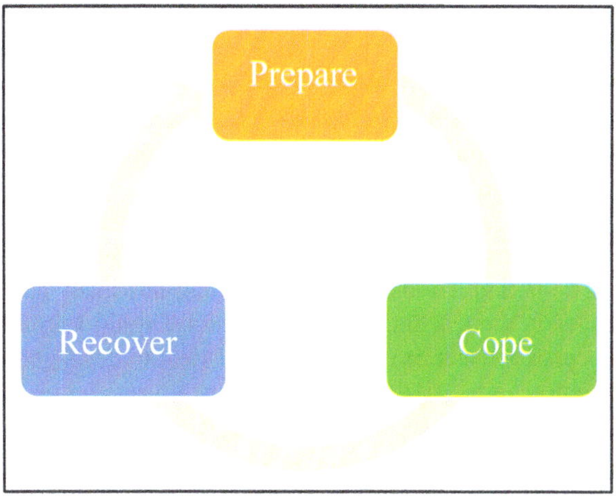

Fig. 7.1 A cyclical approach of education in emergencies (**Source:** UNESCO 2020a)

education shall be made generally available, and higher education shall be equally accessible to all on the basis of merit." (Universal Declaration of Human Rights 1948)

The UN Economic and Social Council (1999) describes education as the primary vehicle for economically and socially marginalized adults and children in getting themselves lifted out of poverty and find avenues to participate fully in their communities. Education does not confine itself articulating just the practicality; instead, an educated, enlightened and active mind, in its most pleasing sense, could rejoice the freedom, generosity that is one of the joys and rewards of human existence on this beautiful planet.

The above clarifications sought education as an instrument of socialization, means of livelihood opportunities, and a principle of the nation-building process. However, the whole socioeconomic and ecological complex on the earth is currently facing the COVID-19 pandemic outbreak, which carries the critical threat for the target-based long-run developmental process of any region and the education system is no exception.

The COVID-19 pandemic and lockdown is a phase of the crisis in the educational sector, which is a part of the cyclical process that UNESCO (2020a) mentioned as the "*preparing—coping—recovery*" cycle to mitigate the impact of emergencies on education. "Preparing" is the response in anticipation of a known or unknown crisis. For an unprecedented emergency like COVID-19, most of the world states had no preparation. When amidst the crisis, they move to "coping" and thus, implementing management measures to minimize the shock. During the outbreak, human health has been the priority as the lives of the citizens are most precious to the government. A highly populous state with the absence of fundamental health infrastructure at its vast territory and remote corners made the initial health shock heavier for India. As the emergency phase dissipates, the government moved into a "recovery" mode by implementing policies and measures to regain lost time (Fig. 7.1).

The emerging situation prompted experts and policymakers to rethink the conventional mode of education. The consequence is very prominent - digital technology swallows the education sector, shifting the traditional white chalk and blackboard method of teaching to digital learning methods. In India, while the former method has long been successful concerning the truest objective of education to reach the mass, the latter is yet over-advertized as interesting, personalized, and enjoyable, however, not tested for its long-term effectiveness, instead looks promising in terms of its reach to the audience (obviously, who could afford) even in the crisis period. Globally, digital learning has many advantages as well as disadvantages in the teaching-learning process. Digital learning has more learning engagement experience, no physical boundaries; students get to learn in the confines of their comfort zone. All these are significant advantages on the one hand, and limited face-to-face interaction and difficulty in lab-based practical classes are the major limitations and challenges on the other hand.

The sudden shift of the teaching-learning method leads not only the students in a dilemma but also the parents, teachers, curriculum makers, material suppliers, and the management in the haul. The pandemic situation and the closure of educational institutions create an invisible divide of public-private modes of education and the privileged-unprivileged divide. While private schools are conducting online classes during the pandemic periods, most of the students in government schools find themselves left behind compared to their privileged peers.

The COVID-19 pandemic has an adverse effect on the population associated with the education system. The worst sufferers are the learners. UNESCO (2020b) estimates that about 32 crores of students are affected in India, including those in schools and colleges.[1] The hard-hit population is the students in the primary classes (Fig. 7.2).

Now, let us prelude, though brief, about "marginality." *Marginality* is an "experience" that affects millions of people all over the world. While the main motto of development is to create a good, healthy, and creative environment, marginalized populations have relatively little control over their lives and the resources available to them, disabling them to contribute to the nation-building process. Therefore, addressing the issue of marginalization is meaningful when the digital divide is concerned.

In general, the term "marginalization" indicates the overt actions or tendencies of human societies, where people whom the society perceives to be undesirable or without useful function are excluded, i.e., marginalized. These people are marginalized from a group or community for their own protection and integration and are known as *marginalized groups*. This restricts their opportunities and means for survival. From the cultural studies perspective, marginalization is defined as the implicit behavior or behaviors of human cultures where individuals are disqualified,

[1]See the UNESCO website, https://en.unesco.org/COVID19/educationresponse

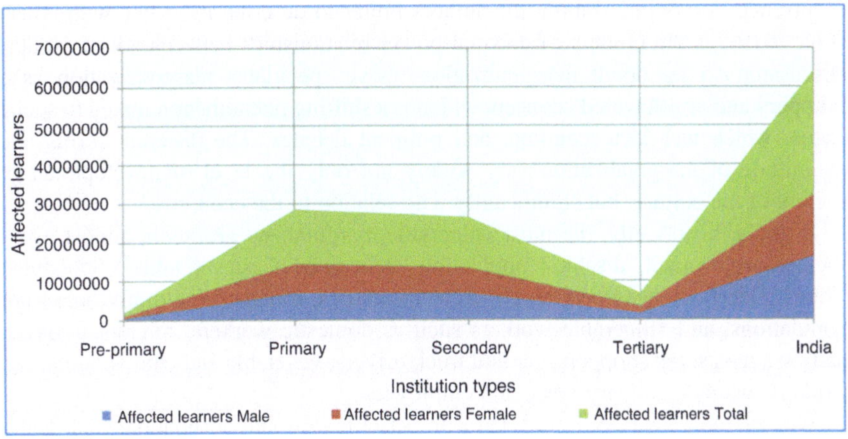

Fig. 7.2 COVID-19 pandemic affected learners in India (**Source**: Prepared by author based on UNESCO 2020a)

i.e., oppressed, viewed as unwanted, or without effective use. Such individuals, who are under oppression, form *marginalized communities*.[2]

While Leonard (1984) describes marginality as "[...] being outside the mainstream of productive activity and/or social reproductive activity", the Encyclopedia of Public Health (2017) defines marginalized groups as "To be marginalized is to be placed in the margins, and thus excluded from the privilege and power found at the center."[3]

UNICEF[4] points out that the marginalized community includes women, the elderly, adolescents, youth and children, persons with disabilities, indigenous populations, refugees, migrants, and minorities. There are ample shreds of evidence which anyone could find in history that marginalized people become more vulnerable in emergencies because their minimal resource base disallows them from coping with the adverse situation. It is not only that the marginalized communities are at risk of coronavirus infection, but they are also susceptible to economic exclusion, joblessness, and lack of food and shelter during this troubled time.[5]

Straightforwardly, this study calls all such people "marginalized," which unconditionally includes all who are excluded from mainstream social, economic, educational, and cultural life. The examples of the marginalized include, but are not limited to, groups excluded due to race, gender identity, sexual orientation, age, physical ability, language, and immigration status. Out of different theories about the

[2] See the details in: https://blog.ipleaders.in/situation-marginalised-sections-amid-lockdown-legal-relief/

[3] For details: https://www.lawctopus.com/academike/problems-marginalized-groups-india/

[4] See the details in: https://www.unicef.org/ukraine/en/documents/COVID-19-how-include-marginalized-and-vulnerable-people-risk-communication-and-community-engagement

[5] See the details in: https://www.gofundme.com/c/blog/marginalized-communities

occurrence of marginalization, the authors prefer to describe it, in line with (Baah et al. 2019), as the product of unequal power relationships between social groups. Discussion so far about marginalization above speculates marginalization as a "slippery and multilayered" concept which is a shifting phenomenon linked to social status, which has socioeconomic and political debates. The different forms and magnitude of marginalization vary widely not only the level of development of society or the culture; but significantly with relation to the economy.

In India, historically, the most vulnerable marginalized groups in almost every society are women, disabled population, Scheduled Castes (Dalits), Scheduled Tribes, elderly or aged people, children, minorities, refugees, culturally backward populations, and vulnerable workers such as domestic workers, migrant workers, daily wagers, scrap shoppers, car and truck drivers, vegetable and fruit vendors, sex workers, waiters, delivery boys, and employees.

Women have long been excluded from specific jobs and occupations and facing gender inequality. People with disabilities have long been battled against biased assumptions, harmful stereotypes, and irrational fears. Socially and economically disadvantages Scheduled Tribes (ST) and the Scheduled Castes (SC) have been witnessing structural discrimination in Indian society. The elderly population's vulnerability appears from increased incidence of illness, old age disability, and economic dependency upon their spouses and kins. Children of low-income families face nutrition deficiencies, lack of access to healthcare, unhygienic living ambiance, and exclusion from quality education. In India, a girl child has ample pieces of evidence of being treated discriminately and differential access to nutritious food and gender-based violence. Growing numbers of children trafficking to the neighboring countries also pose challenges to Indian society and governance (UNICEF 2020).

The pandemic has impacted individuals at the lower end of the spectrum. The laborer working in the informal and the unorganized sectors and micro, small, and medium enterprises (MSMEs) are worst shocked as they lost their temporary job to face deprivation of their livelihoods. The pandemic carries with it the long-term livelihood shocks and occupational displacement that the poor require to adjust newly—economically and socially (Singh et al. 2020). In addition, the financial crises of the population living in the lower end of the spectrum have led to a substantial increase in anxiety and depression, drug addiction, and violence against women and children (World Health Organization 2020a; Ratnam 2020). The marginalized have long been the worst sufferer in terms of socioeconomic, historical, cultural, legal, political, and religious dimensions; and, they are most vulnerable to the shock of the pandemic—not in terms of the fatalities due to infection, rather manifold sufferings beyond the mere health hazard, which carries long-term plight for these people.

With this backdrop, this chapter explores the potential consequences of the COVID-19 pandemic and nationwide lockdown on India's education system, which subsequently shapes the socioeconomic well-being of the population. The present study is also vital to describe how the COVID-19 pandemic is responsible for exacerbated digital divides not only in its geographical perspectives but also in underprivileged and marginalized sections in India. Finally, the study finds some policy approaches and suggests repairing the damage of the education system in rural India and a future strategy.

The research design is exploratory and qualitative in design, based on various reports of national and international agencies on the COVID-19 pandemic. Secondary data and information presented in the current study have been collected from different authentic websites, journals, and e-contents. Required shreds of literature were searched in PubMed[6] and Google Scholar platforms.

Alongside this, the experiences we gained through the socioeconomic survey on selected respondents of the Namasudra[7] community form the ground of validation of what we construct from our secondary sources. We have constructed the binomial ROC from ordinal confidence rating data through the maximum likelihood estimation to validate the situation we investigated from the secondary sources.

We feel it worthy of being mentioned here that the Namasudra community is recognized as a marginalized section by the Government of India. They practice their traditional livelihood status with poor socioeconomic backgrounds. The field data depicts how a marginalized community deals with pandemic circumstances. How one such community has coped with the pandemic conditions; what are their opinions regarding education in the pandemic days? While the field investigation is local, we believe that how such marginalized and traditional communities respond to the pandemic is similar globally.

7.2 Same Shock and Different Impacts: Marginalization Matters

The World Health Organization (WHO) (World Health Organization 2020b) declared the crazy spread of COVID-19 as the global pandemic on 11 March 2020; Wuhan province of China, while witnessing the earliest surge of the pandemic, hundreds of thousands of Indians were protesting all over the country against the CAA.[8] Amidst the news of sporadic cases of infection coming from different metros after the first cases had been recorded way back in January in Kerala, the authorities of the government of India were negating the presence of community transmission of the virus.[9] All of a sudden, on 24 March 2020, the government of India (GoI)

[6]PubMed is a free search engine accessing primarily the database in the health and medical sciences.

[7]Namasudra, A Scheduled Caste community in the province of Bengal, who was earlier known as Chandala or Chandal, a term usually considered as a slur. They are traditionally depressed and marginalized community occupied the lowest position in the rank of caste hierarchy.

[8]The Citizenship (Amendment) Act, 2019 was passed by the Parlament of India on 11 December 2019. The main purpose of this act is to give Indian citizenship to illegial migrants of Hindu, Sikh, Buddhist, Christian, Parsi and jain Community belong to Bangladesh, Pakistan and Afghanistan.

[9]According to the World Population Prospects: The 2012 Revisions, which is archived under the URL: https://population.un.org/wpp/Publications/Files/WPP2012_HIGHLIGHTS.pdf

Table 7.1 Trajectory of shutting down of schools in India in response to the first wave of COVID-19

Date of official closures of the institutions	Indian States/UTs
9 March, 2020	Karnataka (Shutdown of Kindergarden, pre-primary and primary schools)
11 March, 2020	Kerala
13 March, 2020	Chhattisgarh, Manipur, Odisha, Punjab, Puducherry, Karnataka (shutdown of universities and colleges)
14 March, 2020	Rajasthan, Himachal Pradesh, Madhya Pradesh, Maharashtra, West Bengal
15 March, 2020	Goa, Gujarat
17 March, 2020	Telangana, Sikkim, Uttar Pradesh, Chandigarh
18 March, 2020	Andhra Pradesh
20 March, 2020	Tamil Nadu
21 March, 2020	Delhi
22 March, 2020	Bihar
24 March, 2020	Haryana

Source: Compiled by the authors from respective state government's web portals

declared a 21-days nationwide lockdown "for public interest and to control the spread of the COVID-19 pandemic." The community transmission already began to grow by that time, which subsequently appeared as profound health, economic, and social crises to India's population.

In the mid of May 2020, the country has taken to partial lockdowns and unlocking in specific sectors of economic involvement, but the education sector in India is continuing to close (Table 7.1). As a result, the future human resource of India, which is directly or indirectly associated with the education sector, has been pushed to a state of complete mayhem.

As the Government of India (GoI) has decided to shutdown the educational institutions to control the spread of COVID-19, the temporary closing of educational institutions has shifted the teaching-learning method from paper to digital. However, while the government had started to develop some measures for learning to continue through digital/online platforms, like television and radio: a most far-reaching experiment in the history of the education system (Giannini et al. 2020) and started to develop the quality of education, the access once again has been put into question.

The education sector in India is a matter of great concern as the youth population, who contributes a significant part to the population structure, stays away from the direct-contact-based teaching-learning system for a prolonged time ever. The age composition of India is the key element that could reflect the source of anxiety (Fig. 7.3). Census of India (2011a) enumerated that children aged between 5 and 9 years contribute 10.22%, and the young population aged between 10 and 19 years

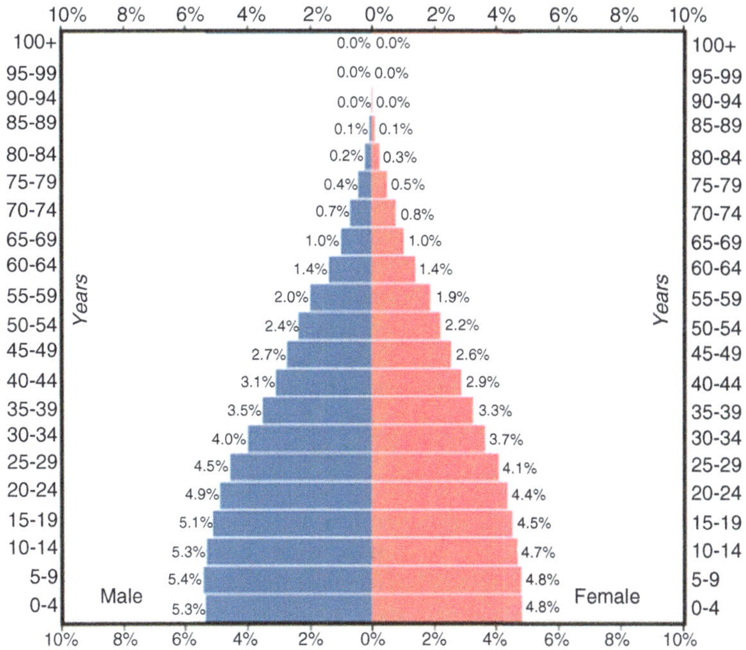

Fig. 7.3 Age-sex structure of India (**Source**: Based on the Census of India 2011a datasets)

share nearly one-fifth (19.64%) of the total population of India. This figure is estimated to have reached 34.43% of the total by the end of 2020.[10] This plenty of youthful population guarantee India's potential growth if given proper implementation and exposure of education and employment. However, the likelihood of the youth population remaining underrepresented and staying inactive from the education sector amidst this prevailing pandemic may challenge future growth quotients.

Preceded by China, India has the world's second-largest school system (Trines 2018). The Indian education system has been witnessing a prolonged closure in its classical "chalk and talk" teaching method due to the emerging pandemic-induced nationwide lockdown that is estimated to have adversely affected 320 million learners in India due to the system is transitioned to e-learning mode (Modi and Postaria 2020).

India has long been dismal to the issues of economic inequalities and social disparities, which set the pretext of the digital divides where the present pandemic deepens the digital education divide in India. The rapid shift to e-learning from the traditional teaching method prompted by the pandemic has resurfaced longstanding regional and household disparities in access to the internet and technology and also

[10]The figures have been mentioned in Youth of India, 2017 a MoSPI Report, p 13, which is available under the URL:http://mospi.nic.in/sites/default/files/publication_reports/Youth_in_India-2017.pdf

created a digital divide in education. However, the pandemic does not confine itself, affecting the learning outcomes in the short run; instead, it brings forth a considerable loss of human capital due to extended school closers and reduced economic opportunities in the long run (Modi and Postaria 2020).

NSSO (2014) statistics confirm that 32 million children had already been out of school in pre-pandemic India. What is more shocking is that the majority of those belong to the socially disadvantaged class in the country (NSSO 2014). Availability and accessibility of technology is a critical issue for learning education to socially disadvantaged students. However, the majority of the students from the marginalized and disadvantaged sections have no technology to support them take advantage of online education. Besides this, little or no availability of electricity is a significant challenge to them. A recent 2017–18 survey conducted by the Ministry of Rural Development, GoI, observed that only 47% of Indian households receive more than 12 hours of electricity, and more than 36% of schools in India operate without electricity (Kundu 2020). Is not it sufficiently evident that a student from an underprivileged background would fail to take advantage of online education, which a student from a family with better means of living can easily carry?

The COVID-19 pandemic has affected the whole nexus of the population associated with education. India boasts of having more than 1.4 million schools and more than 227 million enrollment, and also, there are more than 36,000 higher education institutes (Jha and Shenoy 2016). The COVID-19 pandemic is a threat to world populations, specifically in developing countries like India, where a larger population lives in rural areas. This situation makes it sensible why analyzing the consequences of the pandemic on education takes the rural-urban perspectives. Notably, 68.84% of people of India still live in villages and only 31.16% in urban areas, out of which 17.4% live in the urban slums; along with, a colossal disparity exists in literacy rate (with 85% literacy in urban areas and only 68.9% in the rural counterpart) (Census of India 2011b).

The rural-urban distribution of the total number of educational institutions reflects the win-win situation (Table 7.2). At the end of the 2016 academic session, 180 million students enrolled in rural schools at all levels, contributing almost 72% of overall enrollment, are facilitated through 1,247,492 schools that share 85% of total schools of all levels. On the other hand, 28% of overall enrollment is associated with 15% of total schools that are concentrated in urban areas (U-DISE 2016). In addition, the rural schools have six million teacher representation, whereas two million teachers are working in urban schools. In the case of higher education, there is no clear distinction with Gross Enrolment Ratio (GER) as low as 26.3% (AISHE 2019).

The pre and post-pandemic equations are entirely different. The alternative mode of imparting education through leveraging communication technology through connecting students with their teachers through remotely accessible virtual platforms requires considering the digital access across the country. In India, while the urban internet density is 97.9, the rural areas are outright unfit (with an internet density of 25.3) with the new and so-called "smart" education system (Parsheera 2019). The "smart" education manifests the inevitability of smart gadgets (like smartphones and

Table 7.2 Rural–urban distribution of educational institutes in India

Location	Primary schools (No.)	Upper primary Schools (No.)	Secondary schools (No.)	Higher secondary schools (No.)	Total schools (No.)	School ratio	Enrollment (million)	Enrollment ratio
Rural	756,336	136,334	26,877	5970	1,247,492	85%	180	71.71%
Urban	83,903	11,245	7056	5467	220,184	15%	71	28.29%

Source: U-DISE Flash Statistics 2016–2017

Table 7.3 Network strength of rural and urban households in India by residence and sex

	Households that have own computer (%)			Households with internet facility (%)		
	Rural	Urban	Total	Rural	Urban	Total
Residence	4.4	23.4	10.7	14.9	42	23.8
Gender	Persons of age 5 years and above who					
	can operate a computer (in %)		can use internet (in %)		have used internet in last 30 days (in %)	
	Rural	Urban	Rural	Urban	Rural	Urban
Male	12.6	37.5	17.1	43.5	14.6	40.4
Female	7	26.9	8.5	30.1	6.6	26.6
Total	9.9	32.4	13	37.1	10.8	33.8

Source: 75th round of National Sample Survey, conducted between July 2017 and June 2018

tablets), which were strictly prohibited items in the classrooms and examination halls until the *new normal* in the education system started prevailing in the country. However, this is not the end; the smart gadget needs high-speed data services to connect the online classes. Table 7.3 is self-explanatory to the ground realities based on which one needs to presume the future of the digital divides that looms large in the education system of the pandemic struck India.

The Key Indicators of Household Social Consumption on Education in India (which is based on the 2017–2018 NSSO 2014) confirms that fewer than 15% of rural Indian households have internet access (as opposed to 42% urban Indian households). Only 13% of the people (aged above 5 years) surveyed in rural areas, and just 8.5% of females (aged above 5 years) surveyed reported could use the internet. Girls in poor households face increased domestic duties that induce their inability to access online education because of either inadequate access to the internet and gadgets or the family prioritization of male children in receiving education. This silent exclusion of girl children of poor families results in child labor and child marriage. There is a substantial regional disparity to access the internet and computer operations among the states of India (Fig. 7.4). The availability of internet service and network strength in rural and urban India has a crucial role in accessing education in online mode during pandemic situations.

We will conclude this section by citing one of many such examples reported in ample in the newspapers. The availability of internet service in rural areas of Mizoram state is very irregular and sometimes poor. While the University of Mizoram has started online semester examinations for undergraduate students, they have to cross a few kilometers of hills and jungles to get access to the data network and attend online semester examinations.[11]

[11] Indian national daily newspaper, Anandabazer Partika has published articles on at the top of the hill to give the test, circulated on 07 June, 2020, which is available under the URL: https://www.anandabazar.com/india/camps-are-being-set-up-in-hill-tops-so-that-students-could-give-their-examination-in-online mode/cid/1285481

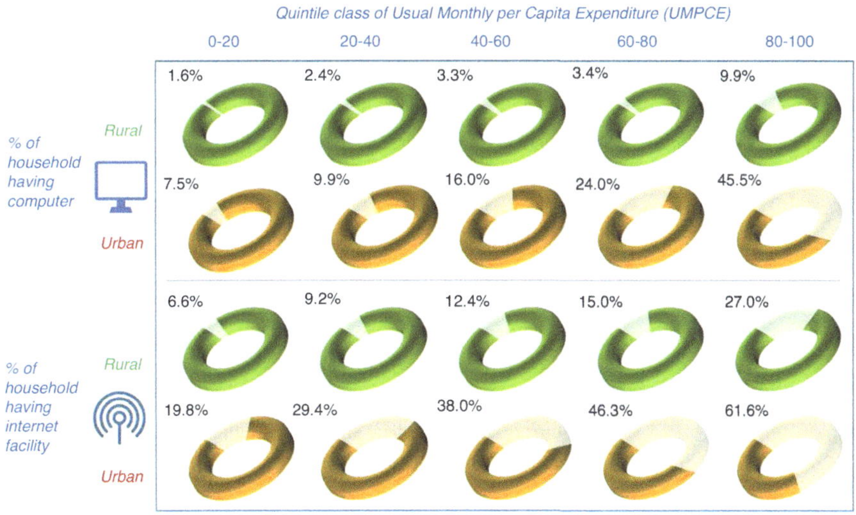

Fig. 7.4 Rural–urban disparity of the households' possession of computers and access to internet facility across the quintile classes of UMPCE in India (**Source:** 75th round of National Sample Survey conducted between July 2017 and June 2018)

7.3 Marginalized are the Fore Precipitates from the Costlier System: Inclusions Needing Priority

Alongside the prevailing swath of differential response in participation, the chances of enrolled students to precipitate out of the system also increases. A study by UNESCO, UIS, and GEM in 2015 reported 49.8 million students as out of school in their upper secondary age in India, out of which 2.9 million were out of school in the primary schooling age; and another 11.1 million in their lower secondary schooling age (UIS and GEM 2017). The National Institute of Educational Planning and Administration (NUEPA, 2016)[12] reported that the average dropout rate in India was 17.06% across secondary schools, and it is higher in rural areas comparing to urban areas. Although the present dropout scenario is disappointing, the overall dropout issues in India have been reducing since the 1970s, and the dropout scenario in primary education has been improving steadily (Table 7.4).

While the country undergoes a thorough move in the right direction, the short-term disruptions in schooling often lead to permanent dropouts among the children (Reddy and Sinha 2010). Thus, the children from the poor and marginalized people are precipitated from the system at the early stages of prevailing adverse situations.

[12]National institute of Educational Planning and Administration is a research institute set up by the Ministry of Education, Government of India.

Table 7.4 Dropout from primary education in India

Year	Dropout rate (% of primary school cohort)	Change
1971	64.1	
1973	61.7	−3.68%
1974	64.6	4.73%
1986	47.3	−26.87%
1987	38	−19.51%
1988	46.6	22.50%
1994	41.4	−11.23%
1995	42.7	3.13%
1999	38	−11.05%
2000	41	8.01%
2001	38.6	−5.77%
2013	18.2	−52.98%

Source: Knoema data atlas (The data atlas is available under this URL: https://knoema.com/atlas)

According to UNESCO,[13] approximately 0.32 billion students in India have been affected by school closures due to the COVID-19 pandemic, out of which almost 84% reside in rural areas while 70% attend government schools (Alvi and Gupta 2020). In the present circumstances, COVID-19 pandemic-induced nationwide lockdown is likely to reduce the earning capacity for many poor households and sometimes loss of parents' employments which subsequently increase the relative cost of sending children to school. Consequently, the COVID-19 pandemic leads the children into the labor market (Bharadwaj et al. 2019) and makes them child labor, especially in rural India.

On the other hand, the dropout rates are more severe for girls (Prakash et al. 2017) whose dropping out from the educational institutions may lead to an increase in child marriages, early pregnancies, domestic violence, and a plethora of other development issues (Birchall 2018).

7.4 Right of Children to Free and Compulsory Education: A Little Deviation may Cost Heavier for the Marginalized

The Constitution of India, in its original enactment, defined education as a state subject. An amendment was added in 1976 Under Article 42 of the Constitution that made education becoming in the concurrent list that enables the central government to legislate it in the manner suited to it. Besides, as India is a signatory to several

[13]The United Nations Educational, Scientific and Cultural Organization is a specialized agency of the United Nations aimed at promoting world peace and security through international cooperation in education, the science and culture.

international covenants (i.e., Jomtien declaration, UNCRC, MDG goals, Dakar declaration, SAARC, SDG Charter for children), it is committed to making education a reality for all children. On 01 April 2010, India, with its 86th Constitutional amendment, makes education a fundamental right that the Parliament passed in 2002. The new law obliges the state governments and local bodies to ensure that every child gets an education in a school in the neighborhood. The Constitution (Eighty-sixth) Amendment Act inserted Article 21A in the Constitution of India, stating education as the Fundamental Right for Children in the age group of 6–14 years:

> "[...] the State shall provide free and compulsory education to all children of the age of six to fourteen years in such manner as the State may, by law, determine."

What is debatable is the two words, "free" and "compulsory," used in the title of the RTE Act. *Free education* implies that "no child shall be liable to pay any kind of fee or charges, or expenses that may prevent him or her from pursuing and completing elementary education." However, it is silent about those children whom their parents have admitted to a school that the appropriate Government does not support. "Compulsory education" obliges "the appropriate Government and local authorities to provide and ensure admission, attendance, and completion of elementary education by all children in the 6–14 age group." So the law confines the state's responsibility within delivering free education disseminated through the state-run schools only by confirming admission and issuing course completion certificates; however, it remains silent about the "quality" of elementary education to be provided as "free." Its enactment will not be capable of taking control of the emerging gradient between the "costs" of education in public- and private-run educational institutions; instead, setting the boundary of the "right" for availing free education seems to offer the private-run system the liberty to treat education as the "purchased products." Like the costlier "commodities," the private-run education system becomes limited for those only whose family economy could afford the cost. The right of all children to elementary education remains elusive to the children from the marginalized in India.

The digital gap starts right from the schools (government schools and private schools). In India, there exist different management authorities of schools: some schools are run by the government (central, state, local government bodies); some by individuals, trusts, and societies, i.e., private sectors. The number of schools managed by central or state governments is almost double that of other bodies taken together (British Council 2014). As per 2016–17 statistics, the government schools shared 71.72% and private schools 19.78% by count. In the rural areas, the government schools dominate the share (78.42%), leaving only 14.4% for the private sector. However, the scenario is the reverse in urban areas where 48.66% are private schools, and the rest are run by the government (U-DISE 2016). However, the number of schools run by private or public authorities does not speak the real stories.

What is the actual scenario of the government-aided "free" primary education system? Only 85% of those primary schools in the country have drinking water facilities; 37% don't have toilet facilities; only 44% have separate toilet facilities for

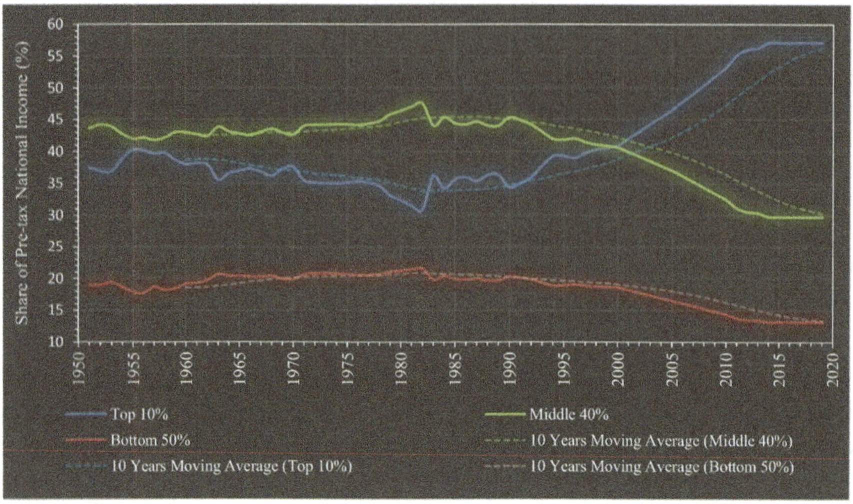

Fig. 7.5 Income distribution over time in India (**Source:** Compiled by the authors from the World Inequality Database [The World Inequality Database (WID) provide open and convenient access to the most extensive available database on the historical evolution of the world distribution of income and wealth, both within countries and between countries. See https://wid.world/data/])

girls, and one-third of them requires major or minor repairs to buildings.[14] While these statistics present the aggregated national averages, the actual picture disaggregated by regions and by social and economic groups of the population must be more disappointing. For example, during pre-pandemic times, only 4.09% of the government schools had an internet connection, where 25.07% of the private schools enjoyed the same (U-DISE 2016).

The long-term trend of the share of national income (Fig. 7.5) does not show any ray of hope to the economically marginalized population. Share of national income by the population at top 10% income group (wealthiest income strata) became larger than that of middle 40% income group (middle and lower-middle-income strata) in post-2000. Moreover, the gap is increasing rapidly with the cost of de-accumulation of income at the lowest income (lowest 50%) strata which is an evil sign indeed. Without getting into the debate about what should be the poverty line, we have focused on the fact itself. If the poverty line is considered as $1.25 (PPP) a day, the UN Millennium Development Goals (MDG) program estimated that about eight million out of 1.2 billion Indians, which is roughly equal to 6.7% of the total population, would live below the poverty line in 2018.

While considering the World Bank (2020) determined the poverty line as $3.2 (PPP) a day for the lower-middle-income countries, the United Nations University (UNU) researchers estimate that the coronavirus pandemic and the economic consequences of an extended lockdown could swell 915 million Indian people, that is

[14] As per the District Information System for Education, published by the NUEPA in 2009–10.

roughly 68% of the total, below the $3.2 (PPP) a day level.[15] The pandemic struck India recorded its lowest economic growth in over a decade. The slowing economy would disproportionately impact the rural areas, the marginalized, the urban slums for perceiving, which probably would not need any official data to perceive. The three factors—high unemployment rate, slowing down consumption expenditure, and stagnating public spending on development, together dictate the well-being of an economy.[16] Amidst this crisis, is not all these facts and figures confirm that the marginalized section would send their students for the "free" education? Moreover, most of those "free education providing" institutions have long been in "hibernation" due to their unfit infrastructure compatible with disseminating digital learning.

The government-run institutions in the remote locations of Himachal Pradesh, Uttarakhand, Jammu & Kashmir (high mountainous areas), Rajasthan (sparsely deserted areas), Madhya Pradesh (forested areas), and the vast areas of Northeast part of India (ironically far from the attention due to their furthest location form the national capital) are shut down. Also, students living in marginalized urban slums and ghettos throughout India are in dire straits in the context of education.

7.5 Validation of the Constructs

We have been discussing so far that the recent outbreak of SARS-CoV-2 and the related shutdown of many systems is sinister for the lives and livelihoods of the world citizens; however, the marginalized are the worst sufferers. Our submission has been orbiting around the idea that economic marginalization has the most prominent role in it. In this section, we will discuss the exploratory survey we made to affirm the constructs we have made for India throughout Sect. 7.2, 7.3, and 7.4, the nutshell of which are:

- The same pandemic shocks differentially where the economically marginalized suffer the most.
- Costlier education systems tend to exclude the marginalized.
- The free and compulsory public education system fails to reach equal opportunities to the marginalized.

The authors have chosen the "Namasudra" communities in the district of Dakshin Dinajpur for these field-based observations as the available secondary sources confirm a high degree of social and economic marginalization prevail within this community that could allow validating our theoretical constructs.

---

[15] See the Business Standard dated 17 April 2020 under the URL https://www.business-standard.com/article/economy-policy/coronavirus-impact-over-100-million-indianscouldfallbelow-poverty-line-120041700906_1.html

[16] See the Down to Earth magazine's specual report dated 07 April 2021 under the URL: https://www.downtoearth.org.in/blog/governance/mass-poverty-is-back-in-india-76348

7.5.1 Marginalization in Scheduled Castes and Scheduled Tribes in India

While the topmost priority of independent India's economic planning is to eliminate poverty, raise the standard of living and quality of life of its weaker sections of the population, Article 46 of the Indian Constitution lays down, as a Directive Principle of State Policy indicate that the state shall promote and protect the marginalized sections from social injustice and all forms of exploitation. The Constitution of India provided special status for such marginalized communities who are now popularly known as Scheduled Castes (SCs) and Scheduled Tribes (STs). However, in practice, SCs and STs face structural discrimination based on ethnicity within Indian society.

The *Brahmins* have traditionally enjoyed the top of the hierarchy within the *Hindu* religious group, and *Shudras* or *Dalits* constitute the bottom. Structural discriminations against these bottom strata frequently appear in the form of physical, psychological, emotional, and cultural abuse, which receives legitimacy from the social structure and the social system. This is well evidenced by the physical segregation of the settlements, which is most common in Indian villages forcing the so-called people of "lower class" to live in the most unhygienic and inhabitable places.[17] These excluded communities form the unorganized labor force in agricultural, plantation, mining, and industrial sectors. The poor socioeconomic background and hardship of poverty make their life miserable specifically, they face low levels of education, poor health, and reduced access to healthcare services. High rates of dropout and mass illiteracy prevail there. Early marriage, trafficking, forced prostitution, and other forms of exploitation are also reportedly high.[18] From the Hindu religious group, the SCs and STs predominate the migrant labor streams, which has long been happening, and the pandemic has placed it under the daylight.

7.5.2 The Namasudra whom we have studied

The term "Scheduled Caste" itself originates from the Government of India (Scheduled Castes) Order, 1936, which contained a list (or Schedule) of castes throughout the British-administered provinces, which was incorporated in the Constitution of India via the Constitution (Scheduled Castes) Order, 1950. Article 341 of the Constitution categorized a section of India's population under the category of Schedule Castes. In the State of West Bengal, one of the major Schedule Caste groups goes by the name of "Namasudra." They are the second-largest Schedule Caste group, constituting about 17% of the total Scheduled Caste population of the

[17] See for details:https://www.lawctopus.com/academike/problems-marginalized-groupsindia/

[18] See for details:https://www.lawctopus.com/academike/problems-marginalized-groupsindia/

state. Several studies on the origin of the "Namasudra" caste group recognized that the "Chandal"[19] caste in Bengal renamed them as Namasudra to escape their lived status of being the lowest of the low in the caste hierarchy (Viswanath 2014; Chandrashekhar 1984; Michael 1999; Biswas 2000; Thapar 2004).

Indian Constitution recognizes the Namasudra community as socioeconomically backward and belongs to scheduled caste groups mainly found in Assam, West Bengal, Orissa, Manipur, Tripura, Meghalaya, Mizoram, and Arunachal Pradesh of India and the six eastern districts of Faridpur, Dacca, Mymensingh, Jessore and Khulna of Bangladesh. These communities have a great history of emergence or evolution, and they are destitute compared to the other caste. James Wise and Herbert Risley (1891) speculates that the word "Nanasudra" was derived either from the Sanskrit word *namas* (i.e., adoration), which means the people who were bound to show obeisance even to the *Shudras* (the lowest strata in the social hierarchy as per the traditional division of labor) or the Bengali word *namate* (i.e., below or underneath) which means a lower grade of *Shudras*.

Bandyopadhyay (2011) claims that the Namasudra community was earlier known as Chandala or Chandals, who mostly inhabited East Bengal (at present Bangladesh) and were forced to migrate to West Bengal during the partition of India in 1947. The peoples of the Namasudra community have settled their habitation in different parts of the State of West Bengal, particularly the then undivided Paschim Dinajpur district (which is later divided into two separate districts, i.e., the Uttar Dinajpur and Dakshin Dinajpur) sharing the Indo-Bangladesh international border (Fig. 7.6). The majority member of the Namasudra community in Dakshin Dinajpur district has settled over here from present-day Bangladesh since the precolonial period, and this process of migration continues today. This migration flow has seen some highs, notably during partition in 1947 and during years preceding the formation of the independent nation of Bangladesh in 1971. However, even after the inception of Bangladesh, the State of West Bengal has a shared historical legacy, good cultural relations, and a close economic alliance. Several causes (which the study has no scope to discuss) had forced the Namasudra community members to move toward India, some of whom have settled in the district of Dakshin Dinajpur. Namasudra community belongs to the lower stratum of the society; unable to fulfill their basic needs mainly due to their lack of education and socioeconomic backwardness occupationally, most of them belong to peasant and fishing communities, practicing subsistence existence.

The district of Dakshin Dinajpur bears a total of 4,82,754 Scheduled Castes population, among which 43,565 (9.02%) belong to the *Namasudra* community (Census of India 2011a, b, c). Most of the *Namasudra* households are in rural areas (88.27%) with their traditional socioeconomic practices, and only 11.74% of Namasudra households living in urban areas of Dakshin Dinajpur district. However, the Namasudra population in this district participates disproportionately lower

[19] A Hindu lower caste community who deals with disposal of corpses, and, long been stigmatized as 'untouchable'.

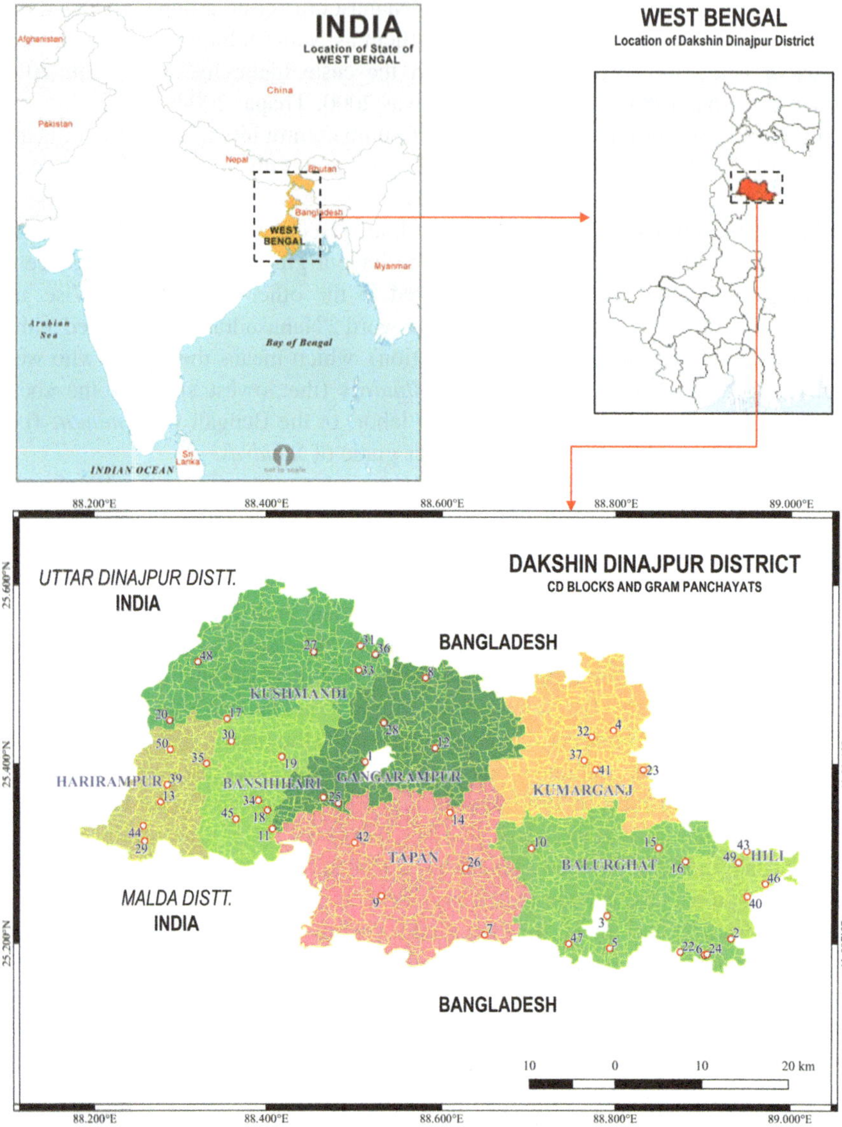

Fig. 7.6 The location of the district of Dakshin Dinajpur where the case study on the Namasudra Community was conducted; the white dots with red circles presents the location of interviews conducted, and the yellow hairlines are the village boundaries (**Source**: Drawn by the authors)

(53.09%) than all other districts of the state (Census of India 2001). With traditional livelihoods and poor socioeconomic backgrounds, most of them live with the poverty line (Roy 2019; Roy and Kumar 2019). Being marginalized community, they are more susceptible to COVID-19 for multiple reasons (Ahmed et al. 2020).

Table 7.5 Expected outcome from expert opinion

Monthly Average HH Income (Pre-pandemic)	Event-1: Likelihood of starvation at least once a week due to hardship?				Event-2: Likelihood of at least one child of the family is dropped out?				Event-3: Likelihood of at least one child of the family taking online classes regularly?			
	E_1	E_2	E_3	F	E_1	E_2	E_3	F	E_1	E_2	E_3	F
> Rs. 20,000	N	N	N	**N**	N	N	N	**N**	Y	Y	Y	**Y**
Rs. 10,000 to Rs.19999	N	N	N	**N**	N	N	N	**N**	Y	Y	Y	**Y**
Rs. 5000 to Rs.9999	Y	N	N	**N**	N	N	N	**N**	N	N	Y	**N**
Rs. 1000 to Rs. 4999	Y	Y	Y	**Y**	Y	N	N	**N**	N	N	Y	**N**
<Rs. 1000	Y	Y	Y	**Y**	Y	Y	Y	**Y**	N	N	N	**N**

Source: Compiled by the authors

7.5.3 Expected Outcomes from Experts' Choice

Advice from three experts having prolonged working experience with similar research fields is approached to inform their expert opinion about which income group in the Namasudra community may be vulnerable to the following three events.

- Event-I: Likelihood of starvation at least once in a week due to hardship;
- Event-II: Likelihood of at least one child of the family is dropped out;
- Event-III: Likelihood of at least one child of the family taking online classes regularly.

For each income group sorted by the events, expert judgment is accepted based on the decision of the majority (Table 7.5).

N.B. HH = Household; Y=Yes (i.e., Coded as "1"); N=No (i.e., Coded as "0"); E_1 = Expert-1; E_2 = Expert-2; E_3 = Expert-3; F=Finalized based on the opinion of the majority.

7.5.4 Recording What We Observed in the Field?

A total of fifty cases were studied between 25 January and 22 February 2021. Pre-set questionnaires were not arranged as we had started with a general idea and tend to use this research as a medium validating our constructs based on the local responses. As this pandemic is unprecedented to the living citizens, this grounded theory approach with interpretive research could answer our questions about "what," "why," and "how" on the digital education divides on the marginalized. The respondents were asked to narrate about his/her family, occupations of the member, family economy, children's education, and their learning experience during the pandemic. The interviewers took written notes and made audio recordings where permission from the respondents was granted. Particularly the interviewer was concerned to note or ask specifically from the respondent. Following are some sample notes we have recorded:

Survey site $25^{0}24'09''$E, $88^{0}30'43''$E (Case 1)
Raju Mondal, a Namasudra respondent (a fisherman), settled in a small hamlet beside the bank of Punarbhava river in Gangarampur block, was describing his experience toward consequences of coronavirus diseases on education and health status on his family and neighborhoods.

> "I earn by catching fishes from the Punarbhava river with the help of traditional instruments and selling them in the local market. My old parents, my three children, and a widow sister with her two children depend on my earnings. In the family, after meat comes mustard. I have no cropland of my own. Fishing activity is my only source of income, and this income is meager. Last year, my elder son dropped from school at class VIII due to the hardship of the economy and started helping me to sell fish. He sells fish in the villages by his bi-cycle. But the sudden outbreak of this pandemic destroyed that sell as the lockdown was imposed. My mother is a chronic patient who requires dialysis and regular treatment. I failed to protect her because of irregular treatment during the pandemic. I couldn't afford the treatment as my earnings were halted. The lockdown and closing of schools introduce a temporary suspension of Mid-day Meal and supplementary nutrition programs. My children remained absent from this part. I don't know whether the online classes are there or even, when the school will open." (Respondent Statement 1, 23rd June 2020, Interviewed by the author, Arabinda Roy).

Survey site $25^{0}12'20''$N, $88^{0}55'55''$E (Case 2)
Pratima Pramanik, a 34 years widow refugee Namasudra respondent, living with her four children in Balurghat block, narrated her experience during the pandemic days.

> "I worked in a house as a servant and get Rs. 2600 per month. My economic condition is not good. With my four children (aged 17 years, 13 years, 8 years and 4 years, respectively), I lived in a small room with one small courtyard. My husband expired one and half years ago who was a farmer. With the increasing number of coronavirus infections, my lord told me to continue my duties as a servant after the lockdown period is over. He did pay the salaries for only those days of the month of April I served; I guess it would be less than half of the month. So I lost my job and the monthly income. I could not afford my children's food. The government's ration was the only means of livelihood. This pandemic totally stopped the education of my children. It is impossible for me and those who belong poor socio-economic background just like me to buy a television or a smartphone during these pandemic days to continue education from home." (Respondent Statement 2, 9 July 2020, Interviewed the author, Arabinda Roy).

Survey site 25⁰19′41″N, 88⁰24′24″E (Case 11)

Bimal Mondal, a 34-year-old Namasudra farmer who lives with his four children in Gouripara village of Banshihari block of Dakshin Dinajpur district, narrated his experiences of living during the pandemic days.

> "I am a small farmer. I own only 1.5 *bighas*[20] of cropland. This little cropland produces minimal crops and vegetables that are insufficient to marketize after feeding all the members. My elder son is physically challenged. In my family, I am the only person who earns. For the last few years, after harvesting, I have been migrating to Kerala for earnings that have subsequently been changing my family economy. This year, when I was in Kerala, the government imposed the lockdown. There, I lost my job in that distant state. All of my savings were gone. After few months, I managed to come back and meet my family. In the meantime, I could not send even a rupee to my family. They spent several days with hunger; no one could help as all of my community neighbors had lost their earnings, and they underwent the same situation. Who could help? My younger son, Ajoy, is reading in class seven, and the little daughter, Kabita, in class five in government high school in the block. The school has been closed. I heard the teachers are teaching online. I could not afford a smartphone for them as managing half a meal every day is a challenge to me. Besides this, the facility of mid-day-mills has also been stopped. My children used to manage their lunch there, but now, they are becoming my responsibility to feed them." (Respondent Statement 3, 9 April 2020, Interviewed by the author, Arabinda Roy).

Survey site 25⁰18′52″N, 88⁰15′29″E (Case 29)

Tarun Pramanik was an assistant headmaster of a private school who was a Namasudra respondent of Santipur village of Chaighara gram panchayat of Harirampur block, described his education and health experience status in the pandemic days.

> "My three friends and I opened a public school named Saraswati Sarada Vidyamandir K. G School in Santipur village of Harirampur block in 2018 for the public interest where most students belong to back warded and marginalized Namasudra community. In our school, we paid attention to our students in their learning and their physical and mental health. The school just began to move; the government declared the nationwide lockdown. The pandemic destroyed everything. Remote teaching and homeschooling are impossible for our students due to their poor socio-economic background. In the last few months, they are entirely confined to their houses. I worried that their mental, as well as physical health, is humiliated. Final examinations, assessment of answer scripts – everything is due." (Respondent Statement 4, 2 August 2020, Interviewed the author, Arabindaspiepr146 Roy).

All the fifty cases were analyzed to record the observation about the likelihoods of the same three events for which the experts judged the expected outcomes based on the monthly household income (as in Table 7.5). The observed outcomes were recorded the following ordinal ratings as below:

---

[20] The bigha (also formerly beegah) is a traditional unit of measurement of area of a land, commonly used in India, Bangladesh and Nepal. There is no "standard" size of bigha. The size of a bigha varies considerably from place to place. The size of Bigha is different in different areas. Normally, one Bigha is equal to 0.6198347106 acre or one Bigha is equal to 17452.0069808 square feet. For details:https://www.magicbricks.com/blog/1-bigha-to-squarefeetacrehectareconverter/115042.html

- Definitely negative (Coded as "1").
- Probably negative (Coded as "2").
- Possibly negative (Coded as "3").
- Possibly positive (Coded as "4").
- Probably positive (Coded as "5").
- Definitely positive (Coded as "6").

7.5.5 Maximum Likelihood Estimation of Binomial ROC Curve from Categorical Rating of Observed and Expected Outcomes

The ROC analysis was employed successfully by the 1960s in experimental psychology and psychophysics (Tanner Jr. and Swets 1954; Swets et al. 1961; Swets 1964; Swets 1973). However, it was successfully used in evaluations of medical decision-making for which Lusted (1960, 1968, 1971) is credited. Subsequently, while increasing attention on the applications of ROC analysis is given in medical diagnosis (Swets 1972; McNeil et al. 1975; Weinstein and Fineberg 1980; Zweig and Campbell 1993), the practitioners of earth sciences made them restricted to use ROC for validating their software-generated models, and the social scientists are rarely interested.

It is widespread in ROC studies to involve human judgments that have employed an ordinal Confidence Rating Scale (CRS) containing a fixed number of discrete response categories. As this space is not suitable for elaborating how the Maximum likelihood (ML) estimation algorithm is used for fitting binormal ROC curves to such categorical data, the readers may follow the works of Dorfman and Alf (1969), Grey and Morgan (1972), Swets and Pickett (1982), Metz (1989) and many more. Figure 7.7 explains graphically how a model is applied to fit a binormal ROC curve to ordinal confidence rating data applying the maximum likelihood (ML) estimation using six-category data (observed outcomes). The decision-variable axis (i.e., the x-axis) is assumed to be divided into I categories by (I- 1) fixed boundaries. "The convergence of the procedure generates ML estimates of the settings of these boundaries along with estimates of parameters that represent the separation and relative widths of the two normal decision-variable distributions, which specify the ROC" (Metz et al. 1998).

The larger values of "Area Under Curve" (AUC) of the three ROCs generated events validate the expert judgment (See Fig. 7.8; also see the calculation in Tables 7.4, 7.5; and also Tables 7.6, 7.7, and 7.8 in the Appendix). The AUC was determined as 0.919 (with an overall accuracy of 88%) for Event-1, 0.940 (overall accuracy of 86%) for Event-2, and 0.974 (overall accuracy of 92%) for Event-3. It is sufficient to state that the impact of COVID-19 on the marginalized Namasudra community in Dakshin Dinajpur district is differential within the community where

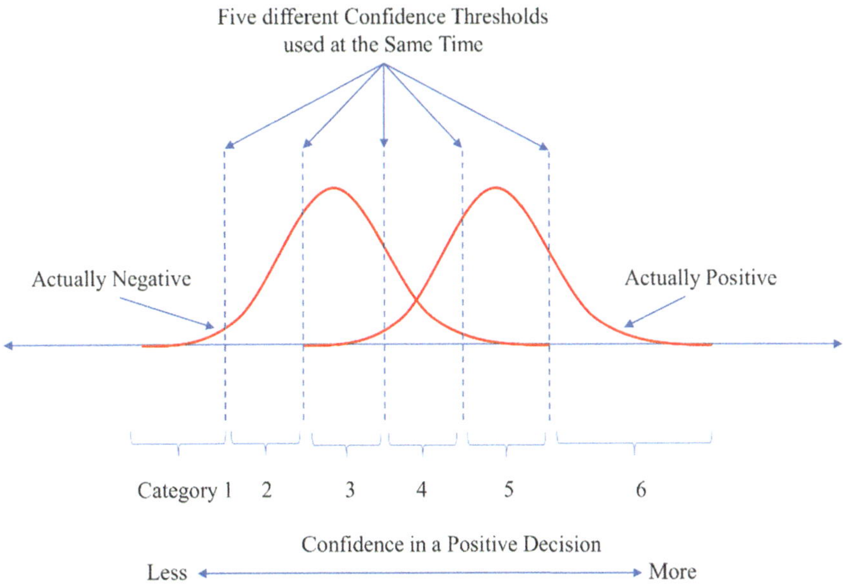

Fig. 7.7 A graphical presentation of the procedure of fitting a binormal ROC curve to ordinal confidence rating data by ML estimation (**Source**: Drawn by the author following Metz et al. 1998)

financial exclusion takes an influential role. It reveals that most of the population suffers from the rapid spread of the COVID-19 pandemic and its association with the digital education divides. The pandemic-induced lockdown has made the life of the lowest economic strata miserable. Thus, marginalized sections are remaining at the margin of the digital divide in the field of education. Lack of socioeconomic affluence of the Namasudra populations remains the main factor for the digital divide, which is more critical for their children who mostly live in outskirt and remote areas and suffer from the home-deficit learning resources.

7.6 The Way Forward

The Prime Minister of India has affirmed that post-COVID India will be strikingly different and that there will be five distinguishing pillars of the new normal. He elaborated on each of the five supporting structures, namely Economy, Infrastructure, System, Demography, and Demand, that will define the new normal (India Today 2020).

Policymakers, governmental organizations, research institutions, and educators are trying their hard to minimize the impression of the disruptions of the pandemic on students across all levels of education. Special focus must be paid to girls and poor children to minimize the long-term disruptions to education as a result of the pandemic. Economically disadvantaged and marginalized students across the nation

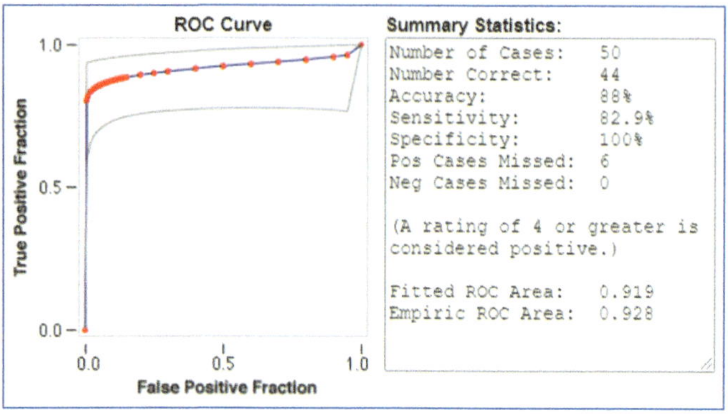

(a)

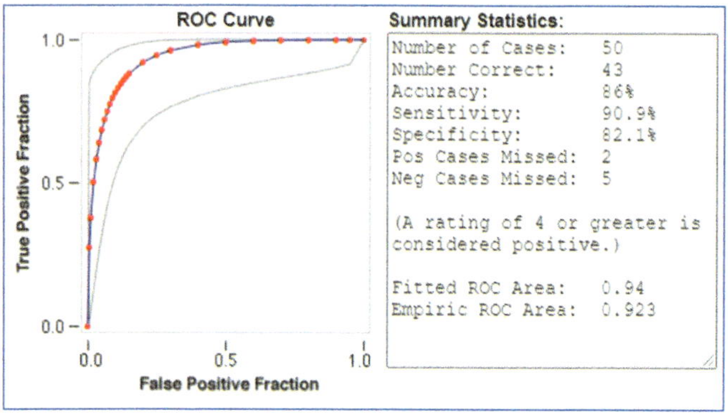

(b)

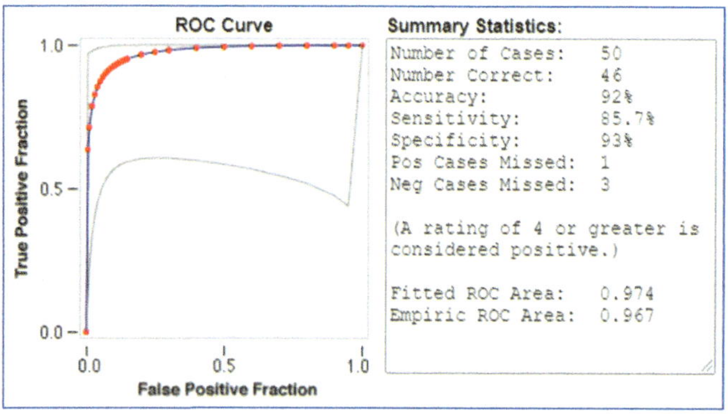

(c)

Fig. 7.8 Maximum Likelihood Estimation of Binomial ROC Curve for **(a)** Event-1; **(b)** Event-2 and **(c)** Event-3. **N.B.** Red symbols and BLUE line show the Fitted ROC curve, and Gray lines show 95% confidence interval of the fitted ROC curve (**Source**: Prepared by the authors) need to be provided with mobile data and computer subsidies during the lockdown periods. Implementation of home-delivered mid-day meals or dry ration to school and Anganwadi children is an effective local way or solution during these pandemic periods, which could reach the benefits millions of children and expecting mothers. Though several southern states (Kerala, Telangana, Karnataka, and Andhra Pradesh) have implemented this program, it should be extended throughout the country.

The decision of the Government of West Bengal to hand out tablet devices to its 9.5 lakh students is grounded on the principle of extending the opportunity of online education to all. The Chief Minister of West Bengal announced the objectives of this government venture as:

> "There are around 14,000 government and government-aided schools and 636 madrasas where approximately 9.5 lakh students will sit for the higher secondary examination this year. Since there is no on-campus class now, most of these students, who don't have access to a smartphone or a computer, are missing out. The government has thus decided to give these students tabs. They can use this during their higher studies also."[21]

Other measures, including TV broadcasted classes and regular SMS/IVR to parents for daily activities with children, are currently underway. In addition, collaborative efforts between PPCs (public sector, private sector, and civil society) would be critical for the educational rehabilitation of affected children. The students from marginalized and vulnerable families should receive practical support in such pandemic situations.

However, there are new and intensified barriers to handling marginalized populations' lives due to the global COVID-19 pandemic. While the marginalized sections are busy maintaining their basic needs, education is so far from them. When most of their kids are first-generation learners, who will whisper to their ears that education is the fundamental instrument to meet livelihoods opportunities and guide socioeconomic development? It needs campaigning, reaching door-to-door to feel the actual scenario evolving at ground levels and record them to frame robust policies in the long run. In the short run, what is most needed is direct and immediate financial support through well-thought channels and adoption of inclusive public policy to meet the needs of the population of India lying at the lowest strata of the economy, particularly the population below and on the Poverty Line, unfortunately, who are marginalized as from economy so from planning.

[21] See the statement of the West Bengal Chief Minister, reported in the Indian English daily, the Times of India on 04 December 2020: https://timesofindia.indiatimes.com/home/education/news/state-to-give-out-tabs-to-9-5l-needy-hs-students/articleshow/79554839.cms

Appendix

Table 7.6 Estimation of binomial ROC Curve for Event-1

```
Maximum likelihood estimation of a binormal ROC curve from
categorical rating data.

DATA CHARACTERISTICS:
  Data collected in 6 categories with category 6 representing
    strongest evidence of positivity (e.g., that abnormality is
    present).
  Number of actually negative cases = 15
  Number of actually positive cases = 35

RESPONSE DATA:
  Category                     1     2     3     4     5     6
  Actually negative cases     10     3     2     0     0     0
  Actually positive cases      3     2     1     8     9    12

OBSERVED OPERATING POINTS:
  FPF:  0.0000  0.0000  0.0000  0.0000  0.1333  0.3333  1.0000
  TPF:  0.0000  0.3429  0.6000  0.8286  0.8571  0.9143  1.0000

INITIAL VALUES OF PARAMETERS:
  A = 1.9680
  B = 0.9797
  Z(K):  0.4303  1.1108  1.6343  1.7343  1.8343
  LOGL = -90.6943

FINAL VALUES OF PARAMETERS:
  Procedure converges after 10 iterations.
  A = 1.4325
  B = 0.2230
  Z(K):  0.4108  1.2536  2.1868  5.2960  8.2439
  LOGL = -67.2093

VARIANCE-COVARIANCE MATRIX:
  A      0.1174  0.0343  0.0279  0.0142 -0.0282 -0.3925 -0.8042
  B      0.0343  0.0260  0.0085 -0.0113 -0.0734 -0.4506 -0.7998
  Z(1)   0.0279  0.0085  0.1111  0.0714  0.0416 -0.0776 -0.1905
  Z(2)   0.0142 -0.0113  0.0714  0.1721  0.1730  0.3321  0.4823
  Z(3)  -0.0282 -0.0734  0.0416  0.1730  0.5978  1.6187  2.5882
  Z(4)  -0.3925 -0.4506 -0.0776  0.3321  1.6187  9.0804 14.9825
  Z(5)  -0.8042 -0.7998 -0.1905  0.4823  2.5882 14.9825 26.3431

CORRELATION MATRIX:
  A      1.0000  0.6206  0.2444  0.0998 -0.1066 -0.3801 -0.4573
  B      0.6206  1.0000  0.1589 -0.1694 -0.5886 -0.9272 -0.9662
  Z(1)   0.2444  0.1589  1.0000  0.5166  0.1615 -0.0772 -0.1114
  Z(2)   0.0998 -0.1694  0.5166  1.0000  0.5394  0.2656  0.2265
  Z(3)  -0.1066 -0.5886  0.1615  0.5394  1.0000  0.6948  0.6522
  Z(4)  -0.3801 -0.9272 -0.0772  0.2656  0.6948  1.0000  0.9687
  Z(5)  -0.4573 -0.9662 -0.1114  0.2265  0.6522  0.9687  1.0000

SUMMARY OF ROC CURVE:
  Area = 0.9190
  Std. Dev. (Area) = 0.0461
```

Source: Calculated by the author on software platform

Table 7.7 Estimation of binomial ROC Curve for Event-2

```
Maximum likelihood estimation of a binormal ROC curve from
categorical rating data.

DATA CHARACTERISTICS:
  Data collected in 6 categories with category 6 representing
    strongest evidence of positivity (e.g., that abnormality is
    present).
  Number of actually negative cases = 28
  Number of actually positive cases = 22

RESPONSE DATA:
  Category                        1      2      3      4      5      6
  Actually negative cases         6      5     12      2      3      0
  Actually positive cases         0      0      2      3      9      8

OBSERVED OPERATING POINTS:
  FPF:   0.0000  0.0000  0.1071  0.1786  0.6071  0.7857  1.0000
  TPF:   0.0000  0.3636  0.7727  0.9091  1.0000  1.0000  1.0000

INITIAL VALUES OF PARAMETERS:
  A = 1.6972
  B = 0.8278
  Z(K): -0.7914 -0.2715  0.9207  1.2420  2.1006
  LOGL = -69.8454

FINAL VALUES OF PARAMETERS:
  Procedure converges after 8 iterations.
  A = 2.3801
  B = 1.1556
  Z(K): -0.7983 -0.2811  0.9319  1.3476  2.3808
  LOGL = -67.4824

VARIANCE-COVARIANCE MATRIX:
  A      0.5316  0.2991  0.0345  0.0453  0.0830  0.0477 -0.1597
  B      0.2991  0.2547  0.0124  0.0153  0.0075 -0.0385 -0.2451
  Z(1)   0.0345  0.0124  0.0707  0.0415  0.0187  0.0148  0.0046
  Z(2)   0.0453  0.0153  0.0415  0.0573  0.0257  0.0207  0.0081
  Z(3)   0.0830  0.0075  0.0187  0.0257  0.0730  0.0641  0.0556
  Z(4)   0.0477 -0.0385  0.0148  0.0207  0.0641  0.0950  0.1183
  Z(5)  -0.1597 -0.2451  0.0046  0.0081  0.0556  0.1183  0.3766

CORRELATION MATRIX:
  A      1.0000  0.8130  0.1782  0.2595  0.4214  0.2123 -0.3570
  B      0.8130  1.0000  0.0925  0.1263  0.0547 -0.2473 -0.7913
  Z(1)   0.1782  0.0925  1.0000  0.6521  0.2598  0.1812  0.0284
  Z(2)   0.2595  0.1263  0.6521  1.0000  0.3968  0.2806  0.0549
  Z(3)   0.4214  0.0547  0.2598  0.3968  1.0000  0.7695  0.3351
  Z(4)   0.2123 -0.2473  0.1812  0.2806  0.7695  1.0000  0.6255
  Z(5)  -0.3570 -0.7913  0.0284  0.0549  0.3351  0.6255  1.0000

SUMMARY OF ROC CURVE:
  Area = 0.9403
  Std. Dev. (Area) = 0.0330
```

Source: Calculated by the author on software platform

Table 7.8 Estimation of binomial ROC Curve for Event-3

```
Maximum likelihood estimation of a binormal ROC curve from
categorical rating data.

DATA CHARACTERISTICS:
  Data collected in 6 categories with category 6 representing
    strongest evidence of positivity (e.g., that abnormality is
    present).
  Number of actually negative cases = 43
  Number of actually positive cases = 7

RESPONSE DATA:
  Category                      1      2      3      4      5      6
  Actually negative cases      23      9      8      3      0      0
  Actually positive cases       0      0      1      2      0      4

OBSERVED OPERATING POINTS:
  FPF:  0.0000  0.0000  0.0698  0.2558  0.4651  1.0000
  TPF:  0.0000  0.5714  0.8571  1.0000  1.0000  1.0000

INITIAL VALUES OF PARAMETERS:
  A = 1.7764
  B = 0.6296
  Z(K):  0.0873  0.6560  1.4778  2.2696
  LOGL = -58.0516

FINAL VALUES OF PARAMETERS:
  Procedure converges after 7 iterations.
  A = 2.5511
  B = 0.8559

VARIANCE-COVARIANCE MATRIX:
  A      1.4259  0.6138  0.0251  0.0346  0.0453 -0.3809
  B      0.6138  0.3536  0.0066  0.0070 -0.0089 -0.3659
  Z(1)   0.0251  0.0066  0.0366  0.0250  0.0162  0.0079
  Z(2)   0.0346  0.0070  0.0250  0.0425  0.0276  0.0173
  Z(3)   0.0453 -0.0089  0.0162  0.0276  0.0856  0.0796
  Z(4)  -0.3809 -0.3659  0.0079  0.0173  0.0796  0.7720

CORRELATION MATRIX:
  A      1.0000  0.8645  0.1099  0.1407  0.1298 -0.3631
  B      0.8645  1.0000  0.0581  0.0575 -0.0509 -0.7003
  Z(1)   0.1099  0.0581  1.0000  0.6329  0.2890  0.0473
  Z(2)   0.1407  0.0575  0.6329  1.0000  0.4573  0.0956
  Z(3)   0.1298 -0.0509  0.2890  0.4573  1.0000  0.3097
  Z(4)  -0.3631 -0.7003  0.0473  0.0956  0.3097  1.0000

SUMMARY OF ROC CURVE:
  Area = 0.9737
  Std. Dev. (Area) = 0.0307
```

Source: Calculated by the author on software platform

References

Ahmed F, Ahmed N, Pissarides C, Stiglitz J (2020) Why inequality could spread COVID-19. Lancet Public Health 5:e240. https://doi.org/10.1016/S2468-2667(20)30085-2

AISHE (2019) All India survey on higher education 2018–2019. Ministry of Human Resource Development, Department of Higher Education

Alvi M, Gupta M (2020) Learning in times of lockdown: how COVID-19 is affecting education and food security in India. Food Security 12:793–796. https://doi.org/10.1007/s12571-020-01065-4

Baah FO, Teitelman AM, Riegel B (2019) Marginalization: conceptualizing patient vulnerabilities in the framework of social determinants of health—An integrative review. Nuring Inquiry 26(1): e12268

Bandyopadhyay S (2011) Caste, Protest and Identity in colonial India; The Namasudras of Bengal, 1871–1 947. Oxford University press, New Delhi, pp 46–51

Bharadwaj P, Lakdawala LK, & Li N (2019) Perverse consequences of well intentioned regulation: Evidence from India's child laborban. J Eur Econ Assoc. 18:1158–1195. doi: https://doi.org/10.1093/jeea/jvz059

Birchall J (2018) Early Marriage, Pregnancy and Girl Child School. Dropout, Resource Centre

Biswas AK (2000) The Namasudras of Bengal: profile of a persecuted people. Blumoon Books

British Council (2014) Indian school education system: an overview. https://www.edcentrum.org/pdf/schooleducation/6_Indian_school_education_system_an_overview.pdf. Accessed 11 Sept 2020

Census of India (2001) Government of India

Census of India (2011a) Government of India

Census of India (2011b) Adolescents and youth in India: highlights from census 2011

Census of India (2011c) Rural-urban distribution of population

Chandrashekhar B (1984) Ethnicity and mobility. Concept Publishing

Dorfman DD, Alf E (1969) Maximum likelihood estimation of parameters of signal detection theory and determination of confidence intervals drating method data. J Math Psychol 6:487–496

Giannini S, Jenkins R, Saavedra J (2020) Reopening schools: when, where and how? UNESCO. https://en.unesco.org/news/reopening-schools-when-where-and-how. Accessed 10 July 2020

Grey DR, Morgan BJT (1972) Some aspects of ROC curve-Þtting: normal and logistic models. J Math Psychol 9:128–139

India Today (2020) PM Modi shares 5 pillars that will make India self-reliant in COVID-19 times. https://www.indiatoday.in/india/story/pm-modi-speech-5-pillars-atm-nirbhar-india-self-reliance-COVID-19-lockdown-1677293-2020-05-12. Accessed 15 June 2020

Jha N, Shenoy V (2016) Digitization of Indian education process: a hope or hype. IOSR J Bus Manag (IOSR-JBM) 18(10):131–139. https://doi.org/10.9790/487X-181003131139

Kundu P (2020) Indian education can't go online – only 8% of homes with young members have computer with net link, published by a newspaper Scroll. in. https://scroll.in/article/960939/indian-education-cant-go-online-only-8-of-homes-with-school-children-have-computer-with-net-link

Leonard P (1984) Personality and ideology: Towards a materialist understanding of the individual. Macmillan, London

Lusted LB (1960) Logical analysis in Roentgen diagnosis. Radiology 74:178–193

Lusted LB (1968) Introduction to Medical Decision Making. Thomas Springfield, IL

Lusted LB (1971) Signal detectability and medical decision-making. Science 171:1217–1219

McNeil BJ, Keeler E, Adelstein SJ (1975) Primer on certain elements of medical decision making. N Engl J Med 293:211–215

Metz CE (1989) Some practical issues of experimental design and data analysis in radiological ROC studies. Investig Radiol 24:234–245

Metz CE, Herman BA, Shen J-H (1998) Maximum likelihood estimation of receiver operating characteristic (ROC) curves from continuously-distributed data. Statist Med 17:1033–1053

Michael SM (1999) Untouchable: Dalits in Modern India. Lynne Rienner Publishers
Modi S, Postaria R (2020) How COVID-19 deepens the digital education divide in India. World Economic Forum.. https://www.weforum.org/agenda/2020/10/how-COVID-19-deepens-the-digital-education-divide-in-india/. Accessed 1 June 2020
NSSO (2014) National sample survey of estimation of out-of-school children in the age 6-13 in India. Social & Rural Research institute, Educational Consultants India Ltd A Government of India enterprice. https://www.education.gov.in/sites/upload_files/mhrd/files/upload_document/National-Survey-Estimation-School-Children-Draft-Report.pdf
Parsheera S (2019) India's on a digital sprint that is leaving millions behind. BBC.. https://www.bbc.com/news/world-asia-india-49085846. Accessed 5 June 2020
Prakash R, Beattie T, Javalkar P, Bhattacharjee P, Ramanaik S, Thalinja R, Murthy S, Davey C, Blanchard J, Watts C, Collumbien M, Moses S, Heise L, Isac S (2017) Correlates of school dropout and absenteeism among adolescent girls from marginalized community in North Karnataka, South India. J Adolesc 61:64–76. https://doi.org/10.1016/j.adolescence.2017.09.007
Ratnam D (2020) Domestic violence during COVID-19 lockdown emerges as serious concern. https://www.hindustantimes.com/india-news/domestic-violence-durincid19lockdownemergesasseriousconcern/storymMRq3NnnFvOehgLOOPpe8J.html. Accessed 29 Apr 2020
Reddy AN, Sinha S (2010) School dropouts or Pushouts? Overcoming barriers for the right to education. http://www.create-rpc.org
Risley HH (1891) Tribes and castes of Bengal. Kolkata: Firma KLM Private Ltd. pp. 186-188
Roy A (2019) Domestic Violence and its consequences on women's health: a study reference to Namasudra community in Dakshin Dinajpur district, West Bengal. Ensemble 1(1):19–30
Roy A, Kumar S (2019) Community knowledge and practices regarding the drinking water and sanitation in Dakshin Dinajpur district, West Bengal. Proceeding of Indian Social Sciences and Humanities Congress 2019 (5th International Summit) held at Jadavpur University, Kolkata, 2019
Sen A (1999) Development as freedom. Anchor Books, New York
Singh AK, Venkataramani B, Ambarkhane D (2020) Post-pandemic Penury of the Financially Marginalized in India: Coping with the New Normal. In: Lee J, Han SH (eds) The future of service post-COVID-19 pandemic, Volume 1, The ICT and Evolution of Work, p 201-225
Swets JA (ed) (1964) Signal detection and recognition by human observers: contemporary readings. Wiley, New York
Swets JA (1972) Signal detection in medical diagnosis. In Jacquez JA (ed) Computer diagnosis and diagnostic methods. Thomas Springfield, IL 1972:8–28
Swets JA (1973) The relative operating characteristic in psychology. Science 182:990–1000
Swets JA, Pickett RM (1982) Evaluation of diagnostic systems: methods from signal detection theory. Academic, New York
Swets JA, Tanner WP Jr, Birdsall TG (1961) Decision processes in perception. Psychol Rev 68: 301–340
Tanner WP Jr, Swets JA (1954) A decision-making theory of visual detection. Psychol Rev 61:401–409
Thapar R (2004) Early India: From the Origins to AD 1300. University of California Press
The Encyclopaedia of Public Health (2017) Marginalization. https://www.egyankoshacinbitstream123456789349231unit3.pdf. Accessed 14 July 2020
Trines S (2018) Education in India. World Education News Reviews (WENR). https://wenr.wes.org/2018/09/education-in-india. Accessed 04 July 2021
U-DISE (2016) School education in India. U-DISE flash statistics 2016–17. National Institute of Educational Planning and Administration
UIS and GEM (2017) Reducing global poverty through universal primary and secondary education. UNESCO. http://uis.unesco.org/sites/default/files/documents/reducing-global-poverty-throughuniversal- primary-secondary-education.pdf. Accessed 30 June 2020

UN Economic and Social Council (1999) General comment No. 13: the right to education (article 13 of the Covenant). UN Committee on Economic, Social and Cultural Rights (CESCR). https://documentsddny.un.org/doc/UNDOC/GEN/G99/462/16/PDF/G9946216.pdf. Accessed 17 Aug 2020

UNESCO (1996) Rethinking Education, towards a Global Commons Good? http://www.unesco.org/new/fileadmin/MULTIMEDIA/FIELD/Cairo/images/RethinkingEducation.pdf. Accessed 04 July 2021

UNESCO (2020a) Education in emergencies. https://en.unesco.org/themes/education-emergencies. Accessed 04 July 2021

UNESCO (2020b) COVID-19 impact on education. https://en.unesco.org/COVID19/educationresponse. Accessed 10 September 2020

UNICEF (2020) A new era for girls: taking stock of 25 years of progress. United Nations Children's Fund. https://www.unicef.org/media/65586/file/A-new-era-for-girls-2020.pdf. Accessed 10May 2020

United Nation (1948) Universal declaration of human rights. UN General Assembly. https://www.ohchr.org/EN/UDHR/Documents/UDHR_Translations/eng.pdf. Accessed 30 May 2020

United Nations (1986) Declaration on the right to development. UN General Assembly. https://www.ohchr.org/Documents/ProfessionalInterest/rtd.pdf. Accessed 8 Aug 2020

United Nations (2015) Transforming our world: the 2030 agenda for sustainable development. Department of Economic and Social Affairs. https://sdgs.un.org/2030agenda. Accessed 9 Aug 2020

Viswanath R (2014) The Pariah Problem: caste, religion, and the social in modern India. Columbia University Press. isbn:978-0-23116-306-4

Weinstein MC, Fineberg HV (1980) Clinical decision analysis. Saunders, Philadelphia

World Bank (2020) The COVID-19 pandemic: shocks to education and policy responses. World Bank. https://openknowledge.worldbank.org/bitstream/handle/10986/33696/148198.pdf?sequence=4&isAllowed=y. Accessed 5 August 2020

World Health Organization (2020a) WHO Timeline – COVID-19. https://www.who.int/news-room/detail/27-04-2020-who-timeline—COVID-19. Accessed 10 June 2020

World Health Organization (2020b) Joint Leader's statement—Violence against children: A hidden crisis of the COVID-19 pandemic. https://www.who.int/news-room/detail/08-04-2020-joint-leader-s-statement—violence-against-children-a-hidden-crisis-of-the-COVID-19-pandemic. Accessed 11 Apr 2020

Zweig MH, Campbell G (1993) Receiver-operating characteristic (ROC) plots: a fundamental evaluation tool in clinical medicine, Clinical Chemistry, 39, 561177, [Erratum published in Clinical Chemistry, p. 39]

Arabinda Roy is presently a faculty member in the Department of Geography, Dr. Meghnad Saha College in West Bengal, India. The college is affiliated with the University of Gour Banga. He is graduated in Geography from the same institution in 2011. He completed his postgraduate studies in Geography in 2013 from the Bilaspur Vishwavidyalaya, India, standing eighth place in panels of merit of the university. Mr. Roy was selected for the Indian Council of Social Science Research (ICSSR) doctoral fellowship by the Ministry of Human Resource Development, Government of India. He had submitted his Ph.D. thesis in the Raiganj University, India, on the theme "Social Well-being of Namasudra Community in Dakshin Dinajpur District, West Bengal: A Study in Social Geography". He has a research interest in social well-being and policy issues, related to the welfare and development of the backward communities and regions in India.

Dr. Mukunda Mishra is an Assistant Professor, Department of Geography, and designated Vice Principal of Dr. Meghnad Saha College in West Bengal, India. The college is affiliated with the University of Gour Banga. Dr. Mishra completed his postgraduate studies in Geography and Environmental Management at Vidyasagar University (receiving top rank in both the B.Sc. and

M.Sc. panels of merit) and held a Ph.D. in Geography from the same University. He was selected for the prestigious National Merit Scholarship by the Ministry of Human Resource Development, Government of India. His research chiefly focuses on analyzing unequal human development, regional planning, and multi-criteria predictive model building. He has more than 12 years of hands-on experience in dealing with development issues at the ground level in various districts of eastern India. Dr. Mishra has in his credit to publish one monograph and three edited research volumes so far from the house of Springer-Nature. Dr. Mishra has more than 30 research articles and book chapters published in the journals and books of international repute. Besides serving as the reviewer of several reputed journals published by Springer-Nature and Elsevier Inc., he is continuing as the Managing and Publishing Editor of Ensemble, a UGC-CARE (India) enlisted journal of repute, since its inception to date.

Chapter 8
Human Resource Management during the Pandemic Period: Emerging Challenges for the Public Sector Employees in India

Ahana Sen

Abstract Almost overnight, the Coronavirus (COVID-19) pandemic has transformed the work and workplaces of the public sector. Public servants are playing a leading role in response to the pandemic. Healthcare workers are keeping medical systems functioning and families safe. In the changed scenario, public sector workforces are being asked to work in new ways and new contexts. During this pandemic, both central government and state governments have announced lots of financial packages for the citizens as well as for the industries to enable them to fight against such unforeseen situations. Banking sectors are continuously working to support the government initiatives throughout this pandemic, even during the nationwide lockdown days. They are relentlessly working towards the interest of the general public in this crunch time. Line Ministries and agencies are learning how to use new technology and tools *on-the-go*, often alongside old procedures and processes. The public sector has become "accidentally agile," with new procedures and protocols governing remote working, accelerated hiring processes, and fast-track mobility programs developed with unprecedented speed. Post-pandemic, governments may review and capitalize on many of the changes introduced and place them on a more sustainable footing. In the more immediate term, public employers' task will be how to get public servants back to work safely and resume non-pandemic related service delivery. This chapter outlines what challenges are being faced by the public servants while rendering services during the outbreak; and how the government backs them to overcome. Moreover, it also aims to highlight the perceptions of these employees in balancing the work-life balance in the crisis.

Keywords COVID-19 · Public servant · Government · Workforces · Need hierarchy

---

A. Sen (✉)
Department of Management, Raiganj University, Raiganj, West Bengal, India

S. Saha et al. (eds.), *Economic and Societal Transformation in Pandemic-Trapped India*, New Frontiers in Regional Science: Asian Perspectives 55,
https://doi.org/10.1007/978-981-16-5755-9_8

8.1 Introduction

The disastrous consequence of World War II had shaken almost the entire world 70 years ago. Apart from the world war, many unfortunate events, including natural calamities, have taken place which had affected either a few regions or communities. However, 2019–20 seems to be a curse for the whole world because of the novel coronavirus. COVID-19 has brought the lives of people to a standstill. The outbreak of Covid-19 has taken a toll on economy, finance and social life of almost all the nations of the world. However, even Italy, one of those countries that provide profound health care facilities, faced immense difficulties fighting this deadly virus. In such a disastrous scenario, each country's government had tried its level best to provide the needed facilities and keep the situation under control as far as possible. To combat this deadly situation, governments have launched different schemes in terms of health care facilities, financial assistance, and other basic facilities. The public servants of different nations have been playing a remarkable role in making these facilities available to the citizens.

The fundamental importance of a welfare state comes into focus through the public sector playing a crucial role in responding to, mitigating, and helping resolve the crisis. States, apart from being expected to craft and provide high-quality public health systems in this pandemic, must also address the strains on ordinarily routine processes of governance. For instance, in economically developed states, welfare administrators are trying to address numerous requests for social support, and officials are framing policies to curb economic damage (Cohen 2020). In the developing world, government agencies coordinate the delivery of essential commodities to the poor, such as food and medicine, as social safety net systems collapse (Kazmin et al. 2020). An effective response to the COVID-19 pandemic thus calls for an efficient administration. Efficient administration, in turn, essentially depends on the hard work and capability of the millions of public sector employees from the front line to central administration. However, for many public sector employees, COVID-19 has fundamentally transformed where and how they work, the demands their job places on them, and the demands they face outside their jobs (Schuster et al. 2020).

In developing nations like India, there is a considerable disparity in the way of living in the urban and rural areas because India has not developed equally across its length and breadth. So, when the government of India had to declare nationwide lockdown on 25th March 2020 (Phase I, for 21 days) to the initial form of the outbreak as the urban health hazard, it was the rural areas of the country which were the worst hit (The Hindu 2020). Food delivery services were prohibited by many states, though there was approval from the central government. As a result of this, there was an acute shortage of foods in different places. In this dire situation, the state needed to provide the daily necessities to anybody in need. Of course, the immediate requirement for universal food and cash delivery was to be dealt with urgency. In such a situation, the public servants have extended their utmost cooperation to deal with the crisis. The responsibility of making the food materials and other basic necessities available to the common people was bestowed on the public servants. In discharging their duties in this Esteem, the public servants had to face immense difficulties. Reaching out to each and every nook and corner of the country was really a challenging task. In the course of serving the society during this pandemic situation, many public servants themselves had fallen prey to this deadly virus, and many had even lost their lives.

8.2 Literature Review

Employees are the backbone of any organization. It is valid for the government as well. All the schemes and facilities that the government had announced have been channelized to the general people by the government employees. The human resource managers are the key players in maximizing the satisfaction level of the employees during the pandemic situation apart from the workplace guidelines and support, access to information, and financial assistance (Elsafty and Ragheb 2020). The expert interview conducted in Georgia revealed that under pandemic situations, the human resource managers should put more emphasis on developing new policies to accommodate the employees into the hybrid working model (Gigauri 2020).

Now, countries need more than ever strong, agile, and well-linked public administration to respond to the COVID-19 pandemic while continuing to provide necessary and day-to-day state services amidst fast-changing conditions (Schuster et al. 2020). The paper argues that to manage a pandemic, governments need dynamic capabilities and capacity—which are very too often missing. These may include the ability to adapt and learn, align public services and citizens' needs, govern resilient production systems, and govern data and digital platforms (Mazzucato and Kattel 2020). By comparing regressions structured according to Maslow's hierarchy to unstructured regressions, a relationship between need satisfaction was found to exist but not necessarily maintaining the order of Maslow's hierarchy (Dodson 2014). A study conducted by Taormina and Gao (2013) attempted to use correlations and regressions to show support for Maslow's hierarchy. Using questionnaire results from 386 adult respondents, they found significant positive correlations between all of the need satisfaction measures, with stronger correlations between adjacent levels. So, their hypothesis holds good and supports Maslow's theory. Moreover, they ran regressions on each need satisfaction using the preceding need satisfactions as well as demographics and exploratory variables as predictors. These regressions provided positive coefficients, indicating a need's satisfaction could be predicted to some extent satisfaction of the lower-level needs, which was their hypothesis (Taormina and Gao 2013).

Many studies could address different issues related to Human Resources management in the public sector in India. However, many studies could be found relating to human resource management in India's public sector during the pandemic and in the post-pandemic period. It is an already known fact that human needs are unlimited, while resources are limited. Whenever different needs remain unfulfilled, human beings may face different challenges. However, the literature survey of this study failed to trace any study, specifically in the Indian context, to address whether the different needs of individuals, specifically as per Maslow's Need Hierarchy theory, are getting satisfied or not for the public sector employees, specifically those, who are in emergency services like health sector, banking sector, and so on, during the pandemic as well as in the post-pandemic period. So, the present study makes a humble attempt to address this gap in the literature.

8.3 Objectives

This study takes into account "Maslow's Need Hierarchy" theory, and it attempts to justify whether the different motivational needs of the individuals as per the theory are works or not for the Indian public sector employees, who are in emergency services during the pandemic post-pandemic period. Hence, the specific objectives of the study are as follows:

- To find out whether different motivational needs of the individuals as per "Maslow's Need Hierarchy" theory are getting satisfied or not for the Indian public sector employees, who are in emergency services during the pandemic period.
- To find out to what extent different motivational needs of the individuals as per "Maslow's Need Hierarchy" theory are getting fulfilled or not for the Indian public sector employees in emergency services during the pandemic.
- To find out whether there is any difference in the need fulfillment pattern across the different sectors like health, banking, municipality, administration, and police.
- To identify the challenges faced by the human resource managers.

8.4 Methods

The present study is exploratory in nature and is based on primary data. Data is collected through telephonic interviews. A brief telephonic interview was possible due to the stringent nature of duties of the public sector employees rendering emergency services in the pandemic period. Here simple random sampling is used. The sample size was 100, which was well comprised of Indian public sector employees from different sectors like health, banking, administration, municipality, etc. For addressing the third objective, a chi-square test for independence was done.

8.5 Overview of Maslow's Need Hierarchy Theory

This five-stage model can be partitioned into deficiency needs and growth needs. The initial four levels are regularly alluded to as lack of needs (D-needs), and the high level is known as growth or necessities (B-needs).

Insufficiency needs emerge because of hardship and are said to spur individuals when they are neglected. Additionally, the inspiration to satisfy such requirements will become more grounded the more drawn out the length they are denied. For instance, the more drawn out an individual abandons food, the hungrier they will turn into.

Maslow (1943) at first expressed that people should fulfill lower-level shortage needs prior to advancing on to meet more significant level growth needs. In any case, he later explained that fulfillment of a requirement is certifiably not an "all-or-none" wonder, conceding that his previous explanations may have given "the bogus impression that a need should be fulfilled 100% before the following need arises" (Maslow and Frager 1987, p. 69).

At the point when a shortage need has been pretty much fulfilled, it will disappear, and our exercises become routinely coordinated towards meeting the following arrangement of necessities that we presently can't seem to fulfill. These at that point, become our notable requirements. In any case, growth needs keep on being felt and may even become more grounded whenever they have been locked in.

Growth needs mostly originate from a longing for personal developments. When these growth needs have been sensibly fulfilled, one might have the urge for meeting the most significant level of need called self-completion.

Each individual is fit and wants to climb the pecking order toward a degree of self-realization. Tragically, progress is regularly disturbed by an inability to meet lower-level necessities. Beneficial encounters, including separation and loss of a job, may make an individual vary between levels of the order.

Consequently, not every person will travel through the progressive system in a uni-directional way yet may move to and fro between various necessities.

Maslow (1943, 1954) expressed that individuals are inspired to accomplish certain requirements and that a few necessities overshadow others.

Our most fundamental need is actual endurance, which will be the principal thing that inspires our conduct. When that level is satisfied, the following level up is the thing that inspires us:

- **Physiological needs** are organic prerequisites for human endurance, e.g., air, food, drink, cover, apparel, warmth, sex, and rest. In the event that these requirements are not fulfilled, the human body cannot work ideally. Therefore, Maslow considered physiological necessities the most significant as the wide range of various requirements become auxiliary until these necessities are met.
- **Safety needs** - Once a person's physiological necessities are fulfilled, security and safety requirements become striking. Individuals need to encounter request, consistency, and control in their lives. These requirements can be satisfied by the family and society (for example, police, schools, business, and clinical

Fig. 8.1 Maslow's Need Hierarchy (Source: Prepared by the author)

consideration). For instance, enthusiastic security, monetary security (for example, work, social government assistance), lawfulness, independence from dread, social solidness, property, safety, and prosperity (for example, safety against mishaps and injury) can be mentioned.

- **Love and belongingness needs** - after physiological and security needs have been satisfied, the third degree of human requirements is social and includes the feeling of belongingness. The requirement for relational connections persuades conduct. Unlike machines, human beings have their own emotion and sentiments. For the sake of sound mental health, every individual needs company of family, friends, and relatives.
- **Esteem needs** are the fourth level in Maslow's chain of command - which Maslow arranged into two classifications: (i) Esteem for oneself (poise, accomplishment, dominance, autonomy) and (ii) the craving for notoriety or Esteem from others (e.g., status, eminence). Maslow showed that the requirement for Esteem or notoriety is generally significant for kids and youths and goes before genuine confidence or pride.
- **Self-actualization needs** are the most elevated level in Maslow's order and allude to acknowledging an individual's latent capacity, self-satisfaction, looking for self-awareness, and pinnacle encounters. Maslow (1943) portrays this level as the longing to achieve all that one can, to turn into the most that one can be.

As per Maslow's Need Hierarchy theory (Fig. 8.1), the employees' needs may be arranged chronologically from the lowest level to the highest level, as discussed above. In the pandemic situation, we found two types of employees in the public sector. One group of employees worked in non-emergency services or in such services where they can discharge their responsibilities virtually (e.g., education, land revenue, motor vehicles). The other group of employees in emergency services had been fighting against the deadly virus in different ways. For example, people in the administration had been making different decisions regarding lockdown, selecting a contentment zone, sending migrant workers to quarantine centers. Health staff was combatting with COVID affected patients in different government hospitals, health centers. Bank employees are also not left with any other option other than visiting their workplace. Similarly, employees in urban municipalities and police departments were committed to providing their services consistently, even in this crisis period that made them widely exposed to the virus. However, the emergency nature of their service still made them continue to provide services even at the risk of their lives.

As per Maslow's theory, people achieve job satisfaction depending on the fulfillment of specific needs. The telephonic interviews reveal whether it was true for the emergency public service sectors in India. The emergency department employees had to reach their workplace either at their regular place of office or the places where they were allotted, though, in few sectors like the police department, employees were visiting their workplaces on a rotational basis. In that situation, employees had to stay at their offices for emergency services 24 $\times$ 7, making it essential to arrange food, drinking water, refreshments, and rest hours in the workplace. Except for the health departments, all the emergency departments' employees had to arrange their food on their own. Only quarters or guest houses were arranged for the employees, but in most cases, if they had to stay for a long time, the employees had to bear the rent of those rooms in some way or the others. Only 20 percent of the respondent, public sector employees, could report that their employers arranged proper food and shelter for them considering COVID-19 protocol. In this crisis time, when offices almost became the home of these employees, it could be observed that, in some cases, even the basic physiological needs of the employees were not getting appropriately satisfied.

During the nationwide lockdown, employees of the municipalities, administrative department, police department, and health department were provided with transport facilities to pick up their employees from specific locations and dropped them at their respective offices or places. Though some of the bank employees were provided with pick up and drop facilities, most of them had to make their own arrangements. In the rural areas, as most of the people have their own bicycles or motorbikes, they were not provided with any transport facility. But during unlocks once the public transports, i.e., buses and local trains started reopening, this pick and drop facilities were discontinued in many sectors. 57 percent of the respondent public sector employees reported about the availability of the pick-up and drop-off facility from their employers.

As these employees may be directly exposed to the COVID affected persons, their need for face masks, gloves, sanitizers, PPE suits, etc., became absolute necessities. In this crisis situation, when maintaining social distance, frequently washing hands with alcohol-based sanitizers, constantly wearing masks was need of the hour, then government sector employees were to some extent getting those safety measures. The employers had supplied masks and sanitizers to their employees, but those were not in adequate quantity and the quality was also not up to the mark. The quality of the sanitizers was very harsh to sensitive skin, which sometimes resulted in skin problems for some employees. To ensure proper quality masks, gloves, sanitizers, employees had to carry these independently. 37 percent of the respondent public sector employees reported about the availability of the safety measures like PPE kit, masks, gloves, hand sanitizers, etc. from their employers and 63 percent of the respondent public sector employees reported that they received the safety measures only to some extent, which was not at all satisfactory.

While performing their duties, these employees had higher chances of getting affected by COVID-19 as they were widely exposed to the virus. Even the infected employees further transmitted infection to their family members, which, in the worst cases, had caused to loss of the lives of the employees along with their family members in some instances. Usually, the public sector employees are provided with some medical facilities from the government, but not many additional medical facilities were offered to them in this crisis situation. However, for the health sector staff, in the hospitals in which they are employed, seven beds are kept aside for the employees and their families in case they face any medical emergency. Moreover, many public banks' employees were provided with medical cards, which the employees and their family members can use if they or their families suffer from COVID-19. 51 percent of the respondent public sector employees reported that they and their families got proper medical facilities from their employer. In case of the demise of the employees, while discharging their duties, different state governments or union governments had announced different financial packages for martyr's family or may offer suitable jobs for the eligible family members. 17 percent of the respondent public sector employees reported that their families got proper financial assistance from their employer in case of the employee's death in this pandemic.

It is worth mentioning that most of the sectors like health, municipality, administration, etc. had taken proper initiatives for providing vaccines like "Covishield" or "Covaxin" to their employees. This pandemic has snatched jobs of different employees, mainly those in the private sector and those engaged in part-time and contractual jobs. However, the public sector employees, particularly those who are employed on a permanent and full-time basis, have job securities. Therefore, many private sector employees had to experience salary cuts during this challenging phase. Unlike the private sector, there was no such salary cut in the public sector in this pandemic period for the full-time employees. However, for the part-time employees of the municipality, there was 50 percent salary cut temporarily, but later on, the entire salary was paid. 79 percent of the respondent, who are public sector employees, reported that they got their wages and salaries timely, without any deduction. So, quite obviously, it can be observed that the safety needs of the employees were satisfied up to a certain limit.

Unlike machines, human beings have their own emotions and sentiments. That is why human beings need the company of family, friends, and relatives. But the public sector employees in emergency services had to stay away from their near and dear ones because of discharging their duties at their workplaces. Unlike the other group of employees, they did not get any "work from home" facility. It was indeed a very stressful situation for the Indian public sector employees who were engaged in emergency services. Even on weekends and off days, many of the employees could not visit their homes, as they may be exposed to the virus, and in that case, visiting home may increase the chances of their family members getting affected by the virus. Some of the employees, especially those in the health sector, were so busy with their work, almost throughout the day, that there was no time for any relaxation. The employees had to bear all their worries on their own, without any option of sharing their stress with their near ones. Almost none of the respondent, public sector employees, could report that they got motivation from their employers as they were away from their families. However, it is worth mentioning that employees have received different kinds of support in some cases, including certain medical support for their family members. It can be clearly observed that the social needs of the Indian public sector employees mostly remain unsatisfied.

Employees of the Indian public sector, especially those who are in emergency services, have undoubtedly played a remarkable role in this challenging situation. But many did not receive enough appreciation of their work, which they deserved. Almost all of the respondent public sector employees reported that no appreciation was received from their employers for their dedicated work towards society. So, it is pretty evident that most Indian public sector employees' esteem needs remain unfulfilled.

Lastly, it may be observed from the telephonic interview that, only for some of the senior-level employees, self-actualization needs may be satisfied to some extent. However, for most of the employees, self-actualization needs remain unsatisfied.

Finally, for a clear understanding, the relevant information gathered from the telephonic interview is represented in Table 8.1.

As far as the first objective is concerned, it would be wrong to conclude that none of the needs are satisfied. It may be noted that some of the needs are satisfied but the needs on the upper side of "Maslow's need hierarchy" tend to remain unsatisfied for the Indian public sector employees, who are in emergency services.

The second objective of this study may also be addressed from the information collected from telephonic interviews. For most of the cases, it can be observed that the physiological needs and the safety needs are getting satisfied to a large extent. However, social needs remain unfulfilled for most Indian public sector employees in emergency services. On the other hand, in the case of esteem needs, only a few employees have been able to satisfy this need, but it remains unsatisfied for most of the others. Lastly, self-actualization needs are met by only some senior employees, but for most of the other employees, this need remains unfulfilled. So, it would not be wrong to note here that some needs are getting satisfied to a larger extent, some needs are getting satisfied to a lesser extent, and some needs are not getting satisfied at all.

For addressing the third objective, the following hypothesis is formulated:

H_0: Department of employment and level of need satisfaction is independent.

H_a: Department of employment and level of need satisfaction is dependent.

The critical value of $\chi 2$ with 8 degree of freedom is 15.507. Since, 59.9979 > 15.507, we reject the null hypothesis and conclude that the need satisfaction level of the public sector employees depends on the department in which they are employed at a 5% level of significance (Table 8.2).

Table 8.1 Employees perception towards satisfaction of different needs

Statements	Yes	To some extent	No
During nationwide lock down employers provided proper transport facilities towards their employees.	57%	3%	40%
During unlocks employers continued transport facilities towards their employees.	0%	0%	100%
During lock down employers provided all the safety measures like PPE kit, masks, gloves, hand sanitizers, etc. to their employees	37%	63%	0%
While performing their duty many public servants got victimized of COVID-19. They and their family got proper medical facilities from their employer.	51%	2%	47%
While performing their duty, many public servants lost their lives. Their family got proper financial assistance from their employer.	17%	23%	60%
During pandemic all the employees got their salaries, wages, etc. timely and no deduction were made from there.	79%	21%	0%
Employees got motivated from their employers since they were away from their families.	0%	0%	100%
Employees got proper appreciation from their employers for their dedicated work towards society.	0%	100%	0%
During lock down many public servants compelled to stay at their offices. Their employers arranged proper food and shelter for them considering COVID-19 protocol.	20%	20%	60%

Source: Prepared by author as per telephonic interview

Table 8.2 Calculation of χ^2**statistics**

(Observed freq. – Expected Freq.)	(Observed freq. – Expected Freq.)^2	$\frac{(Observed\ freq.-Expected\ Freq.)2}{Expected\ Freq.}$
2.2000	4.8400	0.0837
−20.4700	419.0209	6.3039
−15.1300	228.9169	4.6594
1.2100	1.4641	0.0461
32.1900	1036.1961	12.3636
0.2000	0.0400	0.0010
0.2300	0.0529	0.0012
17.1700	294.8089	8.7144
11.1100	123.4321	5.6387
−28.7100	824.2641	14.2829
−2.4000	5.7600	0.0699
20.2400	409.6576	4.3231
−2.0400	4.1616	0.0594
−12.3200	151.7824	3.3491
−3.4800	12.1104	0.1014
$\chi^2 = \Sigma\frac{(Observed\ freq.-Expected\ Freq.)2}{Expected\ Freq.}$		**59.9979**

Source: Prepared by authors

The information gathered from the telephonic interview is also consistent with the result of the hypothesis testing. It is clearly observed that the level of need satisfaction varies across different public sector departments in India.

For the health sector, physiological needs and safety needs are mostly getting satisfied, social needs are getting satisfied to some extent, and the other needs mostly remain unsatisfied.

For the banking sector, physiological needs are mostly getting satisfied, safety needs and social needs are getting satisfied to some extent, and the other needs mostly remain unsatisfied.

For the municipalities, physiological needs and safety needs are mostly getting satisfied, and the other needs mostly remain unsatisfied.

Safety needs are mostly getting satisfied for the administration sector, physiological needs and esteem needs are getting satisfied to some extent, and the other needs mostly remain unsatisfied.

For the police department, physiological needs, safety needs, and esteem needs are getting satisfied to some extent, and the other needs mostly remain unsatisfied.

So, it is clearly observed that there exists a difference in the need fulfillment pattern across the different sectors like health, banking, municipality, administration, and police.

8.6 Challenges for the Human Resource Managers

COVID-19 converts every person's reality overnight. Managers have had to make many decisions in a concise span Esteeming which employees are supposed to carry out work from home and how this could be done. In a survey conducted by Earnst & Young (EY 2020), it was observed that only 20 percent of the executives surveyed believed that their companies were ready to respond to a large adverse risk. The pandemic compelled a major chunk of stressed employees to perform their duties staying at home. Under these circumstances, the managers' jobs become more challenging as they need to communicate and lead from a distance because of work from home. After COVID-19, all recruitment, selection, training, and employee support procedures need to be modified. Uncertain situations put extra stress on the employees. In order to cope up with such stress, employees leverage their dispositional traits and coping responses (Shaffer et al. 2006). Stress restricts employees' ability to empathize with others, consider plausible options remain open-minded, engage in cognitively challenging tasks, and expand one's experiences to learn and grow. When individuals go through a stressful time, they tend to find peace in the familiarity with people, places, and even food that are the most predictable; that is why there are expatriate communities, demographic faultlines, and comfort food in every culture over the world (Caligiuri et al. 2020).

The COVID-19 has produced immense novelty and uncertainty, which is harming most people's mental health all over the globe (World Health Organization 2020). Even as the health risk of the pandemic begins to wane in few nations and the chances of a vaccine appear high, the novel ways of working remotely and the fears around the world recession will continue to produce a state of uncertainty.

Some people may be capable of handling stress and uncertainty in a calm and composed manner, which provides them with an edge to make proper decisions and carry out their work more effectively across different regions and cultures. Employees with a higher degree of ambiguity tolerance are less likely to suspect the adverse impact of stress caused by working in uncertainty. Employees with resilience have the capability to bounce back after stressful situations. Employees having natural curiosity can adapt first to novel situations, thrive in challenging situations, and even be more innovative.

Organizations should select for these three critical cultural agility competencies during this pandemic situation: ambiguity tolerance, resilience, and curiosity for all employees working multiculturally (Caligiuri et al. 2020). Employees, who are working virtually and are dealing with different people will now need these competencies for being effective. So, selection is very vital.

In this pandemic situation, as there is health-related stress all-over cross-cultural, team cohesion must be encouraged. Training to promote the establishment of relationships is the need of the hour as each team member faces the same kind of stress. When shared, stress and anxiety may create ties and can promote cohesion going forward. In this pandemic scenario, all have a common enemy, COVID-19. Cross-cultural training would assist in minimizing ambiguity for cross-cultural differences with the help of skilled development on how to collaborate across cultures, seek similarities with colleagues from different cultures actively, use technology inclusively, etc. Such training also addresses how to add respectful questioning into cross-cultural workgroups and how to do away with the negative aspects of stereotypes. During the COVID-19 pandemic, lack of professional stimulation during work from home encourages more and more self-driven knowledge-seeking to meet up the need to learn, grow and demonstrate competencies (Caligiuri et al. 2020).

An individual's comfort in a known environment can affect their success, and organizational support can also affect adjustment in a novel environment on a positive note. Many employees have even faced health risks while carrying out their work. In this pandemic situation, the issues that call for support may differ from individual to individual depending on their work-life issues, but organizational support is vital. Initiatives must be taken to provide support practices to help mitigate stress in the form of the webinar on resilience, employee assistance program, and virtual counseling services. These mitigating stress offerings would be of immense help for the employees who engaged in virtual work. Presently the entire world is going through a stressful situation. So, it would be wise for human resource managers to foster cohesion in this uncertain situation by applying proper selection, training, and support practices.

8.7 Conclusion

Here a humble attempt is made to fit public servants into Maslow's need hierarchy. The physiological needs of the public sector employees were mostly satisfied. The safety needs of the public sector employees were also more or less satisfied to a considerable extent. Being a public servant, many persons may be ensured with job security to a certain extent, but while serving the society in the pandemic situation, many of them had fallen prey to this deadly virus, and some had even lost their lives. But, in most of the departments, the employees were provided with the required safety measures like hand sanitizers, masks, gloves, social distancing, etc. It is worth mentioning that in many departments, initiatives have been taken for vaccinating the employees against COVID-19. In the third level of the ladder, the social life of the public sector employees has been very badly affected. Employees in the health care department, banking sector, police, administrative department, municipalities, etc. had to be in direct contact with the people, including COVID-19 affected patients, so they had to isolate themselves from their family, friends, and relatives. This had taken a toll on the mental health of the employees. For other employees who are working from home, even their task was a challenging one. Doing office work from home, in online mode, and managing the household works parallelly created immense stress on the employees' minds. So, the social needs of the employees could not be met. The esteem needs of the employees were also not satisfied fully because, in many instances, the employees did not get the proper recognition for the services they provided. At the highest level of the hierarchy self-actualization needs of the employees were also unsatisfied. In this crisis, people were simply struggling for sustainability and survival. So, from Maslow's need hierarchy, only the first two needs from the bottom are getting satisfied to a large extent for the public servants. However, from the third level onwards the needs mostly remain to be unsatisfied. This had created immense stress on the minds of the employees. In dealing with such a stressful situation, the role of the human resources department is very crucial.

This pandemic situation is a very tough and challenging phase in every aspect and more or less, it had its negative impacts on most organizations. However, this scenario can be handled and managed by human resources managers if they are ready to take proper initiatives to address this new challenge.

References

Caligiuri P et al (2020) International HRM insights for navigating the COVID-19 pandemic: implications for future research and practice. J Int Bus Stud 51:697–713

Cohen P (2020) Jobless numbers are "eye-watering" but understate the crisis. New York Times April 23.. https://www.nytimes.com/2020/04/23/business/economy/unemployment-claims-coronavirus.html. Accessed 6 July 2020

Dodson T (2014) Exploring Maslow's hierarchy of needs empirically

Elsafty AS, Ragheb M (2020) The role of human resource management towards employees retention during covid-19 pandemic in medical supplies sector - Egypt. Business and Management Studies 6(2):50–59

EY (2020) Global board risk survey. EY. (n.d.). https://www.ey.com/en_us/news/2020/04/nearly-80-percent-of-board-membersfelt- unprepared-for-a-major-risk-event-like-covid-19-ey-survey. Accessed 15 May 2020

Gigauri I (2020) Implications of COVID-19 for human resource management. SSRG Int J Econ Mgmt Studies 7(11): 25–33. doi:https://doi.org/10.14445/23939125/IJEMS-V7I11P104

Kazmin A, Pilling D, Schipani A, Wheatley J (2020) Threat of catastrophe stalks developing world. Financial Times 12(3)

Maslow AH (1943) A theory of human motivation. Psychol Rev 50(4):370–396. https://doi.org/10.1037/h0054346

Maslow AH (1954) Motivation and personality. Harpers, New York

Maslow AH, Frager R (1987) Motivation and personality. Harper & Row, New York, p 69

Mazzucato M, Kattel R (2020) COVID-19 and public-sector capacity. Oxf Rev Econ Policy 36(1): S256–S269. https://doi.org/10.1093/oxrep/graa031

Schuster C et al (2020) Responding to COVID-19 through surveys of public servants. Public Adm Rev 80(5):792–796

Shaffer MA, Harrison DA, Gregersen H, Black JS, Ferzandi LA (2006) You can take it with you: individual differences and expatriate effectiveness. J Appl Sci 91(1):109–125

Taormina RJ, Gao JH (2013) Maslow and the motivation hierarchy: measuring satisfaction of the needs. Am J Psychol 126(2):155–177

The Hindu (2020). https://www.thehindu.com/news/national/pm-announces-21-day-lockdown-as-covid-19-toll-touches-10/article31156691.ece. Accessed 24 March 2020

World Health Organization (2020) Physical and mental health key to resilience during COVID-19 pandemic. http://www.euro.who.int/en/health-topics/health-emergencies/coronaviruscovid-19/statements/statement-physical-and-mental-health-keyto-resilience-during-covid-19-pandemic. Accessed 15 May 2020

Mrs. Ahana Sen is presently working as an Assistant Professor in the Department of Management at Raiganj University, Raiganj. Her area of specialization is Accounting and Finance, and her research interest lies in Banking and Finance. She is having 5 years of teaching experience. She has presented many research papers in different seminars, conferences at the national and international levels. Mrs. Sen has authored six chapters in different books published from different parts of India.

Part III
New Challenges for the Indian Economy

Chapter 9
Pandemic Outbreak and the Future of Poverty and Inequality Scenario: Indian Perspective

Abhijit Dutta, Partha Mukhopadhyay, Madhabendra Sinha , and Anjan Ray Chaudhury

Abstract COVID-19 has caused an unprecedented pandemic with mass civic lockdown to counter and contain spreads of the virus infections. Besides being a public health challenge, it created a spatial influence on livelihood, economic growth by hampering the entire logistic process. India had been no exception while the country had to deal with a pandemic with a struggling infrastructure in the health care sector; it had to close down all its economic activities to control the spread of the virus. As the period of lockdown got extended, questions of employment and livelihood came up with the partially paralyzed the economic activity over a long period as labour started migrating homewards, spreading the virus on the way. India observed some pessimistic estimations of the economic indicators. RBI estimated the real GDP at 6.2% from 2019 to 2020. It was declared that the GDP had dropped to 3.5%. MSME sector is in a way the backbone of the country since the largest unorganized workforce engaged in it and was estimated to lose 4.5 million US Dollar. To support low- and middle-income labour, about INR 1042500 core was transferred by the government. This chapter tries to understand and explore the linkages between two: pandemic and poverty.

Keywords COVID-19 · Unemployment · Lockdown · GDP · Economic Development

---

A. Dutta
Department of Commerce, School of Professional Studies, Sikkim (Central) University, Gangtok, Sikkim, India

P. Mukhopadhyay
Business Excellence Department, Durgapur Steel Plant, Steel Authority of India Limited, Durgapur, West Bengal, India

M. Sinha (✉)
Department of Economics and Politics, Visva-Bharati University, Santiniketan, West Bengal, India

A. R. Chaudhury
Department of Economics, Durgapur Government College, Durgapur, West Bengal, India

S. Saha et al. (eds.), *Economic and Societal Transformation in Pandemic-Trapped India*, New Frontiers in Regional Science: Asian Perspectives 55,
https://doi.org/10.1007/978-981-16-5755-9_9

9.1 Introduction

The government of India is imperative to focus on both actions, 'Jaan bhi, Jahan bhi' for India's bright, prosperous, and healthy future (Business Standard 2020). In the current situation, so many mathematical models are used to predict the path of COVID-19. The nationwide lockdown of the last three weeks certainly had an adverse impact on agriculture and the allied sector. Major marketing yards are closed, lorry transport is rather limited, and labour is unavailable in adequate numbers. Summer fruits and vegetables are ripened, wheat is ready to harvest, but nationwide COVID-19 lockdown and millions of farmers are staring at huge losses. Even now, Farmers are bound to go for distress selling of their products. Although Distress selling is not new, and farmers were selling their product below the MSP in the past, the circulatory economy has stagnant this time as COVID-19 raises the question of survival of mankind. The nationwide lockdown over the long days certainly had an adverse impact on the agriculture and allied sector. Major market yards are closed, Lorry transport is somewhat limited, and labour is unavoidable in adequate numbers. As market outcomes become uncertain, risk appetite falls, anxiety drive market players to reduce their level of risk. In this context of investing, the behaviour typically involves shifting assets towards more liquid and Government insured assets, which are perceived as safer.

India is the fifth largest economy by nominal GDP ($ 3.202 Trillion) and third-largest by Purchasing Power parity ($ 11.321 Trillion). But as per Centre for Monitoring Indian Economy (CMIE 2021) data, unemployment in India is increasing in the rural sector than in urban areas (see Table 9.1). Hence the burning question arose that what would be the PPP (Purchasing Power parity). The ILO (International Labour Organization) predicts that 25 million jobs will be lost, and the United Nations Development Program (UNDP) expects a loss of USD 220 billion in revenue from developed countries (International Labour Organisation [ILO] 2020).

Table 9.1 Unemployment Rate (%) in India

Month	India	Urban	Rural	Rural- Urban Difference in Percent {(R-U) /R}*100
Dec-20	9.06	8.84	9.15	3.39
Nov-20	6.5	7.07	6.24	−13.30
Oct-20	7.02	7.18	6.95	−3.31
Sep-20	6.68	8.45	5.88	−43.71
Aug-20	8.35	9.83	7.65	−28.50
Jul-20	7.4	9.37	6.51	−43.93
Jun-20	10.18	11.68	9.49	−23.08
May-20	21.73	23.14	21.11	−9.62
Apr-20	23.52	24.95	22.89	−9.00
Mar-20	8.75	9.41	8.44	−11.49
Feb-20	7.76	8.65	7.34	−17.85

Source: Data Source: CMIE (2021)

Table 9.2 Commodity price increases as compared to previous years

Date	Paddy	Potato	Onion	Date	Paddy	Potato	Onion
15-Apr-19	1750	840	1200	15-Apr-20	1815	1700	1600
12-Apr-19	1750	840	1200	14-Apr-20	1815	1760	1600
11-Apr-19	1750	840	1200	13-Apr-20	1815	1720	1600
10-Apr-19	1750	820	1200	10-Apr-20	1815	1640	1600
09-Apr-19	1750	840	1150	09-Apr-20	1815	1600	1700
08-Apr-19	1750	860	1150	08-Apr-20	1815	1560	1600
06-Apr-19	1750	800	1150	07-Apr-20	1815	1560	1600
05-Apr-19	1750	800	1150	06-Apr-20	1815	1500	1600
04-Apr-19	1750	760	1100	04-Apr-20	1815	1400	1600
03-Apr-19	1750	740	1100	03-Apr-20	1815	1440	1800
02-Apr-19	1750	740	1100	02-Apr-20	1815	1440	1800
01-Apr-19	1750	740	1100	01-Apr-20	1815	1440	1800
30-Mar-19	1750	760	1100	31-Mar-20	1815	1480	1800
29-Mar-19	1750	760	1100	30-Mar-20	1815	1500	2000
28-Mar-19	1750	760	1100	28-Mar-20	1815	1520	2000
27-Mar-19	1750	740	1100	27-Mar-20	1815	1560	2000
26-Mar-19	1750	720	1100	26-Mar-20	1815	1600	1600
25-Mar-19	1750	640	1050	25-Mar-20	1815	1660	1800
23-Mar-19	1750	560	1050	24-Mar-20	1815	1520	1800
22-Mar-19	1750	520	1050	23-Mar-20	1815	1420	1200

Source: Data Source: Ag Market Database (2021)

However, as projected by the IMF, the second outbreak of the virus is anticipated to begin in 2021, and financial constraints are anticipated to be tightened twice, leading to an enormous impact on economic activity. It is assumed that there is a non-linear response from financial markets that are scarring (Währungsfonds 2020). This paper explores whether there is a general direction towards 'safety' in terms of sustainability (i.e., economy and the quality of life of Indian citizens). It should be noted that inflation and unemployment are crucial to the recognition of the most recent process of convergence towards purchasing power parity (Juselius 2003). The IMF also projected that India's purchasing power parity was 0.5 in 2020, and the same is expected to be 6.0 in 2021.

On the other side, food prices, such as vegetables and paddy, are steadily increasing in India's high-producing areas/districts. Compared to previous years, there is a high price in Paddy, Potato, and Onion, etc., in this pandemic time. Table 9.2 explains how the commodity price is increased during the lockdown. For example, potato is usually sold at Burdwan Market (in West Bengal, India) at the rate of around Rs. 520/– to Rs. 840/– per quintal, but now it is around Rs. 1700/– (Ag Market Database 2021). Due to the high production of paddy and potato, *'Burdwan'* is called the *'Granary'* of West Bengal, India, and is thus chosen as a sample market.

From the above discussion, it is clear that we are in a crucial position in terms of sustainability issues, particularly Economic and Social factors. Does the economy indicate to another feminine? So, preparing an overcoming infrastructure is of utmost essential.

9.1.1 Risk Management and Challenges of COVID-19 Pandemic

Zoonotic disease COVID-19 is a very high-impact event (Black Swan) and an eclectic approach to risk management. Today, governments, multinationals, and policymakers, as well as researchers from all nations, are committed to alleviating the lacuna of decision-making on future risks.

With rapidly increasing cross-border activities, countries become more interdependent, and the supply chain has been disrupted and re-engineered. In this background, to maintain sustainable social and economic activities, the stable labour supply and buyer concentration policies should be ensured leading to the future of poverty alleviation.

Regarding MSME and the informal sector, millions of jobs have been lost due to sudden lockdown and also for illness. The declaration of the lockdown should be much more calibrated, and there was no clear cut in the sustainable strategy due to the worse effect of spreading COVD-19. The only strategy was to manage the spread of the virus only through surveillance and strict social distancing measures. From the supply side, as employees were unable to go to the workplace due to lockdown and restricted movements, all kinds of businesses are faced with decreased labour supply.

From the demand side, the shock is severe. Negative generation revenue, fear of infection and disease, sudden loss of demand from domestic or foreign customers, termination of aviation, rail, and transport (except emergency duty) weakened business and consumer trust. Overall, it has been observed that overall confidence in the stock market is very low. As a consequence, there will be a dynamic poverty imbalance in the near future. Apart from a lack of financial literacy, a weak conversation with digitalization, the common people were afraid of digital corruption and shifting towards a cashless economy. Besides the existing safety measures, immunity enhancing the healthcare system and opening up a well-formulated economy, with liberalized taxation (in case of direct tax) and imposition of consumption-based indirect tax according to the situation, will withstand immediate demand-side shocks.

9.1.2 Poverty Scenario and Challenges of COVID-19 Pandemic

Although technology has vastly improved in the last 100 years, the world is experiencing sudden stagnation due to the rapid transmission of deadly COVID-19 diseases. COVID-19 induces a total factor productivity shock that produces 'relative poverty' around the world. The biggest challenges of COVID-19 are disruption of incomes for global lockdown, rising unemployment and deflation in demand, and supply chains and, as a result, stagnation of cash flow, food shortages, clean water, healthcare, education, etc., depriving citizens of many quality-of-life activities. Based on social, economic, and health aspects, different types of poverty create challenges in the COVID-19 pandemic era.

9.1.3 Effect of Pandemic on GDP of the Country

The IMF has estimated that India's GDP will contract by 10.3 percent this fiscal year. Also, prior to the pandemic, investments as a share of GDP had been declining in India. This statistic had fallen from 39.6% in 2011 to 29.7% in 2019. In 2021 too, it is expected to decline to 27.8 percent and will not reach the 2018 level of 31.7 percent until 2025 (Währungsfonds 2020).

The domestic economy has been losing the GDP growth momentum. According to MOSPI (2021), the GDP growth rates over last few years are presented in Table 9.3.

9.1.4 Effect of Pandemic on Income Distribution on the Country

Although policymakers are struggling to resolve the implications of lockdowns (in their economies) over the coming months, income disparity is the most pressing economic threat in the nation to the post-COVID-19. The traditional approach to

Table 9.3 GDP Growth Rate in India

Year	GDP Growth Rate (in %)
2015–16	+8.0
2016–17	+8.3
2017–18	+7.0
2018–19	+6.1
2019–20	+4.2
2020–21	−7.7

Source: MOSPI (2021)

pandemic containment (i.e., social distancing and lockdowns) tends to have an enormous impact on unorganized sectors of society because they usually have occupations that cannot be done remotely, such as construction employees, cab drivers, people in housekeeping and maintenance, manufacturing workers, etc. According to a survey conducted by 'India Today' in August 2020, 63% of respondents reported that they had lost their jobs or their businesses had failed, while just 1% of respondents reported that their income had risen during the COVID-19 crisis, and 14% said that their income remained unchanged (India Today 2021).

9.1.5 Effect of Pandemic on Trade of the Country

Some remarkable changes have been observed in the COVID-19 pandemic as follows:

9.1.5.1 Indian Economy Has Transitioned towards More Digitalization

The shock of social distancing in the COVID-19 era forces people to adopt more digitalization in monetary transactions - say, consultation with a doctor to buy face masks or even purchasing groceries.

9.1.5.2 'Work from Home' – Evolve a New Nature of the Job

IT companies and related jobs, such as consulting, etc., are carried out by 'work from home' in the COVID-19 period. Facebook, Twitter, and Google also offer their employees the opportunity to 'work from home' in the long term.

The government of India invoked the Infectious Diseases Act 1897, realizing the very precarious state of its health care infrastructure, and consequently shut down educational institutions and other commercial enterprises just after COVID-19 proclaimed a pandemic. India is one of the first countries in the world to seal its borders to prevent the virus from being transmitted through international passengers. The first lockdown took place between 23 March and 3 May 2020, and the Government of India insisted on taking strict measures with the mass awareness programme to maintain social distance. The government gave some relaxation since the second phase of lockdown, including permitting specified commercial operations in the green zones. However, the burning question was how India would respond to the imminent problems facing the poorer sections of society. According to the Ministry of Labour and Employment, Government of India, about 90% of the working population continues to work in informal sectors (Government of India 2021), migrant workers, and landless agricultural labour. COVID-19 Pandemic and Lockdown are collapsing lives well below the level of financial, social, education and health services.

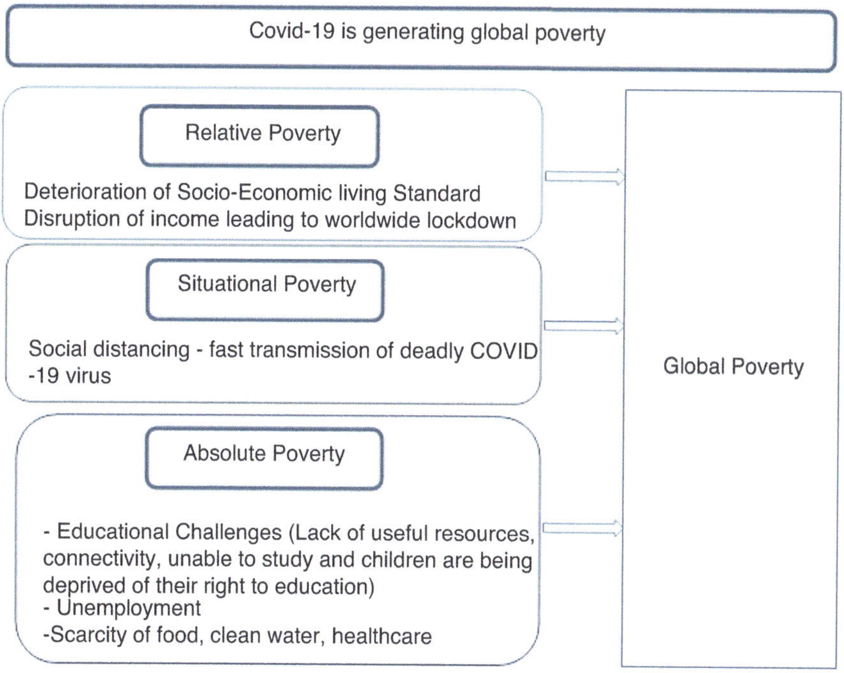

Fig. 9.1 Triggering of global poverty (Source: Authors' own compilation)

Recently, labour law reforms (Government of India 2020) have been introduced by the Government of India with the objective of simplifying and modernizing it. The three labour-bills enacted by Parliament on 23 September 2020 are the industrial relation code 2020, the code on occupational safety, health and working condition code 2020, and social security code 2020 and are specified by the 29th Labour Code.

India is fighting alongside poverty and unemployment with the pandemic and aiming for the future with greater employment in the economy to ensure long-term pathway to sustainable development. The present study carries out some theoretical and empirical investigations to explore the linkages between poverty and pandemic. Figure 9.1 presents a brief scenario of global poverty created by COVID-19.

The rest of the chapter is organized as follows. The next section briefly surveys the available literature, followed by the concise discussions on poverty and inequality in India during the pandemic. The results and subsequent discussions are documented thereafter, followed by the concluding remarks in the final section.

9.2 Literature Survey

The area of studies on poverty and inequality under pandemic outbreaks is not so vast. Dev (2020) addressed an unprecedented challenge to India due to COVID-19 and its effect on productivity, food security, and livelihoods. The study presents the Pradhan Mantri Garib Kalyan Yojona, the activity of the Government of India, to address social safety nets as a bridge between health shock and economic shock. The study reveals that there are nearly 40–50 million seasonal workers in India and hundreds of thousands of migrant workers from many states trudging for miles on highways. India's economy depends on the informal labour market, and the effects of lockdown are highly disruptive.

In the case of India, Singh and Misra (2020) studied the COVID-19 pandemic and identified the use of Chloroquine (CQ), Hydroxychloroquine (HCQ), etc., as a herd mentality. Social media still dominated such conversation. The study concluded that herd activity should be clearly conveyed to all segments of the population, including physicians.

Sahasranaman and Jensen (2020) studied poverty in the pandemic situation. The study focuses on the differential nature of the impact of the pandemic on poor and non-poor communities using the SIR model (Susceptible Infected Recovered). The study finds that under non-intervention situations, regarding infection, there is indifference between the poor and non-poor. Under intervention, poor are consistently worse-off than the non-poor. The study concludes that policies include the development of the entire healthcare infrastructure of developing countries.

Sengupta and Jha (2020) studied the social and economic impact of migrants in the COVID-19 pandemic situation in India. The study describes that about ninety percent of the Indian workforce is employed in the unorganized sector, and their workplaces are placed a great distance from rural homes. The migrant workers faced survival crises at the sudden lockdown through hunger, income loss, destitution, and fearful communities maintaining 'social distance'. The study concludes that India's social policy must rise up to address resurgent poverty.

According to the CMIE report, the unemployment rate jumping from 7.03 percent in May 2019 to 23.98 percent in May 2020. According to a trade union report, nearly seventy thousand people, attached with domestic service and construction work, left Gujarat for their home state just after the announcement of lockdown by the Government of India. (Sharma and Khanna 2020). On the migrant workers' journey, the police forced them to continue walking and not rest or gather anywhere (Pal and Siddiqui 2020). The tragic occurrence is the death of sixteen migrant workers crushed to death by a goods train as they fell asleep on the railway track (Banerjee and Mahale 2020).

Buheji et al. (2020) studied the socio-economic impact on global poverty. The study examined the causes of poverty, the global pandemic crisis in poor communities on four continents and identified vulnerable conditions in the context of restrictive social isolation or lockdown measures. The study concludes that the problems of the poor: their sufferings from hunger, loss of health and livelihoods, and even deaths, are neglected.

Chaudhary et al. (2020) described the effect of COVID-19 on the economy in India. The study explores the impact of all hell breaks on aviation, tourism, capital markets, small and medium-sized enterprises, and the oil industry. The study concludes that multinationals are losing trust in China, and the COVID-19 pandemic aims to implement sustainable development models focused on self-reliance, inclusive and environmentally friendly systems for the Indian economy.

Bloom et al. (2005) reviewed the summary published by the Asian Development Bank in order to assess the economic effects of the avian flu pandemic on Asian economies. Using macroeconomic simulations based on the Oxford Economic Forecasting (OEF) model, which affects both supply and demand sides, the study suggests adapting to a new equilibrium following a shock. The background as mentioned above encourages us to conduct the present study with an emphasis on poverty and inequality under pandemic outbreaks in India.

9.3 Brief Descriptions of Poverty and Inequality in India during Pandemic

The Ministry of Home Affairs (MHA) issued a variety of advice to state governments using a section of the National Disaster Management Act (NDMA) (2005), together with an order to limit the movement of migrants with the provisions that have to be created for them (Ministry of Home Affairs; 29 March and 3 April 2020). A critical gap was the lack of compliance with the minimum standard of relief (including the provision of food, water, shelter, and sanitation) and compensation for the loss of life and living that must be granted to affected persons under the Disaster Management Act (Sibal 2020). In addition, the government has shifted the vital responsibility for disaster relief to states and homeowners, while the NDMA (2005) needs the central government to require some monetary responsibility for immediate disaster relief. In temporary shelters wherever migrants are confined, activists and volunteers have expressed fears about the appallingly bad conditions. The social system of Bharat thrives on the mutuality of families, relatives, and friends, both emotional and economic. Close physical interactions such as living in crowded accommodation and in different locations, pushing and jostle square measures very common and square measures of 'social distance' as decided in this pandemic. Amid the internment, conditions were founded at spiritual places, on road-journey (e.g., 'herds' of refugees on buses), or even while liquor is in the outlet. While 'vertical distance' is the cause of inequalities in India, the position of 'horizontal distance' in the aftermath of COVID-19 has intensified these inequalities. The additional troubling aspect is that the lack of proper provision of safety nets (e.g., food safety) for those most affected by internment, due to the large scale of the matter, remains highly inadequate to government schemes. There is an accrued risk of deficiency disease among the low SES as a result of internment. During the COVID-19 crisis under PMGKAY, around 120 LMT food grains were distributed freely to 80 crore individuals, i.e., approximately two-thirds of India's population (Ministry of Consumer Affairs, Food & Public Distribution, 6 May 2020). Using the epidemic control model, a primary view was that the emergence of sensitive blocks largely caused the inequalities in nature that arose at the time of epidemics by advisers and other regulations.

9.4 Results and Discussion

In order to study the factors for determining the pandemic's effect on the economy, a general regression has been estimated. The basic construct is that the country's GDP is being negatively affected by the onset of the pandemic leading to inequality of distribution. For the purpose of the model, we take GDP at Factor Cost as the dependent factor. The independent factors have been taken on merit. The independent factors include a change in income (depicted by the period on period change in per capita income), the change in consumption (taken as a change in household expenses on spending period on period), and the change in foreign trade (taken as import-export period on period change). Data have been used from 2016 Q_1 to 2021 Q1 in the case of India.

The expected change in the sign is negative for all the independent factors if the pandemic has affected the distribution and positive if it has not affected. The model can be presented as follows:

$$\Delta GDP = \alpha_i + \beta_1\,\Delta I + \beta_2\,\Delta C + \beta_4\,\Delta FT + \varepsilon_i \tag{9.1}$$

In the Eq. (9.1),

ΔGDP = Period on Period change in GDP.
ΔI = period on period change in income.
ΔC = period on period change in consumption (based on Private Final Consumption Expenditure).
ΔFT = period on period change in Foreign trade.
α, β and ε are the constants and coefficient, whereas the last term is the error term in respect of the stochastic model.

It is clear from Table 9.4 that consumption is significant with respect to GDP, which means with the negative growth rate of GDP, their consumption rate is also contrasted.

Table 9.4 Analysis of foreign trade, income, and consumption with respect to GDP

GDP	Coefficients	Linearized Standard Error	t-Statistic	P > t	95% Confidence Interval	
Change in income	−0.033100	0.040870	−0.810000	0.477000	−0.163170	0.096963
FT	−0.000040	0.000008	−4.930000	0.016000	−0.000063	−0.000014
Change in consumption	1.341710	0.094328	14.220000	0.001000	1.041519	1.641906

Source: Authors' Computation in STATA

9.5 Conclusions

Because of COVID-19, a sudden new form of the economic wave has been confronted worldwide, and India is no exception. As a consequence of global lockdown, the stagnant economy has evolved, and the Indian economy has experienced an adverse impact, as focused by negative estimates of some economic indicators. The Indian economy is shifting towards global poverty. The distressed economy saw a large number of labour migrations. To support low- and middle-income labour about Rs. 1,042,500 core was transferred by the Government of India, which was equal to 1.39 million USD. The remittances inflow was expected to fall by 23%, from 83 million USD to 64 million USD. Thus the pandemic affected India's new identity as a nation-state which has long been grappling with poverty and injustice since the beginning. It is necessary to deal with the argument that the nation addresses the pandemic besides poverty and unemployment. This chapter has been a try to explain and discuss the connection between poverty and the pandemic.

References

Ag Market Database (2021) Ministry of Information and Broadcasting, Government of India.. https://agmarknet.gov.in/PriceTrends/

Banerjee S, Mahale A (2020) 16 migrant workers run over by goods train near Aurangabad in Maharashtra. Hindu 8

Bloom E, De Wit V, Carangal-San Jose MJ (2005) Potential economic impact of an avian flu pandemic on Asia

Buheji M, da Costa Cunha K, Beka G, Mavric B, De Souza YL, da Costa Silva SS et al (2020) The extent of covid-19 pandemic socio-economic impact on global poverty. A global integrative multidisciplinary review. Am J Econ 10(4):213–224

Business Standard (2020) Our mantra was 'jaan hai to jahaan hai' but now it is 'jaan bhi jahaan bhi': PM Modi.. https://mybs.in/2YMPoCn

Chaudhary M, Sodani PR, Das S (2020) Effect of COVID-19 on economy in India: some reflections for policy and programme. J Health Manag 22(2):169–180

CMIE (2021) Centre for Monitoring Indian Economy – Unemployment in India.. https://unemploymentinindia.cmie.com/

Dev SM (2020) Addressing COVID-19 impacts on agriculture, food security, and livelihoods in India. IFPRI Book Chapters 12:33–35

Government of India (2020) Labour Law Reforms

Government of India (2021) Ministry of labour and employment.. Employment in Informal Sector and conditions of Informal Employment

India Today (2021) India Today Report 2021

International Labour Organisation [ILO] (2020) ILO monitor: COVID-19 and the world of work (2nd ed.)

Juselius, K. (2003). Institute of economics university of Copenhagen

MOSPI (2021) Ministry of Statistics and Programme Implementation. Government of India

Pal A, Siddiqui D (2020) Special report: India's migrant workers fall through cracks in coronavirus lockdown

Sahasranaman A, Jensen HJ (2020) Poverty in the time of epidemic: a modelling perspective. PLoS One 15(11):e0242042

Sengupta S, Jha MK (2020) Social policy, COVID-19 and impoverished migrants: challenges and prospects in locked down India. Int J Commun Soc Dev 2(2):152–172

Sharma S, Khanna S (2020) India's migrant workers face long walk home amid coronavirus lockdown

Sibal (2020) The Hindu, 3 June, 2020

Singh AK, Misra A (2020) Herd mentality, herds of migrants/people, and COVID-19 in India. Diabet Metab Synd 14(4):497

Währungsfonds I (2020) World economic outlook April 2020, the great lockdown. Washington, DC, International Monetary Fund

Dr. Abhijit Dutta is currently working as the Dean of the School of Professional Studies and Professor at the Department of Commerce, Sikkim (Central) University, India. He was an ICSSR Research Scholar at the Institute of Public Enterprises, Hyderabad. He completed the Masters in Commerce from Utkal University and PGDBM from Symbiosis Pune. He was awarded Ph.D. from Utkal University, Odisha, in 1997. He was awarded the DLit in Management from Sambalpur University. He has taught at several Business schools in India including IISWBM, KSOM, AMITY Business School, and ICFAI Business School. He is represented in the UGC-NAAC Peer Team, academic council of several universities, and the member of the governing boards of government and private collages. He has published in international and national journals including Vikalpa, Vilakshan, Chartered Accountant, Chartered Secretary, Labour and Management Review, Journal of IPE to name a few. He has published extensively in Taylor and Francis, Elsevier, and Rutledge. He has published twelve text and research books in the area of finance, commerce, and ethics. His research interests include capital market, public policy, behavioural finance, infrastructure, and public enterprise. He is a life member of Indian Social Science Congress, Indian Commerce Association, and Indian Accounting Association.

Dr. Partha Mukhopadhyay is an Assistant Professor at the Department of Economics & Politics, Visva-Bharati (University), Santiniketan, India; prior to that he served as an Assistant Professor at the Department of Business Administration, Raiganj University, West Bengal, India. He has been a Senior Research Fellow and received his Ph.D. in Economics at the National Institute of Technology Durgapur, India. He obtained his M.Sc. and M.Phil. in Economics from the University of Calcutta, India, and also qualified UGC NET in Economics. His research interests include international economics, development economics, and applied econometrics. He has to his credit a couple of research publications in several reputed journals including *Journal of Policy Modeling* (Elsevier), *Social Indicators Research* (Springer), *Economic Papers* (Wiley), *Global Business Review* (Sage), *Applied Intelligence* (Springer), *Margin: JAER* (Sage), *Journal of Asian and African Studies* (Sage), *International Journal of Education Economics and Development* (Inderscience), etc. He contributed a number of book chapters published by Springer, Routledge, Emerald, IGI Global, etc. He has participated in several workshops and presented research papers in various conferences and seminars organized by ADBI, IGC, IITs, IIMs, ISI, DSE, IIFT, BIBM & UNCDF, RIS-EXIM Bank, etc. in India, Japan, and other countries. He acted as a reviewer of different journals published by Taylor & Francis, Wiley, Emerald, and Sage. He organized several workshops and seminars on research methodology and other issues and gave invited lectures in different universities.

Dr. Madhabendra Sinha is an Assistant Professor at the Department of Economics & Politics, Visva-Bharati (University), Santiniketan, India; and prior to that he served as an Assistant Professor at the Department of Business Administration, Raiganj University, West Bengal, India. He has been a Senior Research Fellow and PhD in Economics at the National Institute of Technology Durgapur, India. He obtained MSc and MPhil in Economics from the University of Calcutta, India, and also qualified UGC NET in Economics. His research interests include international economics, development economics and applied econometrics. He has credits of a couple of research publications in several reputed journals including Journal of Policy Modeling (Elsevier), Social Indicators Research (Springer), Economic Papers (Wiley), Global Business Review (Sage), Applied Intelligence (Springer), Margin: JAER (Sage), Journal of Asian and African Studies (Sage), International Journal of Education Economics and Development (Inderscience), etc. He contributed a number of book chapters published by Springer, Routledge, Emerald, IGI Global, etc. He has participated in several workshops and presented research papers in various conferences and seminars organized by ADBI, IGC, IITs, IIMs, ISI, DSE, IIFT, BIBM & UNCDF, RIS-EXIM Bank, etc. in India, Japan and other countries. He acted as a reviewer of different journals published by Taylor & Francis, Wiley, Emerald and Sage. He organized a number of workshops and seminars on research methodology and other issues and gave invited lectures in different universities.

Dr. Anjan Ray Chaudhury is a Senior Assistant Professor and the Head of the Department of Economics, Durgapur Government College, West Bengal, India. He has obtained M.Sc. and Ph.D. in Economics from the University of Calcutta, India. He has the credit of 22 years of research and teaching experiences. He received the research fellowship from the University Grants Commission, India. Some of his papers have been published in reputed national and international journals including Social Indicators Research (Springer), Journal of Policy Modeling (Elsevier), Margin: JAER (Sage), Journal of Asian and African Studies (Sage), International Journal of Education Economics and Development (Inderscience) and books published by Springer, Routledge, Emerald, IGI Global, etc. His research interests are welfare economics, development economics, advanced econometrics, etc.

Chapter 10
Healthcare Expenditure and Economic Development Dynamics in India: Experiences from COVID-19 Pandemic

Subrata Saha

Abstract The relationship between health and development is an important topic in the field of economics. India's new Health Policy 2017 also acknowledges this relationship. The COVID-19 pandemic has exposed the chronic underinvestment in India's healthcare system as India's public healthcare system is facing a severe shortage of different resources and infrastructure. The present study aims to investigate the dynamics between healthcare expenditure- public health expenditure and private health expenditure and economic growth for the period from 1999 to 2018. The Johansen cointegration test exhibits that public health expenditures, private health expenditures, and economic growth are cointegrated, implying a long-run relationship between the variables concerned. The result of the Vector Error Correction Model represents that unidirectional causality between public health expenditure and economic growth and causality running from public health expenditure to economic growth exists and another unidirectional causal relationship which runs from economic growth to private health expenditure also exists for the sample period from 1999 to 2018. Block Exogeneity Wald Test has validated the results obtained from VECM. The impulse response functions and variance decomposition reveal weak causality (Granger) link between public health expenditure and economic growth and strong causal (Granger) relationship between Domestic private healthcare expenditure and economic growth during the study period. CUSUM and CUSUM-Square Test confirm that the parameters estimated by the model are stable for the period 1999 to 2018. The study concludes that the impact of public health expenditure on economic growth is relatively small, while economic growth on domestic private healthcare expenditure is relatively strong in India during the period of study.

Keywords Public healthcare expenditure · Private healthcare expenditure · Economic Growth · Johansen Cointegration test · Vector Error Correction Model · Granger Causality

---

S. Saha (✉)
Department of Economics, Raiganj University, Raiganj, West Bengal, India

S. Saha et al. (eds.), *Economic and Societal Transformation in Pandemic-Trapped India*, New Frontiers in Regional Science: Asian Perspectives 55,
https://doi.org/10.1007/978-981-16-5755-9_10

10.1 Introduction

The study of the dynamic relationship between public health expenditure, private health expenditure, and economic growth in India keeps a broad relevance for academicians, researchers, economists, and policymakers. An increase in public health expenditure and private health expenditure has a considerable impact on economic development in any country. The country, which is sincere to their economic development, works on education, health, and other social development of the people. However, there is no consensus among the researchers, academicians, and policymakers whether an increase in both the public and private health care expenditure would accelerate economic growth. Health expenditure in India has been dominated by private healthcare expenditure during the post-independent period. In the last two decades, government spending on health care services has been stuck at around 1% of GDP, which is around 28% of current health expenditure and private healthcare expenditure has been around 72% of current health expenditure (The World Bank Data 2018).

Both health and education are rights under the constitution of India, and education is a fundamental right. As per the constitution of India, health is a state subject, and it is the responsibility of each state government to provide health care facilities to the people. The critical element of any health care system is the excellent service delivery system. Good health care service delivery is playing a critical role and acts as a fundamental input for people's health. Over the last few decades, researchers have revealed the importance of education and health as crucial factors for the country's economic development.[1]

Education gives us the development of human skills and knowledge of the people or labor force. Education is called human capital, and human capital is a significant input of economic development. Like education and other conventional inputs, health is an integral component of economic development because health affects our development or growth by augmenting labor productivity by increasing life expectancy and subsequent human capital formation. Health affects economic growth directly through increasing labor productivity and indirectly through the impact of health on education. Healthy and well-nourished children would perform better in school, college, or any institution, which will positively impact the future income. It is examined that health care expenditure has a positive impact on economic development (Raghupathi and Raghupathi 2020).

So the relationship between health and development is an essential topic in the field of economics, and recent new health policy (2017) also acknowledged this relationship. This study will try to identify the relationship between healthcare expenditure – public healthcare expenditure (GHE) and domestic private healthcare expenditure (DPHE) and economic development as measured by Gross Domestic Product (GDP). Since we are concerned with health care expenditure, which consists

[1]Education in Article 21A of the constitution of India and Health in Article 21A of the constitution of India and Article 39,41,42, and 47.

of public and private spending to provide the quality of health to the people, the present status of the health scenario and the trend of both public and private spending in India have been explored (Sect. 10.4). However, before that, Sect. 10.3 gives a vivid literature review. Section 10.5 deals with the data and methodology of the present study to analyze the short-run and long-run dynamics among public healthcare expenditure and private healthcare expenditure, and Gross Domestic Product. Section 10.6, presents the empirical results along with findings; Sect. 10.7 deals with variance decomposition and impulse response function analysis; and the last Sect. 10.8 is the conclusion of this study.

10.2 Objective of the Study

In the light of the above discussion, the present study attempts to examine the interdependency and dynamic relationship between public healthcare expenditure and private healthcare expenditure and economic development in India during the period from 1999 to 2018. Secondly, the study attempts to measure the strength of the relationship between the variables concerned, and thirdly we will examine the stability of the model.

10.3 Literature Review

Nyasha and Odhiambo (2019) focused on the only objective of assessing the impact of government spending on economic growth. The study was entirely based on secondary data, which carefully evaluates the positive, negative, and no impact of government spending on economic growth from different other studies. The study finds that sample countries have shown the diverse impact of government spending on economic growth—some of the countries showed a positive impact, while some others showed a negative impact, and the rest has no impact of government spending on economic growth.

Varkey et al. (2020) studied the interrelationship between health infrastructure, health outcome, and economic growth in the major India States. They focused on two objectives: analyzing the relationship between infrastructure and health outcomes and analyzing the impact of health infrastructure on economic growth. This study collected data from several secondary sources, including NITI Aayog, Central Statistical Organization, and Ministry of Health and Family Welfare. The investigators studied the above objective for the time period of 1980–2012. Fifteen general category states (i.e., Uttar Pradesh, Rajasthan, Orissa, Madhya Pradesh, Bihar, West Bengal, Tamil Nadu, Kerala, Karnataka, Andhra Pradesh, Punjab, Maharashtra, Haryana, Gujarat, and Goa) having the same funding patterns were considered for the study. The health outcome was measured with Infant Mortality Rate (IMR) and Life Expectancy at birth (LE); economic growth was measured by Gross Domestic

Product (GDP) and Gross State Domestic Product (GSDP), whereas the Health Infrastructure Index (HII) was measured by various variables including primary health centers, community health centers, sub-centers, health workers, and the number of hospital beds in medical colleges. The study used Augmented Dickey–Fuller Test, Johansen Cointegration Test, Fully Modified Ordinary Least Squares, Correlation coefficient, and standard deviation for analyzing the data. The author found that a significant positive relationship exists between HII and LE, whereas a significant negative relationship exists between HII and IMR. Also, at the state level, there is a positive relationship between HII and GSDP and between HII and LE. On the other hand, a negative relationship has been observed between HII and IMR.

Bhat and Jain (2004) applied the time series analysis of private healthcare expenditures and GDP. Their study assessed whether the private expenditure on health and medical care has any long-term relationship with the per capita income. The study collected data from several secondary sources, including National Accounts Statistics, NSSO, and Central government Budget. The study has used cointegration, Fully Modified Ordinary Least Squares (FMOLS). The authors found that spending a high level of private health expenditures results in serious challenges to the policymakers in developing countries. This expenditure may have several undesirable consequences making the health system high cost, unaffordable, and vulnerable to provider payment system and also, it remains ambiguous whether this expenditure is sustainable.

Bedir (2016) studied how economic growth takes on healthcare expenditure in developing countries. The investigator tried to test whether there is any causality between income and healthcare expenditure and whether the health care expenditure is the driving force for economic growth, especially in Europe, Middle East, African, and Asian countries. The study was based on secondary data of per capita gross domestic products and health expenditure, collected from World Bank reports between 1995 and 2013. Nine countries from Europe and Middle East Africa (Czech Republic, [Egypt, Arab R], Greece, Hungary, Poland, Russian F, South Africa, Turkey, and UAE) and seven from Asia (China, India, Indonesia, Korea R, Malaysia, Philippines, and Thailand) were selected. The study used the modified version of the Granger (1969) causality test proposed by Toda and Yamamoto (1995) and Dolado and Lütkepohl (1996). The study found that Egypt, Hungary, Korean Republic, South Africa, and the Philippines support the health view over the income view, whereas, Greece, Poland, the United Arab Emirates, China, Indonesia, and the Korean Republic support the income the health view. The investigator also observed that income is the most important factor for explaining the differences in health care expenditures among countries.

Mohapatra (2019), in her investigation of public health expenditure (PHE) and its effects on health Outcomes in India, focused on three primary objectives - checking the effects of each component of PHE on health outcomes; estimating the effect of the components of PHE on multiple health outcomes including mortality, fertility, morbidity and disease proneness indicators; and, investigating the effects of PHE on health outcomes with different time lags as suggested by Farahani et al. (2010).

The study collected the data from various secondary sources, including the Statistics on State Finances published by the Reserve Bank of India (RBI), different factsheets of the National Family and Health Services (NFHS) reports (NFHS-1, NFHS-2, and NFHS-3) for sixteen major Indian states for the period of 1992–93, 1998–99, and 2005–06. The study used pooled panel data analysis to analyze the panel data of the given states with the Structural Equation Model (SEM). It confirmed that PHE affects the health outcomes, the revenue component of PHE affect the health outcomes more than the capital component, the revenue expenditure of health does not affect CMR, the capital expenditure has a significant positive effect on TFR in no lagged effects model, and the capital expenditure has a significant negative impact on TB.

Maitra and Mukhopadhyay (2012) studied the public spending on education, health care, and economic growth. They focused on examining the role of public spending on education and health sectors in promoting the Gross Domestic Product (GDP) of 12 countries in Asia-Pacific (Republic of Korea, Singapore, Malaysia, Philippines, Maldives, Sri Lanka, Bangladesh, Nepal, Tonga, Fiji, Kiribati, and Vanuatu) for 1981–2011. The study used Augmented Dickey–Fuller (ADF) test, Johansen's Multivariate Cointegration Test, Vector Error Correction Model (VEC), and Vector Auto Regression (VAR) model. The authors observed that for Bangladesh, Kiribati, Malaysia, Maldives, the Philippines and the Republic of Korea, Johansen cointegration tests confirmed the existence of cointegrating relations, while Fiji, Nepal, Singapore, Sri Lanka, Tonga and Vanuatuo do not. For Bangladesh, Nepal, the Philippines, Singapore, and Sri Lanka, the health spending has a positive impact on GDP. There were also countries like Kiribati, Maldives, and Vanuatu in which spending on health showed a negative impact on GDP.

Zaman et al. (2017) tested the association of total health expenditure with GDP and life expectancy in Bangladesh based on the secondary data. The scholars collected Total Health Expenditure (THE) and GDP data from Bangladesh Health Bulletin 2011 and life expectancy data from Sample Vital Registration System 2010. The study considers the period of 1996–2006. The study used descriptive statistics and multivariate logistic regression and found that a direct relationship exists between GDP and THE; and also, there is no significant relationship between total health expenditure and increased life expectancy.

Gangal and Gupta (2013) analyzed the impact of public expenditure on India's economic growth from 1998 to 2012. The study used secondary data from World Economic Outlook, IMF for collecting Total Public Expenditure, and Gross Domestic Product datasets. Augmented Dickey–Fuller (ADF) Test, Cointegration Test, and Granger Causality Test were applied for the data analysis. The study finds that there is a positive impact of total Public Expenditure on Economic Growth. The study observed a unidirectional relationship from total public expenditure to GDP. Basumallik (2017) examined the relationship between health and economic growth in India during the period 1961–2015. The secondary data was taken from the World Development Indicators (WDI) data published by World Bank. The study used Two-Stage Least Squares (2SLS) as the Ordinary Least Square (OLS) failed to find out any significant relationship between health and economic growth. The

2SLS showed a highly significant effect of health indicators on economic growth in India.

Rajeshkumar and Nalraj (2014) tried to examine the causal relationship between healthcare expenditure and economic growth in four Indian states (Madhya Pradesh, Orissa, Kerala, and Tamil Nadu) based on an annual data series for the period of 1991–2010. The study used Cointegration Regression Test like Cointegration Regression Durbin–Watson (CRIW) test, Dickey–Fuller (DF) test, and the Augmented Dickey–Fuller (ADF) test to find that the health expenditure and economic growth are cointegrated in all the four states, and there exists a unidirectional causality from health expenditure to economic growth.

The work of Atuahene et al. (2020) is exceptional, where the investigators explored the impact of economic growth and CO2 emissions on health expenditure and differentiated the situation between India and China from 1960 to 2019. It used a dynamic panel data model estimated by the Generalized Method of Moments (GMM) and found a significant relationship between health expenditure, CO_2 emissions, and economic growth. The CO_2 emission showed a significant positive impact on health expenditure, whereas economic growth negatively impacted health expenditure in both China and India. In the case of health spending, India's population growth rate exhibited a transposed effect, whereas it positively impacted China's case. It was also observed that the strong correlation between health expenditure and economic growth is crucial for economic development.

Verma and Usmani (2019) studied the relationship between health and economic growth in India with the help of Granger Causality (GC) test covering fifteen Indian states from 1985 to 2015. The Infant Mortality Rate (IMR) and Gross State Domestic Product (GSDP) were analyzed by using Vector Auto Regression (VAR) model and found a unilateral causality in each state, which either runs from IMR to GSDP or from GSDP to IMR.

Mohanty and Behera (2020) raised the question about how effective public health care expenditure is in improving health outcomes and tried to answer it through empirical evidence from the 28 Indian States by investigating the effects of public health expenditure on various proximate and ultimate health outcomes during 2005–16. The study found that per capita public health care expenditure adversely affects the infant and child mortality rate and malaria cases and favors life expectancy and immunization coverage across the states.

Gupta and Mitra (2004) tried to analyze the links between economic growth, poverty, and health using panel data of 15 major states of India. The variables like the growth rate of national state domestic product (NSDP), infant mortality rate, life expectancy, per capita health expenditure have been used. The study used a large volume of secondary data for the years considered here are 1973–74, 1977–78, 1983–84, 1987–88, 1993–94, and 1999–2000 for an inter-state cross-sectional analysis and time series analysis. The investigators found that growth tends to reduce poverty, and improvements in health status are necessary to reduce poverty. Economic growth and health status were found as positively correlated and had a two-way relationship, i.e., better health enhances growth by improving productivity, and higher growth allows better human capital formation.

Shafuda and De (2017) focused on analyzing whether there is any relationship between government expenditure on health and economic growth and other development measures and tried to establish the direction of the relationship. The study was conducted with the datasets of Namibia for the period 1980 to 2015. Investigators applied standard Granger Test, Vector Autoregressive (VAR) model, Augmented Dickey–Fuller (ADF) test, Johansen Cointegration test, trace statistic and Max-Eigen, Vector Error Correction Model and Granger Causality test in analyzing data. They found a long-run relationship between government spending on health care and the fertility and infant mortality rate; however, there was no cointegration between government expenditure on health care and life expectancy and adult mortality rate.

Pradhan (2010), in his paper entitled "The Long Run Relation between Health Spending and Economic Growth in 11 OECD countries: Evidence from Panel Cointegration," examined the role of health spending in determining economic growth. The study was based on 11 OECD countries (Austria, Canada, Finland, Iceland, Ireland, Japan, Norway, Spain, Switzerland, UK, and the USA) from 1961 through 2007. The study applied a panel cointegration approach to explore the possibilities of estimating both short-run and long-run impacts of health spending and found that health spending and economic growth are panel cointegrated, indicating the existence of a long-run equilibrium relationship between them. In both the short and long run, the bidirectional causality between health spending and economic growth was confirmed to exist.

Dincer and Yuksel (2019) aimed to identify the causality relationship between health expenditure and economic growth in E7 countries (Brazil, China, India, Indonesia, Mexico, Russia, and Turkey), the biggest emerging economies for 1996–2016. The investigators found a long-run relationship between economic growth with total health expenditure and public health expenditure. It was also found that there is no causality relationship from health expenditure to economic growth. Most importantly, the study succeeded in determining that total, public, and private health expenditure is caused by economic growth; however, this relationship is not valid between private health care expenditure and economic growth.

Seshaiah et al. (2018) studied the general government expenditure and economic growth in India from 1980–81 to 2015–16 based on secondary data, mainly sourced from the Statistical Abstracts published by the Directorate of Economics and Statistics every year, datasets of National Sample Survey Organization, Reserve Bank of India, various reports of Ministry of Finance. The study used Simple Regression Analysis, Breusch–Pagan test for analyzing the data, which finally concluded that the variables, other than the FDI growth rate, pose a positive and significant effect on GDP growth rate. Also, no correlation and autocorrelation were observed among the variables, and the non-development expenditures were found occupying a large proportion of general government expenditure.

Kulkarni (2016) examined the differences in the health care systems of emerging BRICS nations (Brazil, India, China, Russian Federation, and South Africa). The study was based on secondary data, and the data on Infant Mortality Rate, GDP, Adult literacy rate were collected from World Health Organization (WHO) and

World Bank datasets for 16 years (1995–2010) to follow the Grossman model for estimating health outcome. The study found a positive relationship of health outcome with GDP per capita, adult literacy rate, and out-of-pocket expenditure, whereas a negative relationship with that of environmental pollution, represented by CO_2 emissions per capita metric ton and female workforce participation. The age dependency rate and IMR showed a positive elasticity, whereas a negative relation between age dependency and health production was observed.

Ehikioya and Mohammed (2013) studied the determinants of public health care expenditure in Nigeria for the period of 1986–2010. The study used descriptive tools and the analytical tools of the contemporary cointegration test. Investigators also applied the error correction technique. The study concluded that the total population of 14 years and younger and the share of health expenditure in the gross domestic product are the main determinants of the health expenditure. In contrast, the variables of gross domestic product per capita, unemployment rate, population per physician, consumer price index, and political instability were found insignificant.

10.4 Present Status of Health Scenario and the Trend of Healthcare Spending in India

The present study is concerned with health care expenditure, which consists of public and private spending to provide the quality of health to the people, so the status of health infrastructure and expenditure in India during 1999–2018 is analyzed. Figure 10.1 depicts overall spending on healthcare expenditure, both public and private spending, during the sample period, and the domestic private healthcare expenditure was found much higher than public healthcare expenditure during the study period. While the current health expenditure per capita (current US $) in India

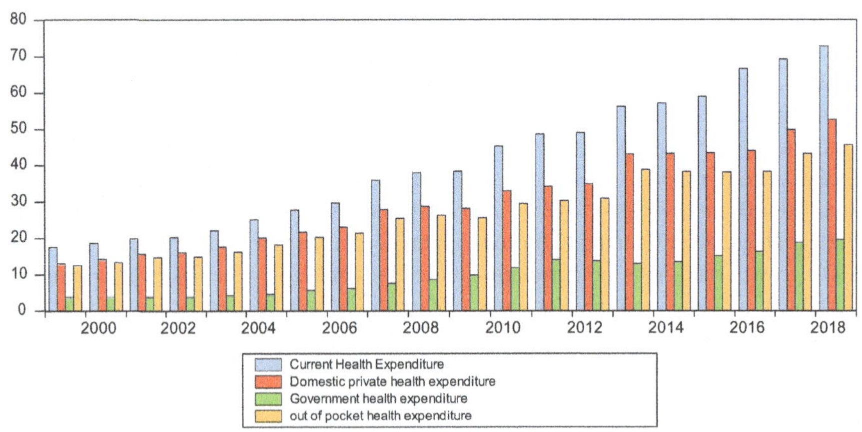

Fig. 10.1 Plots of different health expenditures in India (Data Source: The World Bank, 2018)

Table 10.1 Status of public health Infrastructure in India

Different components of Public health Infrastructure	Existing capacity
Community health worker per 1000 people	0.581
Hospital beds per 1000 people	0.70
Number of doctors per 1000 people	0.857
Nurses and midwives per 1000 people	1.727
Number of ventilators	48,000
Number of ICU beds	95,000
Number of hospital beds	19,000,00
Number of hospitals	69,000
Number of medical colleges	539

has been increasing from 17.54(US $) to 72.83 (US $), which was 40.86 (US $) per capita on an average during the period 1999 to 2018, the government healthcare expenditure per capita was found 9.86 (US $). India's overall spending on the health sector is found low during 1999–2018, compared to any developed country.

The bar diagram (Fig. 10.1) exhibits that domestic private healthcare expenditure has been rising gradually in India, reflecting US$30.23 on an average per capita from 1999 to 2018, while out-of-pocket expenditure in India was US$ 27.11on an average per capita, which puts a burden for healthcare on poor households. This is a very alarming picture because, after 74 years of India's independence, the government spending on healthcare is around on an average US$ 9.86 per capita, private spending US$ 30.23, and external spending US$ 0.45, which is very insignificant. The average health spending is US$40.86 per capita on average, which is still below 4% of the country's GDP. In India, 80% of the poor people live in rural areas, around 28% of the population (364 million according to UNDP), and have limited resources to spend on healthcare. The current pandemic situation has exposed the dismal state of India's health infrastructure, and the public healthcare system is far from satisfactory. People, who could afford the expenditure, prefer to avail of private health services than the public health systems for their medical treatment during emergency health issues; even the former is costlier to manifold extent than the latter.

Sources: The World Bank Data 2018 and CDDEP Survey, 2000.

Table 10.1 presents the bleak picture of Indian health infrastructure, and there is a considerable shortage of hospitals, hospital beds, doctors, nurses, ventilators, ICU beds for 1.353 billion people. The accessibility among the people, particularly in rural areas, is significantly lower than that of in urban areas. In India, where most of the population are rural bound, the mass could not finance their health care expenditure. It grounds the reality of the rural people to rely more on the public health system; however, due to the poor infrastructure of the public health care system, the people need to cline more on private health care providers than public health care providers, compromising the affordability of the familys' economic background.

The reality is that India's neighboring countries like Maldives, Sri Lanka, Nepal, Bhutan spend more on healthcare than that India (Fig. 10.2). On average, while in India, the average public health expenditure contributes 3.68 per cent of GDP, it is

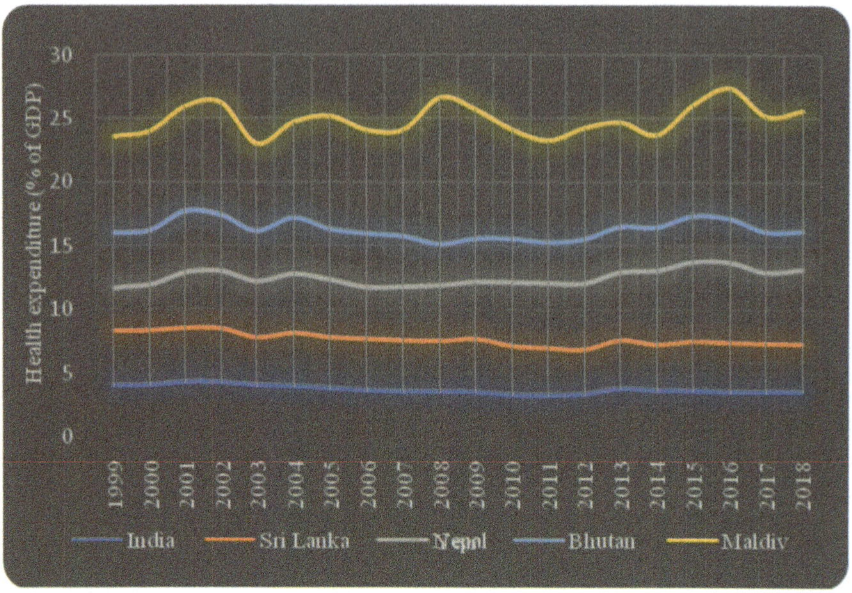

Fig. 10.2 Trends in current health expenditure (% of GDP) of few SAARC Countries (Data Source: The World Bank Data 2018)

3.95 per cent in Sri Lanka, 4.85 per cent in Nepal, and 8.55 per cent in the Maldives; and the developed countries spend more than 10 per cent of their GDP.

10.5 Data and Methodology

The objective of the study is to examine the dynamic relationship among public healthcare expenditure, private healthcare expenditure, and economic development in India using time series data for the sample period 1999 to 2018. To this end, the Augmented Dickey–Fuller test (ADF) and Phillips–Perron test (PP) have been applied to examine the stationarity of the dataset. The estimation methodology used in the present study is the Johansen cointegration test to examine the cointegration among the variables concerned and Vector Error Correction Modelling (VECM) technique to examine the dynamic relationship. The Impulse Response Function (IRF) and Variance Decomposition are applied to measure the strength of the relationships (Granger 1969).

The data on public healthcare expenditure and private healthcare expenditure, and GDP have been collected from the World Bank[2], and all the data have been

[2]World Bank data is available at www.data.worldbank.org

transformed into natural logarithms. Annual data has been used in the present study because the monthly and quarterly data are not available.

10.6 Empirical Results and Findings

The present study is concerned with the examination of the dynamic relationship among three concerned time series data. The public healthcare expenditure series (GHE_t), domestic private healthcare expenditure series ($DPHE_t$), and economic growth (GDP_t) series are used as proxy variables of economic development which is crucial from the policy point of view. The stationarity or non-stationarity of the dataset is examined through two unit root tests: the Augmented Dickey–Fuller test (ADF) and Phillips–Perron test (PP).

The government healthcare expenditure ($lnGHE_t$), domestic private healthcare expenditure ($lnDPHE_t$), and economic growth ($lnGDP_t$) series are integrated of order "1." I(1) implies that GHE_t, $DPHE_{t,}$ and GDP_t series are non-stationary at the level, as evident from both ADF and PP test results with intercept and intercept with trend assumptions. In contrast, the first differenced data on government healthcare expenditure ($lnGHE_t$), domestic private healthcare expenditure ($lnDPHE_t$), and economic growth ($lnGDP_t$) series are integrated of order "0." I (0) implies all the concerned series are stationary, as evident from ADF test and PP test results with intercept and intercept with trend assumptions (See Table 10.2). The lag length used to compute the Augmented Dicky–Fuller test statistics is based on the Schwartz information criterion (SIC), and for Phillips–Perron test statistics, lag length is determined by applying Newey–West using Bartlett Kernel.

In the next step, the Johansen cointegration method identifies multiple cointegrating vectors of any number of non-stationary series. Since all the concerned series, i.e., GHE_t, $DPHE_{t,}$ and GDP_t in the present study, are non-stationary, the Johansen cointegration test is used to detect whether all the three series are cointegrated or not during the sample period. The Trace and maximum Eigenvalue statistics are considered to examine the null hypothesis of no cointegration for these time series data over the period of study. The results of the Johansen cointegration test are reported in Table 10.3.

The null hypothesis of no cointegration between GHE_t, $DPHE_t$, and GDP_t ($r = 0$) based on the Trace statistic and Max-Eigen statistic is rejected at 5% level of significance, while the null hypothesis of no cointegration among GHE_t, $DPHE_t$, and GDP_t ($r \leq 1$ and $r \leq 2$) based on the Trace statistic and Max-Eigen statistic is not rejected at the same level of significance which indicates the existence of only one cointegrating equation among GHE_t, $DPHE_t$, and GDP_t series. Thus, the concerned series of the study are cointegrated, and it implies a long-run equilibrium relationship among these three variables throughout the study period.

In the third step, since the concerned time series data GHE_t, $DPHE_t$ and GDP_t have the long-run equilibrium relationship, there may be disequilibrium in the short-run, and the present study examines whether such disequilibrium converges to the

Table 10.2 Unit root test results

Variable	ADF Intercept		ADF Intercept and trend		PP Intercept		PP Intercept and trend		Degree of Integration
	t -statistic	probability	t -statistic	probability	t -statistic	probability	t -statistic	probability	
$lnGHE_t$	−0.09	0.93	−1.57	0.76	−0.21	0.92	−1.91	0.60	I (1) i.e. non-stationary
$\Delta lnGHE_t$	−2.78[a]	0.08	−2.65	0.26	−2.78[a]	0.08	−2.65	0.26	I (0) i.e. stationary
$lnDPHE_t$	−1.69	0.41	−2.73	0.23	−1.97	0.29	−2.64	0.26	I (1) i.e. non-stationary
$\Delta lnDPHE_t$	−5.45[b]	0.00	−6.00[b]	0.00	−8.23[b]	0.00	−10.61[b]	0.00	I (0) i.e. stationary
$lnGDP_t$	−1.21	0.64	−1.16	0.88	−1.51	0.50	−1.91	0.90	I (1) i.e. non-stationary
$\Delta lnGDP_t$	−3.99[b]	0.00	−3.85[b]	0.03	−3.98[b]	0.00	−4.21[b]	0.01	I (0) i.e. stationary

[a]refers to rejection at10% level of significance and
[b]refers to rejection at 1% level of significance
Notes: $lnGHE_t$ refers to level data on government health expenditure;
$\Delta lnGHE_t$ refers to first differenced data on Government Health Expenditure;
$lnDPHE_t$ refers to level data on Domestic Private Health Expenditure;
$\Delta lnDPHE_t$ refers to first differenced data on Domestic Private Health Expenditure;
$lnGDP_t$ refers to level data on Gross Domestic Product;
$\Delta lnGDP_t$ refers to first differenced data on Gross Domestic Product;
ln represents natural logarithm.
Δ is the first difference operator.
ADF & PP represent Augmented Dickey–Fuller and Phillips–Perron t-statistic, respectively.
Source: Computed by the author using EViews 10 software

Table 10.3 Results of Johansen's cointegration Test

Hypothesized No of cointegration	Eigenvalue	Trace statistic	Critical value at 5%	Probability[b]	Max-Eigen statistic	Critical value at 5%	Probability[b]
None[a]	0.666	33.642	29.79	0.01	19.753	21.131	0.07
At most 1	0.455	13.889	15.49	0.08	10.941	14.264	0.15
At most 2	0.151	2.948	03.84	0.08	2.948	3.841	0.08

[a]Denotes rejection of the hypothesis at the 5% level of significance
[b]Mackinnon–Gaug-Michelis (1999) p values (Source: Computed by the author using EViews 10 software

Table 10.4 Results of VECM Regression

Independent variable	Δ GDP_t	Δ GHE_t	Δ $DPHE_t$
ECT_{t-1} (t-statistic)	$ECT_{t-1}{}^1$ -0.02(0.10)	$ECT_{t-1}{}^2$ -0.14 (−0.56)	$ECT_{t-1}{}^3$ -0.59 (4.00[a])
Δ GDP_{t-1} (t-statistic)	−0.57 (−3.16)	−0.24 (−1.02)	−0.52 (−3.68[a])
Δ $DPHE_{t-1}$ (t-statistic)	0.36 (1.12)	−0.07 (−0.17)	0.50 (1.97)
Δ GHE_t (t-statistic)	0.66 (2.93[a])	−0.34(−1.13)	−0.04 (−0.26)
Constant (t-statistic)	−0.007 (−0.39)	0.007 (0.30)	0.001 (0.13)
Different statistical values	R-squared value-0.64	R-squared value-0.21	R-squared value-0.80
	Adj. R-Squared-0.52	Adj. R-Squared-0.04	Adj. R-Squared-0.73
	F-statistic- 5.46	F-statistic- 0.82	F-statistic- 12.11
	AIC—2.11	AIC—-1.52	AIC—-2.60
	Schwarz SC—1.86	Schwarz SC—-1.28	Schwarz SC— -2.36
	Log likelihood-22.94	Log likelihood-17.98	Log likelihood-27.15

Source: Computed by the author using EViews 10 software
[a]refers the statistical significance of the coefficients

long-run equilibrium or not. Vector Error Correction Model (VECM) is used to generate such short-run dynamics to reach that equilibrium. So, the error correction mechanism is a means to reconcile the short-run and long-run behavior.

The estimation of the Vector Error Correction Model on GHE_t, $DPHE_t$, and GDP_t is reported in Table 10.4. The number of lags in the model has been determined on the basis of the Akaike Information Criterion (AIC), and the lag order of the model is "1."

The estimated coefficient of error correction term ($ECT_{t-1}{}^3$) is statistically significant at 5% level, and the negative sign confirms the long-run causal (Granger) relationship among GDP_t and GHE_t and $DPHE_t$ in India. This causality is unidirectional—running from Gross Domestic Product to Domestic Private Healthcare Expenditure (Table 10.4). The coefficient of error correction term "-0.59" suggests that about 59% of discrepancies between the long-run and short-run domestic private health expenditure are corrected within a year. In the Δ $DPHE_t$ equation, the coefficient of Gross Domestic Product (Δ GDP_t) is significant at 5% level of significance, which shows the short-run causality (Granger) running from Gross Domestic Product toward the Domestic Private Healthcare Expenditure. The coefficients of error correction terms ($ECT_{t-1}{}^1$ and $ECT_{t-1}{}^2$) in Gross Domestic Product and Government Healthcare Expenditure equations are positive and statistically insignificant. In the Δ GDP_t equation, the coefficient of Δ GHE_t is statistically significant at 5% level, which implies that the Government Healthcare Expenditure Granger causes Gross Domestic Product in the short run.

In order to confirm the results of short-run Granger causality between Government Healthcare Expenditure and Gross Domestic Product and between domestic private health expenditure and Gross Domestic Product based on VECM estimation,

Table 10.5 VEC Granger Causality/Block Exogeneity Wald Test

Dependable variable: Δ GDP			
Excluded	Chi-Square	df	Prob.
ΔGHE	8.60	1	0.00
ΔDPHE	1.26	1	0.26
All	8.85	2	0.01
Dependable variable: Δ GHE			
Excluded	Chi-Square	Df	Prob.
ΔGDP	1.04	1	0.30
ΔDPHE	0.03	1	0.85
All	1.09	2	0.57
Dependable variable: Δ DPHE			
Excluded	Chi-Square	Df	Prob.
ΔGDP	13.61	1	0.00
ΔGHE	0.06	1	0.79
All	15.29	2	0.00

Source: Computed by the author using EViews 10 software

the Block Exogeneity Wald Test has been conducted based on Chi-Square Statistic. The results of the Block Exogeneity Wald Test are demonstrated in Table 10.5.

Based on a high chi-square value and low probability value (less than 1%), Government Healthcare Expenditure Granger causes Gross Domestic Product, and Gross Domestic Product causes domestic private healthcare expenditure in the short-run in India during the period of study. Block Exogeneity Wald Test has validated the results obtained from VECM (Table 10.5).

Based on these above causality tests, in the short run, the changes in government healthcare expenditure Granger cause the changes in Gross Domestic Product. On the other hand, the changes in Gross Domestic Product do Granger cause the changes in Domestic Private Healthcare Expenditure in the short run in India under the period of study.

So far as the stability condition of the VECM, all the roots of the characteristic polynomial are inside the unit circle, which is exhibited in Fig. 10.3. Therefore the defined Vector Error Correction model is stable.

10.7 Variance Decomposition and Impulse Response Function Analysis

Variance Decomposition (VD) and Impulse Response Function (IRF) are used to quantify the relationship between the variables concerned, as evident from VECM. In a Vector Error Correction Model, variance decomposition analysis reports the proportion variation in each variable due to a shock in the variable itself and the shocks in other variables.

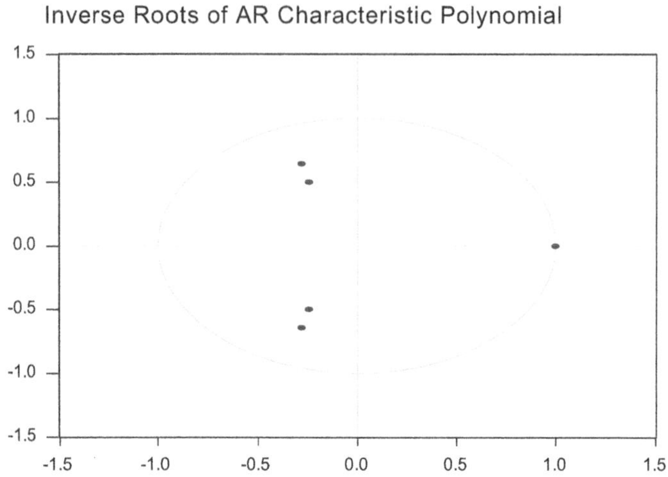

Fig. 10.3 Roots of AR Characteristic Polynomial (Source: Computed by the author using EViews 10 software)

Table 10.6 Variance Decomposition of Gross Domestic Product

Period	S.E	Δ GDP	ΔGHE	ΔDPHE
1	0.074	100.000	0.000	0.000
2	0.123	79.285	19.339	1.374
3	0.129	81.134	17.567	1.297
4	0.152	82.876	16.129	0.993
5	0.170	81.966	16.913	1.120
6	0.179	83.044	15.935	1.020
7	0.194	83.592	15.527	0.880
8	0.207	83.562	15.602	0.832
9	0.216	84.040	15.182	0.777
10	0.227	84.346	14.947	0.706

Cholesky Ordering: GDP GHE DPHE (Source: computed by the author using EViews 10 software)

A variance decomposition for VECM for a period of one year to 10 years is reported in Table 10.6. The results of variance decomposition for GDP show that after the second year, Government healthcare expenditure explains around 15% variations in Gross Domestic Product since $3 \leq t \leq 10$, implying a very low dependence of Gross Domestic Product on Government healthcare expenditure. On the other hand, domestic private healthcare expenditure contributes around 1% variation in Gross Domestic Product since $2 \leq t \leq 0.10$, and GDP itself contributes 84% in its own variation. So, no dependence of Gross Domestic Product on Domestic private healthcare expenditure is found while weak causality running from Government healthcare expenditure to Gross Domestic Product exists.

The results of variance decomposition of Government Healthcare Expenditure for a period of one year to 10 years are reported in Table 10.7. The results of variance

Table 10.7 Variance Decomposition of Government Healthcare Expenditure

Period	S.E	Δ GDP	ΔGHE	ΔDPHE
1	0.100	29.898	70.101	0.000
2	0.116	24.618	73.733	1.648
3	0.125	23.658	74.717	1.623
4	0.142	23.439	75.149	1.411
5	0.154	22.429	75.998	1.571
6	0.164	21.937	76.493	1.568
7	0.175	21.695	76.775	1.529
8	0.185	21.346	77.095	1.558
9	0.195	21.090	77.348	1.560
10	0.204	20.921	77.527	1.551

Cholesky Ordering: GDP GHE DPHE (Source: computed by the author using EViews 10 software)

Table 10.8 Variance Decomposition of Domestic private healthcare expenditure

Period	S.E	Δ GDP	ΔGHE	ΔDPHE
1	0.058	37.186	0.435	62.378
2	0.059	36.604	2.279	61.116
3	0.070	42.891	2.204	54.904
4	0.088	49.817	13.013	37.168
5	0.089	49.162	13.607	37.230
6	0.092	51.199	12.706	36.094
7	0.100	54.475	14.838	30.685
8	0.101	55.278	14.533	30.187
9	0.103	56.798	14.035	29.166
10	0.107	58.734	14.258	27.006

Cholesky Ordering: GDP GHE DPHE (Source: Computed by the author using EViews 10 software)

decomposition for GHE show that the variations in Government healthcare are mainly due to its own shocks since $1 \leq t \leq 10$ and Gross Domestic Product explains around 23 percent variations and Domestic private healthcare expenditure contributes around 2 percent variations in Government Healthcare Expenditure. This result indicates the dependence of Government Healthcare Expenditure on Gross Domestic Product to some extent.

The forecast error variance of Domestic private healthcare expenditure is more explained by Gross Domestic Product, around 59 percent; and nearly 27 percent of the variance of its forecasting error is explained by itself, as evident from Table 10.8. On the other hand, government Healthcare Expenditure contributes the 14 percent variance of forecasting error of Domestic private healthcare expenditure. The above results indicate the higher dependence of domestic private healthcare expenditure on Gross Domestic Product while lowering Government Healthcare Expenditure on Gross Domestic Product. This result is also validated by the VECM Granger causality test, which confirms the long-run causal relationship.

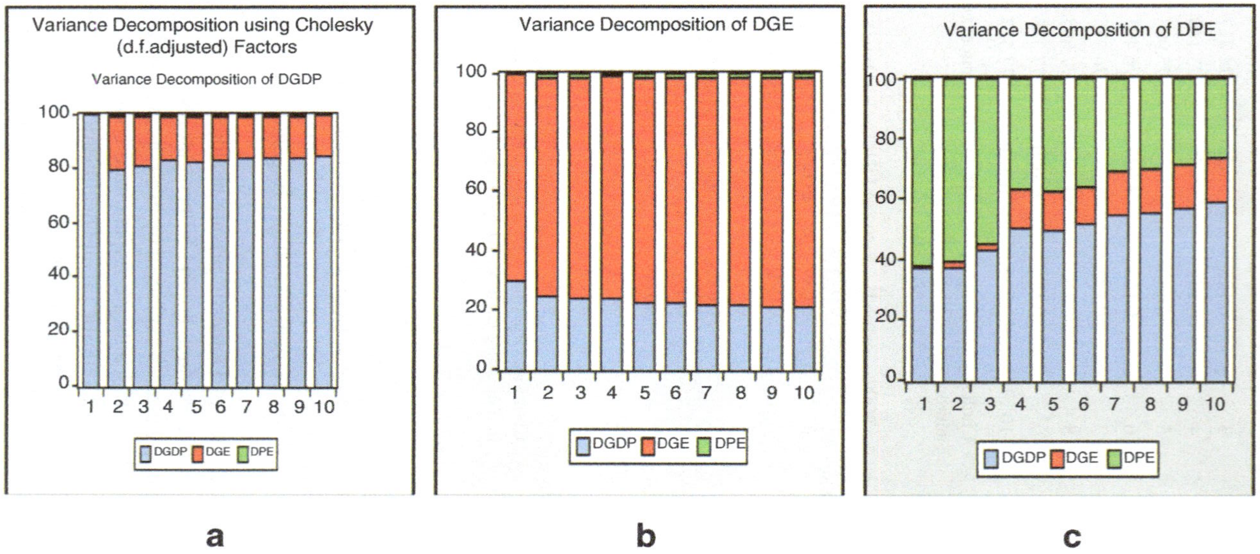

Fig. 10.4 Variance Decomposition of **(a)** GDP, **(b)** GHE, & **(c)** DPHE (Source: Computed by the author using EViews 10 software)

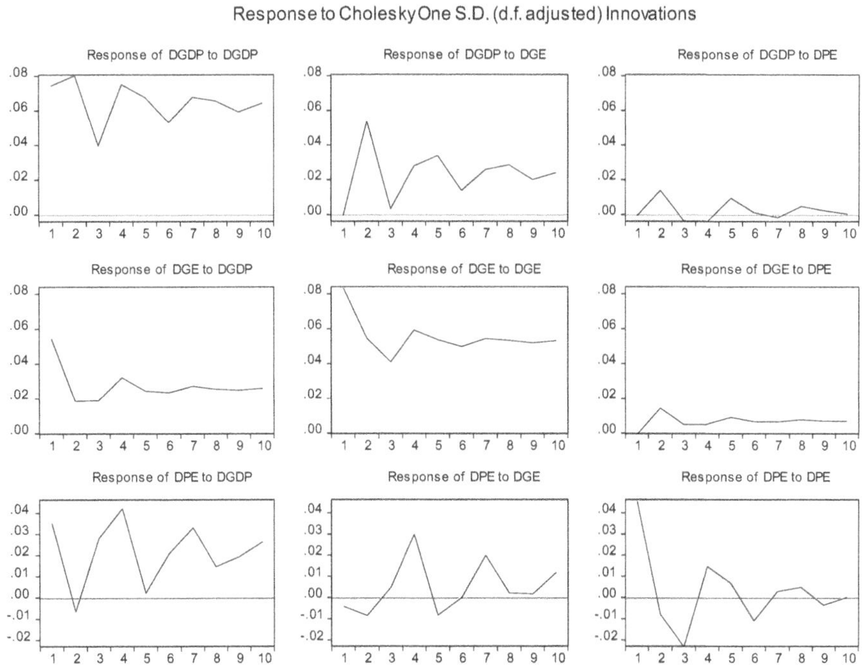

Fig. 10.5 Impulse Response Functions Analysis (Source: Computed by the author using EViews 10 software)

The tacked graphs (Fig. 10.4) on variance decomposition demonstrate that the domestic private health care expenditure is explained more by gross domestic product and the substantial part of the government health expenditure is explained by itself, and a very small part of government health expenditure is contributed by gross domestic product in India during the sample period. In the case of gross domestic product, government healthcare expenditure has little influence, and domestic private health care expenditure has no impact.

The impulse response functions (in Fig. 10.5) show the responses of all variables to an exogenous shock. Solid lines represent the function. Due to one standard deviation, positive shock in government healthcare expenditure, the responses of the gross domestic product are positive from the initial period, and positive impact persists till $t = 10$ while the responses of the gross domestic product are positive at the initial period ($t = 1$) but gradually decline after the second year ($t > 2$) due to positive shocks in domestic private healthcare expenditures and finally asymptotically die out to zero. A positive shock in gross domestic product and domestic private healthcare expenditure was found to have an insignificant positive effect on government healthcare expenditure. One standard deviation positive shock in gross domestic product props up domestic private healthcare expenditure after the second year ($t > 2$), and the positive effects persist till $t = 10$. Domestic private healthcare expenditure responses due to positive shocks in government healthcare expenditure are fluctuating from the first year to fifth year and gradually decline in the lead period. The findings of variance decomposition and impulse response function confirm that Gross Domestic Product (GDP) has more impact on Domestic Private Healthcare Expenditure (DPHE) whereas the Public Healthcare Expenditure (GHE) is relatively less in having an impact on Gross Domestic Product (GDP). These results further validate the results of the VECM Granger causality test, which demonstrates the long-run causality running from gross domestic product to domestic private healthcare expenditure and exhibits another causality working from government healthcare expenditure to gross domestic product in India during the period of study.

During the entire sample period, there might be a possibility of structural changes in the relationship as evident from VECM because of policy changes or institutional changes or external shocks or change in social attitudes and motivation etc. In the presence of structural changes, the estimated econometric relationships between gross domestic product, government healthcare expenditure, and domestic private healthcare expenditure fail to produce a true economic relationship. If any structural change occurs during the period of study, then the values of the parameter of the VECM do not remain the same, and consequently, the estimated causal relationships may undergo changes. The stability of the VECM is examined through CUSUM and CUSUM-Square Test.

The plots of CUSUM (in Fig. 10.6) indicate that the cumulative sum lies inside the two critical lines with 5% level of significance, implying the parameter estimated by the model are stable from 1999 to 2018. While the plots of the CUSUM-Square test show that the cumulative sum square lies inside the two critical lines with 5% level of significance, indicating the stability of parameters estimated by the VECM during the entire sample period.

10.8 Conclusion

Health makes a vital contribution to economic development as healthy people live longer, more productive, and save more resources. Since productive labor can work more efficiently for a longer time, it obviously influences the production process. The present study examines whether any interdependency and dynamic connections among Public Health Expenditure (GHE), Domestic Private Health Expenditure (DPHE), and economic growth (GDP) in India during the period from 1999 to 2018. The current COVID 19 pandemic situation unveils the grim picture of the Indian healthcare system, as evident from different healthcare expenditure data and inadequate public health infrastructure and limited human resources in India during the period of study. The study's overall results strongly suggest that the influence of public healthcare expenditure on India's economic growth is relatively minor. The prime reason is that India's public healthcare expenditure is as low as below 1% of the country's GDP during the period of study. From the study, it is also observed that the country's economic growth has had a significant impact on domestic private

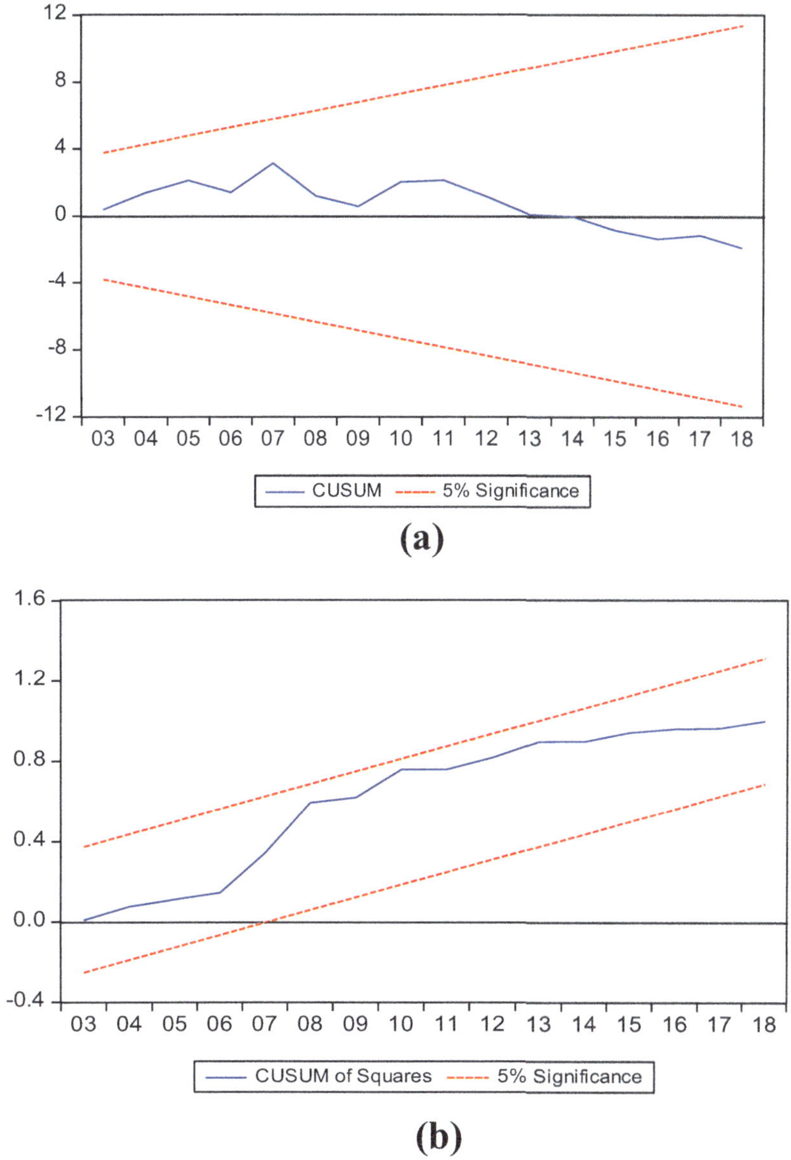

Fig. 10.6 Plots of **(a)** CUSUM and **(b)** CUSUM-Square tests (**Source:** computed by the author using EViews 10 software))

healthcare expenditure. The coefficient of GDP is negative, suggesting that an increase in GDP will decline the private healthcare expenditure, which is extremely necessary to reduce the out-of-pocket expenditure in India. The result of the present study is very much relevant from the policy point of view. Given, such findings, the policymakers and both the central government and state government should invest more in the health sector to strengthen country's health care system which will expedite the India's economic growth. Global evidence on health spending shows that unless a nation spends at least 5–6% of its GDP on health care, basic health care is rarely met.[3] The present study suggests that healthcare reforms should be given utmost priority. It is also recommended that subject of health should be shifted to concurrent list of Indian constitution from the state list and right to health should be declared as a fundamental right.

References

Atuahene SA, Yusheng K, Micah GB (2020) Health expenditure, co2 emissions and economic growth: China vs. India. Doi:https://doi.org/10.20944/preprints202009.0348.v1

Basumallik S (2017) Health and its impact on economic growth in India- an explanation. Int J Creat Res Thoughts 5(4):320–882

Bedir S (2016) Healthcare expenditure and economic growth in developing countries. Adv Econ Business 4(2): 76-86. doi:https://doi.org/10.13189/aeb.2016.040202

Bhat R, Jain N (2004) Time series analysis of private healthcare expenditures and GDP; co integration results with structural breaks. Indian Institute of Management, Ahmedabad

Dincer H, Yuksel S (2019) Identifying the causality relationship between health expenditure and economic growth: an application on E7 countries. Journal of health systems and policies. 1: 157

Dolado JJ, Lütkepohl H (1996) Making Wald tests work for cointegrated VAR systems. Econ Rev 15:369–386

Ehikioya IL, Mohammed I (2013) Determinants of public health care expenditure in Nigeria: an error correction mechanism approach. Int J Busi Soc Sci 4:13

Farahani M, Subramanian SV, David C (2010) Effects of state-level public spending on health on the mortality probability in India. Health Econ 19(11):1361–1376

Gangal VLN, Gupta H (2013) Public expenditure and economic growth: a case study of India. Glob J Mgmt Busi Stud 3(2):191–196

Granger CWJ (1969) Investigating causal relation by econometric models and cross-spectral methods. Econometrica 37:424–438

Gupta I, Mitra A (2004) Economic growth health and poverty: an exploratory study for India. Dev Policy Rev 22(2):193–206

Kulkarni L (2016) Health inputs, health outcomes and public health expenditure: evidence from BRICS countries. Int J Appl Econ 31(1):72–84

Maitra B, Mukhopadhyay CK (2012) Public spending on education, health care and economic growth in selected countries of Asia and the Pacific. Asia-Pacific development journal 9(2):14

Mohanty RK, Behera DK (2020) How effective is public health care expenditure in improving health outcome? An empirical evidence from the Indian states. NIPFP working paper no. 300

Mohapatra S (2019) Public health expenditure and its effect on health outcomes: a new methodological approach in the Indian context. Great Lakes Herald 13(1):148

Nyasha S, Odhiambo NM (2019) The impact of public expenditure on economic growth: a review of international literature. Folia Oeconomicstetinesia Sciendo 19(2):120. https://doi.org/10.2478/foli20190015

[3] Situation Analyses: Backdrop to the National Health Policy-2017, Ministry of Health and Family Welfare. Government of India.

Pradhan RP (2010) The long run relation between health spending and economic growth in 11 OECD countries: evidence from panel cointegration. Int J Econ Perspect 4(2):427–438

Raghupathi V, Raghupathi W (2020) Healthcare expenditure and eco-nomic performance: insights from the United States data. Front Public Health 8:156

Rajeshkumar N, Nalraj P (2014) Public expenditure on health and economic growth in selected Indian states. Int J Sci Res. 3(3):13–204

Seshaiah SV, Koti Reddy T, Sarma IRS (2018) General government expenditure and economic growth in India: 1980-81 to 2015-16. Theor Econ Lett 8:728–740

Shafuda CP, De UK (2017) Upshot of public health expenditure on economic development, MPRA paper no. 101846

The World Bank Data (2018) Data.worldbank.org

Toda HY, Yamamoto T (1995) Statistical inference in vector auto regressions with possibly integrated processes. J Econ 66:225–250

Varkey RS, Joy J, Panda PK (2020) Health infrastructure, health outcome and economic growth: evidence from Indian major states. J Crit Rev 7(11):2394–5125

Verma CS, Usmani G (2019) Relationship between health and economic growth in India. Ind J Human Dev 2(2):1–13. https://doi.org/10.1177/0973703019887601

Zaman SB, Hossain N, Mehta V, Sharmin S, Mahmood SAI (2017) An association of total health expenditure with GDP and life expectancy. J Med Res Innov 1(2):27. https://doi.org/10.5281/zenodo.576546

Dr. Subrata Saha was awarded with the Ph.D. in Economics from the University of North Bengal. Dr. Saha specializes on issues of fiscal policies and effects of Structural changes on fiscal relationships in Southeast Asian Countries and India. The area of research interest is Public Finance and Time Series Econometrics. He is a member of the editorial board of UGC CARE listed journal ENSEMBLE and a life member of the Indian Econometric Society and Indian Economic Association. Dr. Saha is presently working as an Associate Professor of Economics at Raiganj University, West Bengal, India. He has supervised four M.Phil. and one Ph.D. and presently supervising four Ph.D. scholars at Raiganj University. He has contributed several research papers to National & International Journals and edited research volumes. Besides, he has authored two books in his credit. IRDP, India, has awarded him with prestigious Sarvepalli Radhakrishnan Lifetime Achievement National award in 2018 in recognition of his contribution to his activities toward promoting educational excellence.

Chapter 11
Mapping Linkages between the Agriculture Sector, Informal Economy, and Inequality amid Pandemic

Pooja Sharma and Anjan Chakrabarti

Abstract With the onset of the COVID-19 pandemic, a substantial need to introspect the socio-economic systems of specifically the developing nations has emerged. The outbreak of informal sector crises and the health crises have compelled the developing countries to revisit their development process. The status of informal employment and the prevailing inequality amidst rapid economic growth indicate the strong presence of a dual economy in India. After highlighting the economic growth scenario during the COVID-19 pandemic, the chapter examines the status of inequality embedded in India's capitalist structure and the agriculture sector. The chapter attempts to empirically test the plausible linkages between inequality, employment, the agriculture sector, and economic growth by performing a regression analysis to comprehend the agriculture sector's role in explaining inequality. Finally, the chapter witnesses the deep-rooted linkages between the agriculture sector, inequality, and employment. There is a rise in disguised unemployment in the agriculture sector; the corresponding skilled and formal employment is needed to absorb all the displaced and disguised unemployed workers. The focus on skill enhancement by various government schemes and, ultimately, the absorption of people employed in the informal sector is highly recommended.

Keywords COVID-19 pandemic · Inequality · Dualism · Agriculture sector · Informal employment · Disguised employment

---

P. Sharma (✉)
Daulat Ram College, The University of Delhi, Delhi, India

A. Chakrabarti
UGC-Human Resource Development Centre, The University of North Bengal, Darjeeling, West Bengal, India

S. Saha et al. (eds.), *Economic and Societal Transformation in Pandemic-Trapped India*, New Frontiers in Regional Science: Asian Perspectives 55,
https://doi.org/10.1007/978-981-16-5755-9_11

11.1 Introduction

The sudden outburst of the COVID-19 pandemic has marked itself as the biggest catastrophe the world has ever seen in the century. It has affected almost every part of the world and each human being. Human civilization has witnessed a phenomenal transition in their life patterns and their interactions with the environment. The year 2020 became part of History because the entire world is combatting one of the greatest medical holocausts, the COVID-19 pandemic. The last global pandemic occurred in 1918, known as Spanish flu, and it took the lives of around 25 million worldwide. In India, 17 million have died within a short period. Present-day Mumbai was most badly affected; 20,258 people died between tenth September and tenth November 1918. In Africa, 2 percent of the total population lost their lives (Ramanna 2003). One hundred years passed, human civilization is confronting another deadly pandemic. Till 15th February 2021, the number of corona-affected persons reached 1,09, 16, 589 with a death toll of 1,55. 732 and 97 percent corona-affected people have recovered. The death toll remains much higher in the US and Brazil.

The COVID-19 pandemic has revealed the ecological imbalance the world is undergoing. Consequently, the detrimental forces underlying the socio-economic systems emerged and got unveiled. In the Indian context, the country many times has tagged as the fastest-growing country or the second fastest-growing country after China. The average growth rate during this entire period is closer to 6 percent. The moot point is that does growth brings equity or increases income inequality. India's economic growth pattern and change in the degree of inequality do not say so. The research paper of two well-known economists Lucas Chancel and Thomas Piketty, published in 2017, highlighted that India's income inequality reached the highest level since the income tax was introduced in 1922 (Chancel and Thomas 2017).

The prevailing levels of unemployment and inequality indicate the failure of capitalism to deliver equitable income to all. The growth-inequality trade-off is a mixed bag again. The per capita GDP growth rate is highest in China and the lowest in Brazil. At the same time, inequality is highest in Brazil in all 4 years, while lowest in Finland. In rural India, 85 percent of the workforce do not have a salaried job, and for urban areas, the percentage stands at 53 percent. The existing inequality in India exhibits a wide gap between the productivity of agriculture and industry. During the pre-COVID period, the economic slowdown became prominent in 2019. The existing problems of unemployment, low income, rural distress, malnutrition, and income inequality acerbated. The emerging scenario has exposed India's informal sector's glaring reality, which does not receive much attention before the pandemic outbreak. Out of the national total, 465 million workers, around 91% were informal workers, including around 42 percent agricultural workers and 9–10 percent migrant workers from the agricultural sector (International Labour Organization 2020). Though agriculture registered 3.4 percent growth in 2020–2021 due to good monsoon, it faces both supply-side and demand-side shocks and has cascading effects. Besides, the pathetic plight of daily wage-earners, migrant labourers, and most of the agrarian community that has been unfolded during the lockdown period raised few fundamental questions on the efficacy and outcome of economic reforms in India and on the theoretical underpinning of the philosophy of neo-liberal economy. Therefore, the agricultural and informal sectors should be monitored simultaneously, and policy reorientation ensures life and livelihoods have become imperative.

With this backdrop, the chapter examines the prevailing linkages between capitalism and inequality. Finally, having examined India's unemployment and employment situation, the chapter discusses the agriculture sector and its situation related to employment.

11.2 Emerging Growth Scenario during COVID-19

India declared a complete lockdown at the end of March 2020 and continued till May 2020. Supply of all non-essential goods and services reached a near halt. As a result, the economic slowdown became pronounced during 2019. Especially the industrial growth was reduced to a meager 0.9 percent. The overall growth rate has been reduced to 4.2 percent during FY 2019–2020, which is much lower than the economy's average growth rate during the entire reform period. During the last three decades of economic reform, the Indian economy grew by more than 6 percent. Therefore, industrial slowdown, as well as economic slowdown, became prominent since late 2019. The FY 2020–21 or the pandemic year registered a negative growth of −7.7 percent, accompanied by −9.6 percent growth of industry and −8.8 percent growth of services. Only agriculture and allied activities registered a modest growth of 3.4 percent (Fig. 11.1). Quarterly data of 2021–2022 reflected the real economic scenario of the post-COVID scenario. A whopping 23.9 percent contraction of the economy was observed during the first quarter of FY 2021 (Table 11.1). The quarterly GDP growth from the first quarter (Q1) of Financial Year (FY) 2020 to FY 2021 will reveal the deteriorating growth scenario in pandemic and post-pandemic situations.

Between Q2 FY 2020 and Q1 FY 2021, gross value added (GVA) has been declined by 22.8 percent. Construction experienced a significant slide down of 50.3 percent, manufacturing by 39.3 percent, and mining by 23 percent. Gross Fixed Capital Formation declined by 52.9 percent and choked the investment, which has both supply-side and demand-side impacts on goods and services. In addition, core industries received a significant shock. Index of eight core industries (ICI) recorded a massive decline of 38.1 percent during April 2020. Coal, crude oil, natural gas, refinery products, fertilizers, steel, cement constitute two-fifth of overall industrial production, and the output of all these industries has fallen drastically immediately after the countrywide lockdown announcement. Production of steel and cement declined by 84 percent and 86 percent, respectively. Production of automobiles, railway wagons, infrastructure, and construction has also become standstill due to lockdown.

Consequently, the horizontal and vertical linkages of iron and steel industries and cement industries got utterly raptured. Due to the slowdown of industrial activities,

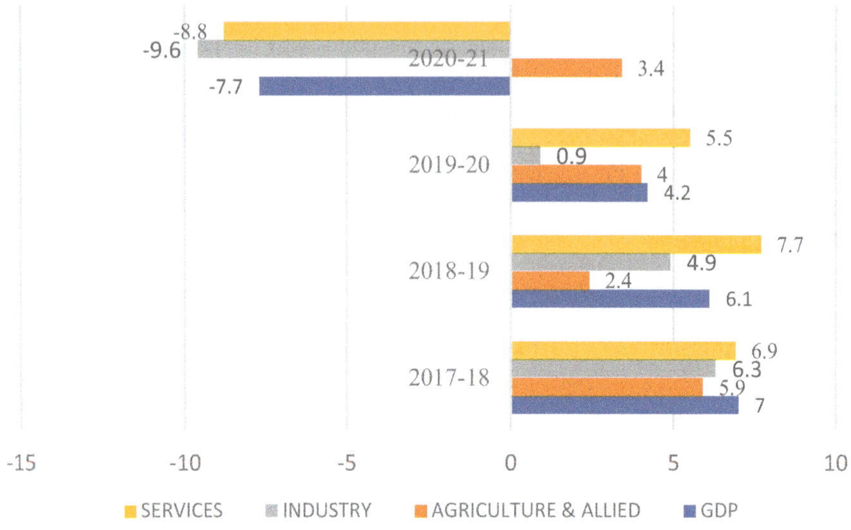

Fig. 11.1 Growth Rates of GVA at Basic Prices from 2017–18 to 2020–21 (Source: Ministry of Statistics and Programme Implementation, Govt. of India)

Table 11.1 Growth rates of in various quarters in 2020–2021

Financial Quarter (Q1, Q2, Q3, Q4) and Financial Year (FY)	GDP Growth (%)
Q1 FY 2020	5.2
Q2 FY 2020	4.4
Q3 FY 2020	4.1
Q4 FY 2020	3.1
Q1 FY 2021	−23.9

Source: Ministry of Statistics and Programme Implementation, Govt. of India

electricity generation declined by 22.8 percent by April 2020, resulting in a fall in coal production by 15.5 percent. During the lockdown, public transport, railways got severely impaired. Civil aviation came to a standstill. The production of crude oil and petroleum products is heavily dependent on domestic demand for diesel, petrol, and aviation turbine fuel (ATF).

Consequently, during May 2020, consumption of ATF suffered a fall of 87 percent and diesel by 38 percent (Shah 2020). Demand for natural declined due to the low production of electricity and fertilizer. Therefore, the revival of core industries in the post-COVID scenario is challenging to achieve in the short run. The first phase of supply shocks created a second-round wave of shocks on demand and supply. The initial supply shock caused a loss in wage, income, and a drastic fall in employment, both formal and informal. As a result, the aggregate demand got severely squeezed. Productive capacity has been declined further and created another wave of a supply shock. As a result, the Indian economy got into a severe recessionary mode.

11.3 Capitalism and Inequality

The famous Kuznets curve (Kuznets 1955) advocated an inverted U-curve, postulating that inequality first increases and then decreases over time. Baumol (1986) and Barro (1991) determined convergence in developed countries while divergence in less developed nations. Romer (1986) observed a lack of convergence across nations and argued for an alternative framework for modern growth. Later, endogenous growth theories modified the neoclassical production function to incorporate factors such as human capital, innovation, increasing returns, and spatial spillover in the production function to determine long-term growth rate.

There are mixed results; one set of literature believes that lagging regions will catch up with the fast-growing regions by adopting balanced regional development strategies such as Borts and Stein (1962) and Needleman et al. (1968) for developed countries as the USA. While Perroux (1950), Myrdal (1957), argued regional incomes would not converge in the long run. Later, Martin and Sunley (1998) argued that policymakers will attempt to intervene with balanced development but will not overcome regional divergence. In China, Aroco et al. (2008) pointed out that income distribution has shifted away from convergence to "polarization".

The growth-inequality trade-off is a mixed bag again. The Per Capita GDP growth rate is highest in China and the lowest in Brazil. In contrast, inequality is highest in Brazil in all four years, while lowest in Finland. The human development index (HDI) is highest in Switzerland, followed by Finland. HDI is lowest in Nepal and Bangladesh, as in Table 11.2. The Indian economy has grown over the last two decades. However, the growth has been unequal substantially when compared to other nations. China has shown a substantial increase in the development of the top 1 percent of the population, followed by India. This gap has been reduced in the Russian Federation and Brazil (World-inequality Database). Thus, there has been an inverse relationship between growth and inequality in the case of India.

Almost three decades passed away since India initiated economic reforms in 1991. The country many times have been tagged as the fastest-growing country or second fastest-growing country after China. However, the average growth rate during this entire period is closer to 6 percent. The moot point is that does growth brings equity or increases income inequality. India's economic growth pattern and change in the degree of inequality do not say so. The research paper of two well-known economists Lucas Chancel and Thomas Piketty, published in 2017 (Chancel and Thomas 2017), highlighted that India's income inequality peaked since the income tax was introduced in 1922. Their study revealed that the gap between rich and poor narrowed between 1951 and 1980, while the trend got reversed from 1980 to 2014.

The Gini coefficient is estimated to be close to 0.50, which would be an all-time high. A general rise in the Gini coefficient indicates that government policies are not inclusive and may be benefiting the rich as much as (or even more than) the poor. The Gini coefficient is one of the most frequently used measures of economic inequality. The coefficient can take any values between 0 and 1. A coefficient zero

Table 11.2 Country-wise inequality, expenditure on health and education as a percentage of GDP

Country	1991			2001			2010			2016		
	HDI	Gini	Per capita GDP growth	HDI	Gini	Per capita GDP growth	HDI	Gini	Per capita GDP growth	HDI	Gini	Per capita GDP growth
Switzerland	0.832	33.90	−2.146	0.889	33.4	0.67	0.932	32.6	1.935	0.943	33.0	0.617
Finland	0.784	22.90	−6.424	0.858	27.2	2.37	0.903	27.700	2.714	0.918	27.100	2.333
Brazil	0.611	60.50	−0.258	0.684	58.4	0.013	0.727	52.900	6.524	0.758	53.300	−4.09
Russian Federation	0.734	48.40	−5.259	0.720	36.9	5.54	0.780	39.500	4.453	0.815	36.800	0.145
India	0.427	32.70	−0.983	0.493	36.8	3.02	0.581	37.500	7.042	0.636	37.800	6.9970
China	0.502	32.20	7.812	0.594	38.7	7.555	0.706	43.700	10.10	0.748	38.500	6.160
Bhutan	NA	NA	−1.120	NA	40.9	6.083	0.566	38.800	10.77	0.609	37.400	6.83
Bangladesh	0.387	27.60	1.106	0.468	33.4	3.113	0.545	32.100	4.39	0.597	32.400	5.950
Nepal	0.387	35.20	3.627	0.446	43.8	3.052	0.529	32.800	4.31	0.569		−0.319
Srilanka	0.625	32.40	3.347	0.685	41.0	−2.243	0.745	36.400	7.27	0.768	39.800	3.338

Source:World Bank Open Data (See here: https://data.worldbank.org/)

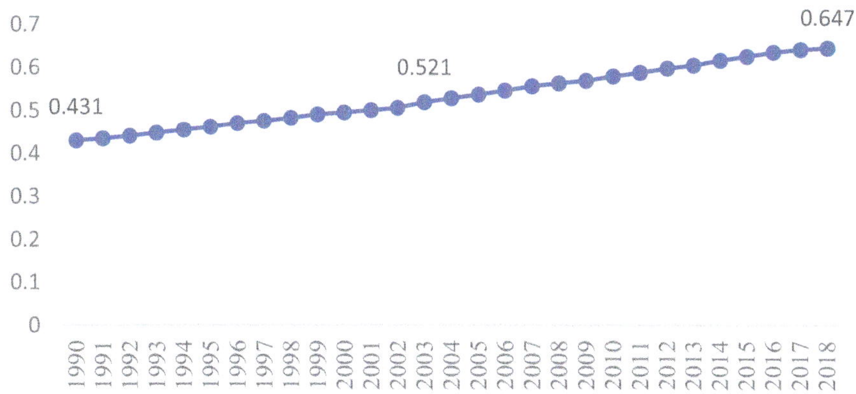

Fig. 11.2 Changing Gini Coefficient for India from 1990 to 2018 (Source:World Bank Open Data [See here: https://data.worldbank.org/])

indicates a perfectly equal distribution of income or wealth within a population. A coefficient of one represents a perfect inequality when one person in a population receives all the income while other people earn nothing. Plotting data of GINI coefficient from 1990 to 2018, it has been observed that there is a secular rise in the value of the coefficient. From 0.431 in 1990, the value of the Gini coefficient reached 0.647 (Fig. 11.2). This clearly reveals that high economic growth during the entire reform period contributed to India's secular rise in inequality.

The survey conducted by Oxfam India in 2018 suggests the rising inequality in India in the recent past. The report stated 73 percent of the wealth generated last year went to the wealthiest one percent (Oxfam International 2018).

It is beyond doubt that the pandemic has created an unprecedented economic crisis. Still, the vulnerability of 90 percent of the workforce, which has now become the point of a significant argument, is the outcome of unfettered faith in market forces which India considers sacrosanct. As Sen (1991a, b, 1992) consistently cautioned that growth does not ensure the well-being of each individual, and to achieve the well-being of every individual, more focus should be given to developing human resources. To him, income cannot be used as a sole measure to evaluate inequality, as it does not entirely ensure the capability or freedom of an individual to achieve what he desires. On the contrary, economic inequality takes into account not only income but also education, health, justice, credit, and other productive resources and opportunities, which provides an incentive to an individual to better his overall situation (Sen 1997).

11.4 Employment and Unemployment

The immediate impact of COVID-19 and lockdown has resulted in an unprecedented spike in the unemployment rate. CMIE estimates stated that unemployment during April–May reached 23.5 percent, and for urban and rural India, figures stood at 24.9 percent and 22.9, respectively (Fig. 11.3). A commensurate fall took place in the labour participation rate (LPR) between March and April 2020. From 22nd March 2020, the LPR had fallen from 42.6 percent to 35.4 percent. During May–June, the unemployment rate started reducing with the rise in LPR. As Mahesh Vyas of CMIE points out, recovery is mostly among the informal, self-employed workers who do not have the choice but to search for livelihoods for survival (Vyas 2020a, b). It is equally important that over 27 million young people in their 20s forced to leave the job during the lockdown period. This is the cumulative impact of job-cut unleashed by the corporate sector since late-2019 and due to lockdown. Now in post-June 2020, a recovery in employment is taking place. It is least expected that all those young minds will be re-employed. Therefore, young India has to suffer for a more extended period (Vyas 2020a).

The average economic growth rate during the entire reform period remains closer to 6 percent. Now growth–employment relationship has become questionable. In 2019–20 Budget, Agriculture and Rural Development together find an allocation closer to 11 percent of GDP, but numerous schemes for these two sectors failed to improve the rural economy. Data of CMIE also demonstrated that the formal sector had not been spared. From April to July 2020, 18.9 million salaried people lost their

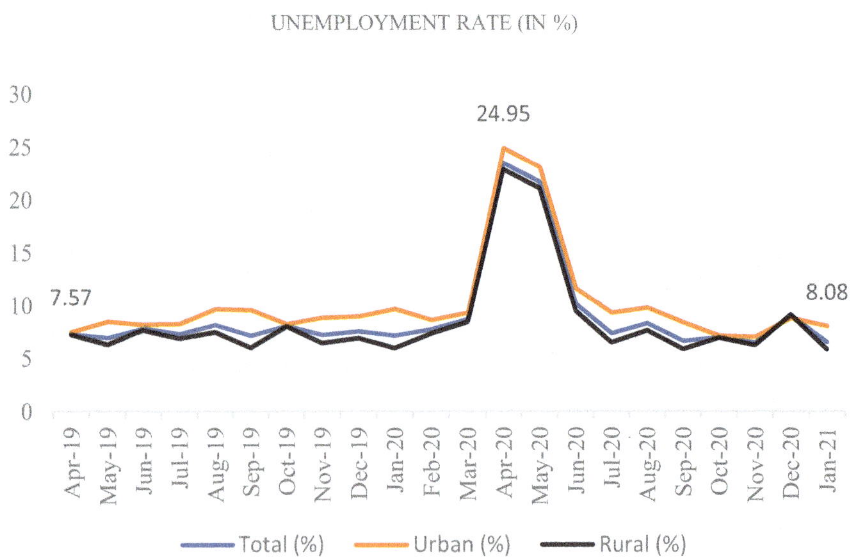

Fig. 11.3 Unemployment Rate (in %) during April 2019 to January 2021 (Source: Centre for Monitoring Indian Economy Pvt. Ltd., CMIE)

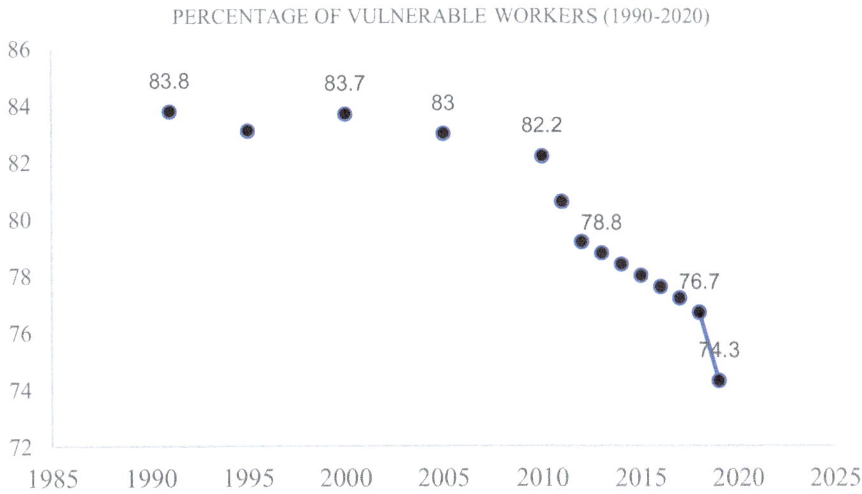

Fig. 11.4 Changing Percentage of Vulnerable Workers in India from 1990 to 2020 (Source: ILOSTAT Database [See here: https://ilostat.ilo.org/])

jobs, and most of them shall find it difficult to get back their jobs, and they may be re-routed from the formal sector to the informal sector (The Wire 2020). The rise in agrarian distress has forced the labour force to join the urban informal sector and migrate away from home. In brief, Economic reforms and expansion of the informal sector went hand in hand. Informal workers do not have any written contract, paid leave, health benefits, or social security. Closer to 90 percent of the workforce is engaged in informal sectors. UNDP estimate reveals 76.7 percent of the workforce can be termed as vulnerable employment. In 2011, the percentage was 80.6 percent, and in 1991, it was 83.3 percent. Vulnerable employment is defined by the International Labour Organisation (ILO) as the percentage of employed people engaged as unpaid family workers and own-account workers. In 2020, the percentage of vulnerable employment stood at 74.3 percent (Fig. 11.4). It clearly manifests the economic reforms have created poor quality jobs with a high degree of economic and social insecurity.

According to the National Statistical Office (NSO), 46 percent have no “paid leave” for those who are in a salaried job, and 70 percent work without any written contracts. According to the annual survey of NSO (46th Round) combined percentage of self-employed and casual workers was closer to 77 percent in 1991. Therefore, it is evident that three decades of economic reforms have expanded the informal sector with a high degree of casualization and vulnerability with constant degradation of the quality of jobs. As per the 2011 Census, 4.1 crores or 8.5 percent of the total population work as migrant workers, and 3.5 percent are temporary migrants. They mostly migrate from the Eastern region and UP to Western and South Western Coastal districts. These indicate that many states, agriculture as well as rural economy fail to ensure year-long subsistence income.

11.5 The Agriculture Sector and Inequality: Conundrum of Capitalism

The Indian economy has a great paradox. India unleashed structural reform and adopted the neo-liberal policy in 1990 without addressing the inherent duality between agriculture and the non-agricultural sector. The situation is such that around 43 percent of the agricultural community is sharing only 14 percent of income. The per capita income from agriculture has been declining consistently. The economic condition of small and marginal farmers and agricultural labourers has been deteriorating over time. Economic reform in India has emphasized fiscal reforms; the major input subsidies have been curtailed. Public investment in agriculture and expenditure on research and extension has been slowed down. Contract farming and export-oriented crop production got priority. The regulated market is on the verge of closure in many states (Ramkumar 2020). The cost of production in agriculture, as a consequence, increased many times. This added further uncertainty to agriculture, abated rural to urban migration and expanded the informal sector. Despite having a plethora of schemes for rural India, it has become gradually clear that rural India's economic condition has started deteriorating much before the outbreak of the COVID-19 pandemic. As per the "Key Indicators: Household Consumer Expenditure in India" conducted by the NSO, an individual's average monthly spending has been reduced to Rs. 1446 in 2017–18 from Rs. 1501 in 2011–11. This implies that there is an overall decline in per capita consumption expenditure by 3.7 percent. But the worrying factor is that in rural India, consumption expenditure has been reduced by 8.8 percent and for urban India by 2 percent. The crucial issue is that Rural India's monthly spending on food has declined by 10 percent between 2011–12 and 2017–18. Rural India spent Rs. 643 per month on food items, and in 2017–18, the spending has been reduced Rs. 580. The figures are inflation-adjusted, and hence the decline is in real terms. Both rural and urban India spent less on essential food items such as oil, salt, sugar, and spices. The former Planning Commission member opined that a fall in food spending, especially in villages, shows that malnutrition has increased along with a rise in poverty level (The Economic Times 2020). It also indicates a lack of job opportunities in the non-farm sector and slows down in the farm sector (Editorials 2020: 7). During the lockdown period and in the post-lockdown period, disruption in the supply chain prices of vegetables and other food items has already been increasing. Lockdown and halt in economic activities will throw minimum employment and income opportunities for both rural people and people associated with the large urban informal sector. The spread of hunger and impoverishment has become imminent.

11.6 Linkages between Inequality, Economic Growth, Employment, and Agriculture Sector: An Empirical Take

The emerging crises of the informal sector unveiled during the COVID-19 pandemic exhibit the underlying linkages between the prevailing income inequalities, economic growth, and employment. Time series analysis is performed by deploying the time series from 1991 to 2018 to explore this linkage. Regressing growth rate, workforce participation rate, and expenditure on the social sector as a percentage of GDP on the level of inequality, we derive some of the significant deductions (Tables 11.3, 11.4, 11.5, and 11.6).

It is evident that the increase in workforce participation significantly reduces inequality. It is also crucial to observe that the increase in expenditure on the social sector has significantly aggravated inequality, confirming that social goods expenditure has not been instrumental in mitigating the inequality. However, such expenditure has enhanced the inequality of income across the country. At the same time,

Table 11.3 Correlation Matrix

	GINI	GDP growth	Work participation rate	Exp on social good as % of GDP	Ln Food productivity
GINI	1				
GDP growth	0.390531	1			
Work participation rate	−0.95394	−0.23507	1		
Exp on social good as % of GDP	0.673507	0.057082	−0.75033	1	
Ln food productivity	0.969827	0.371917	−0.9431	0.681816	1

Source: Authors Calculation

Table 11.4 Regression results

Model		Unstandardized Coefficients		Standardized Coefficients		
		B	Std. Error	Beta	t	Sig.
1	(constant)	−0.981	0.519		−1.891	0.071
2	GDP GROWTH	0.003	0.002	0.091	1.948	0.063
3	Work participation rate	−0.005	0.001	−0.492	−3.535	0.002
4	Exp on social good as % of GDP	−0.004	0.006	−0.042	−0.690	0.497
5	Ln food productivity	0.231	0.062	0.501	3.709	0.001

N.B. Dependent variable GINI
Source: Authors Calculation

Table 11.5 Model Summary of the Regression model

Model Summary[b]

Model	R	R Square	Adjusted R Square	Std. error of the estimate	Change statistics					Durbin-Watson
					R Square change	F change	df1	df2	Sig. F change	
1	.981[a]	0.962	0.956	0.014380	0.962	152.233	4	24	0.000	1.327

[a]Predictors: (constant), ln food productivity, GDP GROWTH, Exp on social good as % of GDP, work participation rate
[b]Dependent variable: GINI
Source: Authors Calculation

Table 11.6 ANOVA Result

ANOVA[a]						
Model		Sum of squares	Df	Mean square	F	Sig.
1	Regression	0.126	4	0.031	152.233	.000[b]
	Residual	0.005	24	0.000		
	Total	0.131	28			

[a]Dependent variable: GINI
[b]Predictors: (constant), ln food productivity, GDP GROWTH, Exp on social good as % of GDP, the work participation rate
Source: Authors Calculation

the increase in growth rate significantly increases inequality. An improvement in agricultural productivity is associated with a further increase in inequality, suggesting that as there is a rise in disguised unemployment in the agriculture sector, the corresponding skilled and formal employment is needed to absorb all the displaced and disguised unemployed workers. If the economy fails to create employment opportunities in the formal sector, the underlying duality will create more informal employment. This will further aggravate inequality in the economy.

The results of regression in Tables 11.4, 11.5, and 11.6 indicate a significant role of economic growth in increasing inequality. It is also observed that the workforce participation rate significantly reduces inequality, emphasizing the role of employment and job opportunities in reducing inequality. Furthermore, the government's role in the social sector as a percentage of GDP in mitigating inequality is substantially insignificant. While improvement in agricultural productivity significantly increases inequality, disguised employment is not sufficiently absorbed in the formal employment offered by the manufacturing or services sector. Around 96 percent of the inequality variations have been explained by the selected explanatory variables, evident from the coefficient of determination that is approximately 96 percent. The overall goodness of fit of the above regression is quite significant, suggesting that the model is a good fit.

11.7 Conclusion

The sudden outbreak of the COVID-19 pandemic has marked itself the most significant health catastrophe of its kind. The COVID-19 pandemic has revealed the ecological imbalance the world is undergoing. Consequently, the detrimental forces underlying the socio-economic systems emerged and got unveiled. Almost three decades passed away since India initiated economic reforms in 1991. India unleashed structural reform and adopted a neo-liberal policy in 1990 without addressing the inherent duality between agriculture and the non-agricultural sector. The situation is such that around 43 percent of the agricultural community is sharing only 14 percent of income. The per capita income from agriculture has been declining consistently. The economic condition of small and marginal farmers and agricultural labourers has been deteriorating over time. The chapter empirically tested the plausible linkages between inequality, employment, the agriculture sector, and economic growth by performing a regression analysis to comprehend the agriculture sector's role in explaining inequality. The rise in agrarian distress has forced the labour force to join in the urban informal sector and migrate away from home. In brief, economic reforms and expansion of the informal sector went hand in hand. Informal workers do not have any written contract, paid leave, health benefits or social security.

Finally, the chapter witnesses the deep-rooted linkages between the agriculture sector, inequality, and employment. There is a rise in disguised unemployment in the agriculture sector, and the corresponding skilled and formal employment is needed to absorb all the displaced and disguised unemployed workers. The focus on skill enhancement by various government schemes and ultimately the absorption of people employed in the informal sector is highly recommended.

References

Aroco PA, Guo D, GJD H (2008) Spatial convergence in China: 1952–1999. In: Guanghua W (ed) Inequality and growth in modern China. Oxford University Press, Oxford, pp 125–143

Barro JR (1991) Economic growth in a cross-section of countries. Q J Econ 106(2):407–443

Baumol WJ (1986) Productivity growth, convergence, and welfare: what the long-run data show. Am Econ Rev 76(5):1072–1085

Borts GH, Stein J (1962) Regional growth and maturity in the United States. A study of regional structural change. Swiss J Econ Stat. 98(3): 290-321

Chancel L, Thomas PT (2017) Indian income inequality, 1922-2015: From British Raj to Billionaire Raj?, WID. World Working Paper Series 2017/11, World Inequality Lab, https://wid.world/document/chancelpiketty2017widworld/

Editorials (2020) Abrupt planning looming hunger, economic a political weekly. LV 15:7

International Labour Organization (2020) Rapid assessment of the impact of the COVID-19 crisis on employment, ILO Brief, June, https://www.ilo.org/newdelhi/whatwedo/publications/WCMS_748095/lang--en/index.htm

Kuznets S (1955) Economic growth and income inequality. Am Econ Rev 45(1):1–28

Martin R, Sunley P (1998) Slow convergence? The new endogenous growth theory and regional development. Econ Geogr 74(3): 201–227

Myrdal G (1957) Rich lands and poorlands: the road to world prosperity. Harper & Bros, New York

Needleman P, Passonneau JV, Lowry OH (1968) Distribution of glucose and related metabolites in rat kidney. Am J Physiol Legacy Cont 215(3):655–659

Oxfam International (2018) 15 shocking facts about inequality in India. https://www.oxfamindia.org/blog/15-shocking-facts-about-inequality-india?gclid=EAIaIQobChMI8fSE9sqW6QIVApCPCh0-qAX7EAAYASAAEgL0ufD_BwE

Perroux F (1950) Economic space: theory and applications. Q J Econ 64(1):89–104

Ramanna M (2003) Coping with influenza pandemic: the Bombay experience. In: Phillips H, Killingray D (eds) The Spanish influenza pandemic of. Routledge, Abingdon Oxon, pp 1918–1919

Ramkumar R (2020) Agriculture and the COVID-19 pandemic: an analysis with special reference to India. Rev Agrar Stud 10(1):6–124

Romer PM (1986) Increasing returns and long-run growth. J Polit Econ 94(5):1002–1037

Sen AK (1991a) Welfare, preference and freedom. J Econ 50(1–2):15–29

Sen AK (ed) (1991b) On indexing primary goods and capabilities unpublished paper. Harvard University, Cambridge, MA

Sen AK (1992) Inequality reexamined. Oxford University Press, Cambridge, MA

Sen AK (1997) From income inequality to economic inequality. South Econ J 64(2):384

Shah Y (2020) Output of core industries plummets in April 2020 31st May. www.cmie.com

The Economic Times (2020) World Bank sees FY21 India growth at 1.5–2.8%; slowest since economic reforms three decades back April 13. https://economictimes.indiatimes.com/news/economy/finance/covid-19-causes-severe-disruption-to-indian-economy-says-worldbank/articleshow/75104474.cms?utm_source=contentofinterest&utm_medium=text&utm_campaign=cppst

The Wire (2020) What Job Losses in the Formal Sector Tell us About the Lockdown's Impact on Economy 19th August. https://thewire.in/economy/job-losses-formal-sector-lockdown-impact-economy-coronavirus-cmie

Vyas M (2020a) Longterm cost of lockdown 12th may. Economic Outlook, CMIE Pvt. Ltd. https://www.cmie.com/kommon/bin/sr.php?kall=warticle&dt=2020-05-12%2010:21:58&msec=653&ver=pf

Vyas M (2020b) Expect a recovery in June 16th June, economic outlook, CMIE Pvt. Ltd. https://www.cmie.com/kommon/bin/sr.php?kall=warticle&dt=2020-06-16%2010:02:45&msec=733

Pooja Sharma is an Associate Professor, Department of Economics, Daulat Ram College, University of Delhi, worked at the Institute of Economic Growth. She is an alumnus of Miranda House, University of Delhi and completed Masters in Economics from Delhi School of Economics with the experience of teaching at the undergraduate level in the University of Delhi and M.Phil. from JNU. She had undertaken a project "Accounting for Monetary Benefits by reducing air pollution in Delhi" and has contributed several chapters for e-Pathshala MHRD project on macro-economics, public economics, and environmental economics. She has several papers at national and international journals related to energy security, energy transition, renewable energy, human capital. Her Ph.D. is from Energy studies program, School of International Studies Jawaharlal Nehru University at Delhi India, titled, "Role of renewables in Energy transition: A Comparative Study of India and Norway". She has been a Ph.D. fellow at the University of Agder, Norway as an exchange program for Ph.D. students.

Dr. Anjan Chakrabarti is a Professor-Director, UGC Human Resource Development Centre, The University of North Bengal, West Bengal. Previously he served as an Associate Professor at UGC Human Resource Development Centre, The University of Burdwan, West Bengal and as an Assistant Professor in Economics at St. Joseph's College, Darjeeling. His research interests include development in northeastern states of India, agrarian relations, agriculture in Bengal, and policy research. He published more than 50 research papers in national/international journals and edited volumes. He completed seven research projects (International and National). He authored "Economic Development and Employment" in Sikkim, Authorspress, New Delhi, 2009, and edited, a volume entitled "Interrogating Development: Perspectives on the Economy, Environment, Ethnicity and Gender" Setu Prakashani, Kolkata, 2017 and a Text Book on Micro Economics in 2019 for under graduate students.

Chapter 12
Impact of COVID-19 Pandemic on Informal Labour Market in India

Anil Kumar Biswas

Abstract The majority of labour forces are employed in un-organized sectors, are poverty-stricken, and vulnerable in the Indian labour market. A large number of workers migrate from less developed areas of the country to industrial areas and megacities. Although the COVID-19 pandemic affected all sections of people throughout the globe, informal workers both in un-organized sectors and organized sectors in India were hit the hardest. They have become jobless, lost their income, exhausted their savings, and aggravated their poverty. Reverse migration of workers has increased the excess burden on the rural economy. The government has initiated fiscal policy and monetary policy measures to overcome the crisis posed by the pandemic. Public expenditure was increased by 10 percent of GDP for boosting activities in different sectors of the pandemic-stricken economy. The RBI reduced interest rates for the smooth functioning of the system. However, informal workers are least benefitted from these policy initiatives. The future of informal workers in India remains totally uncertain as the growth rate reached the lowest level. The scholarly can assess the overall impact of the pandemic only for the time being.

Keywords Economic slowdown · Informal workers · Fiscal policy · Migrant workers · MSMEs

12.1 Introduction

Severe Acute Respiratory Syndrome Coronavirus 2 (SARS-Cov-2), generally known as COVID-19, was first detected in Wuhan, China, in December 2019 (Andrews et al. 2020). The World Health Organization (WHO) declared the menace of COVID-19 as an international Public Health Emergency on 30th January 2020 and coined it as a pandemic on 11th March 2020. In India, the first COVID-19

A. K. Biswas (✉)
Department of Economics, P. D. Women's College, Jalpaiguri, West Bengal, India

S. Saha et al. (eds.), *Economic and Societal Transformation in Pandemic-Trapped India*, New Frontiers in Regional Science: Asian Perspectives 55,
https://doi.org/10.1007/978-981-16-5755-9_12

positive patient was found in Kerala on 30th January 2020, who was a female medical student of a university in Wuhan, China (Hindustan Times 2020). Since then, the number of COVID-19 positive cases in India increased exponentially due to its highly contagious nature.

As a preventive measure, the Government of India announced a country-wide lockdown for a period of 21 days (25th March 2020 to 14th April 2020) known as Phase-1 lockdown. Furthermore, three phases of lockdowns were announced and strictly followed throughout the country. Phase-2, Phase-3, and Phase-4 lockdowns covered the period of 15 April–3 May, 4 May–17 May, and 18 May–31 May of 2020, respectively (Singh et al. 2020a). During this lockdown period, the government adopted restrictive measures to confine people in their current places for avoiding human contact by suspending all kinds of transportations and other business activities except a few essential services.

From the third phase onwards, some sort of relaxation was permitted beyond the red zones (with a high incidence of COVID-19 positive cases). The entire economy was affected by this unprecedented lockdown. Complete lockdown in India and abroad caused the closure of all business and industrial ventures leading to huge unemployment and loss of workers' income, particularly in the un-organized sectors. Most of the workers in India are informal workers without any written contract or insurance, and they generally do not have any savings or other assets to meet the needs during the pandemic. This chapter will investigate the socio-economic impact of the COVID-19 pandemic on informal workers in India. The study is based on available secondary data collected from various national and international agencies. Available literature shows that most of the research works are based on guesses rather than measuring economic impacts. In fact, sufficient data is not available in the issues of interest because the pandemic is still ongoing, and lockdown persists in some critical sectors like foreign trade. Nevertheless, available individual and aggregate level data have been analysed to find the possible impact of the pandemic on the lives and livelihood of informal workers in India.

12.2 The Informal Labour Market

In India, the share of informal labour in the total labour force is the largest in the global labour market. As per the International Labour Organization (ILO 2020a) policy report, the total number of estimated workers in India is 473 million, using population figures of 2020. Around 90 percent of the total labour force in India is employed in the informal sector, which contributes 45 percent of the Gross Domestic Product. In 2020, out of 473 million total labour force, 118 million were casually employed, and 246 million was self-employed. Informal or un-organized sector implies enterprises that are not registered under competent government departments. All the labour force employed in the informal sector is informal labour. The Un-organized Workers' Social Security Act 2008 includes the workers who are home-based workers, self-employed workers, wage workers in the un-organized sector, or workers in the organized sector not covered under social security acts as informal or un-organized workers (Ministry of Labour and Employment 2020). The majority of informal workers are engaged in agriculture and micro, small and medium enterprises (MSMEs) performing jobs like beedi rolling, agarbatti making, papad making, tailoring, and embroidery work.

In a research work based on the 73rd round of the National Sample Survey (NSS), it was observed that the micro-sector is having around 630.52 lakh enterprises, provide employment to 96.96 percent of total un-organized sector employment, while small and medium scale enterprises employ 2.88 percent and 0.16 percent, respectively (Ghosh 2020). Informal workers suffer from the seasonality of employment, absence of formal employer–employee relationships, and lack of social security protections. There is a disparity between formal sector and informal sector workers in India, and the share of women workers is less than that of their male counterparts. Several legislations have been passed for social security measures in protecting the informal workers. The Ministry of Labour and Employment of the Government of India is conducting welfare activities for the informal workers in the fields of health care, housing, and education for the children.

Multiple factors caused a larger share of informal labour forces in India. Historically, the aftermath of British colonial rules in India over three consecutive centuries, starting from the mid-eighteenth century, destroyed domestic industries and unilateral drain of economic resources to England. As a result, the bulk of labour forces were compelled to work in agriculture without any land ownership rights under the Zamindari system introduced by the British in 1793. A section of workers was engaged in small-scale cottage industries producing handicrafts. On the other hand, the population of India started to rise since 1921, known as the year of the Great Divide. Over time, it has led to population explosion and enlargement of the volume of labour. Most of the workers of Indian agriculture and the small-scale cottage industries were informal workers during that time as the British government did not provide any social security. Labour-intensive small-scale industries remained neglected and could not cope with the sophisticated products of large-scale industries of England.

After the independence of India in 1947, development planning was introduced in 1951, which followed an un-balanced development strategy by emphasizing a particular sector in each plan. In the Second Five Year Plan of India (1956–1961) and the Industrial Policy of 1956, priority was given to capital-intensive large-scale industries under government ownership because the Constitution of India adopted a socialistic pattern of society (Datt and Mahajan 2011). Under the planned development, two vibrant factors were highly responsible for the rapid growth of informal workers in India. First is the reservation of small-scale industries with subsidized production of non-durable consumer goods. Production of non-durable consumer goods under reservation of small-scale sector remained a few initially, which increased to 836 in 1990 (ILO 2019). Medium and large-scale industries were debarred from producing these consumer goods, which has incentivized the small-scale sector to remain small without accruing economies of scale. Secondly, there were many mutually inconsistent labour laws for the organized sector workers, which tended the organized sector to adopt technologies to limit the number of workers to avoid the burden of social insurance (ILO 2019). As a result, larger firms have sometimes given birth to smaller enterprises.

Globally, 740 million women work in the informal economy, and for some countries, they are paid 16 percent less than men. Between the age group of 25 to 34, women are 25 percent more likely than men to live in poverty (United Nations 2020). In India, the labour force participation rate (LFPR) for women as defined by the percentage of the population in the labour force is the lowest in the world. As per the Periodic Labour Force Survey (PLFS) conducted by the National Sample Survey Organization (NSO) since 2017, it is evident that in the urban areas, LFPR in Current Weekly Status (CWS) was only 17.2 percent during the quarter October–December 2018 considering the age group of 15 years and above. This rate has marginally increased to 19.0 percent in the quarter October–December 2019. During the same time period, the women's unemployment rate in urban areas of the country decreased from 12.1 percent to 9.8 percent (NSO 2019).

Indian women are generally engaged in the unpaid family works such as cooking, cleaning, shopping, and caring for the elderly and children for which they are not paid. Therefore, they are not included in the employed category. This is one of the major causes of low WPR for women in the country. However, women's participation in unpaid family works is declining over time, and women stay in education for more extended periods. From 2011–12 to 2018–19, women's participation in domestic and allied activities has declined. Among the employed women, the bulk of them are self-employed, and there is no written job contracts, paid leave, and any social security benefits for regular women wage employees (Chakraborty 2020a). The danger faced by these women informal workers is that they are affected by any shock like the COVID-19 pandemic because they do not have any social security, economic assets and hence characterized by poverty and vulnerability.

A significant part of informal workers in India is migrant workers normally migrating from rural areas to urban business centres for less than one year or so for employment or higher wages. They work in urban areas for road construction, building construction, factories, street vending, and various services like hotels, tourism, and house servants. Most of these migrant workers migrate from Bihar, Uttar Pradesh, Madhya Pradesh, Odisha, Jharkhand, West Bengal, and the North-eastern states to Delhi, Mumbai, Ahmedabad, and Kerala. In India, there are 100 million internal migrants who constitute 20 percent of the country's total workforce. They had no choice but to return to their villages due to the COVID-19 pandemic because they had no accommodation, whether owned, leased, rented, or any formal contracts with their employers (ILO 2020a; Singh et al. 2020a).

12.3 Impact of the Pandemic on Informal Workers

Lockdown, that is, closure of all kinds of business activities except the provision of essential services, was globally accepted to contain the spread of the coronavirus, which resulted in a drastic reduction of global GDP. Like many other international agencies, the Asian Development Bank (ADB) has made some assessments of the worldwide impact of worldwide lockdown. The Asian Development Bank (ADB) assesses that the global GDP will be lower by 5.8 trillion dollars (6.4 percent of global GDP) for a lockdown period of three months. It will be lower by 8.8 trillion dollars (9.7 percent of global GDP) for a lockdown period of six months compared to the non-existence of the pandemic (ADB 2020a). The impact was different for different markets of the economy. The effect is critical on the labour market because 158 million jobs (6.0 percent of total employment) will be lost under the assumption of lockdown for three months, and 242 million jobs (9.2 percent of total employment) will be lost if the lockdown is for six months (ADB 2020a).

The Indian economy was slowing down even before the beginning of the pandemic. According to the International Monetary Fund (IMF), the average annual growth rate in real GDP during the period 2012–2018 was 6.96 percent which is obviously a moderate growth rate. However, this real GDP growth rate decreased to 4.2 percent in 2019, while the projected growth rate of real GDP is −10.3 percent for the pandemic year of 2020 (IMF 2020). The mounting unemployment rate is the direct aftermath of this lower level of GDP growth hitting hurt the labour market in general and the informal labour market. Considering youth (15-24 years) unemployment in India, an estimated 4,084,000 jobs (full-time equivalent) will be lost for a lockdown of three months, and 6,113,000 jobs (full-time equivalent) will be lost for a lockdown of six months. Youth unemployment in 2019 was 23.3 percent, while this unemployment was estimated to be 29.5 percent for short containment and 32.5 percent if the containment is for six months (ILO and ADB 2020). During the last week of March 2020, LPR went down to 39 percent, and the employment rate was as low as 30 percent, the lowest after the New Economic Policy (NEP) was introduced in 1991 (Vyas 2020).

Only seven sectors such as agriculture, retail trade, hotel and restaurants, inland transport, other services, construction, textile, and textile products are estimated to contribute 77.3 percent of youth job loss in India in 2020 (ILO and ADB 2020). Figure 12.1 shows the respective share of estimated job loss of youths in these seven sectors.

Figure 12.1 shows that the highest loss of jobs for the youth occurred in the case of the agriculture sector (28.8 percent) followed by construction (24.6 percent). On the other hand, the lowest share of job loss is in the hotel and restaurants sub-sector (1.9 percent). The foregoing discussion indicated that most of the workers are informal workers and migrant workers in these three sub-sectors of the Indian economy.

The COVID-19 pandemic is an enormous labour market shock because it has a significant adverse impact on informal workers. The Food and Agriculture

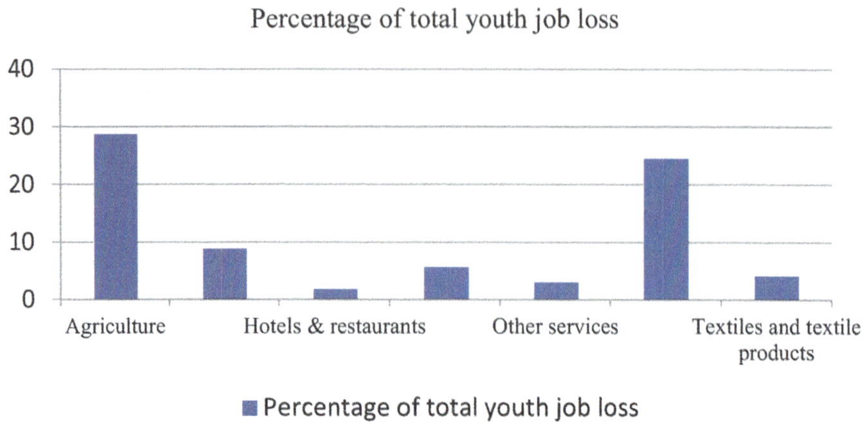

Fig. 12.1 Sectoral distribution of India's youth losing their jobs (Source: Calculated from International Labour Organization and Asian Development Bank (2020) Tackling the COVID-19 youth employment crisis in Asia and the Pacific. https://www.ilo.org/wcmsp5/groups/public. Accessed 11th December 2020)

Organization of the United Nations (FAO) reported that the self-employed and wage workers in the rural areas are badly hurt because of the disruption of agricultural supply chains and markets for farm products. Specific vulnerable groups among the informal workers such as the women, youth, children, indigenous people, and migrant workers in the informal economy will be hit hurt. Some families of these informal workers will resort to distress sale of their assets, borrowing money from moneylenders, or deliver child labour (FAO 2020a). Indian agriculture, which employs more than 90 percent of informal workers of the sector, suffered from lockdown because farmers faced a shortage of labour due to reverse migration of informal workers during the harvest of *Kharif* crops and disruption of transport services to sell the products in urban markets (Singh et al. 2020b). On the other hand, agricultural production was hampered by the short supply of fertilizers, pesticides, pieces of equipment, and veterinary medicines. In addition, due to the closure of hotels, restaurants, sweet shops, tea stalls, and restrictions on marriage and other social ceremonies, demand for agricultural and animal products decreased dramatically.

Next to the informal workers in agriculture and allied activities, urban informal workers in the five sectors, viz. construction, manufacturing, trade, hotel and restaurant, transport, communication and finance, business and real estate consisting of 93 million informal workers are most affected by the pandemic (Mehta and Kumar 2020). 50 percent of these informal workers are engaged in self-employment, 20 percent are casual workers on daily wages, and 30 percent are salaried or contract workers without any social security. Informal workers in the urban sectors are more vulnerable because they are engaged in rag picking, street vending, food stalls, construction, transport, and domestic help, high-risk zones for virus infection. In addition, a large number of informal economy workers are health workers whose lives are easily exposed to health risks during the COVID-19 pandemic. Several informal workers in the economy faced the dilemma of whether to work or starve by losing their income, even the available jobs (ILO 2020b).

Informal workers who migrated from one place to another within the country are vulnerable, and the COVID-19 pandemic accentuated their vulnerability. A large number of migrant workers is employed under informal or casual setup and is susceptible to exploitation, poverty, food insecurity, and inaccessibility to healthcare and social protection. Lockdown retarded normal movement of migrant workers who are employed in the agriculture sector. It will affect agricultural value chains, food availability, and food prices globally (FAO 2020b). The return of migrant workers increased population density in some areas of the country. For example, reverse migration posed risks of the spread of the virus due to high population density, particularly in UP and Bihar (Lele et al. 2020).

Due to the COVID-19 pandemic, there are five types of vulnerability of migrant workers. These are hardships faced during their return to home, loss of their jobs and resultant income flow, overcrowding, problems of maintaining social distance at the origin, lack of maintaining hygiene, and shortage of labour supply at the places from where they have returned. Close to 104 lakh migrant workers returned from urban areas to rural areas within 30 days from first May 2020 to 31st May 2020. Modes of transportation of their return were Shramik (labourer) trains, buses, trucks, auto-rickshaws, bicycles, and last but not least, walking of hundreds of and even thousands of kilometres. Total 4150 Shramik trains were used for this reverse migration to move 55 lakh informal workers (Singh et al. 2020a). Most of these trains had destinations to the states of UP and Bihar (90 percent), and the rest went to Jharkhand, Odisha, MP, Rajasthan, and West Bengal.

Migrant workers became victims of the dilemma between the central government and some state governments regarding the arrangement of transport facilities for their return. Hence, a large number of migrant workers started to return on foot, and they were in deplorable condition on their way to home. In Aurangabad district of Maharashtra, sixteen migrant workers returning to Madhya Pradesh were crushed to death by a cargo train on eighth May 2020 as they fell asleep due to their exhaustion (The Statesman 2020). They were walking along the railway tracks because police hindered their movement. Migrant workers felt that they were undervalued by their employer or the administration and were not ready to take responsibility because the business was closed. This has led to the complete loss of their employment and income, supply of food, and other essentials under the pandemic (Kumar et al. 2021). The psychological impact of the lockdown on migrant workers was massive because there were several cases of suicides and suicide attempts as well as a growing number of domestic violence (Kundu 2020).

The impact of the COVID-19 pandemic on men and women is different. Lockdown due to coronavirus has pushed women towards more vulnerable situations by raising unpaid family works and gender-based violence (United Nations 2020). A research work based on rapid assessment of the Institute of Social Studies Trust (ISST) on 176 women informal workers in Delhi unveiled that 66 percent of the informal women workers reported an increase in unpaid work at home, and

36 percent reported an increase in care works for children and the elderly (Chakraborty 2020b). According to this report, the women informal workers in Delhi were suffering from various problems caused by COVID-19 lockdowns such as food shortage, hike in the prices of food grains and distress sell off their personal assets, or exhaustion of their household savings to finance daily expenses.

In India, women and girls do six times more unpaid family works than their male counterparts. This crisis is more profound for vulnerable women such as pregnant women, breastfeeding women, elderly women, disabled women, homeless and destitute women, women and girls in custody, and women sex workers (Dasgupta 2020). The economic crisis is faced by 94 percent of women workers, mostly in the informal sector like agriculture, construction, and the un-organized MSMEs. Among the daily wage earners and the migrant women workers, the immediate impact of lockdown was the complete loss of their wages. This will aggravate the gap between formal sector and informal sector earnings and between the income of men and women workers. This type of disproportionate impact on women workers is assumed to persist for a long time, even after the complete withdrawal of lockdown.

12.4 Measures Adopted by the Government

Under India's federal structure of governance, several measures were adopted by the government at all levels-central, state, and local governments. The central government initiated fiscal policy and monetary policy measures to control the damage caused by the pandemic. The central government declared aggregate fiscal packages of Rs 21 trillion during March 2020 to May 2020 to support health care, social protection measures for migrant workers, the poor, women, elderly, disabled, and the employment guarantee schemes (ADB 2020b). Under the above package, a financial package of Rs 150 billion ($2 billion) was declared on 24th March 2020 by the Government of India for the development or up-gradation of infrastructure for the treatment of COVID-19 patients. The second financial package of 25 billion US dollars, which is 0.8 percent of GDP, was declared on 26th March 2020 for the in-kind benefit and cash transfers.

The Indian Prime Minister's address to the nation on 12th May 2020 on *Atmanirbhar Bharat*[1] traces five pillars: growing a new economy, creating state-of-the-art infrastructure, technology-driven delivery system, leveraging the young

[1] *Atmanirbhar Bharat Abhiyan* was announced by the Prime Minister of India during the COVID-19 pandemic for the achievement of self-reliance through the incorporation of local resources in the development and for the benefit of labourers, farmers, honest taxpayers, MSMEs, and cottage industry. Structural reforms, marketing reforms, and making Indian companies more competitive in the global supply chains were major objectives of the project. Under this project, space exploration was opened to the private sector. However, *Atmanirbhar Bharat Abhiyan* is criticized on the ground that the ongoing *Make in India* project of the Government of India was designed basically to achieve the same goals, just their names are different.

demography, and exploiting domestic demand (The Telegraph 2020). As part of this address, a large financial package of Rs 20 trillion, which is close to 10 percent of GDP, was declared incorporating short-term, medium-term, and long-term measures (ILO 2020a; Kundu 2020). However, the minimal benefit of these fiscal packages reached the informal workers, migrant workers, women workers, and the poor.

Another stimulus package of 25 billion US dollars was declared covering 800 million people with free-of-cost five kilogrammes of wheat or rice and one kilogramme of pulses per month for a period of three months under Prime Minister Garib Kalyan Anna Yojana. For a period of three months, the Pradhan Mantri Jan Dhan Yojana (PMJDY) account holders received Rs 500 per month. However, 23 percent of PMJDY accounts were non-active in 2018, and hence many rural women might have been excluded from these cash benefits (Kulkarni 2020). Over the same period, Rs 1000 per month was transferred to vulnerable individuals such as poor, senior citizens, widows, and the disabled. An additional amount of Rs 400 billion was sanctioned for the Mahatma Gandhi Rural Employment Guarantee Act (MGREGA) project, and the revenue of workers was raised to Rs 202 per person from Rs 182 per person (ILO 2020c), benefiting 136 million families. Around 80 million migrant workers were provided with shelters either in government-built or privately operated relief camps. They were provided with one kilogramme of grain per person and one kilogramme of pulses per family for a period of two months (ILO 2020a).

The Finance Minister of India appeared in press conferences for five consecutive days to elaborate on various aspects of the fiscal package for *Atmanirbhar Bharat*. On the first day, on 13th May 2020, MSMEs were re-defined to encompass a larger number of these firms in the government's support programmes. The definition of micro, small and medium enterprises (in terms of investment) was revised to a higher level of investments. A collateral-free loan of Rs 3 lakh for the next five months, fully guaranteed by the government, was provisioned for the newly defined MSMEs (Ghosh 2020). But a small number of firms would be eligible for this loan because only the firms already having outstanding loans are eligible to take the loans. Besides, the factors which affect the demand for the products of the MSMEs were simply neglected. Informal workers employed in these MSMEs are benefited very little because of the absence of any support system directly for these workers.

The Reserve bank of India (RBI) introduced expansionary monetary policy measures for the normal functioning of financial markets, the financial institutions and reinforcing monetary transmission under the COVID-19 pandemic. RBI reduced Cash Reserve Ratio by 100 basis points to 3 percent, and 40 bps lowered the Liquidity Adjustment Facility (LAF) than the policy repo rate. The Marginal Standing Facility (MSF) continued to be 25 bps above the policy repo rate under the condition of persistent excess liquidity during March 2020 (RBI 2020). The monetary measures injected total liquidity of Rs 3.74 lakh into the system. The RBI permitted all the commercial banks, co-operative banks, and other financial institutions a three months moratorium on term loans (ILO 2020c). Subsequently, RBI has further lowered the repo rate as well as the reverse repo rate. But under the lockdown and persistent economic slowdown, demand for loans is critically limited. The government decided to provide bank loan facilities of Rs 10,000 as working capital for the street vendors in both urban and rural areas through the commercial banks. However, most commercial banks lack experience in dealing with such small lending.

Respective state governments of the country introduced programmes to tide over the pandemic. For example, the Government of West Bengal distributed and still is distributing free food grains to the majority of people through the rationing system since March 2020, and this food programme has been extended up to June 2021. The children under basic education receive food grains for their mid-day meals despite the fact that schools remained closed even in December 2020. Similarly, the Government of Maharashtra decided to provide food grains to 80 percent of its population for a period of six months. Some state governments decided to change the existing labour laws to ease the functioning of firms and business enterprises of different sizes. But these measures are not a solution to the problem of unemployment and under-employment of the informal workers. In fact, most state governments do not maintain proper data regarding migrant workers. Introducing a skill-based policy for employment and imparting necessary training to the workers is essential. Enlargement of employment opportunities in the organized sector and protecting the interests of informal workers in the informal sectors in a sustained manner may mitigate the far-reaching impact of the COVID-19 pandemic.

12.5 Conclusion

The ongoing COVID-19 pandemic caused the decline of global output and employment tremendously. In India, the growth rate and employment both have reached the three decades' lowest level. The labour market in India consists of a majority of informal workers and migrated workers caused by protective policies of the government for the small-scale consumer goods industries and excessive labour laws applicable to large-scale heavy industries. Due to the current pandemic, these poor informal workers have become unemployed and fall into stress and poverty traps. Reverse migration of informal workers posed an excess burden on the rural economy characterized by the lack of infrastructure and the existence of surplus labour. The government has introduced fiscal and monetary policy measures to overcome the adverse impact of the pandemic on the economy. Public expenditure was increased by 10 percent of GDP for boosting activities in different sectors of the pandemic-stricken economy. In addition, the RBI reduced interest rates for the smooth functioning of the system. However, informal workforces in India hardly accrued any benefit from these measures partly because they are not directly linked to these initiatives or they have been neglected in these policy measures. Therefore, broad-based separate policy measures are essential to secure informal workers' employment, income, and working conditions. Otherwise, overall human development in India would be adversely affected because they represent the bulk of the population of the country.

References

ADB (2020a) An updated assessment of the economic impact of COVID-19. ADB policy brief no. 133. https://www.adb.org/sites/default/files/publication. Accessed 21 Nov 2020

ADB (2020b) Asian development outlook 2020 update: wellness in worrying times. ADB, Philippines

Andrews MA et al (2020) First confirmed case of COVID-19 infection in India: a case report. Indian J Med Res 151(5):490–492

Chakraborty S (2020a) COVID-19 and women informal sector workers in India. Econ Polit Wkly 55(35):17–21

Chakraborty S (2020b) Impact of COVID 19 national lockdown on women informal workers in Delhi. A report. Institute of Social Studies Trust. New Delhi. http://www.isstindia.org/publications. Accessed 25 Nov 2020

Dasgupta J (2020) A gender-responsive policy and fiscal response to the pandemic. Econ Polit Weekly. 55(22):13–17

Datt G, Mahajan A (2011) Datt & Sundaram Indian Economy. S. Chand & Company Ltd., pp 289–290

FAO (2020a) Impact of COVID-19 on informal workers. 7th April. http://www.fao.org. Accessed 21 Nov 2020

FAO (2020b) Migrant workers and the COVID-19 pandemic. 7th April. http://www.fao.org. Accessed 22 Nov 2020

Ghosh S (2020) Examining the COVID-19 relief package for MSMEs. Econ Polit Weekly 55(22): 10–12

Hindustan Times (2020) India's 1st coronavirus case detected in Kerala. Daily Newspaper 31(01): 2020

ILO (2019) Informal employment trends in the Indian economy: persistent informality, growing positive development. Employment Policy Department, Working Paper No. 254, Geneva. http://www.ilo.org/wcmsp5/groups/public. Accessed 22 Oct 2020

ILO (2020a) Rapid assessment of the impact of the COVID-19 crisis on employment. ILO brief, India. http://www.ilo.org/wcmsp5/groups/public. Accessed 22 Nov 2020

ILO (2020b) Extending social protection to informal workers in the COVID-19 crisis: country responses and policy considerations. Policy brief. 8th September. https://www.ilo.org/wcmsp5/groups/public. Accessed 28 Nov 2020

ILO (2020c) COVID-19: labour market measures (India). https://www.ilo.org/wcmsp5/groups/public. Accessed 11 Dec 2020

ILO and ADB (2020) Tackling the COVID-19 youth employment crisis in Asia and the Pacific. Co-publication of the ADB and ILO. https://www.ilo.org/wcmsp5/groups/public. Accessed 11 Dec 2020

IMF (2020) World economic outlook: a long and difficult ascent. October. https://www.imf.org/~/media/Files/Publications. Accessed 22 Oct 2020

Kulkarni S (2020) Locked in crisis: concerns of rural women. Econ Polit Weekly 55(23):15–18

Kumar A et al (2021) The exodus of migrant workers during the coronavirus disease 2019 (COVID-19) pandemic in India: thematic findings on emotional concerns. Letter to the editor. Members copy. Open J Psychiat All Sci 12(1):67–69. http://iproxy.inflibnet.ac.in:2406/ijor.axp?target. Accessed 09 Nov 2020

Kundu S (2020) The era of COVID-19 in India: an economic overview. Ind J Econ Dev 16(2): 313–319

Lele U et al (2020) Health and nutrition of India's labour force and COVID-19 challenges. Econ Polit Wkly 55(21):13–16

Mehta BS, Kumar A (2020) Impact of the lockdown on urban livelihoods. Econ Polit Wkly 55(26): 4–5

Ministry of Labour and Employment (2020) Annual report 2019–2020. Government of India

NSO (2019) Quarterly bulletin. Periodic labour force survey (PLFS) October–December, 2019 Ministry of Statistics and Programme Implementation, Government of India

RBI (2020) Statement on developmental and regulatory policies. Press release on 27th March, 2020. https://rbidocs.rbi.org.in/rdocs/PressRelease/PDFs. Accessed 27 Nov 2020

Singh SK et al (2020a) Reverse migration of labourers amidst COVID-19. Econ Polit Wkly 55(32–33):25–29

Singh G et al (2020b) Economic impact of COVID-19 pandemic: who are the big sufferers? Ind J Econ Dev 16(2):320–326

The Statesman (2020) Train runs over 16 sleeping migrant workers. Daily news paper. Kolkata, 09.05.2020

The Telegraph (2020) Rs 20000000000000. Daily news paper. 13th May 2020

United Nations (2020) Policy brief: the impact of COVID-19 on women. 9 April. https://www.un.org/sites/un2.un.org/files/policy_brief. Accessed 15 Nov 2020

Vyas M (2020) Unemployment rate over 23%. CMIE. 7th April. https://www.cmie.com/kommon/bin. Accessed 23 Nov 2020

Dr. Anil Kumar Biswas has been graduated from ABN Seal College, Cooch Behar, in 1997 with honors in Economics and completed his post-graduation in Economics from North Bengal University in 1999. In 2003, he has completed the degree of M. Phil in Economics. He was awarded with a Ph. D. in Economics in the Centre for Himalayan Studies, University of North Bengal. Dr. Biswas is teaching Economics in P. D. Women's College, Jalpaiguri since 2005. He has a number of publications in national and international journals, edited books, conference proceedings, and presented papers in national and international seminars. His area of interest is international economics and development economics. Dr. Biswas functioned for 3 years as the honorary member of Divisional Railway User's Consultative Committee (DRUCC), Katihar Division of Indian Railway, Government of India.

Part IV
COVID-19, Governance and Policies

Chapter 13
New Pandemic in India: Emerging Challenges to Governance and Responses to Overcome

Jitendra Sahoo

Abstract In context to the ICT-ruled globalized world, the effect of the COVID-19 pandemic situation has been changing the lifestyle like wildfire that lead the new approach of 'work from home', amidst which both the governments at the centre and different states are trying to tackle the situation and through managing each and every aspect of the society including every person for providing the better opportunity to live with peace and safety. In the modern complex world, every society today faces a severe socio-economic and political crisis due to this COVID-19, and there is no exception to India. This chapter focuses on three things: *firstly*, understanding the New Pandemic; *secondly*, governance before and during this Pandemic in India which includes two facets - emerging challenges during this pandemic and the opportunities coming out of this pandemic; and, *finally*, brings forth the author's perception to the changing socio-economic-political dimension during this *new-normal*.

Keywords COVID-19 · Self-reliant India · e-governance · Socio-economic · Political problems

13.1 Introduction

Due to this COVID-19, the governance system has been affected on a large scale in several countries worldwide, resulting in countrywide lockdowns and requires 24×7 monitoring. In federal countries like India, the directions coming from the centre and the uniformity in response, both preventive and curative, are essential. As per the constitutional distribution of powers between the central agency and its units where the local problems are vested in the hands of the states, and other important economic, industrial, commercial, including disaster management are vested with

J. Sahoo (✉)
Department of Political Science, University of Gour Banga, Malda, West Bengal, India

S. Saha et al. (eds.), *Economic and Societal Transformation in Pandemic-Trapped India*, New Frontiers in Regional Science: Asian Perspectives 55,
https://doi.org/10.1007/978-981-16-5755-9_13

the centre. The Government of India has taken various majors, including the shutdown of educational institutions, different workplaces, cancellation of various public gatherings and social events, cancellation of the public transportation system. The pandemic posed several challenges arising out of this in the sphere of governance, and also it explored few opportunities which are coming out of this pandemic not only for the government but also for the common people as a whole (Hannah et al. 2020).

With this backdrop, this chapter discusses the meaning of this new pandemic, how it affected the governance, and brought some opportunities for India. The main objectives are to clarify the inner meaning of this new pandemic and also to explore the real problems that arise in governance not only to introduce new policies but also to convince the people to follow this for breaking the chain of coronavirus and giving a better economic, health facility and also social security to all the people.

While the world community faced many problems, there are also many hindrances coming in front of the government that includes maintaining proper governance systems and providing the safety and security of the citizens through various measures side by side. State disseminate the welfare function for its citizen in the form of free ration, financial package, health assistance, social security, security from cybercrime and cheating in their daily lives.

In social science research, every researcher has their limitations. This chapter explains a few challenges and opportunities arising from this COVID-19 in the sphere of governance in Indian circumstances, which kept vast things out of the discussions if judged from global perspectives.

People have been curious about social events from times immemorial and have been using various methods to understand their significance. These methods fall into two broad categories in social science research: qualitative and quantitative (Sharma n.d.). To complete this analysis, the author has used the former to understand the emergence of this new pandemic and the new challenges and opportunities coming out from this pandemic for governance, especially in India.

13.2 Emerging Challenges from the New Pandemic

The author prefers to call this a 'New Pandemic' as it was very much unknown to us at the beginning of 2020, and the people of the whole world had their curiosity for the unknown. The quest for knowledge and curiosity for new facts and unveiling covert facts is an essential human trait. It is the urge to know that inspires the man to explore his surrounding environment. For example, until the end of 2019, the fundamental ideas about the symptoms and diagnosis related to COVID-19 were unknown. However, the curiosity for this unknown virus inspired the medical science practitioners to investigate the causes and how to prevent COVID-19. Therefore, as far as the present situation is concerned though the number of patients has been increasing day by day, still, we have been successfully fighting against this COVID-19, and the death rate is very low in India in comparison to other countries of the world according to the statement of the Health Ministry, Government of India. It shows that though the COVID-19 is fragile, the recovery rate is quite adequate, and the efforts of the frontline warriors are pretty appreciating.

At the beginning of March 2020, the robust healthcare infrastructure of Europe, especially in Italy, Spain, and many developed countries globally, did not restrict the deaths due to COVID-19. On the other hand, India, where around 1.3 billion people live, wants to control the spread of COVID-19 through the lockdown approach (Chandrashekhar 2020). In the beginning, in India, fragile healthcare facilities could not be able to face the precarious situation. Therefore, the government was in a dilemma to tackle the situation with that context when no specific medicines and vaccines were available side by side.

Out of several challenges in the way of good governance amid this crisis phase, this discussion is specifically focused on some specific crisis with examples and illustrations: the stress from the federal structure of India during the pandemic; the issues of corruption amid the crisis; the issues of price hike as a result of over-stock of goods at both the seller and buyer's end that harms the poor; and, finally the issues of migrant labourers and the role of the government that raised questions even at the international arena.

13.2.1 Federal Structure of India and the COVID-19 Pandemic

The very first article of the Indian Constitution says that India, i.e., *Bharat,* shall be a Union of States, and it is said that India is federal in form but unitary in spirit and to quote Prof. K.C. Where, it is quasi-federal in nature (Ghai 2011). One of the examples is the Disaster Management Act, 2005, and its adoption by the central government from the beginning of this COVID situation brings excessive control of the centre over the states, and the states have minimal functions to do except obeying the lockdown process and then the guidelines during unlocking periods. Few states also criticized the PM CARES, which falls outside the purview of federal auditor, and also criticized the centre's decision regarding declaration of zones, purchase of PPE kits, transportation of migrant labours (Verma and Sughosh 2020).

If one critically analyses the federal principles of India, especially during this pandemic, he or she will find many conflicts arise between the centre and the states that directly impact the governance system.

Due to pandemic and lockdown approaches, the sources of states revenue have been collapsed. A significant part of states' revenue comes from sales duty, stamp duty from property transactions, and sales from petroleum products, which are very poor during this situation. The collection of GST has been severely affected. However, the payment concerning staff salary, social welfare schemes, and interest payments is still the same. Apart from this, the expenditure to revive the health care facilities due to this COVID situation is another great challenge for the states, including health infrastructure, testing, treating, quarantining etc. As far as the FRMB Act, the states have their limitations to borrow from the open market. The suspension of MPLADs and diversion of funds to the consolidated fund of India goes against cooperative federalism. As the Indian states are entirely dependent upon the financial assistance of the centre, it creates a massive challenge for the states to lead normal governance during this COVID situation.

Reforms in the agriculture sector are another severe issue that leads to a hurdle in smooth governance during this pandemic. The introduction of the agricultural bill 2020 by the government of India creates tremendous agitation by farmers from various states, including Punjab and Haryana pretending that these reforms impinge even more state autonomy by dismantling the long-standing agricultural marketing system, which will bring monopolized trade in agriculture within states and prevents the growth of more efficient marketing system. Still, the farmers' agitation is going on in various parts against the cancellation of this agricultural reform (Burman 2020).

The seventh schedule of the Indian constitution deals with the distribution of powers through three important lists: *union list*, *state list*, and *concurrent list*. Here, the union list deals with the major provisions related to the country's governance concerning industries, foreign trade, inter-state trade, and commerce (Fadia 1997) that empower the government of India to control all such things during this pandemic. This control of the centre has often been the point of objection as raised by different states. The most objected phenomenon was somehow the enormous problem for millions of workers when the industries were closed, inter-state trade and commerce were seized, and even road and rail transport was suspended overnight, exercising the power of the centre. While the state list includes the items like public order, public health, and sanitation, hospitals, and dispensaries (Ghai 2011), during the first wave of this pandemic, all the items were guided by the directions of the central government from time to time, leading to problems in the governance process as objected by many states of India. As the concurrent list of the Indian constitution includes the items like social security, labour welfare, employment opportunity, both the central and state governments had always been imposing allegations to each other in connection to labour welfare and their employment opportunities. Again with the constitutional provisions as enshrined in Part-XI-Chapter-I, i.e., legislative relations between centre and states, and chapter-II deals with administrative relations between the centre and states also empowers the centre in many respects to supersede over the states had provided scope to conflict among the two entities (Constitution of India 1998).

13.2.2 Panicked Buyers and Profiteering Sellers: Challenges to the Poor

During this COVID-19, there arise two types of severe problems in cases of buying and selling. We also have observed that there is a change in the behaviours of both the consumers and sellers. The consumers' demand has been raised, and here are many reasons which are responsible for creating panic among the consumers or buyers to change their behaviour for making stocks of essential goods in their homes, such as; (i) frequent visit to markets leads to fear of infection of coronavirus, (ii) panic in the imposition of frequent lockdown resulted in the shutdown of markets and there will be a shortage of essential goods, (iii) more consumption due to work from home and it also leads to purchase of bulk amount of essential commodities, (iv) spread of fake news that the government will at any day shutdown the retail shops of essential commodities due to the spread of coronavirus and people rushed to purchase huge amounts of goods at their homes. Both at the centre and states, the government frequently gives assurances to the people not to panic over this. As doctors and physicians worldwide are always giving suggestions to boost the immunity to fight against the corona virus, the consumers are hoarding to buy the immunity booster medicines and ayurvedic products rather than luxurious and other cosmetics. This shows a change in their lifestyles and behaviours in purchasing varieties of essentials in their daily lives. Again, the doctors are giving suggestions for maintaining the health hygienic by frequent washing of hands and disinfect the surfaces, there is a high demand for purchasing the hand wash items, and other disinfect sanitizers-cleaners also continue even after the COVID situation. By taking this opportunity, there is a price hike in all those essential items by different production companies all over (Kharat 2020).

During the lockdown and the spread of coronavirus, it creates another panic among the people. Problems arise in selling essential commodities like vegetables and fruits, where the small businessmen faced considerable problems in maintaining their livelihood and daily life. When we analyse the buyers' part, we have experienced that the consumers are suddenly forced to purchase maximum essentials online with a fear of infection while going to open market. On the other hand, the companies who never thought about online selling are also forced to start an online delivery system even if the small businessmen and other local brands are now well-versed with this online buying and selling with the common people. Some sectors like household cleaning accessories and frozen foods have seen the rises in consumer demands and have had to navigate the supplies (Rajamani 2020). However, this process brings significant challenges to the poor as they have no idea about these online transactions and have no smart gadgets for taking this opportunity. The pandemic had caused them to lose their jobs, making them have no sufficient money in their hands to stock the essentials in advance. Amidst this, the poor are the worst sufferer. The networks between buyers and sellers have been transformed into a virtual form from which the poor are isolated.

13.2.3 Migrant Labours and the Government

Migration refers to the change of living place by an individual or family from one place to another. There are several causes; however, for the developing countries like India, the causes are primarily poverty situation, lack of employment opportunities, shock from various natural calamities like flood, famine, drought, limited land resources, overcrowding in agriculture, non-affordability of costlier agriculture, and income disparity. Millions of people migrate from rural India to different cities and metropolises to find employment as migrant labours. However, due to COVID-19, their workplaces were shut down, and as they lost their jobs, they returned to their villages, portraying the grim scene of reverse migration ever in the country's history if the migration due to the partition of the country is excluded. Due to the shutdown of factories at different places in India, millions of migrant workers have also lost their income (Sahoo 2020).

To give relief from deep deprivation in rural areas, unemployment, poverty, starvation, even attempt to suicides by the migrant labours and also to instate them, the Government of India has taken several measures and brought policy reforms for the betterment of the migrant workers in three varieties like short-term measures, medium and long term measures (Rajan 2020).

Under the short-term measure, the government announced to provide 5 kg of rice and 1 kg of pulses monthly per head to around eight crores of migrant members. The government has also allocated additional forty thousand crores of rupees to Mahatma Gandhi National Rural Employment Guarantee Act (MGNREGA)[1] programmes to provide job opportunities to the migrant workers at their home states, and also it was announced that the minimum wage would be increased from Rs.182 to Rs.202 per day (Rajan 2020). In addition, accessible credit scheme facilities are also provided to the street vendors to carry on their small business and livelihood.

Under the medium and long term measures Government of India has stated the universal ration card system with a flagship programme, i.e., *One Nation One Ration Card*. Under this scheme, the migrant labours can easily avail the necessary food grains from the ration shops anywhere in India where they are migrated for their livelihood. Besides, under the Pradhan Mantri Awas Yojana (PMAY)[2] scheme, the government will provide rental housing facilities to the migrant labours and their family members at a lower price through a public-private partnership model (Rajan 2020).

[1]MGNREGA is an act passed in September 2005 by the Government of India to provide the guarantee of right to work and aims at enhancement of livelihood security to the rural people with 100 days of wage employment facilities in every financial year.

[2]Pradhan Mantri Awas Yojana (PMAY) is a scheme of Government of India having the objective to provide affordable housing facilities to the poors living both at urban and rural areas in India with toilet facilities and the government has a target of building 2 crores such affordable houses by 31st March, 2022.

13.2.4 E-Governance and the Digital Divides

E-governance refers to the transformation of the functioning of government through the use of Information and Communication Technology (ICT). Governance is broader than the term government because governance is the process of governing or controlling a country by its government. Therefore, e-governance of a country refers to controlling a country by its government through the help of ICT to provide SMART (Simple, Moral, Accountable, Responsive, Responsible and Transparent) government (Civil Service Chronicle 2018). This pandemic thrusts the use of ICT and the process of e-governance, which is often unconsciously confused with e-commerce with an e-delivery system by ordinary people. In India, with its 22% citizens below the poverty line and even as high as 45% in the states like Chhattisgarh and Jharkhand (Sharma 2019), participation in e-governance appeared as a roadblock to many, especially during this pandemic period. Again the relentless spread of coronavirus infections among the citizens and the growing joblessness, especially among the poor, create the situation for the food crisis. At this point, while the poor should have taken advantage of e-governance, they excluded themselves due to their non-affordability of the cost involved in it, arising the ever-witnessed digital divides in the country (Kumar 2018). Another challenge is that a considerable proportion of the population, particularly aged persons, slum dwellers, and remote rural residents, are not e-literate, excluding them from participating in the e-governance process. After all, the vast rural areas in India still have no consistent electricity facilities, i.e., 24×7 hours, along with which there are also network issues and lack of modern accessories to avail this e-governance system. This poor communication and proper understanding of the web portal are well evident with the long queue in the different vaccination centres for getting vaccination appointments, even though the government had made the portal for pre-scheduling the appointment online through the *Arogya Setu App*, *Umang App*, or the dedicated *COWIN*[3] portal.

13.3 Major Steps toward Maintaining a Proper Governance System

13.3.1 Emphasis on the Principle of 'Self-Reliant'

It is often said that if there is hope, there is no despair. Therefore, we should be self-reliant and have the confidence to overcome all the hurdles coming before us. Our present Prime Minister said COVID-19 is a people-centric fight. Therefore, we should maintain *Do Gajki Duri, Bahoot Hai Jaroori* (i.e., maintaining two yards

[3]Government of India's online vaccination scheduling portal for its citizens is available under the URL https://www.cowin.gov.in/

of distance is essential). Therefore, the word '*Atmanirbhar*' or 'Self-reliant' has been added to the work culture of the Indian governance system to boost up the Indian economy through a classical structure of production made of the people, by and for the people.

When we analyse this 'Self- Reliant' principle of the Government of India, we will find that in every sector, the government desires to become self-reliant having a slogan of 'Local should be global', which means we will be self-dependant in every way and not to depend upon the products of other countries, our own products will capture the global markets. Of course, it would not be possible to discuss every aspect. Still, the following sections will explain a few of them.

- *Firstly*, the government of India has allocated an economic stimulus package called the *Atmanirbhar Bharat Abhiyan* (ABA) Package to boost up the Indian economy in various sectors like providing financial assistance to the MSMEs considered being the major employers of the unorganized and migrant labours.
- *Secondly*, to provide a better opportunity to the farmers by facilitating better and predictable prices, measures to strengthen infrastructure, logistics, capacity building, governance, and other administrative changes like Essential Commodities Act, reforms in agricultural marketing, assurance of standard price and support in farmers' additional activities, etc. (Sankar and Gopal 2020).
- *Thirdly*, under the umbrella of Atmanirbhar Bharat Rozgar Yojana, the finance minister of India, has declared in November 2020 to provide a subsidy for the employers to hire new employees, and this scheme is also targeting the people who lost their jobs due to pandemic. In addition, the government will also provide subsidies to all the new employees through Employee's Provident Fund (EPF) with Aadhar seeded EPFO accounts.

The *Atma Nirbhar Bharat* (Self-Reliant India) mission has been launched as a crisis management principle to face the COVID-19 pandemic. One of the significant achievements in the health sector that the Government of India has achieved success in the development of an indigenous supply chain of Personal Protection Equipment (PPE) kits within a short span of sixty days only, which is an example for local manufacturers that 'self-reliant' in most high quality of production. During the beginning of this COVID situation in India on 30th January 2020, the local manufacturers did not have the capability to produce this PPE kit which is very much essential for the frontline warriors like doctors, staff nurses, and other workers, but with the initiatives taken by NITI Aayog, the local manufacturers have achieved the goal successfully (Kapoor and Sandeep 2020).

Finally, the success of our scientists and technicians in the development of two vaccines in India is another milestone in the Aatma Nirbhar Bharat Abhiyan, which is driven by the global well-being, said by the President of India Sri Ramnath Kovind, while addressing the Pravasi Bharatiya Divas Convention on ninth January 2021 (ANI 2021).

13.3.2 Importance of Testing, Tracing, Isolating, and Treating

Some simple steps often result in an excellent outcome; so crucial for testing, tracing, isolating, and treating the infected people, and also random testing is also going on for all the people in asymptomatic. WHO recommends 140 tests per million population per day, whereas India is doing more than the stipulated number because all the states and union territories performing very well, e.g., Andhra Pradesh (1418), Bihar (1093), Odisha (1072), Goa (1058), Jharkhand (994), Jammu & Kashmir (984), Telangana (947), Tamil Nadu (936), and Haryana (863) (Kaul 2020). Now the Government of Delhi has adopted the principle of 5 T's Tracing, Testing, Tracking, Treatment, and Teamwork. In this regard, the Government of India has adopted so many steps, and the *Aarogya Setu App* is one of them to trace the infected people. The total number of samples tested up to 22nd November 2020 was 13,25,82,730, and on 22nd November 2020 total of 8,49,596 was tested in India (DD News November 23 2020).

India has started rapidly developing its healthcare infrastructure all over the country, and both the central and all the state governments are in action to meet the rising demands for hospital beds and ventilators. Now, apart from the general administration department, other departments like armed forces, paramilitary forces, railways department, and various public sector undertakings are trying to meet this health emergency with their independent capacities by setting up medical camps, COVID hospitals by taking schools, colleges, stadiums, inside the railway coaches, and also using many open spaces. Therefore, in India, from every corner, efforts are made to provide better health facilities.

Indian parliament document states that on 21st April 2020, the oxygen-supported beds were 62,458, and it has been increased to 2,47,972 as of 22nd September 2020. Also, the number of ICU beds and ventilators is increased from 27,360 to 66,638 and 13,158 to 33,024, respectively (Rawat 2021). That means the number of oxygen-supported beds rose by 297 percent, ICU beds by 143 percent, and ventilators by 151 percent. As per the Union Health Ministry report, India has 2, 70,710 oxygen-supported beds, 40,486 ICU beds, and 40,627 ventilators (Rawat 2021). On 16th January 2021, during the launch of the COVID-19 vaccination drive, Prime Minister of India, Mr. Narendra Modi, said that India had set an example for the whole world in its fight against the pandemic, and everyone in our country had performed their duties with patience which he did recognize as the people's movement (Rawat 2021).

13.3.3 Strengthening Food Supply Chains

Food and Agriculture Organisation (FAO), in his report entitled *Policy brief: The Impact of COVID-19 on Food Security and Nutrition*, said that 45million people would be pushed into an acute shortage of food due to the pandemic (Civil Service Chronicle 2020a, b, c). In this context, the Government of India has taken several effective measures to better the common people which has been praised globally. For example, through the public distribution system, the GOI has distributed free ration from April 2020 to June 2020, and later it has been extended up to November 2020 (Civil Service Chronicle 2020a, b, c). Furthermore, the GOI has taken another step as the *One Nation One Ration Card System*, and on 1 May 2020, the Central Government included 17 states and Union Territories under the National Food Security Act. (Civil Service Chronicle 2020c).

13.3.4 Strengthening the SHGs

Under the National Rural Livelihood Mission, about 20,000 Self-Help Groups across the country have been engaged to produce thousands of litres of sanitizers, liquid hand wash, masks, etc. which creates enormous numbers of employment opportunities for the poor in rural areas and provide large numbers of livelihood. In addition, as per the report of the World Bank, the SHGs have set up over 10,000 community kitchens across India to feed stranded workers, the poor, and the vulnerable (The World Bank 2020).

During this COVID situation, it is too much hard to avail of banking facilities, especially for women. Therefore, the SHGs women are also working as banking correspondents in the name of *'bank sakhis'* continued to provide doorstep banking services, including distribution of pensions. At the same time, the banks have given them training and orientation in this regard to make them capable of continuing these types of banking works not only to help the banks but also to provide assistance to the needy and aged people during the lockdown. The women SHGs are working under the NRLM programmes under the GOI is financed by the World Bank, functioning across all the states and union territories in India, reaching more than 67 million women (The World Bank 2020).

13.3.5 Role of IT in Pandemic Management

The spread of the corona virus has taken a global pandemic trend becoming a severe threat to the health care system. It is a real challenge for the global community to check coronavirus spread and provide proper health care facilities to every people. Therefore, it needs to share accurate health care information concerning clinical research information, promptly taking action to leverage health information technology to monitor the COVID situation, detection, early warning, prevention-control mechanism, and other information. The Internet of Things (IoT), Artificial Intelligence (AI), intelligent diagnosis with RT-PCR (Real-Time Reverse Transcriptase–Polymerase Chain Reaction), integration and computation of varieties of data including location-based data and sincere learning can enable us to diagnose and detect the coronavirus and also help to develop the invention of remedial measures like vaccines and other medicines to control this virus (Hong et al. 2020).

Lots of other ways by which technology can help and manage this pandemic as follows (Manjunath 2020):

- Finding information regarding COVID hospitals and number of vacant beds, contact numbers of doctors, online diagnosis and treatment facilities, newly discovered medicines, to get the proper knowledge on symptoms of particular diseases, etc.
- Tracing the patients and testing is possible by using government-identified mobile apps, other necessary web portals, etc.
- Technology supported temperature monitoring; by using the wireless thermometer guns and other similar infrared human being temperature measurement devices, which are now most popularly used at the checkpoints of most of the public offices, restaurants-hotels, hospitals, railways stations, shopping malls, and other public places.
- Remote working technologies to support social distancing and maintain business continuity.
- Technology helps contactless movement and deliveries through autonomous vehicles, drones, and robots.
- It also helps control and checks the misinformation regarding the number of fatalities, diagnosis, treatment, vaccines, medicines, etc., which may create panic and anxiety.

The government of India has taken initiatives with the help of IT to monitor and control the spread of the corona virus through smartphones, GPS systems, or Bluetooth. The successful application of the Aarogya Setu mobile app in quarantining and declaring the highly susceptible zones like reading, orange, and green zones is extraordinary evidence (Banerjea 2020).

E-Gram Swarajya App and *Swamitva Scheme* have been introduced by our Prime Minister during his interaction with Sarpanchs of Gram Panchayats throughout the country through video conferencing on the occasion of National Panchayati Raj Day on 24th April 2020 (Civil Service Chronicle 2020c). The *e-Gram Swaraj App* helps prepare and execute Gram Panchayat development Plans, and the portal will ensure accurate time monitoring and accountability. The *Swamitva Scheme* has been launched in 6 states and helps map rural inhabited lands using drones and the latest survey methods. The scheme will ensure streamlined planning, revenue collection and provide clarity over property rights in rural areas.

Another step that the Government of India, including all the state governments, has started for issuing a digital e-pass system for private vehicles to move beyond the permissible limits during the lockdown. These e-passes are issued to the people through online applications by the local administration, considering their urgency and necessities (Gopal 2020).

While many issues are being addressed during this pandemic, education is vital among them, and technology is the only solution to continue the teaching and learning system (Dhawan 2020). Therefore, when all educational institutions are closed for physical attendance since March 2020, the educational institutions have started an online teaching and learning system by using and taking the help of online platforms like Google Meet, Zoom, Webex meet, etc., to prevent hampering the students' academic careers.

13.3.6 Migration Commission

One commendable step that has been taken by the Government of India regarding the solution of the migrants, both internal and external, is the setting up of the Migration Commission. Indian Prime Minister is well-versed with an attitude of delivering *Maan Ki Baat* on the last Sunday of every month, and during his *Maan Ki Baat* in the month of May 2020, he announced about the setting up a 'Migration Commission' or employment of migrant labours and the commission will look into the mapping the skills of these labourers (Mohan 2020). He also spoke of providing self-employment and setting up small-scale industries in villages. Immediately after the announcement, the Chief Minister of Uttar Pradesh announced a migration commission on 26th May 2020 to provide jobs and social security to the workers returning to that state (Mohan 2020).

The main objective of this commission will be to consider the voices of migrants while framing specific guidelines for their betterment. The commission will institutionalize their demands and will be accountable for fulfilling those. The activities of the commission will be focused on whether the migrants are exploited by any organization or agency anywhere in the country and to protect them from such incidents by providing legal shelter (EPW 2020).

For the international migrants, the Government of India, on tenth July 2020, launched ASEEM, i.e., Aatmanirbhar Skilled Employee Employer Mapping portal for facilitating employment opportunities with SWADES (Skilled Workers Arrival Database for Employment Support) Scheme. This project will work jointly with the Ministry of External Affairs, Ministry of Civil Aviation, Ministry of Skill Development, and entrepreneurship with an objective to prepare the database of returned migrants with accurate information about their skills and experience.

13.3.7 Introduction of Shrmik Special Trains for the Migrant Labours

The government of India issued an order on 29th March 2020 for the landlords to not taking any rent from the people staying in their houses on rent. Alongside this, the employers were advised to disburse wages or salaries to their employees without any deduction. Furthermore, the Central Government also stated the Shramik Special trains on and from first May 2020 for the migrant workers to bring them back to their native places (Bhargava 2020).

13.3.8 Pro-Activating the Local Government Machinery

As Lord Bryce opined that, democracy is the best principle for the practice of local self-government. Therefore, as the majority of the countries in the world, including India, believe in democracy, the decentralization of power empowers the local bodies to be active during this COVID-19 situation, and our Prime Minister has emphasized that 'local should be vocal'(Kapur 2016).

The 73rd and 74th Amendment Act, 1992 empowered the local self-governments in India and working as the pillars of democracy. Article 243G of the Indian constitution includes 29 items, whereas the local administration, especially the panchayat institutions, are responsible to the provide the services to the common people concerning minimum basic amenities including health and sanitation, markets and fairs, women and child development, social welfare including the welfare of the physically challenged and the weaker sections (e.g., scheduled castes and tribes) through the public distribution system, poverty alleviation programmes, etc. (Ghai 2011). The local self governmental institutions are working tirelessly as the centre's instruments for decentralization; fiscal discipline and governance come into reality in each district of India. In this dire period of tackling the COVID situation, the local governments are at the frontline to provide better decentralized economic governance in our country (Khosla 2020).

13.3.9 The Power of the Human Spirit

As the pandemic is now a real challenge globally, the most outstanding global crisis has arisen. Still, the world witnesses a great example of the human spirit to defend against the crisis. As we have observed, the doctors, nurses, police personnel, banking sector employees, public administration employees, political leaders, ordinary citizens – all are working tirelessly and helping each other in a selfless manner to face the challenges of COVID-19, and because of this, the recovery rate is high in India despite its a few times larger population size and comparatively weaker public health infrastructure than the developed countries.

In India, many landlords have waived their rent from their tenants who had lost their jobs, and several common people continued to pay the salaries of their domestic helpers, drivers, and cooks without taking service from them. On the other hand, many volunteers are still helping the patients live in isolation by providing free foods, medicines, and few entrainment programmes. These are called the common man's spirit, which humanity sincerely salutes. All these actions toward solidarity and kindness could illuminate the dark periods of the pandemic throughout the last year (Victor 2021).

13.3.10 Opportunity for Indian Startup Apps

The Government of India banned 59 Chinese mobile apps on 29th July 2020 in the first phase, another 47 apps in the second phase in August 2020, and nother118 apps in September 2020 to secure the citizens from the stake of cybercrimes. Indian software companies have been invited to participate in the *Atmanirbhar Bharat App Invitation Challenge,* which will help create an *Atmanirbhar App Ecosystem* in the software industry in India (Education Times 20th July 2020; Business Standard fourth July 2020; PTI 2020).

13.4 Policies at the State Level

Now, this section will focus on a few popular policies taken by some state governments in India like Chhattisgarh, West Bengal, and Odisha for smooth governance during this pandemic as follows:

13.4.1 Godhan Nyay Yojana

As the economic conditions throughout the country have been facing a lot of challenges, the Government of India and various state governments have been taking measures to boost up the economy and give specific employment opportunities to facilitate the livelihood of the common people. For example, on the auspicious occasion of the *Hareli festival* on 20th July 2020, the Government of Chhattisgarh has launched the *Godhan Nyay Yojana* (the term '*godhan*' in English refers 'dairy cattle resources') for reaching economic benefits to the livestock owners through generating new employment opportunities (John 2020). The basic features of this scheme are:

- The government to procure cow dung at Rs.2/per kg from livestock owners;
- Repurposing procured cow dung into vermicompost and other eco-friendly items;
- Selling vermicompost at Rs.8/– per kg to the farmers to promote organic farming.
- The scheme is to protect crops from open grazing, prevent straying of animals on roads, etc.

13.4.2 Sandhane App

Residents of this globalized world now, with smartphones and other electronic gadgets, seamlessly communicate with people across the distance that underpins the success of e-governance where the use of various apps has been coming forward with reaching the solutions on mouse click. Likewise, the government of West Bengal has launched an app, *Sandhane,* to trace suspected COVID-19 patients so that there can be early detection and containment of the coronavirus susceptible areas. Workers under the Accredited Social Health Activists (ASHA) schemes were employed to visit the households and note citizens with prevailing fever cases or cases of cough and cold in different areas. Then, the government personnel fed the data into the portal. The data was monitored from the Nabanna (the state's administrative headquarters) directly (Nag 2020).

13.4.3 WB Prochesta Scheme

While many migrant labours lost their job due to countrywide lockdown and faced acute problems for their livelihood, the government at the centre and various state governments started different measures for the benefits of the daily wage labours, including migrant labours. Likewise, the West Bengal Government's *Prochesta Scheme* had launched through this scheme, through which a monthly Rs 1000 assistance was given to daily wage labourers in the state of West Bengal (Sharma 2020).

13.4.4 Initiatives Taken by Govt. of Odisha

Odisha model is a unique one to be followed by other states in India to prevent the spread of COVID-19. From 22nd March 2020, the Government of Odisha started lockdown at par with the Government of India, and then it was the first state to have the dedicated COVID hospitals in all the 30 districts of Odisha. In addition, the state became the first state in India to start online registration of people returning from outside the state and provide an online pass to come out of the home during the lockdown to avail essential commodities during the lockdown period (Mohanty 2020a, b).

The government of Odisha also executed a remarkable move by empowering the *Sarpanchas* (the heads of gram panchayats) with the power of district collectors, and they will prepare the list of migrants returning from outside and put under compulsory quarantine for 14 days with all expenses borne by the state government. In addition, the Sarpanches will be treated as frontline workers, and their kin will be eligible to receive an *exgratia* of Rs. 50 lakh (Garikipati 2020).

For successful management of this COVID situation, the government of Odisha has engaged an additional 1039 medical graduates with 8000 existing doctors, staff nurses, pharmacists, lab technicians, and other health workers with ASHA, ANM, and Anganwadi workers. In addition, one commendable step taken by the government of Odisha is giving advance four months' salary to the doctors and Rs.50 lakh ex gratia to the kin of all frontline workers, including health workers, support staff, policemen, and others. In addition to this, the state decided that in case of casualties of those staff members due to coronavirus infection, their kin would continue to receive the salary until the person's date of superannuation (Garikipati 2020).

13.5 Concluding Remark

India's success in defeating COVID-19 is grounded upon the Centre-State collaboration in the federal structure. The success of nation-building largely depends on the weakening of all traditional forms of authority, whether based on customs and usages or on religious scriptures, and their subordination in the secular realm to the state agencies that enforce laws and regulations that state legislates. Therefore, during this pandemic situation, the majors have taken by the government, whether at the centre or at the states, are quite commendable to tackle not only the COVID patients but also vulnerable migrant labours, daily wage earners, and poor health infrastructure, also managed natural calamities like *Amphan*[4] in West Bengal and Odisha and floods in Assam and Bihar were the high challenges during this pandemic situation. Now the people are more conscious of following the COVID protocols, and the government is also making strict policies and guidelines in this regard. The research institutions and laboratories in India are working on it. As a result, vaccines like Covishield and Covaxin are now being distributed to the common people through various vaccination centres. However, the dismal socio-economic conditions need time to revive. While social distancing is so essential, policies should also explore the potential of IT for activities ranging from distant learning, implementation of government schemes in efficient, transparent, and accountable manners to secure the social safety nets. Finally, the world citizens need to maintain the formalities strictly what we could call in short 'SMS', where 'S' stands for 'maintain social distance,' M for 'use mask always,' and S for 'sanitizers to carry and use frequently.' So as to keep COVID-19 away, we should remember, 'Great things are done by a series of small things brought together'.

[4]*Amphan* was a supercyclone, made landfall at Bakkhali, West Bengal, India on 20th May 2020 with a wind speed of 185 kilometers per hour, caused heavy damage to the coastal belt of Odisha and West Bengal in India and some parts of Bangladesh.

References

ANI (2021) Development of COVID vaccines is major step for Aatma-Nirbhar Bharat Abhiyan: President Kovind, https://economictimes.indiatimes.com/news/politics-and-nation/development-of-COVID-vaccines-is-major-step-for-aatma-nirbhar-bharat-abhiyan-president-kovind/articleshow/80189485.cms?from=mdr, Accessed 10 Jan 2021

Banerjea A (2020) Govt Launches 'Aarogya Setu,' A Coronavirus Tracker App: All You Need To Know. Livemint, 2nd April, https://www.livemint.com/technology/apps/govt-launches-aarogya-setu-a-coronavirus-tracker-app-all-you-need-to-know-11585821224138.html, Accessed 11 July 2020

Bhargava Y (2020) Railways to run 'Shramik special' trains to move migrant workers, other stranded persons, https://www.thehindu.com/news/national/railways-to-run-shramik-special-trains-to-move-migrant-workers-other-stranded-persons/article31481996.ece. Accessed 11 July 2020

Burman A (2020) https://carnegieindia.org/2020/07/28/how-COVID-19-is-changing-indian-federalism-pub-82382. Accessed 16 Nov 2020

Chandrashekhar V (2020) 1.3 billion people. A 21-day lockdown. Can India curb the coronavirus? https://www.sciencemag.org/news/2020/03/13-billion-people-21-day-lockdown-can-india-curb-coronavirus. Accessed on 11 July 2020

Civil Service Chronicle (2018) E-Governance, Vol. XXX No. 1, UP ISSN.-0971–4073, pp. 107–108

Civil Service Chronicle (2020a) COVID 19 Pushed 45 Million Globally into Food insecurity: FAO Report, Vol. XXXII No. 2, UP ISSN.-0971–4073, p. 10

Civil Service Chronicle (2020b) One Nation One Ration Card Scheme, Atmanirbhar Bharat Abhiyan, Vol. XXXII No. 1, UP ISSN.-0971–4073, pp. 23, 29–30

Civil Service Chronicle (2020c), government Waives off Waterway Usage Charges, Financial Management Index for Rural Development Programmes, Covaxin: Indigenous Vaccine for COVID-19, New Agri-Reforms, E-NAM, Vol. XXXII No. 3, U.P. ISSN.-0971–4073, pp. 23, 26–27, 44, 77, 96–97

Constitution of India (1998) Allahabad Law Agency. Faridabad, pp:130–136

DD News November 23 (2020) At 10.03a.m

Dhawan S (2020) Online Learning: A Panacea in the Time of COVID-19 Crisis, https://journals.sagepub.com/doi/full/10.1177/0047239520934018. Accessed 22 June 2020

EPW (2020) Migration Commission. https://www.epw.in/journal/2020/23/editorials/migration-commission.html. Accessed 20 July 2020

Fadia BL (1997) Indian government and politics. Sahitya Bhawan Publications, Agra, pp 210–211

Garikipati N (2020) Public Health Lessons Odisha's Management of the COVID-19 Pandemic.. https://www.epw.in/journal/2020/40/commentary/public-health-lessons.html. Accessed 14 Oct 2020

Ghai KK (2011), Indian government and politics, Kalyani publishers, New Delhi, pp. 161, 172-173, 399

Gopal BM (2020) Apply Online for vehicle pass. https://www.thehindu.com/news/cities/Visakhapatnam/apply-online-for-vehicle-pass/article31586122.ece. Accessed 22 June 2020

Hannah R, Ortiz-Ospina E et al. (2020) Policy responses to the coronavirus pandemic, https://ourworldindata.org/policy-responses-COVID. Accessed 22 June 2020

Hong W et al (2020) Using Information Technology to Manage the COVID-19 Pandemic: Development of a Technical Framework Based on Practical Experience in China, https://www.ncbi.nlm.nih.gov/pmc/articles/PMC7282474/. Accessed 20 July 2020

John J (2020) First-of-its-kind GodhanNyay Yojana launched in Chhattisgarh.. https://timesofindia.indiatimes.com/city/raipur/first-of-its-kind-godhan-nyay-yojana-launched-in-chhattisgarh/articleshow/77063907.cms. Accessed 21 July 2020

Kapoor A, Sandeep G (2020) India's successful journey to self-sufficiency in PPE kits.. https://economictimes.indiatimes.com/industry/healthcare/biotech/healthcare/indias-successful-

journey-to-self-sufficiency-in-ppe-kits/articleshow/78658109.cms?from=mdr. Accessed 16 Nov 2020
Kapur AC (2016) Principles of political theory. S.Chand & Company Pvt. Ltd, New Delhi, p 651
Kaul R (2020) 12 states conduct more COVID-19 tests per million per day than national average: Data.. https://www.hindustantimes.com/india-news/12-states-conduct-more-COVID-19-tests-per-million-per-day-than-national-average-data/story-FwgHkfM2veRIeKvnaRracM.html. Accessed 16 Nov 2020
Kharat RR (2020) European J Mol Clin Med:7(10). https://www.ejmcm.com/pdf_7463_260f246ad40a0b1fe6d1c4597754fbfd.html. Accessed 07 Dec 2020
Khosla R (2020) In a Post-COVID-19 World, the only way ahead for india is economic federalism.. https://thewire.in/political-economy/COVID-19-economic-federalism. Accessed 03 June 2020
Kumar A (2018) Digital divide: A challenge to E-Governance in India. http://www.academicjournal.in/archives/2018/vol3/issue2/3-2-51 Accessed (03 June 2020)
Manjunath BS (2020) COVID-19: 8 ways in which technology helps pandemic management.. https://cio.economictimes.indiatimes.com/news/next-gen-technologies/COVID-19-8-ways-in-which-technology-helps-pandemic-management/75139759. Accessed 20 July 2020
Mohan A (2020) Modi indicates restricting imports of goods that can be made in India. https://www.businessstandard.com/article/current-affairs/modi-indicates-restricting-imports-of-goods-that-can-be-made-in-india-120053100310_1.html. Accessed 11 July 2020
Mohanty M (2020a) Odisha announces COVID-19-dedicated hospitals with 1,000 beds. https://economictimes.indiatimes.com/industry/healthcare/biotech/healthcare/odisha-announces-countrys-first-COVID-19-dedicated-hospitals-with-1000-beds/articleshow/74832053.cms?from=mdr. Accessed 21 July 2020
Mohanty M (2020b) Odisha to have a COVID19 hospital in each district. https://economictimes.indiatimes.com/news/politics-and-nation/odisha-to-have-a-COVID19-hospital-in-each-district/articleshow/75239975.cms. Accessed 21 July 2020
Nag J (2020) West Bengal government launches app to track COVID-19. https://mumbaimirror.indiatimes.com/coronavirus/news/west-bengal-government-launches-app-to-track-COVID-19/articleshow/75069663.cms. Accessed 21 July 2020
PTI (2020) Chinese app ban: PM urges start-ups to create 'Atmanirbhar' App ecosystem.. https://www.business-standard.com/article/economy-policy/chinese-app-ban-pm-urges-start-ups-to-create-atmanirbhar-app-ecosystem-120070400700_1.html. Accessed 20 July 2020
Rajamani S (2020) The Impact of COVID-19 on Consumer Goods.. https://www.wipro.com/blogs/srini-rajamani/the-impact-of-COVID-19-on-consumer-goods/. Accessed 16 Nov 2020
Rajan SI (2020) COVID-19-led Migrant Crisis.. https://www.epw.in/journal/2020/48/commentary/COVID-19-led-migrant-crisis.html. Accessed 07 Dec 2020
Rawat M (2021) Just before 2nd COVID wave hit India, ICU beds decreased by 46%, oxygen ones by 36%.. https://www.indiatoday.in/coronavirus-outbreak/story/just-before-2nd-COVID-wave-hit-india-icu-beds-decreased-by-46-oxygen-ones-by-36-1796830-2021-05-03. Accessed 03 May 2021
Sahoo J (2020) Understanding the Centre-state relations: areas of conflicts and cooperation during COVID-19. In: Sahoo J, Datta K (eds) Indian Federal System a COVID-19 perspective. Kunal Books, New Delhi, pp 8–9
Sankar G, Gopal N (2020) Atmanirbhar Bharat Abhiyan and Agriculture. https://www.epw.in/journal/2020/35/commentary/atmanirbhar-bharat-abhiyan-and-agriculture.html. Accessed 31 Aug 2020
Sharma S (2019) Around 22% Indians live below poverty line. Chhattisgarh, Jharkhand fare worst. https://www.financialexpress.com/economy/around-22-indians-live-below-poverty-line-chattisgarh-jharkhand-fare-worst/1713365/. Accessed 22 June 2020
Sharma S (2020) Apply for West Bengal's Prochesta Prokolpo scheme to get Rs.1000.. https://timesofindia.indiatimes.com/home/education/news/apply-for-west-bengals-prochesta-prokolpo-scheme-to-get-rs-1000-check-details-here/articleshow/75547872.cms. Accessed 21 July 2020

Sharma RN (n.d.) Methods & Techniques of social survey and research. Rajhans Agencies, Meerut (UP), pp 78–81

The World Bank (2020) In India, women's self-help groups combat the COVID-19 (Coronavirus) pandemic.. https://www.worldbank.org/en/news/feature/2020/04/11/women-self-help-groups-combat-COVID19-coronavirus-pandemic-india. Accessed 11 July 2020

Verma P, Sughosh J (2020) Reaffirm Cooperative Federalism.. https://www.thehindu.com/opinion/op-ed/reaffirm-cooperative-federalism/article31567966.ece. Accessed 20 July 2020

Victor D (2021) Saluting the human spirit A year after the onset of COVID-19, we bemoan the grave misfortunes that fell on us,. https://www.deccanherald.com/opinion/oasis/saluting-the-human-spirit-967695.html. Accessed 29 Mar 2021

Dr. Jitendra Sahoo is currently working as a Professor, Department of Political Science, University of Gour Banga, Malda, West Bengal, India. He has more than nineteen years of teaching experience at the UG and PG levels. He has obtained two Masters, one in Political Science and another in Public Administration, and was awarded a M.Phil. and a Ph.D. Degree from Utkal University, India. He is actively engaged in research work; participated and presented papers in various National and International Conferences, Seminars, Webinars; contributed 15 articles in different National and International Journals and written 22 chapters in Edited Books. Dr. Sahoo has co-authored two Books and he has edited seven books which was published by the publishers of national and international repute. His research areas are Governance at Local and National Level, Public Administration, Society and Polity in India, etc. He is a life member of the Indian Political Science Association and West Bengal Political Science Association. Currently, he is also a member of the Board of Editors of UGC-Care (India) enlisted Journal 'ENSEMBLE'.

Chapter 14
Employment Dynamics and Labor Mobility amidst COVID-19 Pandemic in India: A Critical Appraisal of ILO Recommendations

Siddhartha Sankar Manna

Abstract During the outbreak of the COVID-19 epidemic, the contagion has constantly been pushing humankind into ambiguity and unprecedented problems. The rapid outreach of the COVID-19 epidemic in India accelerated a severe crisis of mobility among the migrant workers in several metropolises and towns seeking to come back to their birthplaces. This chapter would try to analyze the effect of the COVID-19 epidemic on migrant workforces in India. The exponential surge in infection among the people across the state caused the enforcement of extensive lockdown that resulted in the blocking of massive forms of mobility, downgrading of trade and business, and degradation of social relations. This work judges the dynamics of the vulnerability of domestic migrant workforces in respect of their mobility, gender, and health security in India. It is assumed that the COVID-19 infection would form a gender disparity in the production and trade fulfillment. It demands the working class to follow with relentless vigor of the International Labour Organisation (ILO) a legitimate order for social justice. It demands to establish the rights and the requirements, ends, and rights of people to be at the center of economic, social, and ecological strategies. The International Community and the International Labour Organisation have been involved in a concerted effort to challenge the ravaging humanitarian effect of the epidemic.

Keywords COVID-19 · Employment opportunities and Security in India · Labor strategy · Health policies and Community discourse · ILO in India

---

S. S. Manna (✉)
Department of Political Science, University of Gour Banga, Malda, West Bengal, India

S. Saha et al. (eds.), *Economic and Societal Transformation in Pandemic-Trapped India*, New Frontiers in Regional Science: Asian Perspectives 55,
https://doi.org/10.1007/978-981-16-5755-9_14

14.1 Introduction

The migrant workforces in India have been facing several types of adversities and sufferings throughout the outbreak of the COVID-19 epidemic. The exponential surge in infection among the people across the state caused the enforcement of extensive lockdown that resulted in the blocking of massive forms of mobility, downgrading of trade and business, and degradation of social relations. Millions of migrant laborers and workforces in the country faced uncertainty of income, food security, and their basic needs due to the closure of workshops and offices and the lockdown imposed by the state (Singh 2020). As a result, they suffered from starvation, and their family members who depend on their income became uncertain and insecure. India has already been accommodating the largest volume of displaced and homeless people globally, and the execution of the lockdown significantly increased this number. According to the report by the Census of India (2011), it is estimated that 1.7 million people have been internally displaced persons who are absolutely homeless (Abi-Habib and Yasir 2020).

The Central and state governments took several initiatives to assist those home-bound migrants in crisis, including transportation, food, and refreshment, etc. (The Economic Times 2020). M. B. Dhanya, a renowned writer, pointed out the erratic degrees of informal monetary actions in India. She speculated that more than 90% of the Indian economy had been backed by activities organized in informal sectors (Dhanya 2013). This chapter will attempt to explore the impact of the COVID-19 epidemic, particularly how countrywide lockdown did affect domestic migrant people who had long been the principal internal workforce in the country. This work plans to understand the dynamics of the vulnerability of domestic migrant workforces in respect of their mobility, gender, and health security in India. Furthermore, it significantly studies the limitations of public strategy in referring to the migrants and advocates some recommendations.

14.1.1 Attributes of Migrant Labors and Employment Opportunities in India

The labor market throughout the world has been highly affected and disrupted by the emergence of the widespread infection of the COVID-19 epidemic. The labor markets have been shrunk due to the extensive community spread of contagion that seriously disrupts both the supply chain and demand. In the context of supply and demand deficiency, the production systems have been affected by the arrangement that results in multilayered shocks and susceptibility. Subsequently, the commercial firms were forced to block up, and the temporary termination of trade and business resulted in unemployment in different sectors, and the internal migrant workers have been highly affected by the labor market shock.

With the announcement of lockdowns and restrictions, the millions of migrant workforces became jobless and fell into subsequent depression. *The Labour Union leaders and activists* stated that migrant laborers might once again be looking at the downfall of incomes and the absence of social security in the wake of emerging COVID-19 infections (Nath 2021). The migrant laborers were forced to return to their homes at the bucolic side of the state as their workplaces in the metros shut the doors for their employees due to the lockdown. The labor forces at unorganized sections faced impediments with high risk because of the inaccessibility of wages. On the other hand, the lockdown and the subsequent downturn have been expected to hit initially and harmed contractors (i.e., the middleman in the labor supply chain) and workforces across several industries in the country.

It is found that the *Small, Micro, and Medium Enterprises (MSME)* have been severely impacted and downgraded due to the frequent impositions of lockdown in different states. The Association of Indian Industry revealed that many tourism and hospitality businesses had been turned down, which had been producing a great demise of occupations (Das 2020). This would as well concern the unorganized labors and self-employed laborers. The *Retailer Association of India* reported that around 40 million workers and six million workers were involved in the informal and formal sectors. The association also reported that approximately four million workers had involved on a contractual basis (Nahata 2020). Therefore, approximately 50 million workers engaged in both formal and informal sections. Most of them had been migrant employees who were at immediate risk of losing their current occupation, provisionally or permanently, on account of the upsurge of COVID-19 infections across the country.

The perilous threats to workers employed in the informal sector or whose contracts were nearing completion had been the primary concern of anxiety. Losing employment means uncertainty of human security and human development that affects the gamut of social progress and produces the risk-society (Khanna 2020). The burden of lockdowns and following stagnation initially hit contractors and contract laborers across several industries in the country. The precipitous range of the infection provoked the India Government to impose the countrywide lockdown, and mobility and transport linkages become isolated overnight with accurate restrictions and restraints (Gettleman and Raj 2020).

The rapid extension of the epidemic in India accelerated a serious *crisis of mobility* among the migrant workers in several main metropolises and towns seeking to come back to their birthplaces (Dutt 2020). Their anxious efforts to come back home by arranging the transport measures became significant. On the one hand, implementation of social distancing policies and lockdowns was desiccating occupations, incomes, and the opportunities of wages for laborers, while they would have been likely to upset agrarian products, transport mechanisms, and mobilization on the another (Rajan et al. 2020). On the other hand, this has created a challenge of securing the foods for life and provokes extensive starvation, especially among children who could suffer from malnutrition, and it would increase the infant and mortality rate of children (Kone et al. 2018). In this regard, the national migration

Table 14.1 Migrant Workforces excluding agricultural workers in India (percentage)

Job types	Total	Male	Female
Legislators, senior officials professionals and managers (α)	9.69	11.42	6.02
Technicians, associate professionals and clerks (β)	11.44	12.13	9.98
Service workers and shop and market sales workers (γ)	15.92	19.3	8.8
Skilled agricultural and fishery workers (δ)	6.91	2.31	16.59
Craft and related trades and plant and machine operators/assemblers (ε)	48.34	50.22	44.4
Elementary occupations and workers not classified by occupation (η)	7.7	4.62	14.2
Total percentage	100	100	100
Number of migrant workers (non-agri) (in million)	64.96	44.04	20.92

Source: Khanna 2020

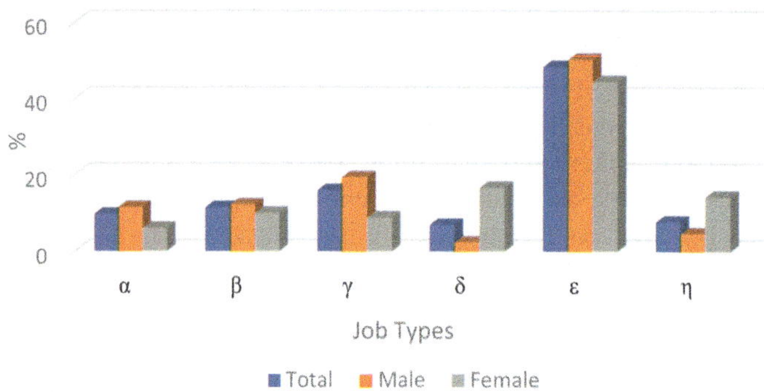

Fig. 14.1 Migrant Workforces excluding agricultural workers in India; Refer Table 14.1 for details about "Job Type" (Source: Prepared by the author, based on Khanna 2020)

strategies need to be relooked to protect migrant laborers as these strategies could have to safeguard them from returning to their homes or providing health facilities.

Correspondingly, there has been a necessity to arrange robust food systems that could diminish food insecurity and uncertainty. According to the Indian Census report 2011, around 65 million migrant laborers had been employed in several occupations and enterprises, excluding agrarian laborers and plows (Table 14.1) (Khanna 2020). According to the work-related category, the study of non-agricultural migrant workforces presents that a huge number of workers have been engrossed in the craft and linked trades and business or herbal productions and engine machinists and machinery parts assemble processes (Fig. 14.1).

The Census of India (2011) report on migration also revealed that a strong pool of migrants had been coming to attain several works at various urban regions, main metropolises from several states in the county. For example, in New Delhi, the capital city of India, more than half of the contributing labor force in different sectors had been sourced from Uttar Pradesh (39%) and Bihar (12%) along with from some other Indian states like Haryana, Uttranchal, Rajasthan, Punjab, and West Bengal.

Table 14.2 Subsector-wise industrial employment trends in India

Sector	Number (Million)	Percentage
Manufacturing sector		
Food and beverages, tobacco products	9.2	16.3
Textiles, wearing apparel	18	31.9
Leather, wood, paper products, printing	5.8	10.2
Petroleum products, chemical products	1.5	2.6
Rubber and plastics, non-metallic products	5.1	9.0
Basic metals, fabricated metals	4.6	8.1
Machinery equipment, electric and electronic machinery	4.0	7.1
Motor vehicles and other transportation	1.5	2.6
Furniture, recycling, jewellery, and sports goods	6.8	12.1
Manufacturing total	*56.4*	*100.0*
Non-manufacturing sector		
Mining and quarrying	2.0	3.4
Electricity, water, gas	2.8	4.4
Construction	54.3	92.2
Non-manufacturing total	*59.1*	*100.0*

Source: Khanna 2020

These migrant laborers had left their villages to attain works in metropolitans and cities to support their families in the homeland financially.

A recent analysis by Mehrotra and Parida (2019) assessed that approximately 115.3 million workforces had been involved in industrial services, of which 56.4 million workers were connected in the manufacturing divisions and another 58.9 million in the non-manufacturing service (Table 14.2), whereas the highest section of employment in the engineering subdivision has been in textiles and foods and drinks. It is important to state that a large number of workers, almost 92% have been involved in the non-manufacturing service especially in the construction and building works (Fig. 14.2).

The migration tables in the Census of India (2011) also revealed that about 143.4 million persons had been working in the service sector itself. Table 14.3 exhibits that 37.3 million workers (which is approximately one-fourth of the total) had been involved in the *retail trade and business*, another 27.7 million (which is nearly one-fifth of the total) in *transport and communication,* and another 14.5 million (comprising 12.3% of the total) in the *education* sub-sector (Khanna 2020). Besides, another 16.5 million workers, nearly 11.5% of the total service sector workers, had been serving in the different social services that numbers an enormous 16.5 million people in those workspaces.

All the discussions so far through different tables confirm that there had accumulated a large volume of migrant workforces in different metropolitans and megacities in India at the pre-lockdown period who had received the shock of the initial joblessness due to pandemic-induced lockdown. What is more worrying is that almost 260 million people had been working in the non-formal sections only

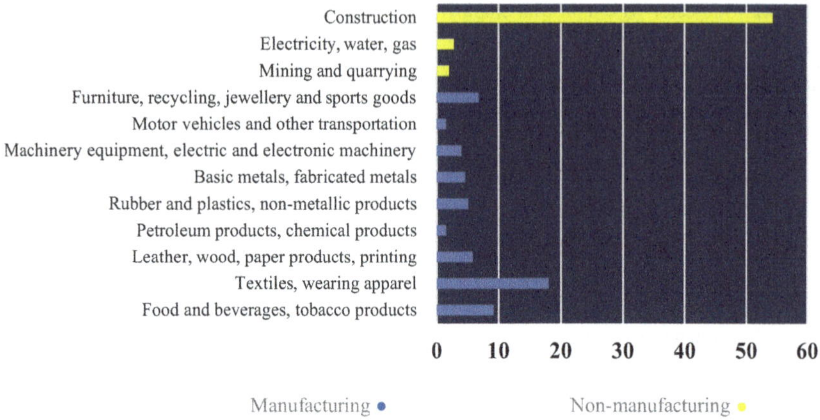

Fig. 14.2 Subsector-wise Industrial Employment Trends in India (Source: Developed by author, based on Khanna 2020)

Table 14.3 Sub-sector-wise service

Sub-sector	Number (million)	Percentage
Sale, maintenance, and repair of motor vehicles	3.6	2.5
Wholesale trade except motor vehicles	6.1	4.2
Retail trade except motor vehicles	37.3	25.8
Hotels and restaurants	8.7	6
Transport and communication	27.7	19.3
Financial intermediation, insurance and pension funding, activities auxiliary to financial intermediation	4.8	3.5
Renting of machinery/equipment, computer, and related activities	3.5	2.4
Other business activities	4.6	3.3
Public administration and defense	7.5	5.3
Education	17.5	12.3
Health and social work	5.6	3.9
Other social services (art, entertainment, R&D)	16.5	11.5
Total service sector employment	143.4	100

Source: Khanna 2020

(Census of India 2011). Table 14.4 indicates that about 101.3 million labors have been employed in the unorganized sections, accounting for almost 71 percent of the whole population engaged in the private sectors or non-farm sections. These vast number of people employed in the *unorganized*, *informal*, and *private sectors* became "surplus" overnight due to the imposition of the lockdown regardless of the matter that in which sector and in what terms they had been contracted with their employers to serve (Khanna 2020).

Tables 14.1, 14.2, 14.3, and 14.4 find the following facts that can be presented in a summarized way:

Table 14.4 Types of employment in non-farm sections (in million)

	Organized/Unorganized		Formal/Informal		Government/Private	
Sector	Organized	Unorganized	Formal	Informal	Government	Private
Manufacturing	18.1	38.4	8.7	47.7	1.2	55.3
Non-manufacturing	15.4	43.5	3.1	55.9	6.3	52.7
Service	43.2	101.3	31	113.4	26.4	118.0

Data Source: Khanna 2020

- Total male laborers employed larger in absolute count than that of the female laborers;
- Averagely a male laborer is engaged based on per person-day salary on a contract which is averagely 9% higher than a female laborer with the same skills involved in the same sector;
- The woman and contractual laborers are waged less than other workforces. Though many measures have been taken to eliminate the gendered difference and incongruity in salaries, the gaps still occur;
- For the other workers, the wage per person-day is higher for males than the female laborers in the workplaces.

Shah and Lerche (2020) speculated that the family members of the migrant laborers constantly support them by upholding their households; however, while the regular works are not available or they are unable to work for income, the family members take care of them. They pointed out five essential facts (Shah and Lerche 2020):

- *Firstly*, trade, commerce, and heavy industries have been depending on migrant workforces who are paid a lesser wage in exchange for long hours of services per day. Although in several regions of the world, such unskilled migrant workers cross the ways of state boundaries. In India, enormous domestic migrant workers crossing state borders for colloquial agreements work in more advanced regions where they have been considered *second-class* citizens. Subsequently, they have been beleaguered by their employers and companies due to their inability to speak the language of where they migrate to. Moreover, they are not allowed to represent the union and institutional sectors in particular regions. They have categorically been controlled and regulated due to their vulnerable condition;
- *Secondly*, millions of migrant laborers have been working seasonally and propagating between their rural homes and their workplaces negotiating hundreds of kilometers distances in between;
- *Thirdly*, the most unskilled laborers have been involved in several regions like *Chhattisgarh, Odisha*, and *Jharkhand*, where strangers extricated the natural resources of aboriginal people. In addition, they who belonged to the retrograde groups, largely from the *Dalit* and the *Adivasi* peoples (laborers), have been lowest waged for the toughest and vigorous works and functions;
- *Fourthly*, the laborers from underprivileged minority sections, i.e., Dalits and Adivasis who live in the worst circumstances, perform the hardest work in India.

They have been exaggeratedly embodied as sporadic migrant laborers. They may be 25% of the total population but comprise more than 40% of the sporadic migrant labor forces like tea plantations of Kerala, biochemical manufacturing factories of Tamil Nadu, or road construction in several areas of India;
- *Fifthly*, periodic migrant laborers can merely be workforces on account of all the work that has been done and assumed across generations at family, as well as care of wives, kids, brothers, and aged parentages. This is the labor of relatives who will take care of the migrant labors in the periods while there has not worked, who will look after the residential places of migrant people so they would have a home-place to come back. They would not be able to work for the reason that employers overused them in practicums or workplaces. In this context, the periodic migrant workforces have not only been exploited or oppressed but also super-exploited physically in the working sectors.

While there were 450 million migrant laborers employed in India, accounting for approximately 37.7% of the total population, a recent survey report ascertained about 600 million internally migrant laborers involved in several formal and informal sectors in 2020. Gupta (2020) rightly pointed out the trends of domestic migrant laborers, saying (Gupta 2020):

> The 2011 census showed that around one-third of all internal migrants are inter-state and inter-district migrants, which makes it almost 200 million. Of these 200 million, about two-thirds are estimated to be workers. This gives us a migrant worker population of about 140 million as of today.

14.2 Migration

The public transport was shut down due to lockdown constraints, and the *'no work and no pay'* policy was executed in the working places. Consequently, a large number of migrant laborers had been returning from their workplaces to their aboriginal villages and families (Pandey 2020). They were hungry while they were coming back to their residential places (Jaiswal 2020). At that time, social isolation was impossible for these migrant laborers as they were collectively traveled together. On 14th September 2020, Mr. Santosh Kumar Gangwar, the Labour and Employment Minister, reported in the Parliament of India that almost ten million migrant workers, as per reports received from the state governments, had tried to come back to their homes due to the increasing effect of COVID-19 infections and subsequent lockdown (Sharma 2020). His statement further included that 71% of these migrant laborers had been sourced from four different states - Uttar Pradesh, Bihar, West Bengal, and Rajasthan. Interestingly, on 15th September 2020, he itemized in the Parliament that no information had been made on the domestic migrant laborers who either expired or turned out to be jobless as an outcome of the epidemic or increasing infection (Nath 2020). UNDP and UN Women (2020) reported that the COVID-19 infection had amplified the scarcity and poverty rate for females and developed the

Table 14.5 The unemployment ratio

Month	Unemployment rate	Month	Unemployment rate
May, 2020	21.7	November, 2020	6.5
June, 2020	10.2	December, 2020	9.1
July, 2020	7.4	January, 2021	6.5
August, 2020	8.3	February, 2021	6.9
September, 2020	6.7	March, 2021	6.5
October, 2020	7	April, 2021	8

Source: Adopted from Beniwal 2021

disparity between men and women who have been living in poverty or under poverty levels in society.

Various people were detained for disobeying the lockdown, crossing the inter-state boundaries, crossing forests between states, or cross the rivers by ships (Babu et al. 2020). Several migrant laborers died because of enfeeblement and enervation (Elsa 2020); several fatalities happened in road accidents. They had faced problems like lack of sufficient vehicles, passenger cars, or railway passenger transport facilities during their journey from workplace to birthplace or hometown (Shantha 2020). Some of them were exterminated at the time of returning to their destination (Nandi and Bhaskar 2020; Auraiya road accident...2020) After coming back to their homes, they engaged in their agrarian works and resumed several small works under MGNREGA. About 3.5 million migrant laborers recorded their engagement under the MGNREGA scheme between first April and May 2020 (Chauhan 2020). This increased number indicates the migrant laborers' returning from different parts of the state.

The unemployment rate has been increasing due to the lockdown, and more than seven million workers lost their occupations. *Centre for Monitoring Indian Economy* (CMIE), a reputed research firm, reported that the rate of joblessness in India had been increased by approximately 8% in April 2021 (Beniwal 2021).

The CMIE report exposed that the unemployment or joblessness doubled in the workweek ended 23rd May 2021 and accentuated the circumstance that the local lockdowns and restrictions to fumigate the community transmission of the disease acutely affected and ruined the community employment opportunities, trade, and business (Beniwal 2021). Table 14.5 shows that the unemployment rate in India has risen to 14.7% on 23rd May 2021. It has been higher than forgoing weeks or years. As per the CMIE report, the weekly unemployment in urban India percentage increased at 14.4%, which was higher from 13.5% on 23rd May 2021 (The Times of India 2021). The issue of unemployment exists in rural parts of the country also; however, the people had been much more jobless in urban areas of the country (Beniwal 2021). When India enforced the national lockdowns, and social isolation plans in April and May 2020, the unemployment and joblessness percentage doubled and augmented the uncertainty of jobs in both urban and rural areas (Vyas 2021). Mr. Mahesh Vyas, the Managing Director, CEO of CMIE stated (Vyas 2021):

Table 14.6 Prickle in urban regions

Date	Month	Village	Urban	Country
04 April 2021	April	8.58	7.21	8.16
11 April 2021	April	8	9.81	8.58
18 April 2021	April	7.31	10.72	8.4
25 April 2021	April	6.37	9.55	7.4
02 May 2021	May	7.4	10.1	8.2
09 May 2021	May	7.3	11.7	8.7
16 May 2021	May	14.3	14.7	14.5
23 May 2021	May	13.5	17.4	14.7

Source: The Times of India 2021

> The double-digit unemployment rate seen in recent times indicates that even these restrictions are taking a toll on the economy. The unemployment rate touched 14.5% in the week ended...

He further supposed that the urban unemployment percentage has been rising since the beginning of April 2021 (Table 14.6). The urban unemployment range in urban areas was 7.21% during the first week of April 2021, but during the first week of May 2021, unemployment had been 10.1%. Mr. Vyas said that the increase of unemployment in rural India had been the latest phenomenon, and he also concentrated on the execution of the MGNREGS in villages. He mentioned that the weekly exceeding of double-digit unemployment in villages and urban areas is a worrying and threatening factor for social solidarity (The Times of India 2021).

Although there have not been nationwide lockdowns to prevent COVID-19 infections, there are strict state restrictions and local lockdowns in different states during the 2nd wave of coronavirus outbreaks. It was expected that the adverse effects would be felt in the job market and employment sectors. Combined with that fear, the unemployment rate in the country reached double digits again on 16th May 2021. According to *CMIE*, it reached 14.50%. In the week ending 23rd May 2021, it raised to 14.70%. The unemployment in rural areas has declined slightly in the past week, but it has risen at a rocket speed in urban areas. In urban areas, unemployment has increased from 14.60% to 17.40%. In rural areas, it decreased to 13.50% from 14.50%. The unemployment rate (Table 14.7) of a few states has also been staggering. Some have crossed the approximately 20% limit. While many people lost their jobs, causing their lives uncertain and hikes social insecurity, the transport system and supply chain in different states have been disrupted lesser than last time. Despite that, the unemployment rate has risen. Though the 100-days project has been at the center of trust in the rural regions, these works are not properly executed

Table 14.7 The Unemployment ratio in different states of India

States	21-Jan 2021	21-Feb 2021	21-Mar 2021	21-Apr 2021
Andhra Pradesh	4.5	3.7	5.9	4.9
Assam	1.5	1.6	1.1	0.2
Bihar	10.5	11.5	14.6	11.5
Chhattisgarh	6.4	6	2.7	3
Delhi	12.5	8	9.4	27.3
Goa	16	20.6	22.1	25.7
Gujarat	3.2	3.2	2.2	1.8
Haryana	17.6	26.3	27.6	35.1
Himachal Pradesh	11.6	15.6	14.2	11.1
Jammu & Kashmir	21.9	14.2	9.5	11.4
Jharkhand	11.3	12.2	12.8	16.5
Karnataka	3.3	2.5	1.2	2
Kerala	5.5	4.3	5.9	7.5
Madhya Pradesh	6.2	2	1.5	1.4
Maharashtra	3.8	3.8	3.5	5.5
Meghalaya	3.3	3.8	1.3	1.4
Odisha	3	2.5	1.6	1.9
Puducherry	7.6	5.8	1.4	2.7
Punjab	8.1	7.2	7.3	5.3
Rajasthan	17.7	25.6	19.7	28
Sikkim	0	4.3	1.7	1.8
Tamil Nadu	4	4.8	3	2.3
Telangana	4.3	5.6	3.7	5
Tripura	18.1	11.1	13.9	17.3
Uttar Pradesh	4.9	4.1	4.1	6.3
Uttarakhand	4.5	4.7	3.3	6
West Bengal	5.2	6.2	7.5	7.6

Source: CMIE, 2021 (Dataset is available at: https://unemploymentinindia.cmie.com/kommon/bin/sr.php?kall=wsttimeseries&index_code=050050000000&dtype=total)

14.3 The ILO's Recommendations

UN Labour Agency said (UN News 2020) that the rapid expansion of the COVID-19 contagion has been affecting the economic mobility that accelerated the unemployment of approximately 200 million workers worldwide. In 2019, the *International Labour Organization (ILO)* assumed the *Centenary Declaration for the Future of Work* supported by the associates of ILO. These associates were from 187 Countries. This declaration made demands on citizens to follow (ILO 2020a):

> with unrelenting vigour its [ILO] constitutional mandate for social justice by further developing its human-centered approach to the future of world.

The downturn caused by the emergence of COVID-19 seems to be worse and more harmful for human beings than what we have experienced during the economic crisis and depression in 2009. The ILO assessed that almost 25 million people turn out to be jobless globally because of the effect of COVID-19, extending between 5.3 million unemployment on a "*low scenario*" and 24.7 million on a "*high scenario*" (ILO 2020b).

The effect of COVID-19 has been diverging from state to state. Between January and February 2020, roughly five million people in China lost their employments in Wuhan of China with the expansion of the COVID-19 (Cheng 2020). The *Angus Reid Institute* pointed out that the different types of unemployment were seen up to 44% in Canadian households (Macleans 2020). Approximately 900,000 workforces lost their works and functions in Spain during the lockdown due to the extensive widespread infection in March 2020 (Keeley 2020). It is essential to state that ten million American people applied for governmental support due to joblessness (Weissmann 2020). *The Federal Reserve Bank of St. Louis* projected that the 47 million workforces lost their works in the USA that affected 32% (Cox 2020).

In this way, ILO called for setting the right of laborers that ensure the entitlements of working people. Also, it tried to fulfill the goals and objectives of all people at the sentiment of socio-economic and eco-friendly strategies. However, the extensive expansion of COVID-19 infection among people all over the world plunged the unprecedented crisis and insecurity. The *international community* and the associates of *ILO* are involved in a cooperative endeavor to confront the contagion on the human effect. Dreze (2020) further claimed that (Dreze 2020):

> As the monsoon advances, there is an urgent need to consider what can be done to prevent hunger during the rainy season — the hardest time of the year for poor families in large parts of rural India. The monsoon is expected to be good, but the Kharif harvest is many months away. Meanwhile, with their reserves depleted by the lockdown, millions of people will find it hard to keep on going. The National Rural Employment Guarantee Act (NREGA) may help, but large numbers are likely to remain exposed to food insecurity.

The fact of reverse migrant workers owing to the widespread infection would affect several states in diverse ways.

UN News (2020) reported that the workforces in four different sectors that have undergone the most severe effects of the infection and deteriorating the production and manufactures are Food and Accommodation, Retail and Wholesale, Business Services and Administration, and Manufacturing (Table 14.8).

According to ILO, several sectors across the world have been affected by the extreme widespread contagion (UN News 2020). It is essential to explicate that the lockdown courses imposed by the different states in India have ominously affected the laborers who have been forced to come back to their rural places, and they face the risk of employment opportunities

In this regard, Dreze (2020) argued that the food grain must be distributed adequately among the poor and marginal people with special attention in this difficulty of corona infection across the country. There have been a few reasons to be very concerned about their problems. Dreze (2020) postulated that:

Table 14.8 Four sectors with most severe effects of the infection and deteriorating the production and manufactures

Sl. No.	Affected sectors	Number (in million)
1.	Food and Accommodation	144
2.	Retail and Wholesale	482
3.	Business Services and Administration	157
4.	Manufacturing	463

Source: UN News 2020

- The poor states in India would become extremely poor, and its population are surviving in a marginal way or below poverty levels;
- India's poorest states would have been likely to bear an uneven burden of the existing economic emergency. Moreover, the working people would be worse off in these states. For the reason that local laborers would be benefitted in advanced states due to leaving of migrant laborers, but laborer groups in weaker states will swell due to the coming back of migrant laborers that reduce the job opportunities for local laborers;
- the poorest states in India have been severely affected by COVID-19 infections. The preparation of the health arrangement, particularly in highly populated areas of the weak states in India, has not been satisfactory or can provide a healthy atmosphere to tackle the contagion among the people.

ILO estimated that around 400 people had been stricken with poverty due to epidemic and lockdown across India (ILO 2020c). The ILO quickly realized that the COVID-19 epidemic is a threat to health and a perilous menace to the financial and marketplaces of employment. The lockdown measures have been assumed in several states in India to stop the extensive widespread of the epidemic that limited the economic functions. Obviously, India has faced troubles in trade and business, generating negative fiscal growth, but not development. In India, the lockdown has dealt a serious setback to the economy by merging with headwinds around the world. ILO assessed that macro- and micro- trade and businesses had been depressingly affected by the emergence of lockdown that produces unemployment. It is assessed that 77.1% of jobs in India have been temporary, self-employed or informal workforces (ILO 2020c). The UN assessed that the 364 and 473 million workforces would have been a threat and trouble owing to the lockdown. In 2020, about ten million migrant laborers were jobless in India on account of lockdown (Sanghera 2020). In this situation, several types of migrant laborers have come back to their villages and hometowns.

In the context of extensive widespread of coronavirus infection across the world, the ILO (2020c) responded to four different pillars— job creation, rights at work, social protection, and social dialogue, with gender equality as a cross-cutting objective of labor standards that would ensure the equitable and inclusive strategies in the social phenomenon (Fig. 14.3).

The protections of the workforce in the workplaces acclimatize work agreements and preclude discrimination and segregation. ILO is recommended to provide timely

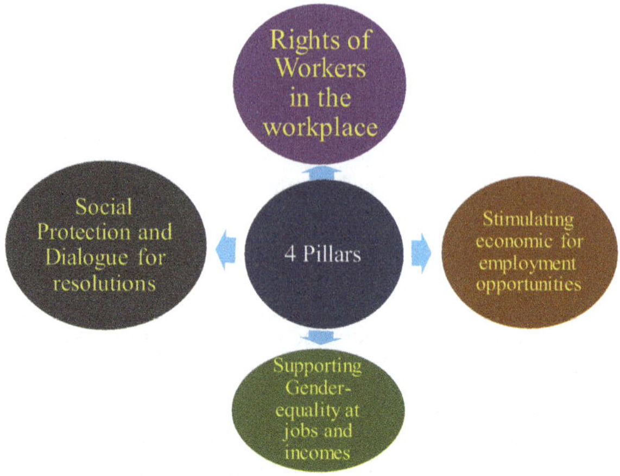

Fig. 14.3 ILO recommended four pillars to respond during the pandemic (Drawn by the author, based on ILO's recommendation)

health services for all laborers in the workplaces and extend adequate paid leave for the workers. The practical and accommodative economic strategy must be stimulated and accelerated to develop public health and sanitation. The employment and incomes must be socially safeguarded. Social discourses between workforces, employers of organizations/enterprises, and government need to be prolonged for the resolutions.

14.4 Strategies and Plans for Inclusive Migrant Labors

In India, the migrant workforces contribute 10% of India's *GDP* and consider it as the backbone of various financial areas, i.e., building, textile productions, domestic job, angling, and fish processing, excavating, mining and quarrying, and even cultivation (ILO 2020d). Migrant workforces have not been counted and unfamiliar at the *'local, regional and national levels'* (ILO 2020d). As no reliable records for migrant laborers are found, it is difficult to provide opportunities for public arrangements and facilities at different stages of governance. And, the lack of formality and documentation weakens the effectiveness of the *local self-governments (LSGs)* and branches of *labor departments*. So, the informal segment seems to be an enormous unaccounted threat and insecurity for migrant workforces across jobs and services. On the other hand, the woman migrant workforce has also been extremely vulnerable to pugnaciousness and persecution. The state of population changes in various areas in the state and its consequences on the workers have gradually been protuberant. Migration has appeared as an essential factor in the dividend of population. So, the inclusion and amalgamation of the migrant laborers and their families have been more significant for sustainable human development in the domain of contagion.

14.4.1 *Five Policy Visions*

The extensive effect of the lockdown because of the COVID-19 contagion on migrant workforces have expectantly left undoubtedly in strategy attentions that this group desires exclusive responsiveness. In the context of contagion, India's socio-economic backbone can be structured by the amalgamation of planned actions with regard to diminishing the suffering of migrant laborers and endorsing decent employment opportunities for migrant workers that mobilize the welfare of their families in the domain of crisis. One of the fundamental objectives of this paper is to lay out suggestions that can generate extensive transformation in the field of work, employment, *and living conditions of migrant* (ILO 2020d) laborers in India. Consequently, the strategy must be holistic and universal in its characterization and application; otherwise, it will be insufficient for long-term transformation demands. In this article, five essential strategies for the inclusiveness of laborers in India need to be considered that affect the schemes to national and state governments (Fig. 14.4).

14.4.1.1 Focusing on Informality

In spite of the uninterrupted economic growth over the past two decades, the informal economy has remained a relentless challenge and persisted a more priority for regimes, employers of organizations, and trade unions in India. It is significant to state that many people have been compelled to join in the informal economy due to

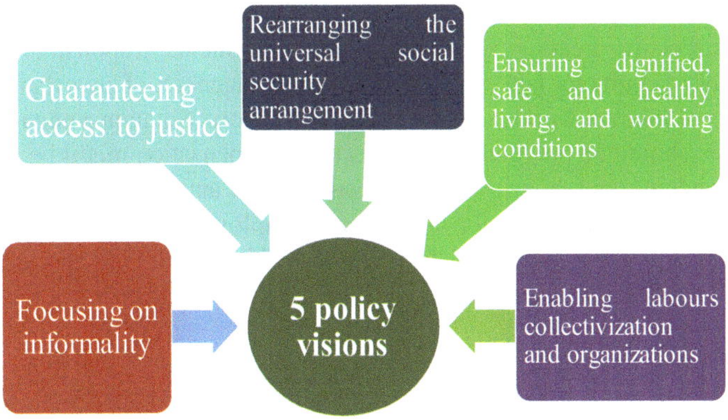

Fig. 14.4 Five Policy Visions (Source: Drawn by author)

the deficiency of chances in the formal economy sectors. In this context, ILO hypothesized that (ILO 2020d):

> "The limited employment creation in the formal economy, despite high rates of economic growth, means that for many, the only alternative is to seek employment in the informal economy, including those who migrate for work. Therefore, one way to address the dire situation migrant workers are in is by addressing the informality and segmentation in India's labor markets."

So, India has taken several types of measures to address vulnerability between unorganized laborers and their households. The measures, for example (ILO 2020d): (1) *formalization of recruitment processes*; (2) *wage fixation and payments*; (3) *extension of social security and adherence to mandatory benefits* through the whole value process, would also play an essential function in contributing to the resolution of informalities. Furthermore, the government, especially in the destination areas, should corroborate and guarantee the registration of migrant workforces. In addition, the formalization facilities need to be enhanced for a large number of self-employed labors as well as homegrown and piece-rate labors in India. In this way, mandating the formal procurement agreement and arrangement can bring transparency and clarity to the retailers' supply chains. In this respect, ILO (2020a, b) correctly pointed out that (ILO 2020d):

> "This will be a way in which these retailers can be held accountable for the welfare of self-employed migrant workers who supply to them. It is also necessary to ensure that labour standards are complied with and made applicable to all workers, including those in non-standard forms of work. Formalization is a powerful tool to enhance labour productivity, which, in turn, will bring far greater value to the economy."

14.4.1.2 Guaranteeing Access to Justice

The access to justice for migrant laborers would help to offset the incalculable dearth of earnings or wages, prevent frauds and disavowals, and oppose their suppression and exploitation, as well as aggression, maltreatment, and persecution. An active legal support arrangement must have to be included to protect the labor rights of migrant workers by creating responsible labor departments, responsible employers, subdivisions of courts for labors, and administrative agencies of laws that can defend the benefits of migrant workers labors in the workplaces. So, unjust engagement and conscription should be prevented, and effective compensation might be provided in case of accidents and injuries. The effective legal and redressal system would postulate a common platform for laborers and employers to resolve disagreements and objections within the ambit of the rule of law. The guaranty to access justice has been a very substantial factor for the rights of work of laborers.

14.4.1.3 Rearranging the Universal Social Security Arrangement

In COVID-19 infection, the social security process needs to be strengthened to protect and safeguard migrant laborers at national and state levels. Two ILO took two different initiatives regarding the standards of labor in the world to ensure social protection, i.e., (1) the *ILO Social Security (Minimum Standards) Convention, 1952 (No. 102)*; (2) *the Social Protection Floors Recommendation, 2012 (No. 202).* These guidelines outlined the standards of labor for the development of social protection arrangements in the state that would mobilize the coverage of inclusive standards for social security in different states. In respect of high informality in India, the comprehensive social protection grounds need to be intensified that would fulfill the minimum requirement of migrant laborers.

Migrant laborers who enter into the state's labor markets or outside of the states in India would have lawful access to basic social security schemes. In the framework of basic security framework, India has a variety of social security programs (ILO 2020d): (1) *the Public Distribution Systems, the world's largest subsidized food distribution program*; (2) *the nutrition supplementation programs,* (3) *health benefits and medical insurance*; (4) *social security and pension schemes;* (5) *housing;* (6) *public employment programs and other social assistance benefits,* etc. A clear strategy has been needed to integrate and amalgamate the migrant laborers into the social security structure so that they can realize their entitlements of citizenship in the state legitimacy.

14.4.1.4 Ensuring Dignified, Safe and Healthy Living, and Working Situations

Migrant laborers must have access to a healthy environment, secure, and non-precarious risk-free workplace. ILO recommended that regional authorities and employers ensure and execute survival logistics and security policy for migrant laborers in social orders. In this regard, ILO articulated a different set of parameters for the protection of migrant laborers (ILO 2020d):

- Right to Adequate Housing;
- right to education;
- a reasonable proportion of income;
- adequate sanitation facilities;
- enjoyment of the workers' fundamental human rights, including the freedom of association;
- safety, health, and protection against the vagaries of nature;
- security of tenure and protection against forced evictions; and,
- proximity to livelihood opportunities.

These rudimentary requirements would be a precondition for guaranteeing adequate livelihood securities. The newly projected *affordable rental housing complexes* (ARHC) made by ILO for migrant laborers in metropolises and towns will emphasize rising unoccupied or empty government housing campuses into hire-lodging. The *ARHC* system would be constructed accessible to the workers who come from pastoral zones or small-scale cities for better opportunities in industrial, hospitality, health, building, and, among others, occupations. It is important to state that access to housing would be a significant degree for livelihood. But India does not endorse the degrees included in *ILO's OSH (Occupational Safety and Health Convention, 1981 (No. 155).* OSH measures have not been as until now well-developed for an extensive field of informal works and fields. A cross-sectoral, pan-Indian, comprehensive plan that ensures all workers' occupational safety and health needs to be formulated. Subdivisions that significantly engross migrant laborers must have to be prioritized in formulating the strategies and directives.

14.4.1.5 Enabling Labors Collectivization and Organizations

It is important to state that the Labor movements organized at the state and national levels of India have played a dominant role in the endorsement and execution of labor. Yet, the unplanned or temporary work has formed a large informal worker that reduced bargaining in the workplaces. Moreover, many laborers working in the informal divisions have been made up of migrant laborers considered a portion of contractual chains, appointed by numerous employers, and belong to pie-cerate and home-based *workers* (ILO 2020d). This has resulted in the continuous decline of the labor movement at several states as well as national levels.

A significant objective of any type of labor strategy has been to facilitate the consolidation and association of migrant laborers in various sectors where they have been recruited or appointed irrespective of caste and gender. ILO (2020a, 2020b) expounded the existence of working people that (ILO 2020d):

> There are successful regional examples of trade unionism in sectors such as garments, domestic work, construction, brick kilns and mining...... Discrimination between local workers and migrants over wage rates and other work conditions creates additional entry barriers for existing trade unions.

14.5 Inclusive Framework for Migrant Workers

The hostile effects occurred by the extensive widespread of the COVID-19 infection has been multidimensional. Therefore, in India, the framework for developing a strategy for the inclusion of migrant laborers requires a large number of stakeholders (ILO 2020d), i.e., from GoI to state governments, from employers to trade unions, from CSOs to development aid agencies and the UN, to work in coherence and unison to comprehensively address the various issues that have been identified. It is important to state that strong socio-economic strategies can diminish the harmful impacts of contagion. Based on the adverse lessons understood at the time of the

Fig. 14.5 *Inclusive* Framework for Migrant Workers (Source: Drawn by author)

COVID-19 contagion and particular practical observations, five significant measures (*Civil Society, Employment opportunities, Social security, Health services, and governance and administration*) can be arranged for recovering the adverse effects and management on internal migrant workers in the socio-economic development (Fig. 14.5).

14.5.1 Civil Society

Civil societies should have a role to promote the service delivery that would help to develop the community in the state. The joint partnership between local self-government and government at national and state levels needs to have appeared in the distribution of services for migrant laborers. The restraints of government structures in providing effective services can be addressed through collective projects of the Civil society organizations (CSOs) with the government or Local Self-Governments. Such corporations can facilitate the livelihood opportunities at the source regions, develop the skill training, confirm suitable work conditions at the destination regions, and safeguard the social security of migrant laborers and it helps the forge source–destination cooperation for safe migration. Another important thing is that the migrant laborers would be empowered by the development of *community-based organizations (CBOs)*, which can assist their right to speak, endorse economic inclusion, and market connections on the basis of community empowerment.

14.5.2 Employment Opportunities

In this effect, in the COVID-19, there is a need to restructure the employment opportunities for migrant laborers. It is found that the low wages at the locality relegate and marginalize the migrant laborers. The accessibility of employment in local areas extensively affects out-migration in spite of the higher incomes or wages available outside the communities and villages. There is a need to fortify the availability of job opportunities for migrant laborers in local areas to bridge and diminish the gap between income at source and at the destination. In the context of crisis, the nodal executive bodies, which would ensure the inclusion and welfare entitlement of migrant laborers, should be established at the local and national levels. The definite roles of these nodal bodies would comprise:

- Safeguarding orderly collection of information on flows of migrant labors to frame the future decisions;
- Make the recommendations for significant migrant labors at industries, with a special emphasis on their vulnerability;
- accelerating national harmonization between states of origin and destination; and,
- Safeguarding portability of several rights interrelated to supply of food, healthcare services, early childcare, and education. The local income generation has come closer to reassurance and protection. This could be further attained by increasing Mahatma Gandhi National Rural Employment Guarantee Scheme (MGNREGS) or increasing government functions at local levels.

14.5.3 Social Protection

Social protection needs to be strengthened. The *Public Distribution Systems* need to be distributed regardless of migrant conditions for the duration of a national emergency. The "One Nation, One Ration Card Scheme 2021" project has been taken for the public distribution systems. In addition, it is significant to reinforce the organizational arrangements for supporting the migrant laborers by monetary assistance, particularly cash deposits in the bank account and increasing the food subsidies. The *National Commission for Enterprises in the Unorganized Sector* (NCEUS) recommended two categories of measures, i.e., (1) *Social Security Measures*; (2) *Social Promotional Measures*. The social security measures incorporate the registration of unorganized labors at villages and towns with the help of panchayats and municipality governments, respectively; adaptation of registered labors under the social security systems; establishment of a welfare board for supervision; formation of the national fund; minimum wage as per states and minimum conditions of works; and unemployment insurance. The *Social Promotional Measures* incorporate skill development; ensure the scope of coverage of work; strategies for the marginal cultivators; banking system for agriculture; financial arrangement for the unorganized sector; adoption of modern technologies for the development of productions. In this context, a nationwide helpline for migrant laborers must have to be established, where they can contact to enroll themselves and seek to redress the defilements and abuses of their rights in the workplaces and other complaints. Additionally, it is important to state that special attention needs to be taken to the well-being of female migrants because they have been a part of vulnerability in the context of the pandemic. So, the program and strategies should be assumed to promote social dialogue.

14.5.4 Health Services

The COVID-19 infection has been affecting the urgencies of health arrangements around the world. India has taken the initiative for resistant and adjustable health care systems. With restricted periods and resources, prevailing obstructions to health facilities and disorganizations in health arrangements have been increased and extended. However, India as well as several states, have been translating the emergencies urgency around COVID-19 into possibilities to form more strong, flexible, and resilient health arrangements. During the lockdown situation, the migrants will create extremely vulnerable circumstances as the maximum number of the urban *Primary Health Centres* have been available for migrants because of several grounds. Furthermore, the accommodation and hygiene services accessible for migrant laborers will make them more vulnerable and helpless. On the other hand, the mental health of migrant laborers needs to be carefully analyzed because they have been facing a lot of stress, nervousness, and mental pressure during normal situations.

So, public health treatment has been required mainly to recover from the effectuates of the existing epidemic. The structure of Health arrangements needs to be rearranged to facilitate the citizens in the state because of the lack of insufficient health care systems. Nationwide Health Protection Action Strategy needs to be prepared for health upliftment. "*Technical initiatives for telemedicine*", "*m-health systems*", and "*digital platforms or mobile apps*" should be incorporated for healthcare facilities and services in remote areas all over the country. Information regarding public healthcare must be improved with the development of technologies and manpower skilled in epidemiology to confront the health crisis.

14.5.5 Governance and Administration

The mechanisms/apparatuses for governance must have to be extensively rearranged for migrant laborers as well as working people in society that can confront the epidemics and provide health care facilities and services in different areas of the state. The conflict between urban growth and migrant subjugation must have to be addressed for the resolution. The arrangements of local self-government for governance should be boosted and improved for management and supervision of health services and other employment opportunities irrespective of caste, creed, ethnicity, and gender in different areas of states around the country. So that, the employment at local levels will be more worthwhile and suitable. Also, governance can only be effectually executed with an exact concept of the scope of the crisis. So that, the exact information has been essential for a substantial period. In this context, the digital apps or central database must be improved and adapted to gather facts that would help track the trends of transfer and mobility of migrants and restrictions in a more exact method.

14.6 Conclusion

There has also been a vital necessity to rejuvenate and reinforce multilateral procedures and arrangements in manufacturing zones and industries, where a huge number of migrant laborers would be engaged and employed. Evaluating the status of the internal migrant laborers in terms of gender, health services, and social security arrangements, I attempt to stimulate the perception hooked on the existing migrant crisis caused by the COVID-19 epidemic. A large number of migrant laborers in different unorganized sectors have been excluded and deprived of several services and arrangements of government because of the status of their mobility and unpredictability of engagement in their occupations.

In specific response, India as a quasi-industrial social order needs a comprehensive legislative regulation that can provide minimum standards of wages, adequate environment for the job, social protection, job-related healthcare services and security, and so on. Moreover, in pursuit of collective welfare and interests, the alternative administrative mechanism for migrant laborers would endorse the rights of trade federations and their deeds to stipulate a structure for the resolution of industrial conflicts. In this regard, it has also been an essential requirement to make an alternative and uniform administrative mechanism for managing migrant laborers in different parts of the country.

References

Abi-Habib M, Yasir S (2020) India's Coronavirus lockdown leaves vast numbers stranded and hungry. The New York Times. Published on- 29th March 2020. https://www.nytimes.com/2020/03/29/world/asia/coronavirus-india-migrants.html. Accessed 19 May 2021

Auraiya road accident: Two more migrant workers die, toll rises to 26, (2020). Hindustan Times, published on- 17th May 2020, https://www.hindustantimes.com/india-news/auraiya-road-accident-two-more-migrant-workers-die-toll-rises-to-26/story-UclLZQfU1mIjSEEn74d3bP.html. Accessed 25 May 2021

Babu V, Saini S, Swaroop V (2020) Across the country, migrants still forced to walk thousands of miles, published on- 09th May, 2020. https://www.hindustantimes.com/india-news/similar-scenes-in-several-states-despite-new-trains/story-YL9qNcF315SA5o645aj1DP.html. Accessed 25 May 2021

Beniwal V (2021) Second coronavirus wave leaves another 7 million people jobless in India, *Business Standard*, published on- 3rd May, 2021. https://www.business-standard.com/article/economy-policy/second-coronavirus-wave-leaves-another-7-million-people-jobless-in-india-121050301553_1.html. Accessed 26 May 2021

Census of India (2011) *D-series: Migration tables. Government of India.* https://censusindia.gov.in/tables_published/D-Series/D-Series_link/DS-0000-014-D09-2001.pdf. Accessed 24 May 2021

Chauhan C (2020) 3.5 million new enrolments under MGNREGA, as 'distressed' workers return to villages. *Hindustan Times. published on- 22 May 2020.* https://www.hindustantimes.com/india-news/3-5-million-new-enrol ments-under-mgnrega-as-distressed-workers-return-to-villages/story-aDJHYz0vz1tSeLleIhVT 7I.html. Accessed 30 May 2021

Cheng E (2020) Roughly 5 million people in China lost their jobs in the first 2 months of 2020. *CNBC*. published on- 16th March 2020, https://www.cnbc.com/2020/03/16/china-economy-millions-lose-their-jobs-as-unemployment-spikes.html. Accessed 01 June 2021

Cox J (2020) Coronavirus job losses could total 47 million, unemployment rate may hit 32%, Fed estimates. *CNBC*. Published on- 30th March, 2020, https://www.cnbc.com/2020/03/30/coronavirus-job-losses-could-total-47-million-unemployment-rate-of-32percent-fed-says.html. Accessed 01 June 2021

Das G (2020) 136 Million jobs at risk in post-corona India. *Livemint*. Published on- 31st March 2020. https://www.livemint.com/news/india/136-million-jobs-at-risk-in-post-corona-india-11585584169192.html. Accessed 25 May 2021

Dhanya MB (2013). Fundamental Principles and Rights at Work and informal economy in India: Trends, Initiatives and Challenges, Ideas, VV Giri National Labour Institute https://ideas.repec.org/p/ess/wpaper/id5580.html. Accessed 30 May 2021

Dreze J (2020) Averting hunger during monsoon calls for bold food security measures. *The Indian Express*. Published on- 9th June 2020. https://indianexpress.com/article/opinion/columns/nrega-funds-migrant-workers-monsoonpds-scheme-6449293/. Accessed 29 May 2021

Dutt B (2020) There is a humanitarian crisis in India. Lift the lockdown, now. *Hindustan Times*. Published on- 15th May, 2020. https://www.hindustantimes.com/columns/there-is-a-humanitarian-crisis-in-india-lift-the-lockdown-now/story-RHG3Mjv7B3VrNszdbTZ1UI.html. Accessed 25 May 2021

Elsa E (2020) Coronavirus lockdown: 12-year-old Indian migrant worker walks 100 km, dies just 11km away from home. *Gulf News*. Published on- 21st April, 2020. https://gulfnews.com/world/asia/india/coronavirus-lockdown-12-year-old-indian-migrant-worker-walks-100-km-dies-just-11km-away-from-home-1.1587462168019. Accessed 25 May 2021

Gettleman J, Raj S (2020) Lionhearted' Girl Bikes Dad Across India, Inspiring a Nation. *The New York Times*. Published on- 22nd May 2020. https://www.nytimes.com/2020/05/22/world/asia/india-bicycle-girl-migrants.html. Accessed 25 May 2021

Gupta S (2020) 30% of migrants will not return to cities: Irudaya Rajan. *Times of India*. Published on-1 June 2020. https://timesofindia.indiatimes.com/india/30-of-migrants-will-not-return-to-cities-irudaya-rajan/articleshow/76126701.cms#:~:text=NEW%20DELHI%3A%20About%2030%25%20of,smart%20cards%2C%20reports%20Surojit%20Gupta. Accessed 19 May 2021

ILO (2020a) Almost 25 million jobs could be lost worldwide as a result of COVID-19, says ILO. Published on- 18th March 2020 in Press release. https://www.ilo.org/global/about-the-ilo/newsroom/news/WCMS_738742/lang%2D%2Den/index.htm, Accessed 01 June 2021

ILO (2020b) ILO Policy Brief on COVID-19. Published on- 07/05/2021. https://www.ilo.org/global/topics/coronavirus/impacts-and-responses/WCMS_739033/lang%2D%2Den/index.htm. Accessed 29 May 2021

ILO (2020c). ILO's response to the impact of the COVID-19 pandemic on workers and enterprises. Published on- May, 2020, https://www.covid19-evaluation-coalition.org/documents/ILO-COVID19.pdf. Accessed 12 June 2021

ILO (2020d) Road map for developing a policy framework for the inclusion of internal migrant workers in India. Published on- December, 2020. https://www.ilo.org/wcmsp5/groups/public/%2D%2D-asia/%2D%2D-ro-bangkok/%2D%2D-sro-new_delhi/documents/publication/wcms_763352.pdf. Accessed 12 June 2021

Jaiswal P (2020) Coronavirus update: A long walk home on empty stomachs for masked migrants, *Hindustan Times*. Published on- 27th March, 2020. https://www.hindustantimes.com/india-news/caught-in-the-middle-of-an-epidemic/story-kJhANZhDiU7SkU5OYUKJbO.html. Accessed 25 May 2021

Keeley G (2 April 2020) Spain sees historic rise in unemployment as nearly 900,000 lose jobs since coronavirus lockdown, *The Independent*. Published on- 2nd April 2020. https://www.independent.co.uk/news/world/europe/spain-coronavirus-lockdown-unemployment-job-loss-update-a9442146.html. Accessed 01 June 2021

Khanna A (2020) Impact of migration of labour force due to global COVID-19 pandemic with reference to India. J Health Manag. Published on- 11th August 2021. https://journals.sagepub.com/doi/full/10.1177/0972063420935542. Accessed 19 May 2021

Kone ZL, Liu MY, Mattoo A, Ozden C, Sharma S (2018) Internal borders and migration in India J Econ Geography 18(4): 729–759https://ideas.repec.org/a/oup/jecgeo/v18y2018i4p729-759.html. Accessed 19 May 2021

Macleans (2020) COVID-19: Canada layoff tracker. *Macleans*. Published on- 30th April 2020. https://www.macleans.ca/economy/covid-19-canada-layoff-tracker/. Accessed 01 June 2021

Mehrotra S, Parida JK (2019) India's employment crisis: Rising education levels and falling non-agricultural job growth, (CSE Working Paper 2019–04), Centre for Sustainable Development. Azim Premji University

Nahata P (2020) Coronavirus impact: Fear of contract job losses prompt cash transfer calls. *Bloomberg/Quint*. Published on- 19th March2020, https://www.bloombergquint.com/business/fear-of-contract-job-losses-prompt-cash-transfer-calls. Accessed 25 May 2021

Nandi S, Bhaskar U (2020) Migrants' deaths on the tracks a wake-up call for India. *Live mint*. Published on- 08th May 2020. https://www.livemint.com/news/india/migrants-deaths-on-the-tracks-a-wake-up-call-for-india-11588958792629.html. Accessed 25 May 2021

Nath D (2020) Govt. has no data of migrant workers' death, loss of job. The Hindu. Published on-14 September 2020. https://www.thehindu.com/news/national/govt-has-no-data-of-migrant-workers-death-loss-of-job/article32600637.ece, Accessed 25 May 2021

Nath D (2021) Migrant workers vulnerable again, say activists. The Hindu. Published on- 15th April, 2021. https://www.thehindu.com/news/national/migrant-workers-vulnerable-again-say-activists/article34330006.ece. Accessed 21 May 2021

Pandey V (2020) Coronavirus lockdown: The Indian migrants dying to get home. *BBC News*. Published on-20th May 2020. https://www.bbc.com/news/world-asia-india-52672764, Accessed 25 May 2021

Rajan S Irudaya, Sivakumar P, Srinivasan A (2020) The COVID-19 pandemic and internal labour migration in India: A 'Crisis of Mobility'. Indian J Labour Econ. Published on-20 November 2020. https://link.springer.com/content/pdf/10.1007/s41027-020-00293-8.pdf. Accessed 29 May 2021

Sanghera T (2020) Hungry, desperate: India virus controls trap its migrant workers. *Al-Jazeera*. Published on 2nd April 2020. https://www.aljazeera.com/economy/2020/04/02/hungry-desperate-india-virus-controls-trap-its-migrant-workers/, Accessed 19 May 2021

Shah A, Lerche J (2020) The five truths about the migrant workers' crisis. *Hindustan Times*. Published on-13 July 2020. https://www.hindustantimes.com/analysis/the-five-truths-about-the-migrant-workers-crisis-opinion/story-awTQUm2gnJx72UWbdPa5OM.html. Accessed 23 May 2020

Shantha S (2020) Gujarat Police to Probe Allegation That Migrant Workers Were Forced Into Container Trucks. *The Wire*. Published on- 07th April 2020. https://thewire.in/law/gujarat-police-inquiry-migrant-workers-container-trucks. Accessed 25 May 2021

Sharma YS (2020) Labour minister Gangwar clarifies his response on migrant workers in Parliament. *The Economic Times*. Published on- 16th September 2020. https://economictimes.indiatimes.com/news/economy/policy/labour-minister-gangwar-clarifies-his-response-on-migrant-labourers-in-parliament/articleshow/78142699.cms. Accessed 25 May 2021

Singh K (2020). Coronavirus outbreak: Ensuring water, hygiene facilities for migrant labourers can safeguard millions stranded during shutdown. Firstpost. published on- 4th April 2020, https://www.firstpost.com/india/coronavirus-outbreak-ensuring-water-hygiene-facilities-for-migrant-labourers-can-safeguard-millions-stranded-during-shutdown-8228331.html, Accessed 19 May 2021

The Economic Times (2020) More than 21,000 camps set up for over 6,60,000 migrants: State governments. The Economic Times. published on- 01st April, 2020. https://economictimes.indiatimes.com/news/politics-and-nation/more-than-21000-camps-set-up-for-over-660000-migrants-state-governments/articleshow/74920798.cms. Accessed 19 May 2021

The Times of India (2021) Weekly jobless rate has inched up to 14.7%: CMIE, published on- 25th May 2021. https://timesofindia.indiatimes.com/business/india-business/weekly-jobless-rate-has-inched-up-to-14-7-cmie/articleshow/82927861.cms. Accessed 26 May 2021

UN News (2020) COVID-19: impact could cause equivalent of 195 million job losses. says ILO chief. Published on- 08th April 2020. https://news.un.org/en/story/2020/04/1061322. Accessed 30 May 2021

Vyas M (2021) Double-digit unemployment rate makes a return. *Business Standard*. Published on 24th May, 2021. https://www.business-standard.com/article/opinion/double-digit-unemployment-rate-returns-121052400643_1.html. Accessed 30 May 2021

Weissmann J (2020) 6.6 Million Americans Filed for Unemployment Last Week, SLATE, puBlished on- 02nd April, 2020. https://slate.com/business/2020/04/unemployment-jobs-numbers-economy-coronavirus.html. Accessed 19 May 2021

Dr. Siddhartha Sankar Manna is an Assistant Professor in the Department of Political Science at the University of Gour Banga. The focal theme of his research is the analysis of international relations, political theory, democratic politics, and human development. He published more than twenty papers in reputed national and international journals.

Chapter 15
India's Tryst with the Second Wave of COVID-19: Politics and Policies at the Crossroads

Madhuri Sukhija

Abstract The world has been a witness to a series of viruses, be it the Spanish Flu, the Avian Flu or the 2009 H_1N_1 (Swine Flu), the AIDS Epidemic, the Ebola, and the Nipah Virus, so that the question today is not whether the world will see the outbreak of a new pandemic or not, but rather when COVID-19 virus or SARS-CoV-2 is an acute respiratory syndrome coronavirus that has taken the world by storm. In India, it is the second wave that has brought the country down on its knees. The explosive number of cases in the second wave raises an obvious question mark on handling the pandemic this time. Many reasons can be cited for the hike - from the double mutation of the virus to the politics and policies engaged by the government in coping with the second wave. It also becomes imperative to take a good look at the existing management responses and think of the road ahead.

Keywords Virus · Mutation · Spike · Politics · Policies · Coping strategy

15.1 Tracing the Trajectory of the Second Wave of COVID-19 Outbreak in India

There can be no finality of data and statistics in undertaking this particular exercise since it is a work in progress. Health care is serious business, and reframing the management and organization of health care is a task that needs urgent attention (Mintzberg 2017). The second wave of COVID-19 in India made its debut in the middle of February 2021 and was prevalent in interior Maharashtra in Akola and Amravati (Menon 2021a, b). Surprisingly, entire families were being infected. A few days later, Punjab followed suit with cases associated with the so-called UK variant of the coronavirus, more technically named "B.1.1.7" (Menon 2021a, b). Delhi became the next victim, and soon case numbers began to spiral across the country.

M. Sukhija (✉)
Mata Sundri College for Women, Delhi University, Delhi, India

S. Saha et al. (eds.), *Economic and Societal Transformation in Pandemic-Trapped India*, New Frontiers in Regional Science: Asian Perspectives 55,
https://doi.org/10.1007/978-981-16-5755-9_15

Early April saw close to one lakh cases, and towards the end of April, they were more than four times the value at the first peak. The month of May also proved to be a nightmare, as in the first half, one saw close to four lakh people getting affected daily, and nearly 4000 deaths per day.[1] The entire COVID-19 management system in India has left many questions unanswered. In recent weeks, five state elections in the preceding weeks violated all norms of COVID appropriate behavior, with massive political rallies, roadshows, and campaigns.

During the first wave last year, India witnessed some difficult decisions like a centralized lockdown and that too at a four-hour notice. This centralized management did prove useful and helped to arrest the spread and garner the requisite capacity to handle the caseload during the process of unlocking. Thus, it did save many lives. However, there was a change in strategy in groping with the second wave for political and economic reasons. Moreover, health happens to be a State subject in India.

15.2 How Is India's Second COVID Wave Different from the First?

Unlike the first wave, this time, the virus spread like wildfire. Since 15th February, the number of cases jumped multifold. From 10,000 daily new cases to 80,000, the spread occurred in 40 days during the second wave. During the first, this jump took double the time, that is, 83 days (Shetty 2021). Officially, by mid-May, about 23 million infections had been confirmed, and roughly two and a half lakhs people had lost their lives. India seemed to be the epicenter, with a caseload amounting to 400,000 a day (Loke 2021). As mentioned earlier, this can be attributed to newer variants that are more infectious than the original virus. For instance, variant B.1.617 is sometimes called "the double mutant." In Punjab and Haryana, almost 80% of the infection is caused by the UK variant. However, in Delhi, there is the UK variant and the Double Mutant, as well. In Maharashtra, there is the latter (Ghosh 2021).

Viruses mutate, and COVID-19 is no different. There are variants from the UK, South Africa, Brazil, and Japan, among others. They follow different patterns, are more contagious, and evoke different efficacy rates from the vaccines being administered. On 24th March, the government made an alarming announcement that 771 variants of concerns (VOC) were prevalent across the country (Shetty 2021).

In addition, the process of unlocking after the first wave was casual. For economic reasons, unlocking was a natural process, but the rapid pace at which it was carried out coupled with a degree of casualness proved fatal. Markets, malls, and marriage avenues were buzzing with activity. Experts say that nearly 2/3rd of the population is

[1] News coverage in Indian news channel News 18, available under this URL: https://www.news18.com/news/india/delhi-reports-more-COVID-cases-deaths-in-april-may-than-since-the-beginning-of-pandemic-3751346.html.

asymptomatic, and they are the largest carriers of the virus, and they transmit efficiently in a closed indoor setting because these people do not isolate themselves in a home setting in the current wave, the nature of marking the containment zone has changed (Oran 2020). Earlier, if one floor were infected, the entire building would be cordoned off, but this time the micro-containment zones are less strict and with more casual monitoring.

The carelessness exhibited on the part of Central and State Governments and the blame game of passing the buck further worsened the situation. The election rallies were not far behind. The Election Commission of India's delayed response to restrict campaigning for the impending two phases of the West Bengal Assembly election was certainly not becoming of the EC. The extended election cycle in the State during the pandemic had already done the damage. The Calcutta High Court had to intervene, thereby asking for an action report on the safety measures adopted in the context of the spreading pandemic. The Election Commission did respond a little late by ordering the cancellation of all rallies and roadshows and restricting meetings to 500 people, a considerable number by all standards. Besides, approval was given to the Hindu festival (Kumbh Mela), which attracted millions of worshipers. Between April 10 and 14, more than 1600 positive cases were detected at the Kumbh Mela (Jaffrelot 2021).

In all the election-bound states, the election campaign had a direct bearing on the spike in cases from late March (Jaffrelot 2021). Winning Election surpassed all else. In March, false assurances to the public that India had reached the pandemic's "end game" further derailed the strategy to fight the pandemic.

What was evident was the inadequacy of the system in the management of the crisis. COVID-19 management was not carried in unison between the national and state level, resulting in leakages and non-availability of resources to patients requiring it most. From mid-April onwards, the condition of some states like Maharashtra, UP, and Delhi was awful. Precious lives were lost due to shortages of medical equipment like oxygen, hospital beds, ventilators, and other appliances. Cases of hoarding/black-marketing of health appliances, Oxygen cylinders, and medicines resulted in the needy being deprived.

The demand for medical oxygen had also shot up by 18% across 12 states, which accounted for 83% of India's active cases (Kaunain 2021). What is disturbing is that people were dying because of the shortage and exorbitant prices of ambulances in Delhi (Asthana 2021).

This is the first mode of help that can be given to a critical patient, and this can be easily arranged with flatbed vehicles with movers and packers and makeshift ambulances by putting few mattresses and oxygen cylinders inside provided the RTOs are tasked for it. Since the government did not foresee the intensity of the second wave, the supply of vaccines fell short. Ironically, on 12th April, India administered 3.7 million doses of vaccine, and two weeks later, it came down to three million doses a day. Ever since the vaccine drive was expanded to include the18 plus category on 1st May, the maximum number of daily doses daily administered has further come down to 2.4 million. This is despite the fact that the daily new cases approximated 380,000 and deaths close to 4000 a day. Unlike the first wave, the pandemic caught on in rural India (Pandey 2021).

India is the world's largest pharmaceutical industry and is also one of the world's leading vaccine manufacturers, but its citizens have had a raw deal for vaccines. The second COVID wave could certainly have been managed better if the vaccine policy was in place. The Vaccine procurement Policy adhered to by the central government has not aligned with India's explosive population. What is absurd, since the central government was mindful of India's population, it should have placed the orders earlier. Besides, it placed orders in such small installments that made it impossible for the companies to plan production. To begin with, requisition for ten million doses were ordered from the Serum Institute of India at Rs 200 per dose, which was later revised to 200 million doses that brought the price down to Rs 150 per dose. The third order was only placed only as late as 26th April, and that too for another 150 million doses (Rao 2021a, b).

The policy of the government could hardly be justified in the thick of the deadly second wave. Furthermore, the government's policy of resorting to a digital mode for acquiring vaccination is certainly a good option, but then it leaves out a voluminous number of the population who do not have access to digital technology. On 1st May, the first day of the new policy (registrations mandatory for the 18–44 age group), 133 million individuals did register, causing overcrowding at the registration points with COVID safety measures thrown to the winds. What was disappointing was, at several centers, there were no vaccines. As a result, a few states diverted the vaccine to the younger age groups when the center had delineated them for the 45+ group (Rao 2021a, b). Amidst the chaotic second wave, the Central government shifted responsibility for vaccination largely to the States, and that too opened up vaccine pricing to market forces, at a time when vaccines ought to be available for free to one and all. Right now, the production is 60 million doses per month. Both the companies [Serum Institute of India (SII) and Bharat Biotech (BB)] would be able to raise their productions from July to115 million doses per month (Thacker 2021).

Ironically, Serum Institute of India (SII), which has received the technology free of cost from AstraZeneca, as well as funding from the Gates Foundation to expand manufacturing capability,[2] is now hankering for profits and indulging in differential pricing, which is certainly not in tune with expenditure incurred on research and development R&D expenditure. In the same vein, Bharat Biotech also put forth that it got favorable treatment in getting all the raw materials and everything from the National Institute of Virology, Pune. The ICMR also chipped in. Even without full submission of data, both these companies were given tremendous relaxations of emergency authorization. Moreover, pricing the vaccines at such a high rate is

[2]Indian English daily, The Hindu's Special Correspondent reported on seventh August, 2020 on "Serum Institute to make up to 100 million COVID-19 vaccine doses for India, low-income countries by 2021," available here: https://www.thehindu.com/news/national/serum-institute-to-make-up-to-100-million-COVID-19-vaccine-doses-for-india-low-income-countries-by-2021/article32295855.ece.

certainly not encouraging for the states and makes it the reserve of the rich. Furthermore, the differential pricing policy of the vaccines has proved problematic for the states, which now have to compete not only with each other but also with private providers of vaccines. At the end of the day, vaccines are a public good.

Abiding by the principle of Vasudhaiva Kutumbakam (i.e., the *world is one family*),[3] India opted to support the UN-backed COVAX program and thereby provide 2 billion vaccine doses to low- and middle-income countries. This was followed by a magnanimous goodwill gesture by the PM in January 2021, that Mongolia, Oman, Myanmar, Philippines, Bahrain, Maldives, Mauritius, Bhutan, Afghanistan, Nepal, Bangladesh, and Seychelles would get vaccines free of cost (Chandna and Basu 2021).

According to Christopher Jafferlot, in 4 months, in tune with the "Vaccine Maitri" (i.e., *Vaccine Friendship*)[4] initiative, India exported 64.4 million doses of vaccines, the break up being 35.7 million on a commercial basis, 18.2 million through the COVAX program and 10.4 million as donations. "At the beginning of April, only 120 million people had been vaccinated in India" (Jaffrelot 2021).

India could not do justice on both fronts. India had exported vaccines to 80 countries, but ironically, it did not have the vaccines to counter the deadly second wave for itself. So, as a corollary, India banned vaccine exports which have slowed down the vaccination drive in several African countries, which were banking on doses produced by the Serum Institute of India. Unfortunately, this reversal decision has not sent the right message and has affected the image of India, with many countries realizing that India's new policy would impact their vaccination plans.

15.3 Where Do We Go from Here

15.3.1 Public Spending on the Social Sector, Especially Health, Has to Increase Drastically

India spends a little over 1% of the GDP on health, and that is why our crisis management system vis-à-vis a medical emergency has fallen short. Universal health care is a distant dream (Mondal 2021). The second wave of COVID-19 is one of a kind and required a different response altogether. To begin, more stringent controls were required in areas designated as "hot-spots." The situation got aggravated due to unreliable testing and under-reporting of cases. Data inputs have been manipulated, and the loss of human lives has been phenomenal. Strategies such as testing and

---

[3] *Vasudhaiva Kutumbakam* is a Sanskrit phrase found in Hindu texts such as the Maha Upanishad, which means "the world is one family."

[4] *Vaccine Maitri* is a humanitarian and commercial initiative undertaken by the Indian government to provide COVID-19 vaccines to countries around the world. The government started providing vaccines from 20 January 2021.

contact tracing have to be more vigorous, focusing on gathering data on the numbers of those infected in a standardized way so that these numbers are comparable across time and space. More importantly, the Central and State governments must cooperate in gauging how many cases are to be expected per million population. A strong surveillance system continuously reporting the number of cases can lead to early identification of an impending wave, and thereby timely measures can be taken to prevent the pandemic from spreading to other areas. "Concurrent genomic sequencing in real-time in the fixed proportion of samples will give us an idea of the likelihood of the variants causing several outbreaks" (Asthana 2021).

The outbreaks in Kerala, Punjab, and Maharashtra were warning signs. If these outbreaks were noticed from genomic sequencing results, India could have expedited local lockdowns and imposed harsh restrictions in high-burden areas to stop the deadly second wave from spreading. Excessively splurging on projects like Central Vistas must be rethought of, keeping in mind the risks and costs involved. The health care system has to be given the top priority. In a democracy, effective policies require reaching out to the broadest pool of expertise available. The more daring question is getting the governance right.

15.3.2 A Proactive Vaccination Policy

India has earned the title of being the largest producer of vaccines globally, but for a large majority of the population, vaccination is a far cry. Early this year, the central government should have got in touch with the vaccine manufacturing firms in the West so that they could have collaborated with Indian firms under the "Make in India" program. Closer home, India needs to speed up manufacturing all vaccines that have got the green signal from various regulatory authorities through a single-window clearance. Close to seven companies obtained contract manufacturing licenses from Sputnik [Russian vaccine] for production. The vaccine is ready for use in India, and its production should also be facilitated at the earliest. It is not only a question of manufacturing the vaccines but also administering the vaccine as quickly as possible in an organized manner. Keeping in mind the volatility of the virus and with new variants emerging, it is imperative to update the vaccines.

Since the deadly second wave also impacts the young, the government needs to go all out to provide vaccine equity to all irrespective of age (18 years onwards). To add to this, vaccine pricing cannot become the monopoly of vaccine manufacturers or the central government; this defeats the very purpose of universal vaccination and creates major obstacles for the state government to undertake the urgent task of immunization of nearly 600 million people. Interestingly, the Union government has opted for vaccinating the rest of the population which is close to 300 million (Menon 2021a, b). Soon, there is every possibility that children too will need to get vaccinated, and then the financial burden will fall on the states again. With vaccination being the safest way to end the pandemic, undertaking any exercise that leaves a large population unprotected will enormously cost the country enormously in terms of lives and livelihoods.

"Union government seems to be quite satisfied with the vaccine rollout drive as well as with allocating 35000 crores for COVID-19 vaccination in this Budget" (Chakravarty 2021). According to the health secretary, Mr. Luv Agarwal, "the Centre had provided about 17.01 crore vaccine doses for free to the States, and the people had already got them. In addition, in collaboration with the States, 25.59 lakh doses were also given to the 18-44 age group" (Pandey 2021).

Factually speaking, the central government will spend less than 10,000 crores to take care of free vaccination for all above 45 years (Chakravarty 2021). By doing so, can the central government absolve itself of its responsibility? Pricing has to become more transparent and accountable, and the States have to be given leeway to collectively bargain for a lower price as well as determine timelines to receive supplies.

Since 2/3rd of the population has not even got the first dose, so one cannot deny that somebody who starts vaccination today will take at least 2 months to develop protective immunity. This means the vaccination drive has to be sustained over months; only then can the surge be reversed. If this world is to be a healthy family of nations, then vaccine production has to be decentralized at a fast pace. A handful of countries cannot make vaccines for everyone. Vaccine equity amongst the nations of the world is another major challenge. Countries like the USA have enough reserves to vaccinate five times their population, whereas vaccination is still a luxury for many others. The sooner the world understands that no one is safe until everyone is safe, the better it would be. The only way to breaking the transmission is extremely important is through regional lockdowns, citywide lockdowns, and local lockdowns. The fact that the Honorable Supreme Court of India appointed National Task Forces to control the allocation and distribution of oxygen justifies the need for better management l of scarce resources.

15.3.3 Robust Public Health Workforce

The rapid spread of SARS-CoV-2 shows us the importance of timely and efficient public health responses. We can only fight better when we have a battle-ready public health workforce. Unfortunately, our health system is on the verge of collapse, thanks also to the minuscule sum of the GDP that we are spending on the health sector. Doctors and nurses are overburdened, and the public health workforce is understaffed. Front-line workers in public health can contribute immensely in contract-tracing, mobilizing people to avail primary health care services, including vaccination. However, the front-line public health workforce is hardly visible in urban areas, what to talk of rural areas. "While critical care capacity (oxygenated beds, ICU's) is limited in rural areas, the urban-rural divide apart, the country needs to reconfigure the health systems to ensure that one Accredited Social Health Activist (ASHA) worker is hired for every 1,000 people, an Auxiliary Nurse

Midwife (ANM) and nurse practitioner are hired for every 5,000 people and a hospital with at least 100 beds, including beds with emergency and critical care services, caters to a population of 30,000" (Giridhara 2021)

15.3.4 Grappling Efficiently with the Health Emergency

A few measures could have helped, for instance, declaring it as a "National Health Emergency" and revamping the existing system and infrastructure of the National Disaster Management Authority (NDMA) under Prime Minister. The emergency required plugging of loopholes, greater transparency, and better coordination of policy between the center and the states to improve capacity building, medical resource generation, and framing policy guidelines. Currently, various agencies are doing their bit but in silos; without any visible unified approaches, the formation of a National Crisis Management Committee, bringing all stakeholders on board, would have been a step in the right direction. This would involve bringing in representatives from the health, and Home Ministries, Intelligence agencies, expert professionals from various fields, medicine, manufacturers of vaccines, administrators, public and private players, and Defense Services involved in COVID-19 management. It is heartening to see that the three wings of the armed forces are contributing immensely to the national effort in the fight against the COVID-19 pandemic. The Air Force and Navy have utilized their good services in transportation of health-related equipment from abroad and within the country. In the early stages of the second COVID wave, the Indian Army has opened many additional makeshift COVID hospitals, where civil patients are also attended to. Retired medical personnel, too, have joined in and add on to the effort. Trained to tackle grave emergencies, the armed forces have assisted in incorporating logistics expertise of services and improving the supply chain management during such instances.

15.3.5 Taking a Cue from Kerala and Karnataka Experience

Learning best practices from shared experiences can also help in drafting adequate policy responses. Many states like Karnataka and Kerala have managed the second wave better than the rest of India. The Karnataka government ordered private hospitals above a certain size to reserve 2/3rd of their beds for COVID-19 patients, and expenses would be taken care of under a public scheme. Kerala did fairly well in controlling the epidemic of Nipah through local action in the past. States are the best repository of political experiments. Kerala is an example to be emulated as it combines the values of both humanity and science. It holdss the position of being number one as far the human development index ratio is concerned, or the accomplishment of SDGs is concerned.

The creation of an Indian National Health Service has to be given immediate consideration to deal with pandemics of such magnitude. Modalities can be worked out like funding, universal health care entirely from general taxation. In certain rural areas, the doctor-population ratio is over mind-boggling. A lot of money is spent on medical expenses and is one of the reasons for personal debt in India. The current crisis is reason enough for introspection, to reassess, publish spending on social sectors like health.

How best to minimize shortages, build upon reserves, and improve overall efficiency are questions with no easy answers. Centralized planning with decentralized execution at the State level along with the digitized allocation of meager resources is the need of the hour. Hoarders/black marketeers have to be taken to task. Foreign help is certainly welcome. However, the increasing caseload in the future can only be met with self-reliance in capacity building to counter the pandemic at an unprecedented speed. We need to learn from the crisis, and our crisis management policies have to be right so that we are better equipped to meet the third wave.

15.3.6 Planning Sensibly for the Future

It would have made a difference if the central government had made provision for a meticulously detailed "COVID-19 Emergency" Budget for the year 2021–2022, after taking into account previous years' experience with the first wave and paying heed to the advice rendered by experts. This also would have determined our domestic vaccine policy and Vaccine Maitri (the extent to which India could extend a helping hand to other developing nations). Various Indian states fell short of vaccines amid the second COVID wave. Presumably, the Finance Minister should have factored in some details before finalizing the 2021–2022 budget. Keeping in mind the size of India's population, the costs of the vaccines could have been multiplied by the total number of people to calculate the total amount needed for vaccinations.

In the early stages itself, it would have proved beneficial in the long run if the Health Ministry had procured the vaccines directly and worked out one uniform price for each type of vaccine. Thereafter the vaccines could have been distributed to the States who would have administered them in the best possible way.

There is an opinion that is fast emerging that emerging infectious diseases are zoonotic in origin. Being mindful of the boundaries of animals and preserving the ecosystem has to be seriously considered. Adopting a uniform Health' agenda wherein environmental health and animal health are to be given as much priority as human health will help in the long run.

15.4 Ray of Hope

It also becomes imperative to mention that the government undertakes some positive steps. Home Ministry Additional Secretary Piyush Goyal said, "There had been a seven-fold increase in the supply of liquid medical oxygen, from 1,320 MT in March to 8,943 MT. On 9th May, Jumbo hospitals with 12,400 beds were being set up, with the supply of gaseous oxygen from refineries and power/steel plants, while one lakh oxygen concentrators were being procured under the PM Cares Fund" (Pandey 2021).

However, following a surge in the cases in April, the loss of lives in the second wave compelled the center to monitor the allocation process to ensure rational distribution. This would prevent the hospitals from directly procuring oxygen from suppliers. On 7th May, positive cases amounted to 4 lakh 14 thousand. Three weeks later, there were 2 lakh, eleven thousand cases. As a result, the mortality rate has significantly come down. The recovery/positivity rate has crossed 90% from 81 as of third May.[5]

The Prime Minister has laid stress on expediting the National Digital Health Mission to benefit from this digital technology. External Affairs Minister Jai Shankar's visit to the USA in the last week of May reaffirmed a collaborative approach of both countries on the issue of COVID-19. The Indian government has also been in talks with Moderna and Johnson & Johnson about the vaccines.[6]

The central government, off late, has decided that it alone will procure the vaccines from the manufacturers and will provide them free of cost. However, the vaccines would also be available in private hospitals to those who wish to avail them.

15.5 Conclusion

Debora MacKenzie, in her book *"COVID-19: The Pandemic that Never Should Have happened,"* points out, "Fighting this pandemic and preventing the next one will take political action of all kinds, globally, from governments, the scientific community, and individuals-but it is possible" (MacKenzie 2020). India needs to have a health system that can respond to pandemics as they emerge because even war is not as big a threat as pandemics are. Pandemics can completely wipe out populations. Building reactive systems for each wave and each pandemic and the

[5] See the Press Information Bureau (PIB) Bulletin on COVID-19 dated 10th June 2021, available under the URL: https://pib.gov.in/PressReleasePage.aspx?PRID=1727010.

[6] Indian English daily Hindustan Times' Correspondent reported on 3rd June 2021 on "In talks with Pfizer, Moderna, J&J on manufacture of vaccines: Foreign Secretary." URL: https://www.hindustantimes.com/india-news/in-talks-with-pfizer-moderna-j-j-on-manufacture-of-vaccines-foreign-secy-101622723651469.html.

ensuing waves will be next to impossible. The legitimacy of governance and leadership is very important in countering the pandemic.

New Zealand did fairly well under a young and determined prime minister, acted with foresight. The United States mobilized vast resources and set a lofty goal of 100 million vaccinations in 100 days.[7] Surprisingly, it went much beyond its target. In a democracy operating within checks and balances, decisions can be slow, deliberative, polarized. China took barely 70 days to tide over and was back on track. However, India cannot catch up that fast, but a strong leader can certainly undertake hard decisions, and a hard state does not mean an illiberal democracy. Timely foresight on the part of democratic institutions can undoubtedly prove helpful. The elongated and unreasonable schedule of the Bengal election during the second wave of the pandemic could undoubtedly have been avoided. The Vaccine Maitri campaign, under which over 64 million vaccine doses have been distributed to 85 countries across the world, has been projected in official statements as a government initiative to combat vaccine nationalism and provide global leadership in combating COVID-19 (Basu 2020).

The rise of the second wave and the public backlash when it became known that more vaccine doses had been exported than used domestically forced the government to stop exports. The stopping of exports has also brought supplies to poorer countries under the COVAX program, for which the Serum Institute is the major supplier, to a halt. The abrupt reversal of policy did hamper our image as the vaccine Guru of the world, but then India had its domestic compulsions.

The pandemic has exposed the fissures in our governance structures, and it is the moral imperative of the State to look to the welfare of its citizens. In this COVID era, all democracies need to renegotiate the relationship between the State and civil society. National resources should be used to revamp public delivery systems to fulfill citizen entitlements with a sense of fairness and also to bring the private sector and informal sector on board in the domain of the crisis management system. Yet again, it is the civil society, voluntary organizations, community-based interventions, and psychosocial interventions that have proved to be a silver lining during the second wave.

In the current situation, both speed and scale are important. The supply chain shortage has to be countered by bringing in domestic and international vaccine manufacturers in the loop, and additional health care personnel should be employed, shortage issues are resolved. In the current scenario, the second wave is receding, but the shortage issues are stark, and till the time vaccines meet the requirement, the next few weeks should be utilized by the states to focus on better micro-planning and mobilization efforts as well as fill up existing vacancies of health care personnel so that dedicated human resources are available.

The entire private sector has to be incentivized to partner with the government when vaccination sites should be increased substantially, and daily vaccinations

[7] See BBC News report on 9th December 2020 entitled "COVID: Biden vows 100 m vaccinations for US in first 100 days." URL: https://www.bbc.com/news/world-us-canada-55238092.

should exceed ten million from the present average of three million. The central government needs to get realistic and put peoples' welfare uppermost. Vaccines are a public good and should be the fundamental right of every citizen. The sooner the central government revises its policy, the better it would be.

Richard Horton poses an important question: at the beginning of January 2020, when the world knew that a dangerous new virus was causing a devastating human tragedy in China, why did it ignore the warnings (Horton 2020).

The result was that even the most developed countries paid dearly in terms of the loss of human lives. In India, it is the second wave of 2021 that has changed India seminally and irreversibly. One could say that the second wave of COVID-19 in India is a moment of realization and is a significant departure from the illusionary confidence that was nurtured in the first wave. The lessons learned from the crisis in the second wave are unforgettable ones and present an opportunity to ward off impending waves.

References

Asthana SB (10 May 2021) India's COVID-19 Challenge: A Crisis Management Perspective, Financial Express

Basu R (ed) (2020) Democracy and public policy in the post-COVID-19 world choices and outcomes. Routledge, New Delhi, p 204

Chakravarty P (12 May 2021) COVID mishandling foretold in the Budget. The Hindu. Available in https://www.thehindu.com/profile/author/Praveen-Chakravarty-1330/. Accessed 24 May 2021

Chandna H, Basu N (18 January 2021) India to export Covid vaccines free of cost to its neighbours as a 'goodwill gesture. The Print https://theprint.in/health/india-to-export-covid-vaccines-free-of-cost-to-its-neighbours-as-a-goodwill-gesture/587874/

Ghosh B (03 May 2021) COVID-19 second wave: can India find a way out of its health nightmare? Business Standard. Available in https://www.business-standard.com/article/current-affairs/india-s-second-coronavirus-wave-what-went-wrong-and-is-there-a-way-out. Accessed 24 May 2021

Giridhara RB (02 May 2021) Battleplan: Priority actions for second wave and beyond, Times of India. Available in http://timesofindia.indiatimes.com/articleshow/82350218.cms?utm_source=contentofinterest&utm_medium=text&utm_campaign=cppst. Accessed 24 May 2021

Horton R (2020) The COVID-19 catastrophe: What's gone wrong and how to stop it happening again. Polity Press, Cambridge, p 24

Jaffrelot C (27 April 2021) India's Second Wave: A Man-Made Disaster? BLOG - Available in https://www.institutmontaigne.org/en/blog/indias-second-wave-man-made-disaster. Accessed 24 May 2021

Kaunain S (5 May 2021) What has changed in the second wave of COVID-19 in India? Indian Express. Available inl https://indianexpress.com/article/explained/explained-whats-changed-in-second-wave-7289002/. Accessed 26 May 2021

Loke A (07 June 2021) What to know about India's coronavirus crisis. The New York Times. Available in https://www.nytimes.com/article/india-coronavirus-cases-deaths.html. Accessed 9 June 2021

MacKenzie D (2020) COVID-19: the pandemic that never should have happened, and how to stop the next one. The Bridge Street Press, London

Menon AK (10 May 2021b) COVID-19 Vaccination Drive: Many Challenges. India Today. URL: https://www.indiatoday.in/magazine/cover-story/story/20210510-vaccine-vagaries-1796994-2021-05-01

Menon G (19 May 2021a) COVID-19 and Indian Exceptionalism. The India Forum. Available in https://www.theindiaforum.in/article/COVID-19-and-indian-exceptionalism. Accessed 29 May 2021

Mintzberg H (2017) Managing the myths of health care: bridging the separations between care, cure, control, and community. Berrett-Koehler, Broadway

Mondal D (02 January 2021) India spends just 1.26% of GDP on public healthcare. Sunday Guardian. URL https://www.sundayguardianlive.com/news/india-spends-just-1-26-gdp-public-healthcare. Accessed 20 July 2021

Oran D P (01 September 2020) Prevalence of asymptomatic SARS-CoV-2 infection. Ann Intern Med doi: https://doi.org/10.7326/M20-3012

Pandey DK (12 May 2021) Coronavirus Cases rising in 15 States; positivity rate at 21%, says ICMR chief. The Hindu. Available in https://www.thehindu.com/news/national/coronavirus-cases-rising-in-15-states-positivity-rate-at-21-says-icmr-chief/article34538262.ece. Accessed 24 May 2021

Rao S (14 May 2021a) Vaccine conundrums. The India Forum. Available in https://www.theindiaforum.in/article/vaccine-conundrums. Accessed 24 May 2021

Rao S (14 May 2021b) Vaccine conundrums. The India Forum. Available in https://www.theindiaforum.in/article/vaccine-conundrums. Accessed 24 May 2021

Shetty S (12 April 2021) What is driving India's second COVID-19 wave. Available in https://transfin.in/what-is-driving-india-second-COVID-19-wave-how-is-it-different-from-the-first-wave. Accessed 29 May 2021

Thacker T (12 April 2021) Bharat Bio-Tech to raise Covaxin's production to 12 million a month by July. The Economic Times

Dr. Madhuri Sukhija is an Associate Professor in the Department of Political Science at Mata Sundri College for Women (Delhi University). She has had a consistently brilliant academic record. She topped the Delhi University in M.Phil. Political Science. She was the recipient of the ICSSR Fellowship for going abroad for her doctoral research. Dr. Madhuri has published a book titled "*Gandhi Between Tradition and Modernity.*" Besides, she has a number of publications to her credit in both National and International Journals, the latest one being on the National Education Policy 2020: A Perspective on Higher Education, published in the *Bihar Journal of Public Administration*. She has spoken widely on various forums and has been a keynote speaker at various Conferences and Faculty Development Programs.

Part V
Sectoral Impact and Responses

Chapter 16
Beekeeping Livelihood at Stake Amidst the Pandemic Outbreaks: A Study on the Migratory Beekeepers in West Bengal

Anil Bhuimali, Sanghamitra Purkait, and Manish Baidya

Abstract The COVID-19 pandemic accelerates several risks among human civilization as the virus separates lives from livelihoods. COVID-19 induced lockdown affected modern beekeeping, which is predominantly migratory in nature and also created many problems of beekeepers' lives and livelihoods of West Bengal during the lockdown period and after that. They had faced varieties of problems in their migrated fields. Thus, this study seeks to analyze the widespread problems faced by the beekeepers of West Bengal during the lockdown and analyze the situation before lockdown, mainly based on primary data collected through the snowball sampling technique. Statistical tools have been applied for the analysis of the data. The study also offers suitable possibilities to convalesce the beekeepers during the post-lockdown period. Loss of migration period is a big issue for beekeepers regarding future production. It is essential to realize that migration is part of the beekeeping profession and awareness to all human beings for these eco-friendly and agriculture support activities that the beekeeping industry.

Keywords Beekeeping · Colony contraction · Hive-holding capacity · Honey flow · Lockdown · Migration pattern

---

A. Bhuimali
Raiganj University, Raiganj, Uttar Dinajpur, West Bengal, India

S. Purkait
Department of Geography, Diamond Harbour Women's University, Diamond Harbour, South 24 Parganas, West Bengal, India

M. Baidya (✉)
Department of Commerce, Shyampur Siddheswari Mahavidyalaya, Ajodhya, Howrah, West Bengal, India

S. Saha et al. (eds.), *Economic and Societal Transformation in Pandemic-Trapped India*, New Frontiers in Regional Science: Asian Perspectives 55,
https://doi.org/10.1007/978-981-16-5755-9_16

16.1 Introduction

The usefulness of honey bees and beekeeping are known to men from ancient times. Modern beekeeping is on its track that is blessed with discovering the movable frame hive in 1851 by Lorenzo Langshoth (Baidya 2017). In India, beekeeping was introduced in 1882 in the then undivided Bengal (Baidya 2017). Morden beekeeping at present is an essential component of agriculture, rural development, and socio-economic development all over the World. There is a silent and mutual relationship between bees and the beekeepers. So, the beekeepers migrate to different places in different seasons to collect honey. They are not migrant workers. Their profession is migratory in nature. Generally, traders, cooperative societies, different companies through their agents and intermediaries collect the honey from the beekeepers throughout the year. Large numbers of bee colonies are moved among different crops in regional and national areas. These migratory colonies of beekeepers are depended on the distance traveled, and the crops visited. In most cases, colonies are transported by trucks to a series of monoculture crops for different months, except the rainy season over the years.

The floral sources and their blooming period are the leading parameter and tools for modern beekeeping in India. The bee plants are grouped into forest plants, horticultural/agricultural plants, and pastures and hayfields. The nectar and pollen-yielding plants are significant in certain areas due to their abundance and their blooming period. The bee floral sources like Litchi (Nephelium litchi), Mustard, Rapeseed (*Brassica sp.*), Sunflower (*Helianthus annus*), Khair (*Acacia catechu*), Jamun (*Syzygium cumini*), Plectranthus rugosus, Rubber (*Hevea brasiliensis*), Shisham (*Dalbergia sissao*), Eucalyptus (Eucalyptus sp.), Berseem (*Trifolium alexandrinum*), Karanju (*Pongamia pinnata*) contributes to abundant nectar flow, hence migratory beekeeping is practiced to exploit honey production. The migratory routes for the exploitation of several honey flows that take place in different locations at different periods are the key factor for commercial beekeeping (Thomas et al. 2001).

However, this year, due to the COVID-19 pandemic, the beekeepers of India, including West Bengal, faced numerous problems. Like other migrant workers, They are also poorly treated in different parts of India during the lockdown. The bee-keepers of other parts of the World, especially China, France, and Canada, had also faced similar problems (The Economist 2020; Green 2020; Priyadarshini 2020). The lockdown hampered the entire migration plan of the beekeepers of West Bengal. It became challenging to maintain their life and livelihood simultaneously.

16.2 Objectives of the Study

Finding out the problems and prospects of apiculture activity as a whole from the pre-to-post lockdown period is the main objective of this study. Therefore, the current study is focused on the following objectives:

- To analyze the situations of beekeepers before the lockdown.
- To analyze the overall problems faced by the beekeepers of West Bengal due to the pandemic induced lockdown.
- To offer suitable possibilities for convalescing the beekeepers during the post-lockdown period.

16.3 Study Area

Two separate survey zones were chosen for collecting the samples to examine the beekeepers' situations in the state of West Bengal. The first one, which comprises the three districts in the Southern Part of West Bengal (geographically, the South Bengal), namely West Midnapore (or Paschim Medinipur), Bankura, and Purulia districts, will be called South Bengal Study Zone (SBSZ) herein forth. The second zone, with three districts in the Northern part of the state (geographically, the North Bengal), namely Malda, North Dinajpur, and South Dinajpur districts, comprises the North Bengal Study Zone (NBSZ).

The above SBSZ had been chosen as eucalyptus flowers blossomed there. The eucalyptus plants give plenty of honey with uncommon taste and are highly valued as present markets prefer "Ayurvedic" products. More importantly, the trees are devoid of pesticides or fertilizers. As such, the honey produced is organic in character. As far as the honey flow is concerned, it starts from October and ends around January the following year. Therefore, the beekeepers get an average of 3 months of honey flow periods every year, and many beekeepers from different parts of West Bengal flock to these regions during this period in order to collect this unique quality of honey.

On the other hand, the NBSZ had been selected as those regions fall under the major honey flow zones of West Bengal. In this region, the honey flow period starts from September and ends around April the following year. Furthermore, the region is renowned for mustard, mango, and Litchi honey, and the beekeepers get an average of 4 months of honey flow periods from this zone every year. Therefore, many beekeepers visit the region every year during this time to collect honey.

16.4 Limitations of the Study

The current study only includes data and analysis ranging from December 2019 to September 2020 and not beyond. Furthermore, the study is limited to only six districts of West Bengal, namely West Midnapore, Bankura, Purulia, Malda, North Dinajpur, and South Dinajpur districts. Furthermore, due to the lockdown, the fieldwork was largely impacted, and surveys after lockdown had to be taken through telephonic interviews. Furthermore, all beekeepers of West Bengal were not taken into account for the concerned study. Lastly, all investigations and analyses were done based on responses furnished by the respondents (beekeepers) who were active in both phases, before and during the pandemic.

16.5 Methodology

The current study is primarily based on primary data, and a range of secondary data have also been used where essential. Primary data was collected through the field survey that was collected in two phases. The First Phase of the survey (Phase I) was conducted before the COVID-19 outbreaks (without any knowledge and anticipation of the same in India), during December 2019 and January 2020. At that time, the objectives were to gather personal and professional information of the beekeepers and find out the problems and prospects of apiculture activities. Unfortunately, India (and the World) witnessed the COVID-19 pandemic that leads to a significant alteration as far as the scope and objectives of the study were concerned. Like the other migrant workers, the migrant beekeepers also faced problems with a new dimension demanding a fresh approach to the study. Thus, a second field survey was conducted during September 2020 to measure the impacts of the lockdown on this sector and to conduct a comparative study during the pre- and post-lockdown period.

16.5.1 Phase I Survey

There are many honey flow zones in West Bengal. The beekeepers of the West Bengal venture are divided into different fields as per their personal schedules. As many beekeepers of West Bengal come to those two zones every year, the said six districts were selected to make the sample as representative as possible. Consequently, in December 2019, three districts of South Bengal were surveyed, and the field works in the other three districts in North Bengal were conducted in January 2020. Habitually, beekeepers place their hive haphazardly according to their preferences, but they maintain a particular gap between two apiaries.

Moreover, the beekeepers do interestingly not belong to a recognized sector, making their migration unrecorded to the government authorities. Hence, to tackle this problem and to reach the maximum beekeepers, a snowball sampling technique was employed in the study. A total of 446 beekeepers were interviewed (out of whom 196 are from SBSZ and 250 from the NBSZ) during the Phase I survey.

16.5.2 Phase II Survey

The face-to-face interview could not be conducted owing to the COVID-19 restrictions. Hence, Phase II was dependent on the telephonic interviews to collect data. The contact information collected from the beekeeper respondents during the Phase I survey was utilized in the Phase II survey that was conducted during the first week of September 2020. Since all the respondents were not available (and some were not interested in the telephonic interview), 300 respondents were interviewed in this Phase. The First 150 interested telephone receivers were taken from each zone, using a lottery without replacement method, to perceive the nature of problems they were facing during the lockdown and to study the major effects of those problems.

In both Phases of the survey, pre-structured questionnaires were used, which was consisted of multiple-choice questions, mainly close and open-ended questions. Descriptive statistics and graphical presentation have been applied for the analysis of the data.

16.6 Results and Discussion

16.6.1 Migration Pattern of the Beekeepers during Pre-Pandemic West Bengal

The Phase I survey revealed that in the said two survey zones, the beekeepers of 12 districts of West Bengal had come to relocate their hives for collecting honey and taming bees (Fig. 16.1). In the migration activities, the beekeepers of North 24 Parganas (74 out of 300 beekeepers) and Malda (64 out of 300 beekeepers) had participated more extensively than other districts. Alongside, the beekeepers of South 24 Parganas, West Midnapore, and Bankura districts were also significant. Except for North and South Dinajpur, East Burdwan, Purulia, and Hoogly, the remaining seven districts' beekeepers had actively migrated to both survey zone.

In the SBSZ, where 150 beekeepers were taken as respondents, it was found that the beekeepers of North 24 Parganas (33 beekeepers) had actively participated in the honey flow zone of Eucalyptus fields during December 2019. Besides, North 24 Parganas district, Bankura (31 beekeepers), West Midnapore (27 beekeepers), and South 24 Parganas (21 beekeepers) districts had also taken an active part in the migration activities of the zone. In the North Bengal survey areas, where the same 150 numbers of beekeepers were taken as a sample, it was found that the beekeepers of Malda (53 beekeepers), North 24 Parganas (41 beekeepers), and South 24 Parganas (25 beekeepers) had active involvement in the honey flow zone of Mustard fields during January 2020 (Fig. 16.1).

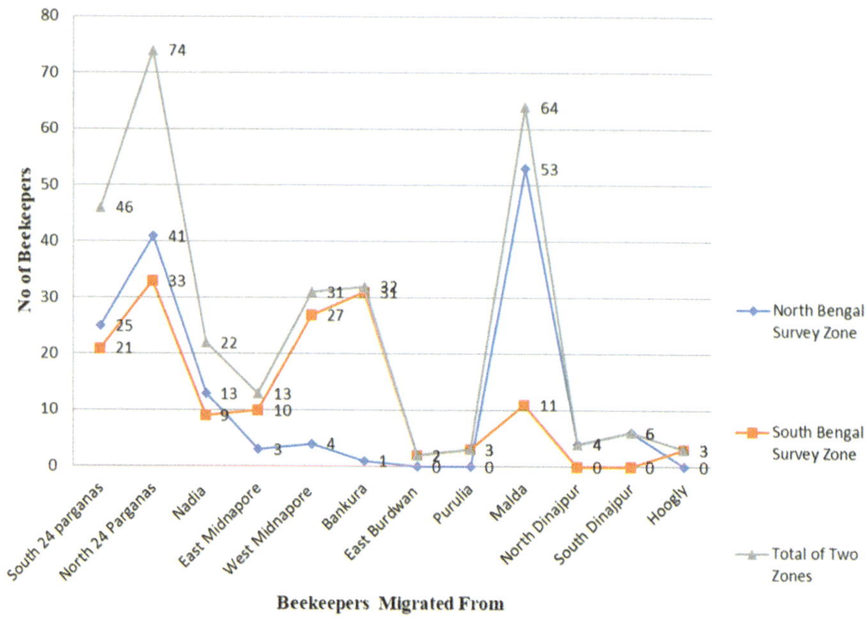

Fig. 16.1 Migration pattern of the beekeepers of West Bengal in two study zones before the pandemic (Source: Drawn by the authors based on field survey datasets)

16.6.2 Hive-Holding Capacity During the Pre-Pandemic Session

In West Bengal, there are different sizes of apiaries. From the Phase I survey before the lockdown, it was found that, among the 300 beekeepers, three beekeepers held below 20 beehives each and seven beekeepers held above 300 beehives each. In that survey it was also found that averagely each beekeeper held 135.37 beehives. But in hive-holding capacity there existed variation between beekeepers to beekeepers as in there standard deviation was 60.49 and coefficient of variation was 44.68%. In the case of the SBSZ, it was 145.50, with a standard deviation of 67.13 and coefficient of variation 46.14%, and in the case of the NBSZ, it was 125.23, with a standard deviation of 51.05 and coefficient of variation 40.76%. Those high variations revealed that hive-holding power was highly varied with one beekeeper to other. Maximum beekeepers held 120–139 beehives per head (Fig. 16.2). They were 68 in total out of the 300 respondents. Out of the 68, most (41 beekeepers) beekeepers migrated with their colonies to the Mustard fields of Malda and both Dinajpur districts of the NBSZ. The large-scale beekeepers (73 out of that 300 beekeepers) had more than 159 beehives per head. Most of them (51 beekeepers) had migrated to the three districts of the South Bengal survey zone to collect eucalyptus honey. The small-scale beekeepers (9 beekeepers) had less than 40 beehives per head; most

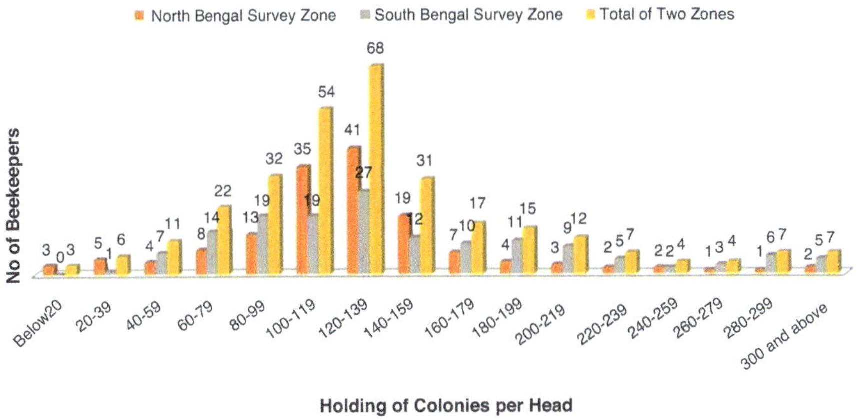

Fig. 16.2 Per head colony-holding capacity of the migratory beekeepers at pre-pandemic months (Source: Drawn by the authors based on field survey datasets)

(8 beekeepers) migrated in the three survey districts of the NBSZ. It was also found that the medium-scale beekeepers (153 beekeepers) holding 100–159 colonies each, maximum of them (95 beekeepers) migrated to the said mustard fields. In all the cases, it was found that there was a minor negative co-relation between the holding of bee colonies per head and the number of beekeepers. In respect of the SBSZ, the NBSZ, and the total of the two zones, those correlations were –0.30 (lower moderate), –0.20 (slight), and –0.30 (lower moderate), respectively (Table 16.1).

16.6.3 Professional Experience of the Migratory Beekeepers

The Phase I survey reveals that the beekeepers of West Bengal had gathered a range of work experiences. Though maximum beekeepers had five to below 10 years of work experience (89 beekeepers out of 300), 59 beekeepers had below 5 years of experience, and only18 beekeepers had above 30 years of experience among the said 300 beekeepers respondents. It was also observed that the beekeepers of mustard honey fields were lesser experienced than the beekeepers of eucalyptus honey fields (Fig. 16.3). The average years of experience of the beekeepers among the total six districts survey areas, which they had accumulated, was 12.70 years, with a standard deviation of 8.78 and coefficient of variation of 69.13%. It was 12.80 years, with a standard deviation of 9.49 and coefficient of variation of 74.14% for the SBSZ, and it was 12.60, with a standard deviation of 8.01 and a coefficient of variation of 63.57% for the NBSZ (Table 16.1). As deviations were high, it means the maximum had not nearer to the average. In the total study areas, beekeepers who owned 260 to 279 bee colonies each reported holding the maximum average experience (23.8 years of experience) was acquired by those beekeepers. In the SBSZ case, the maximum

Table 16.1 Statement relating to hive-holding capacity and professional experience of the beekeepers before the lockdown and effect of the lockdown on their colonies and migration periods

	Study zones	NBSZ	SBSZ	Two zones together
Holding of Beecolonies	Mean (in number)	125.23	145.50	135.37
	Standard deviation (in number)	51.05	67.13	60.49
	Coefficient of variation (in %)	40.76	46.14	44.68
	Co-relation between holding of beecolonies per head with no. of beekeepers	−0.30	−0.20	−0.30
Beekeeping experience	Mean (in year)	12.60	12.80	12.70
	Standard deviation (in year)	8.01	9.49	8.78
	Coefficient of variation (in %)	63.57	74.14	69.13
	Co-relation between mean experience in year with holding of beecolonies	0.85	0.85	0.88
% decreased in Beecolonies for the lockdown	Mean (in %)	22.20	29.67	25.93
	Standard deviation (in %)	27.52	20.96	24.74
	Coefficient of variation (in %)	123.96	70.64	95.41
	Co-relation between mean % decreased in beecolonies with hive-holding capacity	−0.56	−0.28	−0.64
	Co-relation between mean % decreased in beecolonies with their beekeeping experiences	−0.47	−0.48	−0.46
Loss of migration period for the lockdown	Mean (in month)	2.59	3.57	3.08
	Standard deviation (in month)	1.05	0.75	1.02
	Coefficient of variation (in %)	40.54	21.01	33.12
	Co-relation between mean losses of migration months with hive-holding capacity	0.61	0.44	0.72
	Co-relation between mean losses of migration months with their beekeeping experiences	0.85	0.68	0.80

Source: Field Survey by the authors

average experience (25.8 years of experience) was acquired by those beekeepers, who owned 200–219 bee colonies each. Similarly, in the case of the North Bengal study, the maximum average experience (28.8 years of experience) was acquired by those beekeepers, who owned 180 to 199 bee colonies each. In all the cases, it was found that there was an almost high degree of positive correlation between the experience of the beekeepers and their hive-holding capacity. In respect of the SBSZ, the NBSZ and the total of the two zones were 0.85, 0.85, and 0.88, respectively (Fig. 16.4).

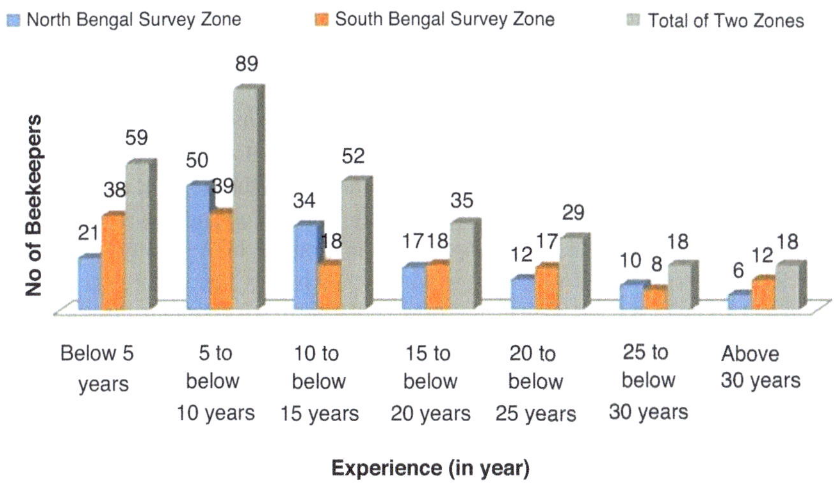

Fig. 16.3 Number of migratory beekeepers and their professional experience (Source: Drawn by the authors based on field survey datasets)

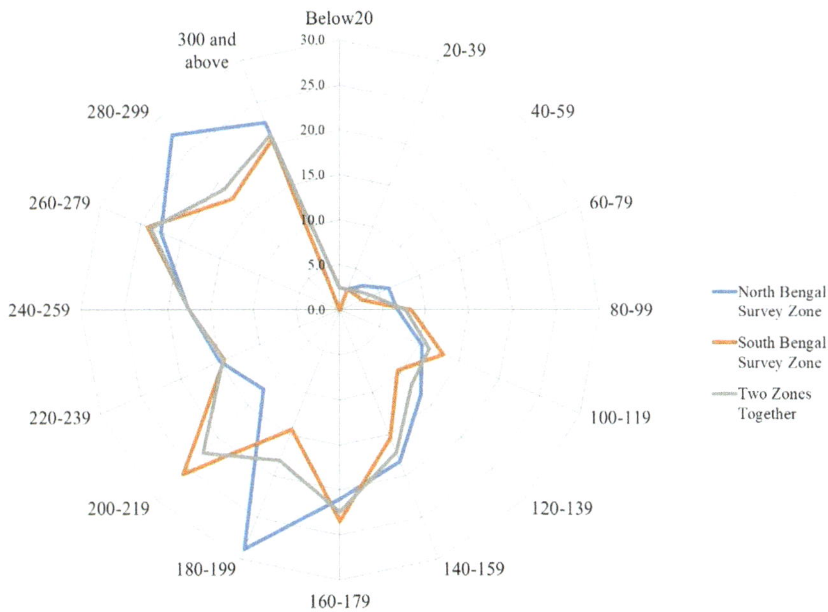

Fig. 16.4 Status of professional experiences of the migratory beekeepers in the perspective of hive-holding capacity (Source: Drawn by the authors based on field survey datasets)

16.6.4 Migratory Beekeepers and the Lockdown: Indian Scenario

The lockdown has crippled the migration activities of the beekeepers. The beekeepers faced various problems when they had blocked into their different migrated fields. Though different governments issued orders in that regard, yet problems remained unresolved. In order to avoid the field level contagious spread of the coronavirus in agricultural activities, the Indian Council of Agricultural Research (ICAR) had prepared (ICAR 2020) state-wise advisories for all 29 states. It had circulated through Krishi Vigyan Kendras (KVKs), ICAR research institutes, and State Agricultural Universities (SAUs) to farmers in their local languages to follow the necessary precautions. However, it was found that only the state of Himachal Pradesh and Uttarakhand had concrete guidelines on beekeeping. Most of the states, including West Bengal, had issued guidelines on agriculture, treating it as an essential activity. However, most of the stakeholders of the society were not aware of beekeeping. They were confused to treat beekeeping as an essential agricultural activity or not. Hence, the West Bengal government had notified in the order dated 18 May 2020 that beekeeping was free from restriction in movement. However, it had been found that the migratory beekeepers had faced various problems in different parts of India during the lockdown.

The North Indian beekeepers (Priyadarshini 2020) were unable to relocate their colonies on seasonal flowers of mango and litchi trees from February to July due to the lockdown. Consequently, the bees were starving to death. Similarly, the beekeepers of Uttarakhand (UN India 2020) were unable to relocate their colonies on seasonal flowers of apple and litchi trees from March to May as they had already migrated to Uttar Pradesh before the lockdown. Consequently, expansion of colonies, honey production as well as crop pollination was affected. The beekeepers of Punjab (Joshi 2020), Himachal Pradesh (HT Correspondent 2020), and Bihar (Khan 2020) also faced problems of selling, storing, producing honey, and relocating bee colonies. Here, it must be mentioned that traders purchase honey from the beekeepers of Odisha (Express News Service 2020) from May to July, which is the harvesting season of the honey in the state. Due to the lockdown, finding a buyer and transporting honey had emerged as a significant challenge for the beekeepers. The COVID-19 pandemic had also made the situation of Tamil Nadu's beekeepers (Yazhiniyan 2020) miserable. The lockdown restrictions had worsened their business as there were no buyers for honey neither in the state and nor outside. The lockdown had also impacted (Priyadershini and Menon 2020) the beekeeping sector in the state of Kerala. It prevented the migratory beekeepers from harvesting honey from colonies in different parts of the state and in Karnataka, for which their production went down.

In West Bengal, there are at least 60,000 beekeepers, mainly in North and South 24-Parganas, East Midnapore, Nadia, East Burdwan, Bankura, Hooghly, Purulia, and Malda. These beekeepers take bees across the state throughout the year, depending on the flowering season (Chaudhuri 2020). Since the beekeepers could not go anywhere due to the lockdown, a considerable number of bees died in the absence of nectar. During the lockdown, the beekeepers had tried to feed the emaciated insects the sugar syrup. However, it was not a viable option owing to its cost. Influenced by the plight of dairy farmers and cows, the West Bengal government had allowed sweet shops to open for limited hours during the lockdown so that milk could be used. But the beekeepers and the beekeeping industry faced totally different complications. For example, the officers of Hemnagar Coastal and Hingalganj police stations disallowed beekeepers to enter the Sunderbans during the lockdown (Chaudhuri 2020). Hence, many beekeepers were compelled to let bees in their hives starve to death, and they sacrificed a part of their total hives.

Moreover, another loss due to the lockdown was that many raw honey buyers did not remit the dues of beekeepers for earlier consignments of honey as they were detached from them. In addition to that, the beekeepers could not sell all the honey due to the closure of markets (Singh 2020; Das 2020). In this regard, the general secretary of the West Bengal Beekeepers' Association (Trust of India 2020) contended that the beekeepers of North-24 Parganas, South-24 Parganas, and Nadia had been reeling under financial crisis since March 2020, when the lockdown was imposed, as none of them could tend to the farms or sell the produce.

16.6.5 The Plight of Migratory Beekeepers Amidst Lockdown: West Bengal at the Crossroads

The sudden lockdown had stopped the migration of beekeepers by arresting them into their different migrated fields. In the Phase II survey, it was revealed that the beekeepers had faced various problems. These are described as under.

16.6.5.1 Stuck at the Fields Last Attended

There exists an innate relationship between honey bees and flowers. Flowers need insects like honey bees for their pollination. Similarly, bees are entirely dependent on flowers for their food, nectar, and pollen. Round the year, a particular flowering plant cannot blossom. Since every plant has its own habitat, flowering seasons vary from one location to another or one season to another season in the same location, depending upon the different geographical regions and agro-climatic zones. The plants those yield both nectar and pollen are collectively called "bee pasturage." When a good number of plants providing nectar and pollen are available for bees, such a period is called the "honey flow period."

Similarly, the days devoid of honey flow period is called as "dearth period" (Abrol 2002). Accordingly, the beekeepers always are concern about honey flow periods in different areas. The West Bengal state has well "bee pasturage" with different "honey flow periods." The significant honey flows of West Bengal are (Table 16.2) from Eucalyptus, Mustard, litchi, Mangrove, Coriander, Cumin, Black

Table 16.2 Significant honey flows of West Bengal and its features

Flower	Scientific name	Available in the following Districts of West Bengal	Migrate mainly for	Honey flow periods	Pick season
Eucalyptus	Eucalyptus globulus	West Midnapore, Bankura and Purulia	Honey collection	September to January	November to December
Mustard	Brassica juncea	Mainly- Malda, Uttar Dinajpur, and Dakshin Dinajpur districts; besides- Murshidabad, Hoogly, Nadia, Birbhum, Burdwan, and north 24 Parganas	Honey collection	October to march	November to February
Mango	Mangifera indica	Malda, Murshidabad, Nadia, and both 24 Parganas	Honey collection	February to march	March
Cumin, black cumin, and coriander	Cuminum cyminum Nigella sativa; Coriandrum sativum	Murshidabad, Birbhum, Hoogly, Burdwan, Nadia, Bankura, Purulia, and west Midnapore	Honey collection	January to march	February to march
Litchi	Litchi chinensis Sonn	Mainly- south 24 Parganas, Murshidabad Besides- Uttar & Dakshin Dinajpur, north 24 Parganas and Malda	Honey collection	March to may	April to may
Mangrove	Rhizophora mangle	Sunderban region of both 24 Parganas	Honey collection	February to may	March to may
Sesame or Til	Sesamum indicum	Murshidabad, Nadia, Hoogly, Burdwan, and Birbhum	Honey collection	May to July	May
Khesari (Almorta)	Lathyrus sativus	Southern part of West Bengal	Honey collection	January to march	January to march
Corn	Zea mays	Areas adjacent to Bihar	Feeding	July to august	July to august
Pumpkin	Cucurbita	Hoogly	Feeding	July to august	July to august
Cucumber	Cucumis sativus	Almost all but in little range	Feeding	July to august	July to august
Gurmi	Cucumis melo	Areas adjacent to Bihar	Feeding	August	August
Paddy	Oryza sativa	Almost all	Feeding	September	September
Indian jujube (Kul or Ber)	Ziziphus mauritiana	North 24 Parganas	Honey collection &feeding	September	September

Source: Baidya (2017) and Field Survey by the authors

Cumin, Sesame, Mango, khesari (almorta), Indian Jujube, Corn, Pumpkin, Cucumber, Gurmi (one type of melon) and Paddy. Mustard, Mangrove, Eucalyptus, and Litchi are the major honey flora in West Bengal, etc. (Baidya 2017). Maximum beekeepers of West Bengal collect honey from Mustard fields of the state. Therefore, the beekeepers of West Bengal have chosen different migration periods and paths for their bees. It is observed that minimum migration periods of beekeepers of West Bengal are 6 months to a maximum of 10.5 months. Though the main intentions of beekeepers are to collect honey and to reduce the cost of feeding bees by migrating in various fields where the honey flow period is operational, the migration activity depends on their hive-holding capacity, experience, residence, financial condition, risk-taking behavior, and entrepreneur skill.

Before the lockdown, most of the beekeepers (243 Beekeepers out of 300) stayed at the honey fields of Malda, North and South Dinajpur, Murshidabad, and Nadia districts for collecting mustard honey (Table 16.3). Since the production of the mustard honey was greater than the preceding years, most of the beekeepers stayed more than usual in the mustard fields. Although 25 Beekeepers already migrated to the cumin, black cumin, and coriander honey fields, 11 Beekeepers to the litchi fields, 9 Beekeepers to the mango fields, and 9 Beekeepers to the mangrove and 3 Beekeepers to the almost fields.

The beekeepers, who ultimately stayed at mustard fields, wanted to leave the fields within two or 3 days. According to their migration calendar, they were preparing to migrate at the different honey fields (Table 16.3). However, they had compelled to detain there for the lockdown. Since the dearth periods had come, they faced various problems in those fields, for the sudden stop due to government restrictions and low awareness among society about beekeeping. Only the nine beekeepers who could migrate to mangrove fields and the 11 beekeepers who could migrate to litchi fields took the benefits of full honey flow periods up to May 2020.

16.6.5.2 Shortage of Field Money

It is evident that the beekeepers faced various problems at different phases during the lockdown. The most immediate among them was the shortage of field money due to the sudden declaration of the lockdown. The economically poor beekeepers do not have an adequate financial backup. Most of them operate by borrowing money from informal sources such as relatives, friends, money lenders, cooperative societies, or intermediaries. In many cases, traders (actually middlemen) supply the money to the beekeepers as advances for their products at different migrated fields. In fact, there is no price (even no minimum support price) or no cash sale of honey in the fields. Throughout the season, the beekeepers take advances from the middlemen, and those advances are adjusted at the end of the season when the middlemen's groups have declared the prices of different floral kinds of honey. As such, the beekeepers are always in this vicious cycle. Lately, both the state and central governments have taken initiatives in this regard, with no evident effects. After that sudden lockdown,

Table 16.3 Migration Paths of the Migratory Beekeepers and their detention fields

Beekeepers of the SBSZ (Total 150 Respondents)	Beekeepers of the NBSZ (Total 150 Respondents)	The forced detention site (due to imposition of lockdown 2020)	The year-round trajectory of the migratory beekeepers				
			April to May (Mainly for honey collection)	May to July (Mainly for honey collection)	July to August (Mainly for feeding)	August (Mainly for feeding)	September (Mainly for feeding)
109	134	Mustard	Litchi / mangrove	Return at home / sesame	Return at home / corn/ pumpkin / cucumber	Return at home / Gurmi,	Return at home / Paddy / Indian jujube (Kul or Ber)
0	9	Mango	Same as above	Same as above	Same as above	Same as above	Same as above
3	0	Khesari (Almorta)	Same as above	Same as above	Same as above	Same as above	Same as above
21	4	Cumin, black cumin, and coriander	Same as above	Same as above	Same as above	Same as above	Same as above
9	0	Mangrove	Already they had come	Same as above	Same as above	Same as above	Same as above
8	3	Litchi	Already they had come	Mangrove / return at home / sesame	Same as above	Same as above	Same as above

Source: Field Survey by the authors

Table 16.4 Problems faced by the migrant beekeepers due to the lockdown

	Beekeepers of the SBSZ (Total 150 Respondents)		Beekeepers of the NBSZ (total 150 respondents)		The two zones together (total 300 respondents)	
Problems faced by the beekeepers due to the lockdown	No of the respondents faced	In %	No of the respondents faced	In %	No of the respondents faced	In %
Shortage of field's money	103	68.67	123	82.00	226	75.33
Shortage of daily needs	121	80.67	96	64.00	217	72.33
Transport problem	150	100	150	100	300	100
Colonies kept without supervision	6	4.00	8	5.33	14	4.67
Problem to get technical support	23	15.33	3	2.00	26	8.67
Lack of traders or middlemen	150	100	150	100	300	100
Problem of overstocking	150	100	150	100	300	100
Extra and high cost of transport	56	37.33	12	8.00	68	22.66
Compel to sell the produce	121	80.67	134	89.33	255	85.00
Local people directed to leave fields	12	8.00	22	14.67	34	11.33
Stopped by the local people to enter into new fields	126	84.00	113	75.33	239	79.67
Stopped by the local bodies to enter into new fields	34	22.67	41	27.33	75	25.00
Extra charge imposed by land owners	4	2.67	16	10.67	20	6.67
Harassments by bee-keeper's neighbors	44	29.33	72	48	116	38.67
Harassment by the bodies of beekeepers' own village	36	24.00	13	8.66	49	16.33

Source: Field Survey by the authors

the supply of working capital by the traders/middlemen had been stopped. As a result, they could not procure food supplements like sugar, sugar syrup, vitamins, or medicines for bees and could not purchase their own daily needs. Out of the 150 beekeeper respondents of the SBSZ, 103 respondents, and 150 beekeeper respondents of the NBSZ, 123 respondents expressed this problem as crucial (Table 16.4).

16.6.5.3 Shortage of Daily Needs

It was found that except for a few, maximum beekeepers were far away from their residence. On the one hand, they had exhausted their field's money, and on the other hand, as they were migrants, it became difficult for them to procure goods on credit. Moreover, the beekeepers suffered from the lack of daily needs such as foods, vegetables, soap, cooking fuels, medicines, etc. Like other migrants, many were also deprived of relief measures provided by the local bodies or NGOs. Out of the 150 beekeeper respondents of the SBSZ, 121 respondents and out of the 150 beekeeper respondents of the NBSZ, 96 respondents suffered from a scarcity of daily needs (Table 16.4).

16.6.5.4 Transport Problem

The order (No 67-CS/2020 dated 31 Marh 020) of the West Bengal Government stated - "some vehicles have been allowed movement for maintenance of essential services, emergency duties and movement of goods to ensure that citizens do not face any problem during this lockdown period." However, the people, including traffic police, truck owners, truck drivers, rural authorities, could not understand the category of beekeeping service (Field Survey). Hence, the beekeeping service failed to be enlisted in the essential services that received few relaxations. Consequently, the beekeepers faced the transport problem. Due to the shortage of field money and the non-availability of food supplements for the bees, they could not supply sugar in the bee colonies; the bees gradually starved to death. This would not have occurred if the beekeepers had transportation facilities so that they could relocate their colonies in the honey flow fields, mainly in the litchi or mangrove fields. However, for transport problems as mentioned above, they were unable to do so.

The order of the West Bengal Government (Memo No 177-CS/2020 dated 18 May 2020) in the third week of May 2020 stated that "restriction in movement of people and non-essential items between 7 PM to 7 AM. However, there shall be no restriction in movement of essential commodities including agricultural produce." In that order, it was also declared that agriculture, agriculture marketing, floriculture, apiculture, horticulture, and associated activities would be permitted even in Buffer Areas (Category—B) and in Clean Areas (Category—C) of this state. [1] However, the problem was that the transport facilities were only available

[1] For implementing lockdown measures, the department of family welfare, Government of West Bengal in consultation with the home department, Government of West Bengal had classified the state's areas into three zones—Red, Orange, and Green, taking into consideration of some prescribed parameters. In there, ward in urban areas and gram panchayat in rural areas had been treated as the basic unit for zoning and polling station (booth) in each ward and gram panchayat had been considered as the unit for containment planning. Farther, each Red and Orange zone had been classified into three categories—Category A: Affected Area (containment zone), Category B: Buffer Area and Category C: Clean Area (Memo No 177-CS/2020 dated 18/05/2020).

to the beekeepers in the middle of June 2020, when the honey collection period almost ended (Table 16.2). Then only a few fields were vacant for the natural feeding of bees but not suitable for honey collection. In such circumstances, the beekeepers have to manage alternate sources to feed the bees like sugar, sugar syrup, and vitamins; this again is taxing for the beekeepers. Generally, to reduce the feeding cost, the beekeepers stay at those natural feeding fields that fall on their return trips. However, due to the lockdown, the beekeepers suffered in this regard; they could not avail of natural feeding benefits as they were detained at the distant fields in different parts of the state. Out of the 300 beekeeper respondents of both survey zones, all respondents suffered from this particular problem (Table 16.4). Moreover, when situations became almost normal, the transport costs were hiked disproportionately (Field Survey), which impacted them negatively.

16.6.5.5 Colonies Kept Without Supervision

A few beekeepers went to their homes just before the lockdown, keeping the colonies alone in their migrated fields. Generally, the beekeepers relocate their colonies in new honey fields after coming back from home. However, due to the pandemic, they could not do it, as they could not come back to their last migrated fields. As a result, their colonies had been kept without the supervision of the beekeepers. Though the event occurred in few fields, problems were very crucial. Out of 150 beekeeper respondents of the SBSZ, only six respondents and out of 150 beekeeper respondents of the NBSZ, only eight respondents suffered from this problem (Table 16.4).

16.6.5.6 Lack of Technical Support

During the off-season, the beekeepers get technical supports from their cooperative societies, KVIC, different associations, and organizations. However, for any technical problem in migrated fields, they often take suggestions and get help from the intermediaries or traders, as the middlemen/traders are continuously attached with them. Since the intermediaries or traders were detached, the beekeepers were rendered helpless and could not get any technical support for a considerable period due to the lockdown. Twenty-three beekeeper respondents of the SBSZ and three beekeeper respondents of the NBSZ wanted few technical supports for their bees (Table 16.4). As they could not get those supports during that period, mainly from the end of March 2020 to the first week of June 2020, their colonies had faced few problems.

16.6.5.7 Unavailability of Traders and Intermediaries

The lockdown led to the communication gap between the migrant beekeepers and the traders/middlemen. As mentioned earlier, the migrant beekeepers are very much dependent on the traders/middlemen. Usually, these traders/middlemen provide different benefits to the beekeepers, such as interest-free loans, advance money, transport facilities at low cost, quick technical supports, clearing of honey stock, supplying empty containers, arranging field support systems, and even providing mental as well as physical supports in fields. As a result, the poor beekeepers become dependents entirely on those middlemen. However, due to the lack of traders or intermediaries during the lockdown, the beekeepers faced many problems as they did not receive the usual benefits. All the respondents faced this particular problem while they were confined to different migrated fields of the state (Table 16.4).

16.6.5.8 Problem of the Overstocking of Honey

Due to the lockdown, the traders or the middlemen could not purchase honey from the migrated fields. Furthermore, the beekeepers lack adequate storage facilities as generally there is no need for it. However, the pandemic raised this peculiar problem. As a result, the beekeepers had to delay the harvest and hence could not proceed with their new harvest. This particular problem was faced by even those beekeepers who were able to relocate their colonies in new honey fields. This ultimately reduced the production of honey. All respondents suffered from this problem. When the transport facilities had been more or less accessible by the beekeepers, they were compelled to pay an extra amount for transporting their produce to the cooperative societies, trading houses, middlemen's warehouses, or their homes. Thus, on the one hand, the beekeepers paid high transportation charges, which had generally been taken care of by the traders/middlemen in previous years. On the other side, bulk and request supply of the beekeepers reduced the price of honey. Out of 150 beekeeper respondents of the SBSZ, 56 respondents, and 150 beekeeper respondents of the NBSZ, 12 respondents experienced this problem (Table 16.4).

16.6.5.9 Lack of Cooperation from Local People and Local Bodies

Though the beekeepers generally are migrants from different parts of the state, they usually stay in a particular migrated field for the period of 1–4 months. Generally, they often converge at the same migrated fields during the same season every year. For this reason, they become familiar with the local people of migrated areas. However, during the times of pandemic, the locals often viewed the migrants from a suspectable perspective, and the beekeepers were not an exception. Many respondents highlighted the kind of harassment they faced at the hands of the locals.

Furthermore, the respondents contended that the locals imposed restrictions by framing few self-made quarantine rules on the migrated beekeepers. Many beekeepers alleged that they were even directed to leave the area. Twelve beekeepers from the SBSZ and 22 beekeepers from the NBSZ had been directed to leave the area by the local people. During a partially normalized situation, when few beekeepers wanted to enter into new fields for relocating their beehives, they were also stopped by the local people. Among 300 beekeeper respondents of the total survey areas, 239 respondents had faced this problem, out of which 126 beekeepers from the SBSZ and 113 beekeepers from the NBSZ. The local authorities and clubs also united to create some troublesome activities. Thirty-four beekeepers from the SBSZ and 41 beekeepers from the NBSZ reported having been harassed by the local authorities and clubs, and the local people (Table 16.4).

16.6.5.10 Misbehavior by Land Owners

There is a tradition to pay some money or honey to landowners as the beekeepers keep their colonies on their land. Not only that, local associations also charge the same as they feel the flowers of their village are utilized for these beekeeping activities. Where in the developed countries of the World and in some parts of India, beekeepers receive some amount for their pollination services, but in West Bengal, they are made to pay some charges to landowners and local associations. Despite the lockdown and despite the lack of yearly produce, many beekeepers were made to pay the yearly charges. Furthermore, since the beekeepers stayed longer than their contracted period due to lockdown, many landowners imposed extra charges.

Consequently, the beekeepers were compelled to pay those charges for the fields where dearth periods had come. Even during the partially normalized situation, when few beekeepers tried to enter into new fields, the fields' owners also charged extra costs for keeping their colonies. Lack of awareness about beekeeping and COVID-guidelines is the major problem in our society. Four beekeepers from the SBSZ and 16 beekeepers from the NBSZ faced the problem of landowners (Table 16.4).

16.6.5.11 Misbehavior by Their Neighbors

The beekeepers also reported harassment by their neighbors when they returned to their homes after the migration. Forty-four beekeepers from the SBSZ and 72 beekeepers from the NBSZ had faced the problem created by their neighbors. Along with the neighbors, other associations of the beekeepers' villages also created the same problem. Thirty-six beekeepers from the SBSZ and 13 beekeepers from the NBSZ had been harassed by the authorities and clubs of the villages along with their neighbors (Table 16.4). Furthermore, the beekeepers have also imposed some restrictions by framing few self-made quarantine rules by their villagers as they had come home from other regions of the state staying there for a considerable amount of time.

16.6.5.12 Exploitation by Intermediaries

It has already been mentioned that intermediaries play a vital part in the marketing activities of the beekeeping industry. The middlemen use the beekeepers for their interest by creating "benefit traps" for the beekeepers and harassing and frightening the beekeepers. Since the beekeepers already has availed the benefits from middlemen like season's advance, off season's loan, tranport and storing facilities, maximum beekeepers are usually forced to sell their produce to the middlemen at a cheaper rate. Moreover, intermediaries control the price and weight of honey for their vested interest and prevent direct and close touch between the beekeepers and the other buyers. Since the middlemen have their own storing and transport facilities, they can collect the entire honey of beekeepers from different fields. They also have good access to the marketplace. Sometimes intermediaries transfer honey from beekeepers' tents to warehouses of different companies by taking commissions.

Due to the lockdown, the beekeepers were unable to get any benefit from intermediaries. However, when the situation became partially normal, the middlemen procured the honey at a meager price. The reason furnished by the middlemen were many, the most common one being that the pandemic badly impacted the international market of honey and that no export house or company would take that honey. In addition to this, the beekeepers contended that the middlemen took around Rs 4 per kg for cold storage and transport charges. Under this situation, maximum beekeepers were compelled to sell their honey to the middlemen at the prices offered by the middlemen as the honey was overstocked, and at the same time, they naturally were in a crisis of money (Table 16.4).

16.6.6 How Lockdown Affects the Migratory Beekeeping in West Bengal

The beekeeping profession is continually heading a downward trajectory due to the problems and issues resulting from the lockdown and thereafter. The production of honey as such has been hampered to a large extent. With that colony, expansion had been totally stopped, which are hindering the current year's production also. Here it is imperative to mention that beekeeping production depends on – number of bee colonies and migration periods. These two were, in fact, the major affected areas of the beekeeping industry during the pandemic.

16.6.6.1 Destruction of Beecolonies

From the survey after lockdown, it was found (Table 16.1) that, in the total survey areas of South and North Bengal zones, the beekeepers who had an average 135.37 number of beecolonies per head with an average of 12.70 years of experience before lockdown, from March 2020 to September 2020, had destroyed their number of colonies by 25.93%. This was a recurring phenomenon even in the SBSZ; their number of colonies was contracted by 29.67%. In the case of the NBSZ, their number of colonies contracted by 22.20%. Standard deviations and coefficient of variations relating to the reduction of colonies in the two separate study zones and in a total of the two zones were also high (Table 16.1). It indicates that huge variations in the reduction of colonies existed. Total 72 beekeepers, nine beekeepers were from the SBSZ and 63 from the NBSZ, who kept their colonies without any contraction. Only a few were able to migrate into new honey flow fields just before the lockdown and could spend huge amounts for supplying artificial foods for their bees. Fig. 16.5 shows the mean decrease (in %) of colonies according to the various beekeepers' groups as per hive-holding capacity before the pandemic. It was also revealed that in all the cases (i.e., the two zones separately or in total), the beekeepers who had less than 80 colonies before lockdown had suffered more (over 40% reduction in colonies). In addition to this, the beekeepers' group of the SBSZ, who held colonies from 240 to 279, had also destroyed more than 40% of their colonies, which they held just before March 2020. The beekeepers' group, who held a number of colonies "below 20″ in the NBSZ and the beekeepers' group, held many colonies "from 20 to 39" in the SBSZ before lockdown, had totally closed their beekeeping activities. The study also showed that beekeepers who held fewer bee boxes (colonies) had contracted more. In the two separate study zones and in a total of the two zones, co-relations between mean decreased (in %) in beecolonies with hive-holding

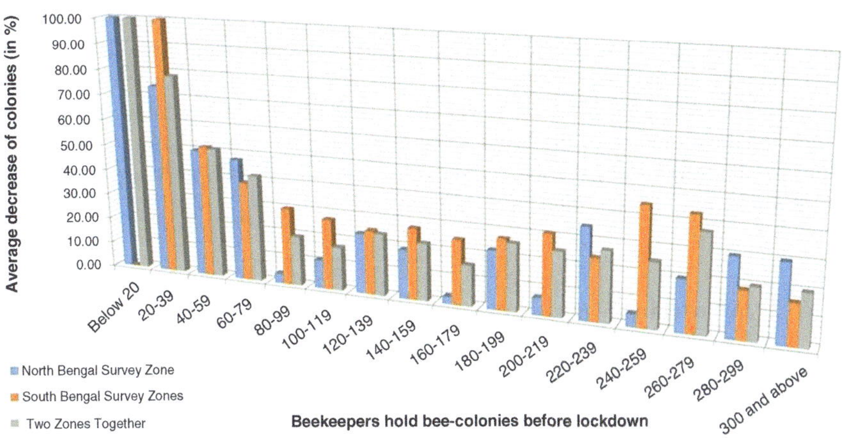

Fig. 16.5 Average contraction of colonies (in %) due to the lockdown as per hive-holding Capacity-wise class of beekeepers (Source: Drawn by the authors based on field survey datasets)

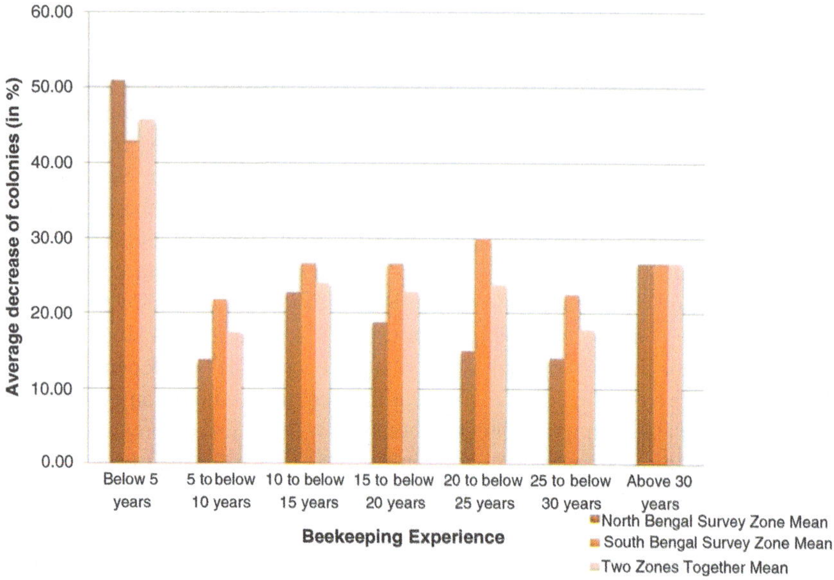

Fig. 16.6 Average decrease of colonies (in %) due to the lockdown as per experience-wise classes of beekeeper Groups (Source: Drawn by the authors based on field survey datasets)

capacity of beekeeper were negative. Those were −0.28 (slight) for the South Bengal zone, −0.56 (moderate) for the North Bengal zone, and −0.64 (moderate) for the total study areas.

Figure 16.6 shows the mean decrease (in %) of colonies according to the various beekeepers' groups as per their beekeeping experiences. It also revealed that in all the cases (i.e., the two zones separately or in total), the beekeepers those having beekeeping experience of below 5 years had suffered more (over 40% reduction in colonies). In addition to this, it was also observed that in all the cases, the beekeepers who had beekeeping experience above 30 years also contracted more than 25% of their colonies. In the two separate study zones and a total of the two zones, it was found that co-relations between mean decreased (in %) in beecolonies with the professional experience of the beekeepers were negative. Those were −0.48 (moderate) for the South Bengal zone, −0.47 (moderate) for the North Bengal zone, and −0.46 (moderate) for the total study areas.

16.6.6.2 Shortening Migration Periods

It is already said that the beekeepers always want to relocate their beecolonies in various fields where honey flows come. Usually, beekeepers of West Bengal have migrated a minimum of 6 months and a maximum of 10.5 months in a year. However, for around 2.5 months' lockdown, the relocation activities had been hampered. They were compelled to minimize their migration months to those

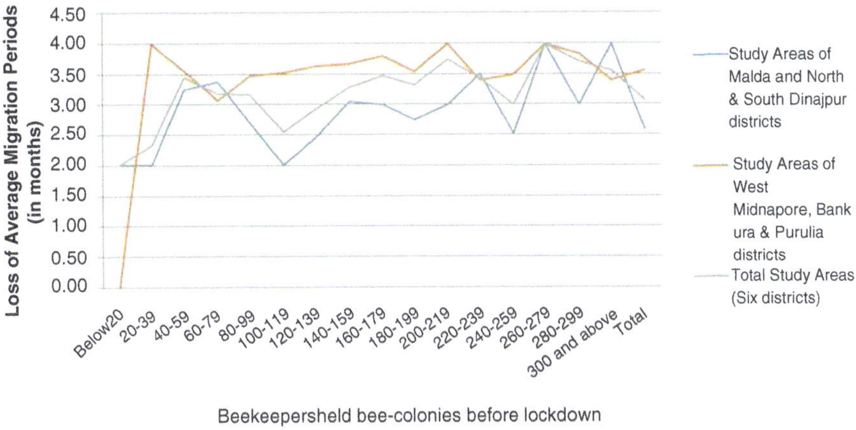

Fig. 16.7 Loss of average migration months due to the lockdown by the different classes of beekeepers according to their past hive-holding capacity (Source: Drawn by the authors based on field survey datasets)

restricted months and over and above the months. From the phase II survey after lockdown, it was found (Table 16.3) that, in the total survey areas of South and North Bengal zones, averagely they had minimized their migration period by 3.08 months. In the case of South Bengal survey areas, the beekeepers had lost their migration period by 3.57 months, and that was 2.59 months to the beekeepers of the NBSZ. Standard deviations and coefficient of variations relating to the loss of migration months were also high (Table 16.3) in the two separate study zones and in total. It indicates variations in loss of migration months among one beekeepers' group to another. Only three beekeepers among the 300 did not lose any migration month. Actually, they had shorter migration periods according to their plan, and they sifted before lockdown in a new field, where the honey flow was just starting. These three beekeepers were from the NBSZ.

Figure 16.7 shows the loss of migration periods (in months) according to the various beekeepers' groups categorized by their hive-holding capacity. It revealed that the beekeepers of SBSZ had suffered more. The study also showed that the beekeepers those held maximum bee boxes had lost more migration months. Since they had faced more troubles and anxieties to move, it was also found that co-relations between mean losses of migration months with hive-holding capacity were moderately positive in the two separate study zones and in total areas. Those were 0.44 (moderate) for the South Bengal zone, 0.61 (moderate) for the North Bengal zone, and 0.72 (moderate) for the total study areas. Similarly, Fig. 16.8 shows the loss of migration periods (in months) according to the various beekeepers' groups categorized by their beekeeping experiences. It also revealed that the beekeepers of SBSZ had suffered more. The study also showed that the beekeepers who had experienced more had lost more migration months. The co-relations between

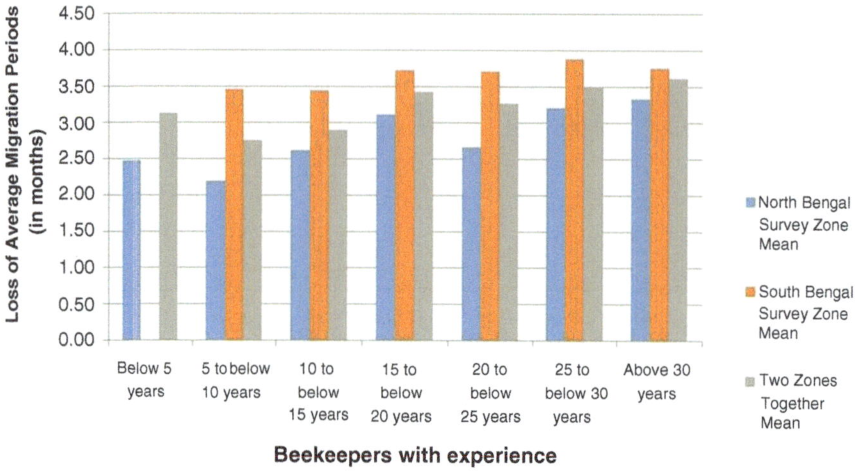

Fig. 16.8 Loss of average migration months due to the lockdown by the different classes of beekeepers according to their professional experience (Source: Drawn by the authors based on field survey datasets)

mean losses of migration months with beekeepers' experiences were highly positive (Table 16.1). Those were 0.68 (moderate) for the South Bengal zone, 0.85 (high) for the North Bengal zone, and 0.80 (high) for the total study areas. As they were more experienced, they had taken more risks. They were migrated for more periods than the lesser experienced beekeepers, and they were more away from their native villages. Hence, that year they suffered more. It was found in all the cases that more experienced beekeepers suffered more.

16.6.6.3 Impact of Lockdown on the Promises of Beekeepers

The tamed bees pay honey, other by-products, and pollination services as gifts to the beekeepers. As a return gift, the beekeepers keep the promises to their bees to ensure a regular supply of food throughout the year, either by relocating their colonies in the honey flow zones or by giving artificial food in the colonies. However, the lockdown-induced problems hampered all migration activities as the beekeepers had blocked at their respective fields. Hence beekeepers had to keep their second option - supply of sugar during the end of the honey flow periods (Dearth Period). However, the buying of sugar daily was becoming too much financial involvement to bear. Hence the second promise also was not kept by the beekeepers. For this scarcity of nectar sources and insufficient sugar supply, the bees were starving to death. For this reason, many beekeepers had destroyed their beehives (Fig. 16.5 and Fig. 16.6).

16.7 Major Findings

- The lockdown posed challenges to the beekeepers of West Bengal severely.
- Mustard crops are the major honey flow fields in West Bengal. All beekeepers have their migration plans to move their colonies in different honey flow fields round the year to collect more honey and minimize feeding costs. The beekeepers of North 24 Parganas, Malda, South 24 Parganas, West Midnapore, and Bankura districts participated more effectively in migration activities.
- The Phase I survey (pre-lockdown) revealed the average beehives of a beekeeper were 135.37 units in West Bengal. While, in the NBSZ, the average was 125 beecolonies per head, the SBSZ had 145.50 beecolonies per head.
- All beekeepers of West Bengal had gathered a range of experiences. The average years of experience of the beekeepers were 12.70 years. The beekeepers in the SBSZ and the beekeepers in the NBSZ were 12.80 years and 12.60 years of average experience, respectively.
- Before the lockdown, most of the beekeepers stayed at the honey fields of Malda, North Dinajpur, South Dinajpur, Murshidabad, and Nadia for collecting mustard, cumin, and coriander honey. However, the lockdown obstructed the entire plan and blocked them at their respective fields where the dearth period already had come.
- The Phase II survey (post-lockdown) found that the beekeepers had lost their colonies by 25.93% on average. While, in the South Bengal study area, it was 29.67%, it was 22.20% in NBSZ. The study also showed that those beekeepers who held fewer bee boxes and lesser experience had lost more.
- The minimum migration periods of the beekeepers of West Bengal are 6 months, and the maximum is 10.5 months. However, the lockdown had hampered more for 2.5 months.
- The Phase II survey estimated that the beekeepers had lessened their migration period by 3.08 months on average in West Bengal due to lockdown. While they lost 3.57 months on average in South Bengal, in the NBSZ, they had lost 2.59 months. Also, the beekeepers holding a higher number of beehives and more extended beekeeping experience had lost more migration months.
- Due to the lockdown, the migratory beekeepers of West Bengal faced several problems; however, out of all problems of shortage of field's money, shortage of daily needs, transport, absence of traders or intermediaries, overstocking and selling of honey were crucial.
- The losses of migration periods and the interruption in the adequate food supply to the bees due to scarcity of working capital resulted in the bees starving to death. This could be a long-term challenge for the apiculture sector in its supply-side in West Bengal.

16.8 Some Recommendations

- The lives and livelihoods of the migratory beekeepers must be given serious considerations. We have witnessed epidemics in the past and are facing the current pandemic. During the lockdown and after that, most of West Bengal's beekeepers were treated like migrant workers in different parts of India. Here, it is essential to understand that the beekeepers help the farms and the farmers through pollination by keeping hives in the agriculture and horticulture fields. Hence, the beekeepers should be given their due, and their essence should be recognized and valued. However, only "black and white" on papers is not sufficient. Political will and seriousness are a must if we address the problems faced by the beekeepers meaningfully. Probably a bottom-top approach is the need of the hour.
- Loss of migration period is an important issue that needs attention. Hence, to avoid migration loss and to minimize its impacts, the governments should provide seamless transport facilities for the beehives and raw honey stocks. Furthermore, special subsidies have to be granted to the beekeeper for maintaining the bee boxes during this pandemic situation or any other emergency that could arise in the future. Therefore, both the central and state governments should understand the gravity of the matter and should be prepared for future pandemic-like situations.
- Throughout the year, traders, cooperative societies, agents of different companies, and intermediaries collect the honey from the beekeepers. However, during the lockdown, the beekeepers are invariably exploited by them. The government and the authorities should give this issue utmost importance. Furthermore, an effective mechanism should be devised, wherein the beekeepers get their due and are not exploited by the agents/middlemen like the field level benefits and supports have to be provided by the government directly or through their societies, government has to purchase honey directly and government has to declare the minimum support price of honey.
- It is also imperative that the government provides bee boxes and financial supports to the affected beekeepers free of cost or at a concessional rate. Though the Central Government declared Rs. 500 crores funds for the development of beekeeping (Hindustan Times 2020), but the said amount can never be sufficient to address the problems of all the beekeepers of India. It is important to note that India has more than 2.5 lakhs active beekeepers, and the newly trained beekeepers are around 30 lakhs. Only in West Bengal, there are around 60,000 beekeepers (both registered and unregistered). These numeric have to be taken into account while allocating funds and reliefs for these beekeepers.
- Maintaining only one colony of bees during the dearth period requires 500 grams of sugar every day as food for the bees, which could help assume the extent of loss that the beekeepers have had to incur during this pandemic. Keeping in mind the importance of apiculture sectors supply-side, the government should arrange for interest-free loans to help beekeepers recover their losses and continue with their profession in the coming year.

- Lastly, we must understand that beekeeping is integral not only to the beekeepers but also to humanity. Without bees and pollination, the whole ecosystem will suffer. Hence, awareness programs have to be conducted to disseminate information to the common people what these eco-friendly and agri-support activities the beekeeping sector offers to us.

16.9 Conclusion

Since early times, there is an association between honey bees and human beings. At present, this modern migratory beekeeping is an indispensable factor of ecological and socio-economic development in many parts of the World. The beekeepers are not migrant workers, but their occupation is migratory in nature. A vast number of bee colonies are moved by trucks to a series of monoculture crops in different parts of regional and national areas for different months. For this reason, the COVID-19 induced lockdown stood challenges to the beekeepers of West Bengal severely. The lockdown hindered the whole migration plan of the beekeepers. It became challenging to maintain their life and livelihood simultaneously. This study helped to know the common migration paths of the Beekeepers of West Bengal. The beekeeping experience and hive-holding capacity of the beekeepers of West Bengal had been investigated vividly. Detention of the beekeepers at their last migrated fields, shortage of field's money, shortage of daily needs, transport problem, unavailability of traders or intermediaries, the problem of overstocking and exploitation by intermediaries appeared as the major problems which mainly impacted on the contraction of colonies and loss on migration periods. These could be a long-term challenge for the apiculture sector in its supply-side in West Bengal. Government officials, people's representatives and general people have to take this issue seriously to address the problems meaningfully.

The loss of migration period of the beekeepers is an important issue that needs attention. Hence, to avoid migration loss and to minimize its impacts, the governments should provide seamless transport facilities for the beehives and raw honey stocks. Instead of traders, agents of different companies, and intermediaries, the field level benefits and supports have to be provided by the central and state governments directly or through their societies to the beekeepers. Similarly, governments have to purchase honey directly from the beekeepers, declare the minimum support price of honey, extend financial supports like subsidies, and interest-free loans to the affected beekeepers. Without bees and pollination, the whole ecosystem will suffer. We must understand that beekeeping is integral not only to the beekeepers but also to humanity.

This study was a small effort where many issues remain unexplored, which can be studied in the future like analysis of the internal and external market of honey, the trend of honey productions, the trend of honey esports, new challenges of beekeeping during new-normal periods, and present situation of the beekeepers of India and its constituting states.

References

Abrol DP (2002) Beekeeping an Indian perspective. Vinod Publishers and Distributors, Jammu & Kashmir

Baidya M (2017) Apiculture Marketing through Cooperatives in West Bengal with Special Reference to South 24 Parganas and North Dinajpur Districts. PhD Thesis, North Bengal University

Chaudhuri S (2020) Untold honey trap of lockdown. https://www.deccanherald.com. Accessed 12 Sep 2020

Das S (2020) West Bengal govt develops safe, lucrative way of collecting honey for Sundarbans villagers. https://www.deccanherald.com. Accessed 12 Sep 2020

Express News Service (2020) COVID-19 strikes huge blow to apiculture in Odisha's Kendrapara. https://www.newindianexpress.com. Accessed 1 Sep 2020

Green K (2020) Lack of bee imports due to COVID-19 will affect honey supply, agriculture. https://calgary.ctvnews.ca. Accessed 1 Sep 2020

Hindustan Times (2020) Rs 500 cr for beekeeping initiatives, will help 2 lakh beekeepers: FM Nirmala Sitharaman. https://www.hindustantimes.com/india-news. Accessed 2 Aug 2020

HT Correspondent (2020) HP wants beekeepers stuck in Punjab, Haryana to return. https://www.hindustantimes.com. Accessed 28 Aug 2020

ICAR (2020) Note on Advisories to farmers during lockdown due to COVID 19. https://www.icar.org.in. Accessed 1 Sep 2020

Joshi V (2020) Beekeepers failed to recover cost this season on exportsuspension. https://www.hindustantimes.com. Accessed 30 Aug 2020

Khan MI (2020) COVID-19: Bihar beekeepers find it tough to relocate bee boxes amid lockdown. https://www.downtoearth.org.in. Accessed 10 June 2020

Priyadarshini S (2020) Honey bees starve in COVID-19 lockdown. https://www.nature.com. Accessed 15 Aug 2020

Priyadershini S, Menon A (2020) Bee keepers to the rescue. https://www.thehindu.com. Accessed 23 Aug 2020

Singh G (2020) Embracing beekeeping to stop fatal tiger attacks in Sundarbans mangrove forest. https://www.lifegate.com/sundarbans-beekeeping. Accessed 12 Sep 2020

The Economist (2020) China's beekeepers feel the sting of COVID-19Lockdown. https://www.economist.com. Accessed 2 Sep 2020

Thomas D et al (2001) Bee flora and migratory routes in India, in: 37th international apicultural congress, 28 October – 1 November, 2001, APIMONDIA, Durban

Trust of India (2020) Amphan Stings Beekeeping Business in Bengal. https://www.republicworld.com. Accessed 25 Aug 2020

UN India (2020) Bees, biodiversity and COVID-19. https://in.one.un.org. Accessed 30 Aug 2020

Yazhiniyan (2020) Virus effect turns beekeeping trade sour in state. https://www.dtnext.in. Accessed 2 Sep 2020

Chapter 17
Struggle of Apiculture Sector in West Bengal, India During COVID-19 Pandemic: Analyzing the Demand and Supply Sides

Sanghamitra Purkait and Anindya Basu

Abstract Honey is the main product of apiculture. Apiculture is a part of Argo practice. Beekeeping is not only limited to honey production but also plays a vital role in providing pollination services and maintaining bio-diversity. Modern bee-keeping is migratory in nature. Across India, during the lockdown phase of the COVID-19 pandemic, the beekeepers faced many difficulties with transporting, increasing, or even maintaining their production. It is challenging to balance life and livelihood. On the other hand, the government issued COVID guidelines to boost immunity through a healthy lifestyle and a nutritious diet. Honey has been highlighted again. Though advisories were to increase honey intake, the level of consumption astonishingly dropped in the national and international markets. In the present study, through a primary survey of both the parties, i.e., the beekeepers and the common people, across the state of West Bengal, the stocktaking of the ground situation was done. Likert scaling has been taken to analyze the problems faced by beekeepers. The Chi-square Tests and paired t-test have been used for testing the hypothesis. Finally, certain feasible mid-term planning and policies have been chalked out in the study to smoothly run the apiculture sector, gunning for higher production in the "new normal phase."

Keywords Beekeepers · Consumption · Immunity booster · Pandemic · Migration · New normal

17.1 Introduction

COVID-19 pandemic has presented an unmatched, complex phase in India. Few catchphrases have been generated in these trying times like "people's curfew," "lockdown," "only if there is life there will be livelihood," "both, lives and

S. Purkait (✉) · A. Basu
Department of Geography, Diamond Harbour Women's University, Diamond Harbour, West Bengal, India

S. Saha et al. (eds.), *Economic and Societal Transformation in Pandemic-Trapped India*, New Frontiers in Regional Science: Asian Perspectives 55,
https://doi.org/10.1007/978-981-16-5755-9_17

livelihood matter equally," "from an individual to the whole of humanity," "unlock period," "new normal," "vocal for local," and many more. It is a challenging task to balance life and livelihood under this pandemic condition. In fact, the economy of India has been greatly affected by this lockdown due to the coronavirus pandemic, and India has started with a negative growth rate in the financial year 2020–2021. India Ratings and Research had demoted the financial year 2020–2021 estimate to 3.6% (The Hindu, PTI 2020). On 12 April 2020, World Bank had reported focusing on South Asia said that India's economy would expect to grow 1.5–2.8% for the financial year 2020–2021 (The Hindu, PTI 2020). Confederation of Indian Industry (CII) had estimated that India's GDP for the financial year 2020–2021 would be between 0.9% and 1.5% (Kumar 2020). Executive Summary Data (Department of Economic Affairs 2020) for the April–June quarter has reflected a significant worldwide year-on-year contraction of output, narrating the actual situation of the world economy under the COVID-19 pandemic environment. India's GDP contraction at 23.9%, which is slightly higher compared with other advanced nations.

Apiculture is a kind of agricultural activity involved in beekeeping, and, apart from providing honey, it helps provide pollination services and bio-diversity maintenance. Since early times, honey has been valued as a sweetener (Jones and Sweeney-Lynch 2011), which is a high-calorie food with an extensive market, has innumerable utilities. It is used as medicine, preservative, cosmetics, food item, media in laboratories, etc. (Hall and Lothrop 1934). The medicinal importance of honey has always been a highlight as it fights germs and maintains the blood's proper alkalinity (Singh 1958; Ghosh and Ghosh 1999).

In India, agriculture is the only sector that has recorded modest growth of 3.4% in year-on-year terms (Jebaraj 2020). As countries unlocked in the quarter starting in July 2020, gradual recovery was underway globally. A sharp V-shaped recovery has been found in the case of India also from May 2020, with agriculture being the brightest spot in the restoration of growth (Department of Economic Affairs 2020). Beekeeping with honey as the prime product is referred to as the "golden wonder" of nature. It is most interesting to know that honey is the first biological sweet of the world (Nagaraja and Rajagopal 2009), and it was only sweet of ancient times (Root 1980). Since the Vedic age, honey has persistently been in use as a holy item and in our household or for religious purposes. The value of honey can be measured from the fact that it has multi-purpose usages right from religion, health, nutrition, beauty-care, to medicine. Ironically, the bees and beekeepers always remain under the radar of planning and development, even during this pandemic situation.

Moreover, in India, beekeeping is mostly migratory in nature (though stationary beekeeping is also operated in some states). For this reason, like migrant workers, beekeepers had also been harassed in different parts of India during this lockdown period. The beekeepers of other parts of the world, especially China, France, and Canada, also face challenges relating to the COVID-19 pandemic (The Economist 2020; Green 2020; Priyadarshini 2020). Alongside the production problems, the marketing of honey in international and national markets has also been an area of concern as transportation is restricted at large.

17.2 Literature Review

The lockdown affected honey production across the globe. Green (2020) reported that how shipping restrictions hampered the movements of bee boxes in Western Canada, leading to low honey production and inefficient pollination services. The situation was analogous in France, where the beekeepers were victims of unsold honey due to the closure of farmers' markets, and many were forced to leave their profession (Trtworld 2020).

The Indian apiculture scenario was similarly dismal. In Tamil Nadu, more than 50,000 families were involved in the profession of beekeeping (Yazhiniyan 2020). Unfortunately, there were no honey buyers within and outside of the state due to the COVID-19 pandemic situation. Since the beekeepers could not sell honey this year, they were left with no liquid cash to invest in the next season. Due to the pandemic, cooperative societies had not procured honey from the beekeepers as the complete and partial lockdowns hampered the usual movements.

Every year beekeepers from Kanniyakumari travel all the way to Kerala, Karnataka, and Goa to sell honey, but this year these trips too had to be canceled. Nearby 1000 farmers of Kendrapara, Odisha, depend on apiculture for their livelihood. In Express News Service (2020), it was reported that every year, traders purchase honey from the farmers from May to July, which is the harvesting season of the honey, but last year was an exception as, due to the lockdown, finding a buyer and transporting honey has emerged as a significant cause of concern for the farmers. Due to the lockdown, the demand had gone down, and the local price of honey had come down from Rs 600 to Rs 250–300 per kg. Priyadarshini (2020) said that during the summer months between February and July, farmers, especially in northern India, go from one state to the other with their bee boxes to feed the bees on seasonal flowers of mango and litchi trees. However, the transport restrictions had made beekeepers unable to move their massive number of beehive boxes from one state to another and even one field to another within the states. As a result of the scarcity of nectar sources, the bees had starved to death. Priyadarshini and Menon (2020) narrated how the migration of the beekeepers from Kerala to Karnataka was thwarted due to the pandemic as relocation of the colonies was not possible, and with the increasing number of unattended boxes, the production goes down.

Khan (2020) reported that around 9000 beekeepers in Bihar were unable to relocate lakhs of bee boxes to neighboring Jharkhand to give their bees easy access to nutrition amid the nationwide lockdown. The flowering season for the litchi was over by the first week of April in Bihar, and then usually every year, around 2 lakh bee boxes were relocated from Bihar to Jharkhand. The lockdown has kept trucks off the roads leading to huge losses. A similar dismal situation was faced by the beekeepers of Uttarakhand in northern India, where nearly 7000 beekeepers are tending to 70,800 colonies, producing over 15,000 quintals of honey each season (UN India 2020), but during this year's spring season, the entire interplay between pollination, the honey collection was greatly affected in terms of quantity and quality. The apiculture business cycle in Punjab (Joshi 2020), having around has

2.7 lakh bee colonies, being the third-largest producer of honey in the country, suffered badly. The much-coveted highly-priced mustard honey did not find any importers as export was closed and local markets offered meager prices, and the rolling business came to a halt.

Singh (2020), Das (2020), Chaudhuri (2020) depicted the situation of Bengal apiculture where stories of unsold honey, inability to transfer bee boxes, incurring high maintenance costs were the major issues.

17.3 Objectives

The study has three primary objectives:

- To investigate the problems faced by the beekeepers of West Bengal during the sudden lockdown period due to the COVID-19 pandemic.
- To determine whether there has been any remarkable change in the extent of honey consumption by the general public, taking into different factors like gender, dwelling place, annual family income, the total number of members in a family before and during the pandemic situation, awareness about beneficial apiculture products.
- To find the ways forward, which would help maintain the better livelihood of the beekeepers and increase the demand for honey, helping in the growth of the apiculture industry at large.

17.4 Methodology

Though secondary data have been used, this study is mainly based on primary data. The *snowball* sampling technique has been applied here to collect the primary data under this lockdown situation. Mainly this survey has been done in two phases. In the first phase (First week of September 2020), considering this present situation relating to lockdown, the telephonic interview was done with the beekeepers-respondents of several parts of West Bengal, who were migrating to different mustard fields of Malda, North Dinajpur, and South Dinajpur districts during entire honey flow session before lockdown. The opinions of 150 beekeeper respondents were considered in the study. Questionnaires were made, mainly consisting of close-ended questions on a 5-point Likert Scale to study the severity of the problems faced by the beekeepers of West Bengal during the sudden lockdown period due to the COVID-19 pandemic. In the second phase, a structured questionnaire has made under Google Forms and forwarded through WhatsApp and Facebook to few known people. These few people have filled up the form and then forwarded the link to their friends/relatives. The study has been conducted between August 2020 and October 2020. The total number of respondents, i.e., the common people, is 137. Among the total respondents, 68 are male and 69 are female, out of which 40 numbers of respondents live in rural areas, and 97 numbers of respondents live in urban areas. The Chi-square Test and paired t-test have been used through SPSSv23 for testing hypothesis, involving several factors (e.g., gender, dwelling place, annual family income, the total number of members in a family, and prior knowledge on benefits of apiculture products) to study whether there is any significant difference between mean consumption of honey before the pandemic and that of during pandemic situation.

17.5 Hypothesis

Several hypotheses have been constructed to analyze the honey consumption scenario -

- Gender and honey consumption are independent.
- The nature of the dwelling place and honey consumption is independent.
- Annual family income and honey consumption are independent.
- The number of family members and honey consumption is independent.
- Percentages of children present in a family and honey consumption are independent.
- Prior knowledge about the benefits of apiculture products and honey consumption is independent.
- There is no significant difference in the mean consumption of honey before and during the pandemic situation.

17.6 Results and Discussion

Beekeepers of West Bengal always follow the "honey flow period" for their migratory beekeeping and utilize flowering plants for their bee colonies for the collection of nectar and pollen. The lockdown hampered the entire plan of the beekeepers and restricted them at their respective fields where the dearth (with no honey flow) period had come. It brings different problems to those beekeepers stuck in those fields due to the restriction of the transport facilities. Besides being harassed by local people, they were often interrogated by the different government departments. Most of such incidences were generated due to the rumors about the imposition of government restrictions on outsiders in the villages where they were camping in the fields. The subsequent problems came from overstocking raw honey for lack of traders and problems for the scarcity of working capital due to detachment with societies or traders. Amidst lockdown, they were measured with suspicious eyes during their stay at the remote honey fields, during their movement on roads, during their entry into the new villages, and even during their return to the homelands after migration.

Besides production, one of the significant problems is the selling problem. They cannot supply their raw honey to the market as demand in international and national markets has been falling due to pandemics and as transport activities are hampered during this time.

Modern beekeeping is migratory as the central promise of beekeepers has to ensure a regular supply of food to their bees either by migrating the colonies in flowering areas or by giving artificial food like sugar in the colonies as the flowering plant cannot blossom round the year. When there is no honey flow, those days are called the "dearth period," which is the major problem in apiculture. The presence of this floral dearth period results in decreasing and desertion of the bee colonies. Every locality has its own floral dearth periods of different durations. Therefore, beekeepers have to observe and retain information about the dearth period in their present camping area and the area where they need to move their bee colonies. Hence beekeepers of West Bengal migrate in different fields of the state or outside the state.

In West Bengal, before this lockdown, most of the beekeepers were camping at the honey fields of Malda, North Dinajpur, South Dinajpur, Murshidabad, Nadia, Hooghly, Birbhum, Bankura, Purulia, and West Midnapur for collecting honey from Mustard (*Brassica juncea*), Cumin (*Cuminum cyminum*), and Coriander (*Coriandrum sativum*). Since mustard fields give the premium quality of honey, most of the beekeepers opted for staying in the mustard fields. Few migrated to cumin and coriander honey fields. Even very few migrated to Litchi (*Litchi chinensis Sonn*) and Mango (*Mangifera indica*) fields in North Dinajpur, South Dinajpur, Murshidabad, or the neighboring districts in Bihar; and some moved far south to the North and North 24 Parganas. Very few have reported migrating to Almorta (*Lathyrus sativus)* fields to South Bengal, mainly Murshidabad, Nadia, Burdwan, Birbhum, South 24 Parganas, and North 24 Parganas. When it was almost time for the beekeepers to wind up for the season, the pandemic and lockdown struck and hampered all their plans of departing the collecting locations.

The primary survey revealed several facets of the problems faced by the beekeepers of West Bengal during the pandemic. As maintaining the bee boxes, regular feeding the bees involves handsome expenditure, though most of them tried their level best to hold on to their profession, the number of boxes maintained by many beekeepers reduced quite sharply (Fig. 17.1). 11 out of 150 beekeeper respondents have reported leaving the profession entirely.

Due to strict lockdown in the initial pandemic phase, most of the beekeepers (except 2 out of 150 respondents) were forced to sit out of work. The span of loss of migration months ranged from 2 to 4 months in most cases, and after that, they started pursuing their livelihood though in a much-muffled manner (Fig. 17.2).

Apart from the decline in bee boxes and loss of migration months, several other problems were also faced by the beekeepers, which compelled them to accept the losses from production as well as migration cost. 5-Point Likert Scaling was adopted to record the respondents' answers against the given questions where respondents were asked to give appropriate weightage according to their judgment about the severity of different problems; on the basis of which the critical issues were identified (Table 17.1). The problem in staying at migrated honey fields during the

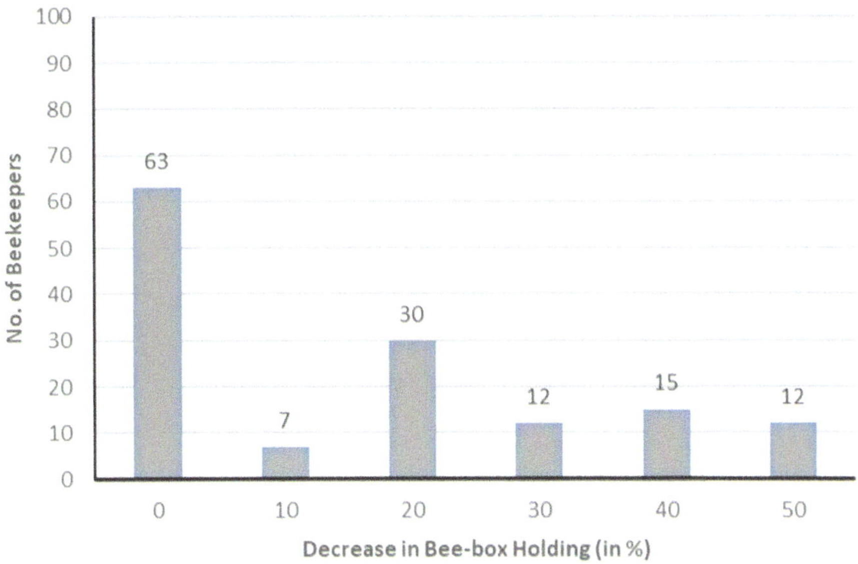

Fig. 17.1 Declining bee box holding by the selected beekeepers in West Bengal from January 2020 to September 2020 (Data Source: Authors)

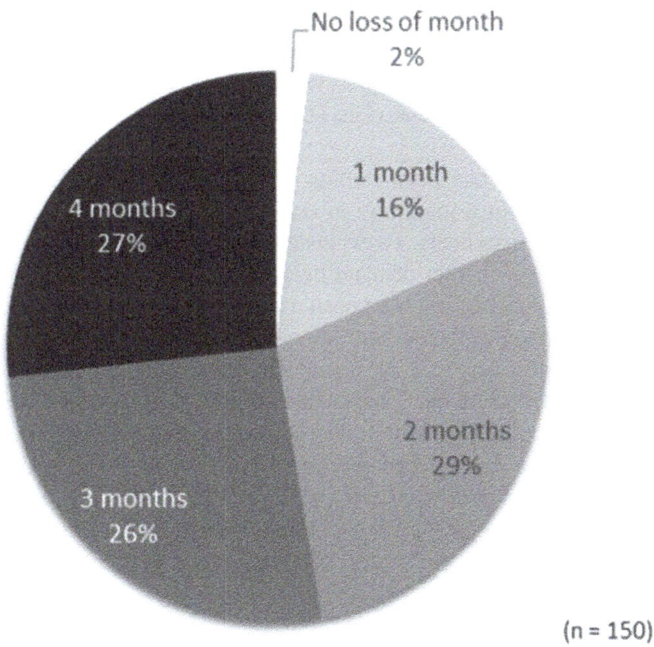

Fig. 17.2 Loss of migration months of the beekeepers due to lockdown during COVID-19 (Data Source: Telephonic Survey, 2020)

Table 17.1 Problems faced by the respondent beekeepers and their respective weight

5-Point Likert Scale	Staying at migrated honey fields during lockdown		Move on the road during lockdown		Entering into new migration fields for honey collection during lockdown		Staying back at home after migration during lockdown	
X	f	fx	f	fx	f	fx	f	fx
Negligible effect (1)	0	0	0	0	0	0	75	75
Minimum effect (2)	0	0	0	0	0	0	46	92
Moderate effect (3)	0	0	0	0	0	0	26	78
Major effect (4)	0	0	0	0	0	0	3	12
Extreme effect (5)	150	750	146	730	150	750	0	0
Total	150	750	150	745	150	750	150	257
Weighted mean		5		4.96		5		1.71

Data Source: Primary Survey, 2020

lockdown and entering into new fields was evident as comparatively more severe. The beekeepers expressed that how helpless they felt when they were stuck to their migrated fields much beyond their schedule with very little monetary resource at hand to fall back and had to even face the unwelcoming attitude of the locals. Since the lockdown continued unabated for long when the time for visiting the next location for honey collection came, they even missed out on that getting further ridden in loans and misery. While making frantic attempts to get back to their homes with their equipment and boxes they were detained by police and had to pay hefty fine or bribe. So, most of the respondents had a sigh of relief when they managed to get back to their homes but were anxious about continuing their vocation and future earnings.

Ministry of AYUSH has been issuing advisories for general people to take nutritious diet for boosting immunity to combat against COVID where honey is one of the significant ingredients. To evaluate the importance of honey consumption and to assess the actual market demand here some hypothesis tests are used. The Chi-Square Test and Paired t-test have been used to find out the relationship between the several socioeconomic factors and honey consumptions. Primary survey data analysis has revealed a very starling result. There has been a common belief that during the lockdown pandemic phase, the general people were more inclined towards honey consumption, but the study showed that there had been no significant change in honey consumption behavior among the public. The beekeepers also reported that the cooperative societies were also forthcoming to collect honey as there were inter and intra-state market restrictions and the sale in general was also not very promising.

17.6.1 Application of Chi-Square Test

17.6.1.1 Test I: Examining the Association Between Gender and Honey Consumption

This test considers a null hypothesis (H_0) and an alternative hypothesis (H_1).

H_0 = Gender and Honey Consumption are independent; and,

H_1 = Gender and Honey Consumption are not independent (Table 17.2).

Here the value of the Pearson Chi-Square test statistic is 25.821a, with 23 degree of freedom, and Asymptotic Significance (2-sided) is 0.309. Since the P-value (2-sided) of "0.309" is greater than α ("0.05"), the Null Hypothesis is accepted, which signifies that gender and honey consumption are independent.
There is no gendered pattern of honey consumption.

17.6.1.2 Test II: Applying Chi-Square Test to Know the Association Between Dwelling Place and Honey Consumption

H_0 = Dwelling Place and Honey Consumption are independent; and,

H_1 = Dwelling Place and Honey Consumption are not independent (Table 17.3).

Here the value of the Pearson Chi-Square test statistic is 18.483a, with 23 degree of freedom, and Asymptotic Significance (2-sided) is 0.731. Since the P-value (2-sided) of "0.731" is greater than α ("0.05"), the Null Hypothesis is accepted. It means dwelling place and honey consumption are independent. The residents of both rural and urban areas have similar honey consumption patterns.

Table 17.2 Chi-Square Test to know the association between gender and honey consumption

Test	Value	df	Asymptotic Significance (2-sided)
Pearson Chi-Square	25.821[a]	23	0.309
Likelihood ratio	33.661	23	0.070
Linear-by-linear association	0.003	1	0.995
N of valid cases	137		

Data Source: Online Primary Survey, 2020

[a]36 cells (75.0%) have an expected count of less than 5. The minimum expected count is 0.50

Table 17.3 Chi-Square Test to know the association between dwelling place and honey consumption

Test	Value	df	Asymptotic significance (2-sided)
Pearson Chi-Square	18.483[a]	23	0.731
Likelihood ratio	21.427	23	0.555
Linear-by-linear association	0.792	1	0.373
N of valid cases	137		

Data Source: Online Primary Survey, 2020

[a]40 cells (83.3%) have expected count less than 5. The minimum expected count is 0.29

17.6.1.3 Test III: Examining the Association Between Annual Family Income and Honey Consumption

H_0 = Annual Family Income and Honey Consumption are independent; and,

H_1 = Annual Family Income and Honey Consumption are not independent (Table 17.4).

Here the value of the Pearson Chi-Square test statistic is 204.075a, with 161 degree of freedom, and Asymptotic Significance (2-sided) is 0.012. Since the P-value (2-sided) of "0.012" is less than α ("0.05"), the Null Hypothesis is rejected. It means annual family income and honey consumption are not independent; with more disposable income at hand, the affluent families can afford the luxury of consuming honey during the pandemic when many were struggling to arrange four square meals a day.

17.6.1.4 Test IV: Examining the Association Between the Total Member in a Family and Honey Consumption

H_0 = Total Member in a Family and Honey Consumption are independent, and,

H_1 = Total Member in a Family and Honey Consumption are not independent (Table 17.5).

Here the value of the Pearson Chi-Square test statistic is 250.326a, with 230 degree of freedom, and Asymptotic Significance (2-sided) is 0.171. Since P-value (2-sided) =0.171 is greater than α = 0.05, the Null Hypothesis is accepted.

Table 17.4 Chi-Square Test to know the association between annual family income and honey consumption

Test	Value	df	Asymptotic significance (2-sided)
Pearson Chi-Square	204.075[a]	161	0.012
Likelihood ratio	141.185	161	0.868
Linear-by-linear association	0.262	1	0.608
N of valid cases	137		

Data Source: Online Primary Survey, 2020

[a]190 cells (99.0%) have an expected count of less than 5. The minimum expected count is 0.02

Table 17.5 Chi-Square Test to know the association between total member in a family and honey consumption

	Value	df	Asymptotic significance (2-sided)
Pearson Chi-Square	250.326[a]	230	0.171
Likelihood ratio	134.943	230	1.000
Linear-by-linear association	1.402	1	0.236
N of valid cases	137		

Data Source: Online Primary Survey, 2020

[a]261 cells (98.9%) have an expected count of less than 5. The minimum expected count is 0.01

It means the total number of members in a family and honey consumption is independent. Since, in most families, the aged and the young are honey consumers, the other family members do not have much bearing on honey consumption.

17.6.1.5 Test V: Examining the Association Between the Percentage of Children in a Family and Honey Consumption

H_0 = % of Children in a Family and Honey Consumption are independent; and,

H_1 = % of Children in a Family and Honey Consumption are not independent (Table 17.6).

Here the value of the Pearson Chi-Square test statistic is 107.301a, with 69 degrees of freedom, and Asymptotic Significance (2-sided) is 0.002. Since the P-value (2-sided) of "0.002" is less than α ('0.05'), the Null Hypothesis is rejected. It means the percentage of children in a family and honey consumption is not independent. All are very conscious about the immunity and health condition of the children of the family and thus take extra care of their immunity which includes having honey too.

17.6.1.6 Test VI: Examining the Association Between Prior Knowledge About Benefits of Apiculture Products and Honey Consumption

H_0 = Knowledge on Benefits of Apiculture Products and Honey Consumption are independent; and,

H_1 = Knowledge on Benefits of Apiculture Products and Honey Consumption are not independent (Table 17.7).

Here the value of the Pearson Chi-Square test statistic is 404.006a, with 184 degree of freedom, and Asymptotic Significance (2-sided) is 0.000. Since the P-value (2-sided) of "0.000" is less than α ("0.05"), the Null Hypothesis is rejected. It means Prior knowledge about the benefits of apiculture products and honey consumption is dependent. The section of the population who are aware of the

Table 17.6 Chi-Square Test to know the association between the percentage of children in a family and honey consumption

	Value	df	Asymptotic Significance (2-sided)
Pearson Chi-Square	107.301[a]	69	0.002
Likelihood ratio	51.939	69	0.938
Linear-by-linear association	0.728	1	0.393
N of valid cases	137		

Data Source: Online Primary Survey, 2020

[a]84 cells (87.5%) have an expected count of less than 5. The minimum expected count is 0.01

Table 17.7 Chi-Square Test to know the association between prior knowledge about benefits of apiculture products and honey consumption

Test	Value	df	Asymptotic Significance (2-sided)
Pearson Chi-Square	404.006[a]	184	0.000
Likelihood ratio	139.167	184	0.994
Linear-by-linear association	3.888	1	0.049
N of valid cases	137		

Data Source: Online Primary Survey, 2020
[a]209 cells (96.8%) have an expected count of less than 5. The minimum expected count is 0.01

Table 17.8a Paired samples descriptive statistics

Test series		Mean	N	Std. deviation	Std. error mean
Pair 1	Before pandemic (grams)	375.58	137	684.324	58.466
	During pandemic (grams)	291.06	137	506.339	43.259

Data Source: Online Primary Survey, 2020

Table 17.8b Paired samples correlations

Test series		N	Correlation	Sig.
Pair 1	Before pandemic (grams) & during pandemic	137	0.645	0.000

Data Source: Online Primary Survey, 2020

health benefits of honey consumption take the initiative to include honey in their diet.

17.6.2 *Application of Paired T-Test*

17.6.2.1 Test VII: Examining Whether There Is Any Significant Difference Between Mean Consumption of Honey Before and During the Pandemic Situation

The test considers the following:

H_0 = mean consumption of honey before the pandemic and that of during pandemic situation are equal; and,

H_1 = mean consumption of honey before the pandemic and that of during pandemic situation is not equal.

T-TEST pairs = Monthly Honey Consumption of Family (in Grams) Before Pandemic with Monthly Honey Consumption of Family (in Grams) During Pandemic Situation (Tables 17.8a, 17.8b, 17.8c, and 17.8d).

Here level of significance $\alpha = 0.05$. The coefficient of relatedness between "Monthly Honey Consumption of Family (in Grams) Before Pandemic" and "Monthly Honey Consumption of Family (in Grams) During Pandemic Situation"

Table 17.8c Paired samples t-test

Test series		Paired difference					t	df	Sig.(2-tailed)
					95% confidence interval of the difference				
		$\bar{x}$	SD	SE	Lower	Upper			
Pair 1	Before pandemic (grams) and during pandemic (grams)	84.518	526.673	44.50	−4.47	173.50	1.88	136	0.062

Data Source: Online Primary Survey, 2020

Table 17.8d Paired samples effects sizes

Test series			Standardizer[a]	Point estimate	95% Confidence interval	
					Lower	Upper
Pair 1	Before pandemic (grams) and during pandemic (grams)	Cohen's d	526.673	0.160	−0.008	0.329
		Hedges' correction	528.130	0.160	−0.008	0.308

Cohen's d uses the sample standard deviation of the mean difference
Hedges' correction uses the sample standard deviation of the mean difference, plus a correction factor
Data Source: Online Primary Survey, 2020
[a]The denominator used in estimating the effect sizes

is significant, as P = 0.000, which is <0.05. The average difference between before and during pandemic consumption is 84.518. The 95% confidence interval for μd is (−4.465, 173.502). In the pair samples test, the value of the test statistics, t = 1.878, with 136 degree of freedom and P-value (2 tailed) is =0.062. Since P-value (2 tailed) =0.062 is greater than α = 0.05, the Null Hypothesis is accepted. That means the mean of consumption of honey before the pandemic and that of during the pandemic situation is equal. Though monthly family consumption of honey has been decreased during the pandemic situation (mean monthly consumption 291.06 grams) from that of before the pandemic (mean monthly consumption 375.58 grams), this difference is not significant. Though respondents were aware of the health benefits of having honey as an immunity booster, many were unable to scale up their consumption due to economic constraints, and lockdown impacted the supply chain of the cooperative honey supply also.

17.7 Findings

The study based on a telephonic survey of the beekeepers found that beekeepers faced myriads of problems during the lockdown phase and thereafter, which severely affected their livelihood. Their entire migratory pattern, bee box holdings, earnings were jeopardized. Usually, they get at least 12 kg of honey from each bee box per month during the full honey flow season; but this year, beekeepers have lost an average 2.5 months migrations period during the lockdown. Hence, beekeepers have lost on an average 30 kg production of honey per box due to this pandemic situation. Moreover, beekeepers have lost their opportunity to sell their closing stock of honey due to the absence of traders in the field due to the pandemic situation. People who bought honey as an immunity booster in this phase opted for branded ones like Dabur India Ltd., Patanjali Ayurved Ltd., and others, keeping the cooperatives at bay. The companies that purchased honey from the beekeepers just before the pandemic were yet to clear the dues for earlier consignments, making lives more miserable. As their profession involves migrating from one place to the other, people feared that providing access to them might lead to virus infection and were often harassed unnecessarily.

Though both central and state governments have been making many efforts to develop the economic conditions of the beekeepers and to improve the beekeeping industry as a whole, the benefits, in many cases, did not reach the grassroots level where poor beekeepers struggled. A beekeeping development committee was set up under Bibek Debroy in 2019 to jot down measures to achieve the target of doubling farmers' income through beekeeping (The Economic Times 2019). Since this plan did not succeed entirely, in May 2020, the Ministry of Agriculture & Farmers' Welfare took initiatives to promote beekeeping as a supplementary income source for the farmers and double their income by 2024; an allocation of Rs.500 crores was done towards apiculture under the *Atma Nirbhar Abhiyan*. India is among the five leading honey-producing countries, having a meteoric rise in both production and export. So, an agency like the National Bee Board (NBB) has been developed as part of the National Beekeeping and Honey Mission (NBHM), and nearly 30 lakh farmers were trained in scientific beekeeping (Ministry of Agriculture and Farmers Welfare 2020. Emphasis was given on organic honey production and its smart branding, which has a robust international market ushering a "sweet revolution" (Ministry of Agriculture and Farmers Welfare 2020). Khadi and Village Industries Commission (KVIC) is another important honey mission in India that strengthens the beekeeping activity with the formation of 25 lakh bee colonies, helps the farmers in adapting to modern apiculture technologies through trainings, provides business loans, and also strengthens the ties with national and international markets (India Filings 2021). This pandemic situation has again pointed out that these schemes and measures have a long way to go realize their goal.

From the sample survey conducted among the general public, it was seen that there is no relationship between gender, dwelling place, and the total number of members in a family. However, the factors like annual family income, percentage of children in a family, and prior knowledge about the benefits of apiculture products show dependency on honey consumption. The mean of consumption of honey before the pandemic and that of during the pandemic situation has not varied. Though different advisories were to increase honey intake, the monthly family consumption of honey had not been improved during pandemic situations from that of prior-pandemic times. During the pandemic phase, many lost their jobs, the income of the family went down it was not possible for many to bear the additional cost of buying a non-essential commodity like honey, although it has got medicinal properties. There is a chance of an increase in consumption once the economy normalcy returns. Greater awareness needs to be generated to promote the essentiality of honey. It was seen from the survey that the respondents who were more aware of the variety of apiary products consumed honey more, so increasing the knowledge base of the consumers might help the cause. Apart from corporate campaigns, government-level initiatives are needed to popularize honey and other apiculture products produced by cooperatives, Non-Government Organizations, and Self-Help Groups.

17.8 Ways Forward

The central government announced a COVID-19 special relief package of Rs 500 crores for beekeeping initiatives (Bloomberg Quint, 2020). The government has announced that a new law will be formulated to provide adequate choices to farmers to sell their produce at attractive prices, barriers on inter-state trade will be abolished, and a framework for e-trading of agriculture products will be established. Jagaran English (2020) reported that this scheme would be aimed at integrated beekeeping development centers, collection, marketing and storage centers, post-harvest, and value addition facilities; this will lead to an increase in income of 2 lakh beekeepers. Kumar (2020) said that the Center's allocation of ₹500 crores for apiculture, or beekeeping, as part of its monetary stimulus package is expected to boost bee farming across the country, and beekeepers say that this fund can be used to provide hands-on training to farmers to make the sector more profitable. But some other measures that could be taken are –

- The government needs to take initiatives to provide bee boxes to the affected beekeepers at a concessional rate.
- As the beekeepers are reliant on rolling money, 1 year's loss will affect the next year's production too. So, arrangements for an interest-free loan for a long period would be much appreciated.
- Government should include beekeeping activities as an essential service. However, the Central Government, as per the recommendation of ICAR, issued an advisory notice to the state (including the local body) considering beekeeping as an agriculture service where beekeeping is treated as an essential one. But Local level government does not know whether beekeeping is an agricultural activity or

not. Moreover, the West Bengal government also issued notices from time to time declaring relaxation of restriction from lockdown to the agriculture activities, where beekeeping was also implied. Hence police and local bodies harassed the poor beekeepers unlawfully, and such boisterousness needs to be intervened by the higher authority.

- To improve the consumption of honey in West Bengal as well as throughout India publicizing the various usages of honey needs to be done by governmental and non-governmental organizations. Awareness and promotional camps might be organized at various places, especially at educational institutions.
- Special awareness programs targeting different strata of the population, training camps can be conducted for the beekeepers by government, forest department, non-government organizations, apiculture cooperative societies can be adopted to make them more proficient and risk resilient.

References

Chaudhuri S (2020) Untold honey trap of lockdown. Available at: https://www.deccanherald.com. Accessed 12 Sep 2020

Das S (2020) West Bengal govt develops safe, lucrative way of collecting honey for Sundarbans villagers. Available at: https://www.deccanherald.com. Accessed 12 Sep 2020

Department of Economic Affairs (2020) Monthly economic review- August, 2020. Department of Economic Affairs, New Delhi

Express News Service (2020) COVID-19 strikes huge blow to apiculture in Odisha's Kendrapara. Available at: https://www.newindianexpress.com. Accessed 1 Sep 2020

Ghosh GK, Ghosh S (1999) Honey as food, medicine, cosmetics and honey recipes. Firm a KLM Private Limited, Calcutta

Green K (2020) Lack of bee imports due to COVID-19 will affect honey supply, agriculture. Available at: https://calgary.ctvnews.ca. Accessed 1 Sep 2020

Hall HH, Lothrop RE (1934) The use of clarified honey in culture media. Available at: http://www.jb.asm.org. Accessed 30 July 2016

India Filings (2021) National Beekeeping Honey Mission (NBHM). https://www.indiafilings.com/learn/national-beekeeping-honey-mission-nbhm. Accessed 30 June 2021

Jebaraj P (31 August 2020) GDP contracts by record 23.9% in Q1 NEW DELHI, AUGUST 31, 2020 IST. Available at: https://www.thehindu.com. Accessed 9 May 2020

Jones R, Sweeney-Lynch S (2011) The Beekeeper's bible. Stewart, Tabori & Chang, New York

Joshi V (2020) Beekeepers failed to recover cost this season on export suspension. Available at: Joshi V (2020) Beekeepers failed to recover cost this season on export suspension. Available at: https://www.hindustantimes.com. Accessed 30 Aug 2020

Khan MI (2020) COVID-19: Bihar beekeepers find it tough to relocate bee boxes amid lockdown. Available at: https://www.downtoearth.org.in. Accessed 10 June 2020

Kumar NV (2020) Beekeepers want training as govt allocates ₹500 crore for apiculture. Available at: https://thefederal.com/news/beekeepers-want-training-as-govt-allocates-%E2%82%B9500-crore-for-apiculture/. Accessed 1 Sep 2020

Ministry of Agriculture & Farmers Welfare (2020) India among world's top five honey producers. https://pib.gov.in/PressReleasePage.aspx? PRID=1626128. Accessed 30 June 2021

Nagaraja N, Rajagopal D (2009) Honey bees- diseases, parasites, pests, predators and their management. MJP Publishers, Chennai

Priyadarshini S (2020) Honey bees starve in COVID-19 lockdown. Available at: https://www.nature.com. Accessed 15 Aug 2020
Priyadarshini S, Menon A (2020) Bee keepers to the rescue. Available at: https://www.thehindu.com. Accessed 23 Aug 2020
Root AI (1980) The ABC and XYZ of bee culture. The A. I. Root Company, Medina
Singh G (2020) Embracing beekeeping to stop fatal tiger attacks in Sundarbans mangrove forest. Available at: https://www.lifegate.com/sundarbans-beekeeping. Accessed 12 Sep 2020
Singh SJ (1958) Honey for health. Nature Cure Research Hospital, Lucknow
The Economist (2020) China's beekeepers feel the sting of covid-19 Lockdown. Available at: https://www.economist.com. Accessed 2 Sep 2020
The Economic Times (2019) Government to give status of farmers to landless beekeepers. https://economictimes.indiatimes.com/news/economy/agriculture/government-to-give-status of-farmers-to-landless-beekeepers/articleshow/69957060.cms?from=mdr. Accessed 30 June 2021
The Hindu PTI (4 December 2020). World Bank sees FY21 India growth at 1.5-2.8%, slowest since economic reforms 30 years ago. Available at: https://www.thehindu.com. Accessed 17 Feb 2020
Trtworld (2020) How the pandemic affects agricultural production across the world. Available at: https://www.trtworld.com. Accessed 12 Aug 2020
UN India (2020) Bees, Biodiversity and COVID-19. Available at: https://in.one.un.org. Accessed 30 Aug 2020
Yazhiniyan (2020) Virus effect turns beekeeping trade sour in state. Available at: https://www.dtnext.in. Accessed 2 Sep 2020

Sanghamitra Purkait is a Research Scholar, Department of Geography, Diamond Harbour Women's University, Sarisha, Pin-743,368, West Bengal, India, since 2018. She did her Master from The Vidyasagar University and M.Phil. from Jadavpur University (Environmental Science). Her field of interest is Environmental Science, Socio-economical geography, Beekeeping, and Beekeepers. She presented few papers in national and international seminars.

Anindya Basu is working as an Assistant Professor in the Department of Geography, Diamond Harbour Women's University since 2017. Previously, from 2011, she served as an Assistant Professor in Women's Christian College, Kolkata. She has been a rank-holder in Bachelor's and Master's degree programme in Geography from the University of Calcutta and UGC-Junior Research Fellow recipient. She has also received PG Diploma in RS-GIS from Jadavpur University. She has keen interest in socio-political geography, urban-environmental issues, and tourism geography. Dr. Basu presented a number of papers in national and international conferences and seminars, apart from publishing several research papers and book chapters.

Chapter 18
The Importance of Public Expenditure on Crop Production and Exports in India: An Econometric Analysis in the Pandemic Trapped Era

Labonie Mukhopadhyay

Abstract Crop yield is a major concern to the farmers in many of developing economies including India. The recent protests by a large number of farmers against the Government of India's farm bill raise the issue of crop production or export. On the other hand, from March 11, 2020, Novel Corona Virus creates global pandemic situation (according to WHO) and India is no exception. This pandemic has a significant impact on health which could emerge in the short- and long-term time horizons, necessitating early detection and careful planning to reduce the socioeconomic burden. The study examined the sector-specific dynamic impacts of crop production, export, and public spending on agriculture in India. We explore and compare the data from 1987–1988 to 2018–2019. Following the Unit root test, the estimation of cointegration for long-run relationships is carried out in this analysis. Empirical findings of VECM imply that Growth in Crop production (LC) is positively related to Export (LEX) and Public Expenditure (LPE). The study concludes that the agricultural sector's emergence as the only area to experience positive growth can be attributed to an increase in agricultural production as a result of a favorable monsoon and targeted government expenditure in this pandemic.

Keywords Crop production · Export · Public expenditure · Agriculture · Cointegration

18.1 Introduction

India's economy is regarded as one of the world's most important emerging economies. Public expenditure encourages capital accumulation and long-run economic growth. Aside from employment, food, and infrastructure, rural sector investment

L. Mukhopadhyay (✉)
Awadh Dental College and Hospital, Jamshedpur, Jharkhand, India

S. Saha et al. (eds.), *Economic and Societal Transformation in Pandemic-Trapped India*, New Frontiers in Regional Science: Asian Perspectives 55,
https://doi.org/10.1007/978-981-16-5755-9_18

enhances overall economic development. Several pieces of research have analyzed the relationship between agriculture and economic development from various perspectives. Divergences in the methodology and variables used in the study may explain the discrepancy in the correlation between public agriculture spending and economic development. Our study investigates the relationship between crop yield, public agriculture spending, and agricultural export from the recent past to determine the sustainability issues that have arisen as a consequence of the pandemic.

18.1.1 Role of Public Expenditure in Agriculture in Motivating Farmers to Increase Production

Infrastructure is necessary for the subsistence and progress of the agricultural segment and the economy. Investment in agricultural infrastructure leads to reduced costs per unit, increased productivity, increased revenue generation, and capacity building. Public expenditure in Agriculture has significant spill-over effects, such as the development of rural areas, the alleviation of poverty, and the proper conservation and management of natural resources. Public expenditure in agriculture includes facilities that aid farmers from sowing to selling: irrigation, road access, electrification, storage, telecommunications, etc. This study explores the relationship between crop production, agricultural exports, and public expenditure in Agriculture to identify the equilibrium model.

18.1.2 COVID-19 and Farmers' Issues

About the fact that the available literature on the effect of COVID-19 on farmers' issues is limited, the existing analytical research has yielded an intriguing finding. Social distancing reduces agricultural productivity and day-to-day activities and disrupts storage and transportation, resulting in a decline in revenue and a cash flow challenge for farmers, food merchants, and distributors.

Concerning farmer health, they have to perform all activities from showing to harvesting in the field and post-harvest processing, storage, and marketing. In the current COVID-19 situation, farmers are having difficulty drying, shelling, and selling their commodities in the market. Farmers, especially those who produce fruits and vegetables, may be forced to make distress sales. Agriculture, Horticulture commissions, and State agriculture universities played a significant role in connecting farmers to markets to sell perishable agri-commodities in these circumstances. NGOs and government bodies play a significant role in finding profitable markets and connecting producers. Another essential task of government agencies is to publicize COVID-19 control measures along with agricultural information.

18.1.3 New Equations in Agricultural Performance During Pandemic

India is the fifth largest economy by nominal GDP ($ 3.202 Trillion) and the third-largest by Purchasing Power parity ($ 11.321 Trillion). However, as per CMIE (Centre for Monitoring Indian Economy) data, unemployment in India is increasing in the rural sector than in urban areas. Hence the burning question arose that what would be the PPP (Purchasing Power parity). The ILO (International Labour Organization) predicts that 25 million jobs will be lost, and the United Nations Development Program (UNDP) expects a loss of USD 220 billion in revenue from developed countries (ILO Monitor Ed-3, 2020). On the other side, food prices, such as vegetables and paddy, are steadily increasing in India's high-producing areas/districts. Compared to previous years, there is a high price in Paddy, Potato, and Onion, etc., in this pandemic time. In such circumstances, it is clear that we are in a crucial position in terms of sustainability issues.

Food demand in India is rising in tandem with the growing population. However, negative consequences of the Green Revolution, such as environmental trade-offs and fundamentalism in staple cereals, have since emerged. So, it is preferable to transition to a more appropriate model with a greater emphasis on nutrition. The study also seeks to assist policymakers in developing better post-COVID strategies for a healthier population based on previous food and agriculture policies.

In terms of sustainability, specifically economical and social factors, this study seeks to determine whether there is a statistically significant relationship between agricultural productivity (Crop Production Index) and Government expenditure on agriculture or export policies implemented by the Government of India so that policymakers and other government implementation bodies can confront issues such as unemployment and inflation, which are crucial to purchasing power parity.

Fortunately, only the agricultural sector's emergence as the only area to experience positive growth can be ascribed to an increase in agricultural production as a result of a favorable monsoon and targeted government expenditure (Singh and Mishra 2021).

The Indian government has taken steps to assist farmers in the wake of the COVID-19 pandemic. Indian Finance Minister N. Sitharaman stated that the first installment of Rupees 2000 would be transferred to the account of 86.9 million farmers (under PM Kishan Scheme) in April 2000 as part of a cash transfer program for all segments of society (Times of India, August 18, 2020).

18.2 Literature Survey

The hypothesis of export-led growth is thoroughly investigated. In the post-liberalization era, two critical issues became prominent. Pro-liberalization economists opine that once the prices are corrected through the free play of the market, farmers could be able to realize better prices for their agricultural produce, which, in turn, would encourage them to produce more. Gulati and Sharma (1997) pointed out that agricultural income would have increased by 16–25% had domestic prices aligned with international prices. Liberalization of agricultural trade has been accepted on the argument that free trade and opening up of the economy would maximize efficiency and gains. It is advocated that India has comparative advantages in producing highly valued export-oriented crops like fruits, vegetables, and flowers, and therefore, farmers should be encouraged to diversify their crops in favor of highly valued crops. Pro-reformists also conclude that input subsidies and output support prices have suppressing effects on domestic prices and, in turn, kept the terms of trade against agriculture. Parikh (1997) pointed out that agricultural subsidies were 'fiscally unsustainable…inefficient, and costly to farmers'. Gulati and Sharma (1997) speculated that a cut down in agricultural subsidy would arrest the long-term declining trend in public investment in agriculture.

On the other hand, Contemporary researches argue that economic reforms primarily emphasized price factors and infrastructure. However, the institutional aspect of agriculture has grossly been neglected, and as a consequence, deceleration in agricultural growth and a decline in growth of output were accentuated since 1991 (Chadha 2002; Bhalla 2002; Kumar 2002). The public sector gross capital formation in agriculture declined to one-third in 1999–2000, compared to the level attained in 1980–1981 (Reddy 2006a, b). Other studies also revealed that a fall in public sector investments in agriculture had adversely affected agricultural productivity and growth (Chand 2001; Gulati and Bathla 2001).

Ekanayake (1999) used data from 1960–1961 to 1996–1997 to conducted a time series study of the causal relationship between export and economic development in eight developing nations from Asia. Except for Malaysia, the study confirmed the assertion of the export-led development hypothesis.

Chandra Parida and Sahoo (2007) investigated the export-led development for four Asian countries, viz., India, Sri Lanka, Bangladesh, and Pakistan, using Pedroni's panel cointegration methodology. Their findings support the legitimacy of the export-led development hypothesis.

Safdari et al. (2011) employed VECM to explore the causal linkage between economic development and export in 13 developing nations from 1988 to 2008. Their findings show unidirectional causality relation among economic growth and export.

Haalisha and Jahufer (2018) studied the correlation between Consumer Price and Gross Domestic Price of Sri Lanka from 1960 to 2016. Their study reveals that no short-run relationships exist but are in the long run.

Dev (2020) addressed an unprecedented challenge to India due to COVID-19 and its effect on productivity, food security, and livelihoods. The study presented the Pradhan Mantra Garib Kalian Yojana, the activity of the Government of India, to address social safety nets as a bridge between health shock and economic shock. It revealed that nearly 40–50 million seasonal workers in India and thousands of migrant workers from many states had been trudging for miles on highways.

India's economy depends on the informal labor market, and the effects of lockdown are highly disruptive.

Mukhopadhyay et al. (2020) examined the correlation between public expenditure on infrastructure with agricultural sustainability in the Indian context. The study investigated trends in data from 1970 to 2015 and revealed a long-run relationship between public spending and agricultural sustainability. The vector error correction model (VECM) outcome of the study implies that public spending significantly affects agricultural productivity with a one-period lag.

18.3 Research Gap and Objective of the Study

Whether crop production impacts the export and public expenditure, the study carried out a comprehensive literature review, summarized in the Literature Survey chapter. Previously no significant studies were conducted regarding the impact of globalization on both exports and public expenditure on Crop Production in India. Researchers mostly confined to explore the factors impacting various policy measures, reduction of subsidies, restriction of exports, trade-distorting effects, incremental change in high-value agro product decline in public investment, etc. This background led to the exploration of emerging issues concerning the role of public expenditure in agricultural performance and exports regarding Crop Production in India during the post-reform period, where agricultural reform policy is one of the main concerns. In the case of India, the study will contribute to the economic background research of the country.

18.3.1 Research Question

Do public expenditure and agricultural exports stimulate crop production in the post-globalization era?

18.3.2 Objectives

As the relationship between Public Expenditure, Agriculture Export, and Crop Production is still unexplored, the study aims to identify the exact correlation between Crop Production, Exports of Agriculture, and Public Expenditure on Agriculture. *Public Expenditure on Agriculture* refers to spending by public authorities to develop the Agricultural Sector, covering major and minor irrigation, rural infrastructure (e.g., road, electricity, agricultural finance), soil and water conservation, etc. *Agricultural Exports* are determined by supply-side factors, infrastructure facilities, and several regulations by which farmers are encouraged by Agricultural exports. *Crop Production Index* includes all crops except fodder crops.

18.4 Hypothesis

Based on the above objectives, the present study conducts several theoretical and observational investigations to form various hypotheses. We structure some basic null hypotheses in this regard, which will be examined using the data collected and the optimization technique.

The null hypotheses categorized as follows:

$\mathbf{H_{01}}$ There is no statistically significant interrelationship between agricultural production (Crop Production Index) and Public expenditure on agriculture.

$\mathbf{H_{02}}$ There is no statistically significant interrelationship between agricultural production (Crop Production Index) and Export.

18.5 Data

Secondary data have been collected from a range of published sources, including the RBI Handbook, World Bank Data, and others. Accessible literature like books, Journals, and data published by the Government of India and state governments on issues such as agriculture, energy, transportation, and education are also important sources of data. Data for trend analysis was obtained from 1987–1988 to 2018–2019. Data on Crop Production Index, Public Expenditure, and Agricultural Exports from the same period are also poised for trend comparison.

18.6 Methodologies

Different techniques or econometric methods are used for assessing the relationship, including public expenditure, export of agricultural goods, and crop production. The independent variables were used as proxies for agricultural export and crop production.

Some variables in simultaneous equation models are treated as endogenous, while others are treated as exogenous. However, Sims (1980) asserts that if there is simultaneity among a set of variables, all of these variables should be treated differently. A proper simultaneous equation system is treated as endogenous in the VAR model. Individually two series may be non-stationary, yet they may be cointegrated. The cointegration equation depicts the long-run relationship between two variables. The error correction model is used to construct a more comprehensive model that combines the short-run relationship with long-run dynamics.

An autoregressive process (or model) of order p is used to ensure stationarity. The AR(p) model is a generalization of the AR(1) model, will have the form

$$Y_t = \phi_1 Y_{t-1} + \phi_2 Y_{t-2} + \ldots . \phi_p Y_{t-p} + \varepsilon_t \tag{18.1}$$

Y_t returns current and past values at time t are involved.

The Augmented Dickey and Fuller (1979) (ADF) test is used to determine the unit root principle. The AR (p) with a deterministic trend is a more appropriate model for evaluating a unit root:

$$\Delta X_t = \alpha + \rho X_{t-1} + \eta_1 \Delta X_{t-1} + \eta_2 \Delta X_{t-2} + \ldots\ldots\ldots + \eta_{p-1} \Delta X_{t-p+1} + \delta t + \varepsilon_t \tag{18.2}$$

It selects an appropriate lag length to examine if the series has a unit root (p).

There are several statistical parameters and testing methods to determine the appropriate lag length in an AR(p) model. The maximum lag length is determined by applying Schwert's (2002) rule:

$$P_{max} = \text{integer part of } \left[12(T/100)^{.25}\right]. \tag{18.3}$$

The AIC is often used to determine the proper lag length. The maximum lag length is found to be 1 when such standard are followed. As a result, our model will be:

$$\Delta X_t = \alpha + \rho X_{t-1} + \eta_1 \Delta X_{t-1} + \delta t + \varepsilon_t \tag{18.4}$$

The ECM model may be interpreted in one of two ways with two variables, Y and X:

$$\ln X_t = \theta + \sum_{i=1}^{p} \pi_i \ln X_{t-i} + \sum_{j=1}^{r} \phi_j \ln Y_{t-j} + v_t \tag{18.5}$$

$$\ln Y_t = \alpha + \sum_{i=1}^{m} \beta_i \ln X_{t-i} + \sum_{j=1}^{n} \gamma_j \ln Y_{t-j} + u_t \tag{18.6}$$

or,

$$\Delta \ln X_t = \theta + \sum_{i=1}^{p} \pi_i \Delta \ln X_{t-i} + \sum_{j=1}^{r} \phi_j \Delta \ln Y_{t-j} + \lambda ECM_{t-1} + v_t. \quad (18.5.a)$$

$$\Delta \ln Y_t = \alpha + \sum_{i=1}^{m} \beta_i \Delta \ln X_{t-i} + \sum_{j=1}^{n} \gamma_j \Delta \ln Y_{t-j} + \delta ECM_{t-1} + u_t \quad (18.6.a)$$

ECM denotes the error correction method, where u_t and v_t are zero-mean, serially uncorrelated, random disturbances.

If the data are I(1) but not cointegrated, valid tests can be conducted using the first differences without the error correction term.

$$\Delta \ln X_t = \theta + \sum_{i=1}^{p} \pi_i \Delta \ln X_{t-i} + \sum_{j=1}^{r} \phi_j \Delta \ln Y_{t-j} + v_t \quad (18.7)$$

$$\Delta \ln Y_t = \alpha + \sum_{i=1}^{m} \beta_i \Delta \ln X_{t-i} + \sum_{j=1}^{n} \gamma_j \Delta \ln Y_{t-j} + u_t \quad (18.8)$$

18.7 Results and Discussion

Since the variables are time series, running a multiple regression directly may produce a spurious outcome if these time series are not all stationary. Accordingly, testing for the stationarity by using the appropriate test, among other factors, is a pre-requisite for validating the estimated regression. The Unit Root Test is being used in the study to evaluate the stationarity and underlying nature of the variables (Crop Production Index (LC), Agricultural Export (LEX), and Public Expenditure (LPE).

18.7.1 Unit Root Test

Table 18.1 summarizes the output of the unit root test. All variables in the original series have unit root at level I(0), but the series has stationarity at the first difference, level I (1).

18.7.1.1 Structural Breaks

The existence of a structural break determines whether or not a stochastic process is stationary. Perron (1989) incorporated structural break in the conventional unit root

Table 18.1 Estimated statistics of Unit Root Tests

	Augmented Dickey–Fuller Test Statistics		Phillips–Perron Test Statistics	
Series	Level	First Difference	Level	First Difference
LC	−0.62	−7.08***	−1.32	−7.21***
LPE	0.59	−4.95***	0.594	−4.95***
LEX	−0.074	−4.14***	−0.11	−4.15***

Source: Computed by the author
[n.b. LC = Log value of Crop Production, LPE = Log value of Public Expenditure, LEX = Log value of Agricultural Export]
Note: *** denotes the level of significance at 1%

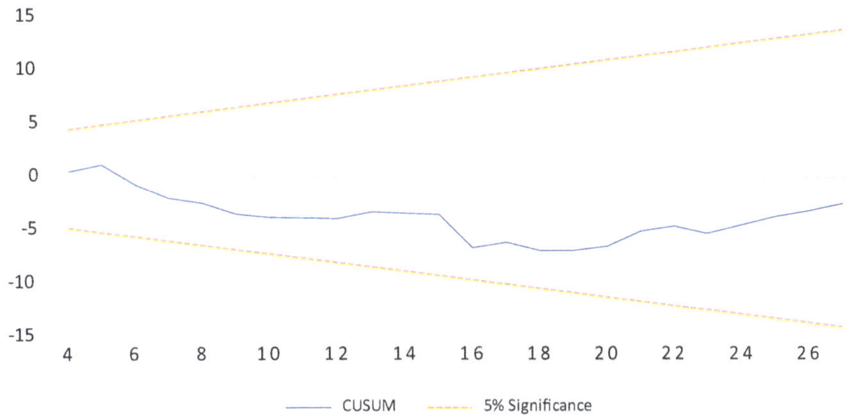

Fig. 18.1 Stability Diagnostics (CUSUM Test) (Source: Prepared by the author)

test (Augmented Dickey Fullar). Bai and Perron (1998) proposed testing procedures for the stability of the equation. This method also has the advantage of consistently estimating the number of breaks and the periods when these occur.

In this study, it is essential to examine if a structural break is present in the series. The study performs the CUSUM Square Test to visualize Structural break and found the plot does not deviate from the 5% significance level boundary (Figs. 18.1 and 18.2).

The CUSUM statistics use the sequence of residual deviations from a model in time series analysis to indicate whether the model is misspecified. In our study, the CUSUM test shows it is within 5% significant level and confirms that the coefficients in ECM are stable.

Another reliability test namely Bai-Perron test confirms the existence of a 'structural break'. Sequential F-statistics determines that the break result is zero, confirming that there is no break in the series (presented in Table 18.2).

The study examined cointegration vectors utilizing Johansen's cointegration tests to determine the dynamic long-term impact of agricultural exports and the various measures of Indian public agricultural expenditure.

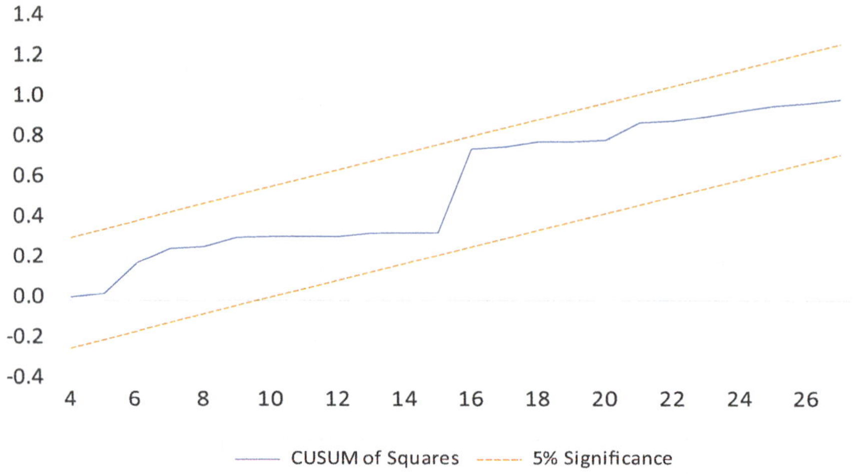

Fig. 18.2 CUSUM Square Test (Source: Prepared by the author)

Table 18.2 Results from Bai–Perron structural break estimation

Sequential F-statistics determined breaks 0			
Break Test	F-Statistics	Scaled F-Statistics	Critical value[b]
0 vs 1	3.651196	10.95359	13.98

Source: Computed by the author
[a]Significant at the 0.05 level
[b]Bai–Perron (Econometric Journal) critical values

Table 18.3 Estimated statistics of Johansen cointegration

Unrestricted cointegration rank test					
Tests	Hypothesized no. of CE(s)	Eigenvalue	Statistics	5% critical value	Prob[b]
Trace	None[a]	0.872973	63.9419	29.79707	0.0000
	At most 1[a]	0.424648	14.42136	15.49471	0.0721
	At most 2[a]	0.046978	1.154811	3.841465	0.2825
Maximum eigenvalue	None[a]	0.872973	49.52054	21.13162	0.0000
	At most 1[a]	0.424648	13.26655	14.2646	0.0721
	At most 2[a]	0.046978	1.154811	3.841465	0.2825

Source: Computed by the author
Trace Test indicates 3 cointegrating eqn(s) at the 0.05 level
Max-eigenvalue test indicates 3 cointegrating eqn(s) at the 0.05 level
[a]denotes rejection of the hypothesis at the 0.05 level
[b]MacKinnon–Haug–Michelis(1999) p-values

18.7.2 Cointegration Test

Table 18.3 depicts the approximate outcomes of Johansen's cointegration studies. The eigenvalues and the trace or LR test statistic are being used to test for cointegration between series and compare the cointegration hypothesis to the alternative hypothesis of maximum rank. Analysis implies three cointegrating equations by employing the trace statistic accounts at the 5% level. Consequently, India's agricultural exports and public expenditure on agricultural yield could have a long-term relationship.

A VECM is an appropriate estimation method when a sequence of variables includes one or more cointegrating vectors. VECM also adjusts for short-term fluctuations in conditions as well as variations from equilibrium. Any cointegrating relationships among the variables which be taken into account by the VEC. The estimated VECM results are listed below in Table 18.4.

With a one-period lag, the findings demonstrate that its agricultural exports and public expenditure substantially influence India's crop production.

$$\begin{aligned}&\text{The Long Run relationship is presented below in equation :}\\&\mathrm{LC} = 3.433 + 0.13466 * \mathrm{LEX}_{(t-1)} + 0.05718 * \mathrm{LPE}_{(t-1)}.\end{aligned} \tag{18.9}$$

Growth in Crop production (LC) is found positively related to Export (LEX) and Public Expenditure (LPE).

18.8 Conclusion

The COVID-19 has emerged as a crisis for Indian farmers. The Indian government has closed down all forms of transportation, closed all institutions, public and commercial offices, factories, and limited mobilization due to an increase in COVID-19 infections. As a result, in this pandemic trapped era, farmers, food merchants, distributors, and all other related agencies are suffering from low earnings and cash flow challenge due to disruptions of supply chain (particularly long time lockdown), maintaining social distancing and a lot of human death. Farmers are having difficulties in drying, shelling, and selling their products at the market. Farmers, especially those who produce fruits and vegetables, may be forced to make distress sales. We conducted this study to investigate whether there is a statistically significant relationship between agricultural productivity (Crop Production Index) and government expenditure on agriculture or export policies enacted by the Government of India. The study intends to expand a deeper understanding of the effect of public spending on agriculture on the crop yield with exports in India by conducting an econometric analysis from 1987–1988 to 2018–2019. Our findings, relying on unit root test results and Johansen-cointegration tests, demonstrate that at least one cointegrating equation exists, as evidenced by trace and maximum

Table 18.4 Vector Error Correction Model (VECM)

Cointegrating Eq:	CointEq1		
LC(−1)	1		
LEX(−1)	−0.13466		
	−0.03061		
	[−4.39904]		
LPE(−1)	−0.05718		
	−0.0358		
	[−1.59747]		
C	−3.433		
Error correction:	D(LC)	D(LEX)	D(LPE)
CointEq1	−0.86884	1.492299	−1.86803
	−0.40059	−1.57774	−1.108
	[−2.16887]	[0.94584]	[−1.68594]
D(LC(−1))	0.143555	−0.71502	1.336476
	−0.33806	−1.33146	−0.93504
	[0.42464]	[−0.53702]	[1.42932]
D(LC(−2))	0.078032	−0.85873	0.611946
	−0.22301	−0.87833	−0.61682
	[0.34990]	[−0.97768]	[0.99210]
D(LEX(−1))	0.058868	0.32108	−0.20556
	−0.07378	−0.29058	−0.20407
	[0.79790]	[1.10496]	[−1.00734]
D(LEX(−2))	−0.03945	0.06392	−0.01299
	−0.0596	−0.23472	−0.16484
	[−0.66202]	[0.27232]	[−0.07881]
D(LPE(−1))	−0.09779	0.371274	−0.2024
	−0.0989	−0.38952	−0.27355
	[−0.98877]	[0.95315]	[−0.73990]
D(LPE(−2))	0.236583	0.239267	0.011528
	−0.09996	−0.39371	−0.27649
	[2.36667]	[0.60772]	[0.04169]
C	0.000453	0.077379	0.119245
	−0.02064	−0.08127	−0.05708
	[0.02195]	[0.95209]	[2.08924]
R-squared	0.626428	0.178978	0.162023
Adj. R-squared	0.46299	−0.18022	−0.20459
Sum sq. resids	0.021269	0.32993	0.162715
S.E. equation	0.03646	0.143599	0.100845
F-statistic	3.83282	0.498273	0.441942

Source: Computed by the author

Eigenvalue statistics. As a result, it is inferred that the variables are in long-term equilibrium. So, findings demonstrate that India's crop production is significantly impacted by agricultural exports and public expenditure with a one-period lag. This analysis should enable policymakers and researchers in India continue improving policy-led growth in the future and the agricultural sector's emergence as the only area to experience positive growth can be attributed to an increase in agricultural production owing to a favorable monsoon and targeted government expenditure in this pandemic.

References

Bai J, Perron P (1998) Estimating and testing linear models with multiple structural changes. Econometrica 66:47–78

Bhalla GS (2002) Globalisation and agricultural liberalisation in India. Government of India, CSRD, Jawaharlal Nehru University, New Delhi, A Study Sponsored by Ministry of Agriculture

Chadha GK (2002), 'Indian agriculture in the new millennium: human response to technological Change', Presidential Address to 62nd Annual Conference of Indian Society of Agricultural Economics, Indian Agricultural Research Institute, New Delhi, Dec 19–21

Chand R (2001) Emerging trends and issues in public and private Investments in Indian Agriculture: a Statewise Analysis. Indian J Agric Econ 56(2):161–184

Chandra Parida P, Sahoo P (2007) Export-led growth in South Asia: a panel cointegration analysis. Int Econ J 21(2):155–175

Dev SM (2020) Addressing COVID-19 impacts on agriculture, food security, and livelihoods in India. IFPRI book chapters. pp. 33–35

Dickey DA, Fuller WA (1979) Distribution of the estimators for autoregressive time series with a unit root. J Am Stat Assoc 74(366a):427–431

Ekanayake EM (1999) Exports and economic growth in Asian developing countries: Cointegration and error-correction models. J Econ Dev 24(2):43–56

Gulati A, Sharma A (1997) Freeing trade in agriculture: implication for resource use efficiency and cropping pattern changes. Econ Politic Weekly 27:A155–A164

Gulati A, Bathla S (2001) Capital formation in Indian agriculture. Econ Politic Weekly 36(20): 1697–1708

Haalisha A, Jahufer A (2018) Influences of consumer price index and gross domestic saving on Sri Lankan economy: a johansen cointegration tactic

International Labour Organisation (ILO) (2020) ILO monitor: COVID-19 and the world of work (2nd ed.). Updated estimates and analysis. Retrieved from https://www.ilo.org/wcmsp5/groups/public/@dgreports/@dcomm/documents/briefingnote/wcms_740877.pdf

Kumar P (2002) Agricultural performance and productivity. In: Acharya SS, Chaudhri DP (eds) Indian agricultural policy at the crossroads. Rawat Publications, Jaipur

Mukhopadhyay P, Sinha M, Sengupta PP (2020) Importance of sustainable rural development through agrarian reforms: an Indian scenario. In: *Environmental and agricultural informatics: concepts, methodologies, tools, and applications*. IGI Global, Pennsylvania, pp 1163–1180

Parikh K (1997) Overview: prospects and retrospect. In: Parikh K (ed) Indian development report 1997. Indira Gandhi Institute of Development Research, Oxford University Press, Mumbai

Perron P (1989) The great crash, the oil price shock, and the unit root hypothesis. Econometrica 57: 1361–1401

Reddy DN (2006a) Changes in agrarian structure and agricultural technology: is peasant farming sustainable under institutional retrogression. In: Radhakrishna R, Rao SK, Dev SM, Rao KS (eds) *India in a Globalising world: some aspects of macro economy, agriculture and poverty, essays in honour of C.H. Hanumantha Rao*. Academic Press, New Delhi

Reddy DN (2006b) Economic reforms, agrarian crisis and rural Distress. In: 4th Annual Prof. B. Janardhan Rao Memorial Lecture, Prof. B. Janardhan Rao Memorial Foundation, Warangal, Telegana

Safdari M, Mahmoodi M, Mahmoodi E (2011) The causality relationship between export and economic growth in Asian developing countries. Am J Sci Res 25(1):40–45

Schwert GW (2002) Tests for unit roots: a Monte Carlo investigation. J Bus Econ Stat 20(1):5–17

Sims CA (1980) Macroeconomics and reality. Econometrica:1–48

Singh RB, Mishra M (2021). COVID-19 pandemic trajectory: challenges and opportunities for India. *COVID-19 pandemic trajectory in the developing world*, p. 3

Times of India (2020). https://timesofindia.indiatimes.com/blogs/agyeya/indian-agriculture-ecosystem-during-covid-19/?

Labonie Mukhopadhyay is presently pursuing her bachelor degree in medical science from Awadh Dental College and Hospital, Jamshedpur, India. Alongside, she has keen research interest on time series econometrics. Till date, she has published two medical papers. One paper regarding different types of malocclusion and the second one is on the management of dentistry during COVID-19.

Chapter 19
A Study on the Impact of COVID-19 on Indian MSMEs

Rajib Lahiri

Abstract Micro, small and medium enterprises (MSMEs) are one of the most vital drivers of the Indian economy as the sector generates employment for more than 110 million people (approximately 23% of the total workforce in India), produces a wide variety of products, utilizes locally available resources, and ensures balanced economic growth. The sector was severely shocked by the sudden outbreak of coronavirus and the nationwide lockdown. This chapter attempts to analyze the changing pattern of the definition of MSMEs, the major problems faced by the MSMEs during the pandemic period, and the remedial measures taken up by the government to safeguard the sector. Lack of finance, supply-chain disruption, dearth of raw materials, and non-availability of workers are some of the major problems faced by the MSMEs during the pandemic. The government has taken some prompt measures like the announcement of special packages, collateral-free automatic loans, revision in the definition of MSMEs under the *Atmanirbhar Bharat* (Self-reliant India) scheme. The latest revision in the definition of MSMEs in July 2020 is likely to increase the sector's competitiveness and encourage the MSMEs to carry out expansion plans. Collective efforts by the government, industry, and all the major stakeholders would be needed to rejuvenate the Indian economy.

Keywords Employment · Efficiency · Identification problem · Manufacturing sector · Service sector · Turnover

19.1 Introduction

Micro, small and medium enterprises (MSMEs) are considered the backbone of the Indian economy. This sector contributes about 30% of Indian GDP and half of the exports; employs more than 110 million people (approximately 23% of the total

---

R. Lahiri (✉)
Department of Commerce, Derozio Memorial College, Kolkata, West Bengal, India

S. Saha et al. (eds.), *Economic and Societal Transformation in Pandemic-Trapped India*, New Frontiers in Regional Science: Asian Perspectives 55,
https://doi.org/10.1007/978-981-16-5755-9_19

Table 19.1 Share of gross value added (GVA) of MSME in all India GDP

Figures in Rs. crores adjusted for FISIM[a] at current prices

Year	Total MSME GVA	Growth (%)	Total GVA	Share of MSME in GVA (%)	All India GDP	Share of MSME in All India GDP (in %)
2014–2015	3,658,196	–	11,504,279	31.80	12,467,959	29.34
2015–2016	4,059,660	10.97	12,574,499	32.28	13,771,874	29.48
2016–2017	4,502,129	10.90	13,965,200	32.24	15,391,669	29.25
2017–2018	5,086,493	12.98	15,513,122	32.79	17,098,304	29.75
2018–2019	5,741,765	12.88	17,139,962	33.50	18,971,237	30.27

Source: Annual Reports, Ministry of MSMEs, Govt. of India 2020–2021

[a]*FISIM* stands for Financial Intermediation Services Indirectly Measured. In the system of National Accounts, it is an estimate of the value of the services provided by financial intermediaries, such as banks, for which no explicit charges are made; instead, these services are paid for as part of the margin between rates applied to savers and borrowers. The supposition is that savers would receive a lower interest, and borrowers pay a higher interest rate if all financial services had explicit charges

workforce in India); produces more than 6000 varieties of products; ensures proper utilization of resources and balanced economic growth by removing regional disparity (Annual Report of MSMEs 2018–2019). Small-scale industries were also an integral part of the ancient Indian economy as the sector was the principal source of income and employment, and products produced by the cottage and small industries were famous for quality and artistic skill. This section attempts to present a brief overview of the MSME sector and the contribution of the sector to the Indian economy.

Table 19.1 shows the gross value added (GVA)[1] and the gross domestic product (GDP)[2] contributed by the MSME sector at current prices from 2014–2015 to 2018–2019. The MSME sector showed an increasing growth trend in total MSME GVA from 10.97% in 2015–2016 to 12.88% in 2018–2019. Similarly, the sector has registered an average 29% share in total GDP in India during the period.

Table 19.2 shows the distribution of MSMEs in rural and urban areas by dividing the sector into four major categories: manufacturing, electricity, trade, and other

[1]*Gross Value Added (GVA):* The estimates of GVA had been prepared at factor cost in the earlier series (base year 2004–2005), while these are being prepared at basic prices in the new series (2011–2012), GVA estimated by production approach: (GVA = Output − Material Inputs) and GVA estimated by income approach: (GVA = Compensation of Employees + Operating Surplus + CFC).

[2]*Gross Domestic Product (GDP):* GDP is derived by adding taxes on products, net of subsidies on products to GVA at basic prices.

Table 19.2 Estimated number of MSMEs (activity wise)

	Estimated number of enterprises (in lakhs)			
Activity category	Rural	Urban	Total	Share (%)
Manufacturing	114.14	82.50	196.65	31
Electricity	0.03	0.01	0.03	0
Trade	108.71	121.64	230.35	36
Other services	102.00	104.85	206.85	33
All	324.88	309.00	633.88	100

Source: Annual Reports, Ministry of MSMEs, Govt. Of India 2020–2021

Table 19.3 Distribution of enterprises (rural and urban area wise)

(Number in Lakhs)					
Sector	Micro	Small	Medium	Total	Share (%)
Rural	324.09	0.78	0.01	324.88	51
Urban	306.43	2.53	0.04	309.00	49
All	630.52	3.31	0.05	633.88	100

Source: Annual Reports, Ministry of MSMEs, Govt. Of India 2020–2021

Table 19.4 Distribution of employment by type of enterprises in rural and urban areas

(Number in Lakhs)					
Sector	Micro	Small	Medium	Total	Share (%)
Rural	489.30	7.88	0.60	497.78	45
Urban	586.88	24.06	1.16	612.10	55
All	1076.19	31.95	1.75	1109.89	100

Source: Annual Reports, Ministry of MSMEs, Govt. Of India 2020–2021

services. MSMEs related to manufacturing activities are primarily concentrated in rural areas, whereas trade-related MSMEs are mostly located in urban areas.

Table 19.3 shows that 51% of MSMEs are situated in rural areas and 49% in urban areas. This distribution shows that the MSMEs have played an influential role in reducing regional imbalances and disparities in India.

Table 19.4 shows that the MSMEs, especially the micro-enterprises, have generated significant employment in rural and urban areas. However, the contribution of small and medium enterprises in employment generation is considerably low. Unemployment is one of the burning problems in India today. The MSMEs can be a vital tool to reduce the problem.

The situation has changed drastically in the first half of 2020 due to the coronavirus outbreak. The virus has taken the entire world in its grip very quickly, causing extreme damage to human life and the global economy. Like other countries in the world, the Indian economy is also badly hit by the crisis as India's GDP for the first quarter of 2020–2021 reduced by 24%, and the share of the manufacturing sector in total GVA reduced by more than 4% in the first quarter of 2020–2021. As a result, the manufacturing sector witnessed a negative growth rate in 2020–2021 (RBI Report 2020).

After the first countrywide lockdown imposed in India in 2020, manufacturing activities almost stopped except for the rice milling sector, where production dropped by half. The major manufacturing industries like metals and chemical products, motor vehicles, machinery and equipment, textiles were adversely affected due to lockdown (United Nations Industrial Development Organization Survey 2020).

The MSME sector was also highly affected due to the pandemic situation. The sector, being sensitive in nature, is still battling to maintain its existence due to lack of finance, supply-chain disruption, dearth of raw materials, price hike in raw materials, non-availability of workers, and lack of reliable data during this pandemic period.

19.2 Objectives of the Study

The concept of MSMEs in India has been changed several times. This chapter's primary objectives are to analyze the changing pattern of the definition of MSMEs and how far such changes have been effective in increasing the sector's competitiveness.

The sudden outbreak of COVID-19 has created a threat to the entire world. Most countries around the world have imposed lockdown measures to restrict the spread of the virus. This has severely impacted economic activities, and the sectors like MSMEs are the worst sufferers. This chapter aims to identify and analyze the problems faced by the MSME sector due to the COVID-19 outbreaks.

In order to protect the MSMEs from the adverse effects of COVID-19, the Indian government has taken some prompt measures. This chapter will analyze the measures taken by the government to safeguard the MSME sector.

19.3 Review of Literature

There has been substantial literature on various aspects of MSMEs, which may form the primary base for the research work. In this section, we have tried to review some essential areas of MSMEs like change in the definition of MSMEs, institutional credit support to MSMEs, the performance of MSMEs, and the impact of COVID-19 on Indian MSMEs.

*Little et al. (*1987*)* analyzed the pattern and relationship between technical efficiency and the size of the small-scale industries (SSI) firms based on information collected from sample SSIs. The number of workers employed or the value of the plant and machinery would determine the size of the SSIs. The technical efficiency analysis revealed that there was no identifiable relationship between size and technical efficiency among the sample firms. The researchers argued that factors like the experience of the labor force and the entrepreneurs, levels of capacity utilization, age of capital stock were the primary reasons for variations of technical efficiency in small firms. In our study, we would try to capture the performance and efficiency of the MSME sector based on the number of MSME units, distribution of such units in rural and urban areas, and employment generated by the sector.

Das (2006) critically analyzed the definitional aspect as per the MSMED Act, 2006. The author argued that the definition of MSMEs as per the MSMED Act, 2006 was based on historical values of plant and machinery and the limits suggested for category determination were useless. The registered micro and small firms enjoy several preferential measures. Once the firm crosses the specified limit of investment, it does not intend to register itself as a medium-scale industry as this would result in withdrawal of preferential benefits. In our study, we would try to determine whether the change in definition of MSMEs has been effective for the sector several times.

Bhavani (2011) argued that though the MSME sector was known to generate large employment, most of such jobs were low-income jobs as the general level of technology employed in Indian MSMEs was poor. Employment generation was high in the quantitative term, but such jobs offered very low remuneration. Therefore, the MSMEs were urgently needed for technological upgradation to create quality employment, improving remuneration, duration, and skill. Due to this structural shift, employment generation might be less in the short run but would ensure a high-income employment generation in the long run.

Biswas (2011) observed that MSME entrepreneurs needed proper training and skill development to exist and shine in the highly competitive market. However, the educational background of MSME entrepreneurs was poor as many people shifting from agriculture to the informal sector were either illiterate or had a low level of literacy. Moreover, the performance of the training institutes, both government and private, was not satisfactory as the training courses lacked job orientation and failed to prepare trainees to set up their individual enterprises. The author opined that a proper mix of formal and informal training techniques might be beneficial for MSME entrepreneurs. In our study, we would highlight the measures taken by the MSME-supporting agencies to safeguard the MSME sector during the COVID pandemic period.

Nair (2011) argued that most government schemes for offering financial assistance to the MSMEs were ineffective as either the schemes were not suitable for the sector or not adequately monitored. Such schemes were used to grant one-time credit without any follow-up measures and failed to build continuing relations with the borrowers and lenders. Moreover, the MSMEs, especially the micro firms, were not the preferred section for investment by the banks and other financial institutions due to poor financial capacity, irregular loan repayment records, high-risk involvement, and lack of credit rating. As a result, the banks became more cautious while sanctioning credits to MSMEs so that such loans did not become non-performing assets (NPAs). The entire process created a "vicious circle," and the percentage of net bank credit offered to MSMEs showed a declining trend. The author believed that a proper blend between the orthodox measures (institutional credit) and unconventional methods (factoring services, venture capital financing) could be helpful to ensure a steady supply of credit to the MSMEs.

Prasad and Mondal (2020) analyzed the impact of COVID-19 on Indian MSMEs and different remedial measures adopted by the government. The authors argue that the business practices during the post COVID era are likely to change significantly as the economy shall be more digitally driven. Therefore, the MSMEs should focus on innovation, unconventional sources of finance, and digital practices during the post-pandemic period.

COVID-19 Relief for MSME Sector (2020) by the Ministry of MSMEs offered problem-based solutions to the MSMEs during the Corona outbreak. The action plan presented the measures taken up by the Khadi and Village Industries Commission (KVIC), National Small Industries Corporation (NSIC) to safeguard the MSME sector. It also identified the short-term measures for three months and a long-term strategy to combat the adverse effects of COVID-19 on the MSMEs.

Pandey and Pillai (2020) observed that due to lockdown, the regular flow of income for most of the MSMEs was reduced significantly, and a majority of such units could not pay the fixed and operational expenses. The paper also highlighted the "Issue of Identification" relating to the quick delivery of relief to the affected MSMEs. Due to a lack of accurate data, it was challenging to identify the MSME units that need urgent support and ensure proper utilization of relief corpus.

Shah (2020) observed that Government relief packages would be helpful for MSMEs to maintain liquidity and equity infusion to help the MSMEs expand their size and capacity. The author suggested that the MSMEs must rethink their product policy, marketing strategy, distribution system to combat the challenging situation. They must stress digital marketing and online channels to increase customer base and reduce costs.

Sahoo and Aswani (2020) studied the impact of COVID-19 on MSMEs and the policy measures initiated by the government. They observed that the impact of the pandemic on trade, manufacturing, and MSMEs was huge as the manufacturing sector, exports, imports, and MSME net value added (NVA) might shrink significantly in 2020 compared to the previous year. They argued that the economy was moving towards a recession, and aggressive fiscal-monetary stimulus measures were required to control the situation. Our study would explore gross value added (GVA), rural-urban distribution of MSMEs in terms of the number of MSMEs and employment generation.

19.4 Changing Pattern of Definition of MSMEs

19.4.1 Concept of Small-Scale Industries (SSIs)

The definition of MSMEs was changed on several occasions. Initially, the sector was known as small-scale industries (SSI) and was defined in terms of the level of investment and the number of employees (Table 19.5). Table 19.5 shows that starting from 1950 to 1959, gross investment in fixed assets and the number of workers were the primary criteria for defining SSIs. However, from 1960 onwards, the employment condition was dropped from the definition, and gross investment in fixed assets continued to be the primary condition. Therefore, Table 19.5 captures a snapshot of defining factors of SSIs during the pre-liberalization period.

Table 19.5 Definition of small-scale industries (SSI)

1950	Gross investment in fixed assets: value of machinery (original)	Employment less than 50 workers per day (with the use of power) or less than 100 workers per day (without the use of power)
1958	Gross investment in fixed assets: Value of machinery (original)	Employment less than 50 workers per day (with the use of power) or less than 100 workers per day (without the use of power) except that the criteria based on the employment "per day" was henceforth replaced by a "per shift" provision
1959	Gross Investment in Fixed Assets: Value of machinery (original)	Employment less than 50 workers per day (with the use of power) or less than 100 workers per day (without the use of power) except that the criteria based on the employment "per day" was henceforth replaced by a "per shift" provision
1960	Gross investment in fixed assets: Value up to re. 0.5 million (re. 0.5 million)	The employment condition was dropped from the definition
1966	Up to re. 0.75 million (re. 0.57 million)	No condition
1975	Up to re. One million (re. 0.58 million)	No condition
1977	Up to re. One million (re. 0.59 million)	No condition
1980	Up to Rs. Two million (re. 0.92 million)	No condition
1985	Up to Rs.3.5 million (Rs. 2.91 million)	No condition
1991	Up to Rs.6 million (Rs. 3.28 million)	No condition

Source: Annual Report of MSMEs 2002–2003

Notes: The figures in the bracket show the approximate real values of investment in fixed assets; the real values have been calculated considering 1960 as the base year

19.4.2 Concept of MSMEs in India as per Micro, Small and Medium Enterprises Development (MSMED) Act, 2006

A significant change took place in 2006 when the micro, small, and medium enterprises were brought under the same umbrella under the micro, small and medium enterprises development (MSMED) Act, 2006. The main objectives of the introduction of the act are to increase the sector's competitiveness and relieve the sector from complying with multiple rules and regulations (Annual Report of MSMEs 2018–2019). The act also advocates the provision of marketing assistance, infrastructural facilities and backward, and forward linkages to all MSMEs.

Table 19.6 shows the limits for investment in plant and machinery/equipment for manufacturing/service enterprises. Under the provision of the MSMED Act, 2006, the MSMEs are classified into two categories:

- *Manufacturing Enterprises* are the enterprises engaged in manufacturing or producing goods pertaining to any industry specified in the first schedule to the Industries. The manufacturing enterprises are defined in terms of investment in plant and machinery.
- *Service Enterprises* are the MSMEs engaged in the service sector who are defined in terms of investment in equipment (Table 19.6).

Table 19.6 Investment limits in MSMEs as per MSMED Act, 2006 (valid up to 30 June 2020)

Manufacturing Sector	
Enterprises	*Investment in Plant and Machinery*
Micro-enterprises	Does not exceed twenty-five lakh rupees
Small enterprises	More than twenty-five lakh rupees but does not exceed five crore rupees
Medium enterprises	More than five crore rupees but does not exceed ten crore rupees
Service sector	
Enterprises	*Investment in Equipment*
Micro-enterprises	Does not exceed ten lakh rupees
Small enterprises	More than ten lakh rupees but does not exceed two crore rupees
Medium enterprises	More than two crore rupees but does not exceed five crore rupees

Source: Annual Report of MSMEs 2018–2019

19.4.3 Concept of MSMEs, Effective from 1 July 2020

As per the latest development, the definition of MSMEs was again changed by the government with effect from 1 July 2020. As per the revised definition, an enterprise shall be classified as a micro, small, or medium enterprise (Table 19.7).

Two essential features of the revised definition of MSMEs may be mentioned (Table 19.7). Firstly, as per the definition suggested by the Micro, Small and Medium Enterprises Development (MSMED) Act, 2006, the MSMEs were divided into two major categories, namely manufacturing and services. However, as per the revised definition, these two broad categories have been withdrawn. Secondly, the old definition was based on the levels of investment, but the revised definition is based on both investment and turnover.

The new definition of MSMEs would encourage the MSMEs to carry out expansion schemes without the fear of losing the benefits of MSMEs as the new definition has significantly increased the ceiling of investment. The last revision in the definition was made in 2006, i.e., about 14 years ago. Thus, a revision was necessary to encourage the MSMEs to expand in size and still be considered as MSMEs. This measure is expected to increase export, the creation of jobs, and the competitiveness of the sector.

19.5 Major Problems Faced by the MSMEs During Pandemic Period

19.5.1 Lack of Steady Supply of Raw Materials and Spare Parts

The MSMEs in India have always suffered due to an unorganized supply of raw materials. This problem has taken a severe shape during the pandemic caused by COVID-19 due to complete lockdown throughout the country and COVID restrictions during the unlock periods. China has been a significant player in supplying raw

Table 19.7 Revised definition of MSMEs (Valid from 1 July 2020 onwards)

Composite criteria: investment in plant & machinery/equipment and annual turnover			
Classification	Micro	Small	Medium
Manufacturing enterprises and enterprises rendering services	Investment in plant and machinery/equipment: Not more than Rs.1 crore Annual turnover: Not more than Rs. 5 crore	Investment in plant and machinery/equipment: Not more than Rs.10 crore Annual turnover: Not more than Rs. 50 crore	Investment in plant and machinery/equipment: Not more than Rs.50 crore Annual turnover: Not more than Rs. 250 crore

Source: The Gazette of India 2020

materials to Indian MSMEs like organic chemicals, mineral fuel, cotton. Imports have been severely hit during the lockdown periods and eroding political relations with China, resulting in a 40% reduction in imports (Prasad and Mondal 2020).

However, India has already taken several measures to reduce dependence on other countries through the *Atmanirbhar Bharat* project announced by the government. *Atmanirbhar Bharat* means "Self-reliant India." It refers to a project announced by Prime Minister Narendra Modi on 12 May 2020 to combat the economic crisis created by the coronavirus pandemic. A financial package of Rs. 20 lakh crore was announced under this project (Atmanirbhar Bharat: Business Including MSMEs 2020). The package would be beneficial for laborers, farmers, honest taxpayers, MSMEs, and the cottage industries.

Atmanirbhar Bharat is based on five pillars: (1) Economy; (2) Infrastructure; (3) Technology-driven system; (4) Vibrant demography; and (5) Demand. It will be implemented in five phases: (1) Business including MSMEs; (2) Poor, including migrants and farmers; (3) Agriculture; (4) New Horizons of Growth; and (5) Government Reforms and Enablers. "Self-reliant India" does not mean cutting off from the rest of the world and advocating a self-centred protectionist system (PRS Legislative Research 2020). On the contrary, this scheme aims to achieve self-dependence, ensure the best utilization of available resources, and reduce dependence on foreign countries.

19.5.2 Lack of Demand

This is also an existent illness of Indian MSMEs as a significant number of MSMEs, and especially the micro firms, do not find adequate customers to sell their products. It is surprising that even after being one of the most middle-class-dominated economies globally, Indian MSMEs fail to generate sufficient demand for their products in the market. This may be due to poor marketing ability, lack of product mix, and inappropriate marketing strategy. During the COVID-19 pandemic, these problems aggravated manifolds. As per the International Monetary Fund (IMF) projection, the Indian economy would experience a sharp decline of 4.5% (World Economic Outlook Update 2020). The consumption of petroleum products, proxy for transport activity and oil demand, from April to July 2020 recorded a decline of 22.5% compared to consumption in the same period in 2019. Due to earning insecurity, loss of jobs, supply-chain disruption, lack of demand, and a stagnating economy, the usual behavior pattern of common consumers has also changed. Consumers have become cautious while purchasing goods and services which adversely affected the growth potential of MSMEs (Rath and Das 2020). NSS 73rd Round Survey (2015–2016) revealed that more than 50% of manufacturing MSMEs had not shown any growth since 2013–2014. Moreover, 37% of manufacturing and service MSMEs face a severe lack of demand (NSS 73rd

Round Survey 2015–2016).[3] The study indicates that lack of demand was already persistent in the economy, which might have aggravated due to the attack of the COVID-19 pandemic.

19.5.3 Poor Supply of Credit

The MSMEs, especially the micro-units, are not the coveted field of investment for banks and other financial institutions due to the low growth rate of the small firms, informal business practices, inability to maintain collateral securities, lack of creditworthiness, high processing cost, and poor flow of information (Haider 2000). Moreover, poor supply of institutional credit results in a dearth of working capital, technological inefficiency, poor managerial performance, irregular payment of loan installments, and reduces loan eligibility of MSMEs.

The Indian MSME sector faces a credit deficit of $240 billion in 2020 (RBI Publication 2020). As per the report, this excessive credit gap has occurred due to lack of proper business operation due to frequent full and partial lockdowns, extremely conservative approach by the lending authorities while sanctioning loans to MSMEs, reduction in cash flow, poor maintenance of business documents, and account statements by the MSMEs, slowdown of the local and global economy, restriction on banks and other financial institutions in terms of operational hours and number of workers.

As per the study conducted by the Reserve Bank of India, intuitional credit offered to the MSMEs showed a decline of 12.7% in June 2020. While loans to micro and small industries shrank by 7.6%, it dipped by 9.4% for medium-sized industries (Reserve Bank of India Publication 2020).

19.5.4 Stoppage of the Steady Flow of Income

Due to strict countrywide lockdown, many MSMEs faced severe crunch in cash flow. As there was a halt in the production and a slump in demand, MSMEs could not pay their creditors, pay regular salaries to the employees, repay loan installments, and many other operational expenses. Full and frequent partial lockdowns have resulted in job losses, revenue losses, and many MSMEs have become defunct. The study conducted by the All India Manufacturing Organization in 2020 among 46,525 MSMEs shows an unfortunate result. As per the study, 35% of sample MSMEs and 37% of self-employed respondents believe that their enterprises would not recover from this shock. Moreover, 32% of respondents feel that recovery would take more

[3]During the 73rd round (July 2015 to June 2016), National Sample Survey Office (NSSO) carried out an all India survey on unincorporated non-agricultural enterprises in manufacturing, trade, non-captive electricity generation and transmission, and other service sector (excluding construction). The previous survey on the similar subjects was conducted during the National Sample Survey (NSS) sixth round (July 2010–June 2011).

than one year (Magazine 2020). According to another study conducted jointly by Magma Fincorp and SPJIMR in 2020 among 14,444 MSMEs, 50% of MSMEs reported reducing their regular income by more than 50% due to the COVID-19 pandemic (Magma-SPJIMR Report 2020).

19.5.5 Non-availability of Workers

The lack of skilled workers has been a significant problem for Indian MSMEs as large enterprises capture most of the skilled workers by offering lucrative compensation packages (Biswas 2011). During the lockdown periods, several restrictions were imposed on transportation facilities and mobility of people to break the chain of infection. The *work from home* concept has become extremely popular during the pandemic days. Unfortunately, in the case of manufacturing and traditional services, the worker's physical presence is more needed. Thus, the work from home concept is not much applicable in MSMEs, especially for unskilled workers. Moreover, Indian MSMEs are much dependent on migrant workers. Due to the pandemic situation, many of such workers returned to their native places. Many migrant workers are at risk of losing their jobs, especially in the unorganized sector and those who do not have any legal contract with their employers (Khanna 2020). At least five million salaried people have lost their jobs in July 2020, taking the total count of job losses in the category to 18.9 million since April 2020 (Pandey and Pillai 2020).

19.5.6 Lack of Comprehensive Data

Lack of reliable data, especially relating to the unorganized MSME sector, has been a major problem for the sector. The last census on the MSME sector was carried out in 2006–2007. After that, some data collection procedures like Udyog Aadhaar Memorandum (UAM),[4] MSME Data Bank, the Goods, and the Services Tax Network (GSTN)[5] were initiated. The first two are the voluntary self-certified type where the authenticity of information is subject to verification, while the GSTN is compulsory only for businesses with a turnover of more than Rs. four million (Annual Report of MSMEs 2018–2019). Therefore, it becomes challenging to identify the sector that needs urgent relief and ensure that the relief reaches the targeted section (Pandey and Pillai 2020).

---

[4] *Udyog Aadhaar Memorandum (UAM)* is a single-page registration that is equivalent to the previous 11 forms protocol for self-certification of the MSMEs.

[5] *The Goods and Service Tax Network (or GSTN)* is a non-profit, non-government organization which will manage the entire IT system of the GST portal. This portal will be used by the government to track every financial transaction and will provide taxpayers with all services—from registration to filing taxes and maintaining all tax details.

19.6 Remedial Measures Adopted by the Government to Combat COVID Scenario

19.6.1 Self-Reliant India

The Government of India undertook several financial and legislative measures under the *Atmanirbhar Bharat* (See 19.5.1, para 2) scheme for the betterment and easy operation of MSMEs.

- *Corpus for MSMEs*: A fund with a corpus of Rs 10,000 crore will be set up for MSMEs to offer equity funding for MSMEs with growth potential and viability. Rs 50,000 crore is expected to be leveraged through this fund structure.
- *Subordinate debt for MSMEs:* This scheme aims to support financially distressed MSMEs where the non-performing assets (NPAs) are high. Under the scheme, promoters of MSMEs will be given debt from banks, which will be infused into the MSMEs as equity. The government will facilitate Rs 20,000 crore of subordinated debt to MSMEs and also ensure credit guarantee of such debt (PRS Legislative Research 2020).
- *Disallowing global tenders:* To protect Indian MSMEs from foreign competition, global tenders of up to Rs 200 crore will not be allowed in government procurement tenders (Atmanirbhar Bharat: Business Including MSMEs 2020).
- *Reduction in TDS and TCS rates:* The rates of Tax Deduction at Source (TDS) and Tax Collected at Source (TCS) will be reduced by 25%. This scheme will be operational from 14 May 2020 to 31 March 2021 and is expected to provide liquidity of Rs. 50,000 crore in the economy (Atmanirbhar Bharat: Business Including MSMEs 2020).

19.6.2 Measures Taken by the Ministry of Khadi and Village Industries Commission (KVIC)

Khadi amd Village Industries Commission (KVIC), a statutory organization, was established under the Khadi and Village Industries Commission Act, 1956. The commission works under the direct control of the Ministry of MSMEs, focusing on strengthening the rural economy. It performs the activities like skill development, transfer of technology, generation of self-employment opportunities in rural areas, production of saleable articles, and building up a strong rural community spirit (Jagriti 2020). KVIC has also played its part in combatting the COVID-19 pandemic:

- *Distribution of food packets:* KVIC has taken the initiative in setting up community kitchens for providing food packets to migrants. Freshly prepared food packets were distributed in Delhi, Jammu, Nagpur, Bengaluru, Thane, Jaipur, Coimbatore, Patna, Varanasi, Cuttack, and Hyderabad (Jagriti 2020)[6].

- *Distribution of face masks and sanitizers:* KVIC has become famous for its Khadi masks. The commission has distributed more than 1,00,000 masks and sanitizers free to local people (Jagriti 2020), police officials, administrators to prevent the spread of the virus and create awareness among people.
- *Release of funds under Market Development Assistance (MDA)*[7]*:* KVIC released Rs. 1000 per month to each registered artisan from the Artisan Welfare Fund Trust and also provided financial support to Khadi Institutions and Artisans under Market Development Assistance (MDA) (COVID-19 Relief for MSME Sector 2020).

19.6.3 Measures Taken by National Small Industries Corporation (NSIC)

The National Small Industries Corporation Ltd. (NSIC) is an ISO 9001-2015 certified Government of India enterprise under the Ministry of MSMEs. The mission of NSIC is to promote the MSME sector by providing marketing, financial, and technological support (Annual Report of MSMEs 2020–2021). NSIC operates through a countrywide network of offices and technical centers. In addition, NSIC has also set up Training cum Incubation Centres managed by professionals. Significant Steps taken by NSIC during COVID-19 are:

- *Raw material distribution:* NSIC plays a vital role in ensuring a steady supply of raw materials at a stable price to the MSMEs. During COVID-19, most of the MSMEs suffered from an acute crisis of raw materials. The role of NSIC has become even more critical during the pandemic period. During the financial year 2020–2021 (up to 31 December 2020), NSIC has distributed raw materials worth Rs. 1070 crore (Annual Report of MSMEs 2020–2021). MSMEs availing raw material assistance against bank guarantee before 1 March 2020 were allowed a moratorium of 3 months to repay outstanding dues (COVID-19 Relief for MSME Sector 2020).
- *Consortia and tender marketing:* MSMEs face tough competition from the large enterprises in procuring bulk orders when they bid for tender on their capacity. NSIC makes consortia with the MSMEs and bids for tenders on behalf of the

---

[6]Jagriti is a monthly journal of KVIC on rural industrialization.

[7]*Market Development Assistance (MDA)* was introduced in 2020 as a part of the comprehensive policy package for promotion and development of MSMEs. The scheme offers funding for (**a**) participation of small and micro enterprises in international trade/fairs/exhibitions, (**b**) sector specific market studies by Industry Associations/Export Promotion Councils/Federation of Indian Export Organisation, (**c**) enhancing export opportunities for the small and micro manufacturing enterprises.

MSME consortia. This process helps NSIC to procure orders in bulk quantities. These orders are then distributed among MSMEs in line with their production capacity. During the financial year 2020–2021 (up to 31 December 2020), the company has participated in 444 tenders valuing Rs. 230.16 crore and executed tenders worth Rs. 74.91 crore (Annual Report of MSMEs 2020–2021).

- *Single point registration scheme:* NSIC operates a single point registration scheme to enable the MSMEs to participate in government tenders. During the financial year 2020–2021 (up to 31 December 2020), 2100 new units were added, and 5141 units were renewed under this scheme (Annual Report of MSMEs 2020–2021).

19.6.4 Collateral-Free Automatic Loans

Special collateral-free automatic loan for Rs. 3 lakh crore was announced for the MSMEs to meet their operational expenses and ensure a steady supply of raw materials. The scheme is applicable for the MSMEs whose outstanding is up to 25 crores and turnover is 100 crores. More than 45 lakh units are expected to benefit from the scheme (Atmanirbhar Bharat: Business Including MSMEs 2020).

19.6.5 Reduction in Employees Provident Fund (EPF) Contribution

The statutory Provident Fund contribution of both employer and employee has been reduced to 10% each from existing 12% for May, June, and July in 2020, keeping in mind the cash crunch during the pandemic situation. This measure is expected to provide liquidity of Rs. 6750 crore to the economy (Employees' Provident Fund Organisation 2020).

19.6.6 Equity Support to the MSMEs

The government has announced an equity infusion scheme for Rs.50000 crore to help the MSMEs to expand and encourage the units to be listed on the stock exchange (Atmanirbhar Bharat: Business Including MSMEs 2020).

19.7 Conclusion

Human civilization has witnessed an unprecedented crisis due to COVID-19 that has resulted in economic activities to come to a standstill. The situation is especially critical for a sector like the MSMEs, which is extremely sensitive in nature. One of the biggest challenges for MSMEs in the current scenario is shrinking demand. The manufacturing and service sectors of MSMEs have shown a negative growth rate during the pandemic situation. The aftershock of COVID-19 is likely to affect the Indian economy in the future.

Atmanirbhar Bharat (self-reliant India) has been introduced to kick start the economy and rejuvenate the MSME sector. However, there exists a broad gap between the declaration of packages and the proper implementation of such packages. Delayed or lack of information from the economic units like banks, corporates, MSMEs, and statistical agencies has become a significant problem for policymakers. The MSME sector will not be benefitted just by announcing economic packages and different policies. Continuous monitoring is needed to assess the ground reality and change the promotional measures accordingly to make such schemes more beneficial for MSMEs. There is an urgent need for up-gradation of technology, development in infrastructure, product diversification, innovative use of human resources, access to finance and market to ensure sustainable development of MSMEs. District Industries Centres (DICs) can play a more meaningful role in MSME registration, creating awareness about the government schemes and motivating the entrepreneurs to continue the business in this challenging time.

The study is based on the immediate effects of COVID-19 on Indian MSMEs and the government's immediate response. The actual damage to the MSMEs and the economy could be much more profound than what has currently been anticipated. This study is expected to open up many future avenues of research like the long-term effect of COVID-19 pandemic on MSMEs, the performance of economic packages announced by the government, role of District Industries Centres (DICs) and MSME-supporting agencies in handling the crisis, effects of revision of the definition of MSMEs in 2020.

However, this challenging situation can be converted into opportunity as several new business types, ideas, and practices (manufacturing of PPE, masks, sanitizers, testing kits, face shields) suitable for MSMEs have emerged. Encouraging exports and integrating Indian MSMEs with the global supply-chain should be a priority for policymakers. MSMEs need to reorient their product engineering, marketing policy, pricing policy, and distribution models to survive; they must ensure better use of digital marketing and online channels to increase customer base at a lower cost and collaborate with academics and engineering institutions for product and process innovation. Collective efforts on the part of the government, industry, and all the major stakeholders can reduce the impact of the COVID-19 pandemic and ensure the revitalization of Indian economy.

References

Annual Report of MSMEs (2002–2003) Ministry of MSMEs, Govt. of India, New Delhi. http://www.dcmsme.gov.in/publications/areports/ar2002-03/ssi-eng0203.pdf

Annual Report of MSMEs (2018–2019) Ministry of MSMEs, Govt. of India, New Delhi. https://msme.gov.in/sites/default/files/MSME-ANNUAL-REPORT-ENGLISH%202020-21.pdf. Accessed 24 June 2021

Annual Report of MSMEs (2020–2021) Ministry of MSMEs, Govt. of India, New Delhi. https://msme.gov.in/sites/default/files/Annualrprt.pdf Accessed 23 December 2020

Atmanirbhar Bharat: Business Including MSMEs (2020) Govt. of India. https://cdnbbsr.s3waas.gov.in/s3850af92f8d9903e7a4e0559a98ecc857/uploads/2020/05/2020051717.pdf. Accessed 27 June 2021

Bhavani TA (2011) Dynamic business environments: what these mean for Indian small enterprises. In: Das K (ed) Micro and small Enterprises in India: era of reforms. Routledge Publication, New Delhi, pp 27–45

Biswas P (2011) System of education, training and skill formation: how relevant for small enterprises? In: Das K (ed) Micro and small Enterprises in India: era of reforms. Routledge Publication, New Delhi, pp 158–184

COVID-19 Relief for MSME Sector (2020) Ministry of MSMEs, Govt. of India, New Delhi. https://msme.gov.in/sites/default/files/ActiononCOVID-19RelieforMSMESector.pdf. Accessed 24 Dec 2020

Das K (2006) Micro and small enterprises during reforms: policy & concerns. http://ideas.repec.org/p/ess/wpaper/id1803.html. Accessed 28 June 2021

Employees' Provident Fund Organisation (2020) Reduction in Statutory Rate of EPF Contributions. https://www.epfindia.gov.in/site_docs/PDFs/EPFO_PRESS_RELEASES/19_05_2020_Press_brief_reduction_in_EPF_contribution.pdf. Accessed 1 July 2021

Haider E (2000) Credit Bureaus: Leveraging Information for the Benefit of Microenterprises. http://www.microfinancegateway.org. Accessed 27 Dec 2020

Jagriti (2020) KVIC. Jagriti 64(1):25–38

Khanna A (2020) Impact of migration of labour force due to global COVID-19 pandemic with reference to India. J Health Manag 22:181–191. https://doi.org/10.1177/0972063420935542

Little I, Mazumdar D, Page M (1987) Small manufacturing enterprises: a comparative study of India and other economies. Oxford University Press, Washington DC

Magma-SPJIMR Report (2020). https://magma.co.in/wp-content/uploads/2020/10/MAGMA-SPJIMR-Report-min-1.pdf. Accessed 25 June 2021

Magazine A (2020) One in three businesses close to winding up. Indian Express (2 June)

Nair T (2011) Credit and microfinance development: a critical appraisal. In: Das K (ed) Micro and small enterprises in India: era of reforms. Routledge Publication, New Delhi, pp 129–157

NSS 73rd Round Survey (2015–2016) National Sample Survey Office, Ministry of Statistics & Programme Implementation, GOI. http://mospi.nic.in/sites/default/files/publication_reports/NSS_581.pdf. Accessed 25 June 2021

Pandey R, Pillai A (2020) COVID-19 and MSMEs: The 'identification' Problem. https://www.ideasforindia.in/topics/macroeconomics/covid-19-and-the-msme-sector-the-identification-problem.html. Accessed 27 Dec 2020

Prasad R, Mondal A (2020) Impact of COVID-19 Pandemic on Indian Economy with Special Reference to Indian MSME Sector. https://www.researchgate.net/publication/341775901_Impact_of_COVID-19_pandemic_on_Indian_Economy_with_special_reference_to_Indian_MSME_Sector/link/5ed372f2299bf1c67d2cc395/download. Accessed 18 Dec 2020

PRS Legislative Research (2020) Summary of announcements: Aatma Nirbhar Bharat Abhiyaan. https://www.prsindia.org/sites/default/files/parliament_or_policy_pdfs/Summary%20of%20Aatma%20Nirbhar%20Bharat%20Abhiyaan.pdf. Accessed 28 June 2021

Rath C, Das P (2020) Impact of COVID-19 on MSMEs. North Orissa Chamber of Commerce & Industry. http://nocci.in/Impact%20of%20Covid%2019%20on%20MSMEs.pdf. Accessed 26 June 2021

RBI Publication (2020). https://rbi.org.in/scripts/PublicationsView.aspx?id=20269. Accessed 27 June 2021

Sahoo P, Aswani (2020) COVID-19 and Indian economy: impact on growth, manufacturing, trade and MSME Sector. Global Business Review 22:1159–1183. https://journals.sagepub.com/doi/10.1177/0972150920945687

Shah N (2020) Covid 19: big opportunity or doomsday scenario for MSMEs? Stage set for firms to decide their future. Financial Express. https://www.financialexpress.com/industry/msme-eodb-covid-19-big-opportunity-or-doomsday-scenario-for-msmes-stage-set-for-firms-to-decide-their-future/2011215/

The Gazette of India (2020) Government of India, New Delhi. https://msmegovin/sites/default/files/MSME_gazette_of_indiapdf Accessed 27 June 2021

United Nations Industrial Developmennt Organisation (2020) India's Manufacturing Reels from the Impact of COVID-19. https://www.unido.org/stories/indias-manufacturing-reels-impact-covid-19

World Economic Outlook Update (2020) International Monitory Fund. https://www.imf.org/en/Publications/WEO Accessed 28 June 2021

Rajib Lahiri presently serves as Associate Professor in the Department of Commerce in Derozio Memorial College of Kolkata in India. He has been awarded a Ph.D. by the North Bengal University in 2015. He has written thirty research papers and presented papers in many national and international conferences. In addition, he has authored three books and completed one UGC-sponsored Minor Research Project (MRP). He has a wide range of interest in accounting, tax laws, micro, small, and medium enterprises (MSMEs), and microfinance.

Chapter 20
Indian Tourism Sector Under Siege of the COVID-19 Pandemic

Arunava Kumar Choudhury and Subrata Saha

Abstract The COVID-19 pandemic has created an unprecedented breakdown in the global economy. While the quarterly growth figure of every country exhibits the impact of this economic lockdown, the intensity does not fall equally in all the sectors of an economy. UNWTO forecasted that the travel and tourism industry could be most affected by this pandemic. This is probably the biggest challenge that the travel and tourism sector is facing and need to survive, revive, and thrive for the large section of people who directly or indirectly depend on this sector. The tourism sector is termed as the *sunrise sector* of India due to its immense potentiality. Tourism has a strong forward and backward linkage to other sectors of an economy. According to the WTTC (2020), travel and tourism are that sectors that recover more slowly than any other economic sector. In the Indian tourism sector, 50 million jobs are at risk due to the COVID-19 crisis. In the short run, establishing a healthy and safe tourism industry is very important. Government should help tourism enterprises and take necessary measures to bring back the confidence of the tourists as well as the stakeholders in this sector.

Keywords Economic lockdown · Forward linkage · Backward linkage · UNWTO · WTTC

20.1 Introduction

The whole world has been witnessing the most horrible pandemic of the century, the COVID-19. During one year how rapidly the virus spreaded and how much fetal it was reported in December 2019 (Page et al. 2021), and till December 28, 2020

A. K. Choudhury (✉)
Department of Economics, Sripat Singh College, Jiaganj, West Bengal, India

S. Saha
Department of Economics, Raiganj University, Raiganj, West Bengal, India

S. Saha et al. (eds.), *Economic and Societal Transformation in Pandemic-Trapped India*, New Frontiers in Regional Science: Asian Perspectives 55,
https://doi.org/10.1007/978-981-16-5755-9_20

the cumulative figure raised to 84 million positive cases and 1.9 million deaths.[1] As a result, most countries choose to impose lockdown and travel restrictions to capture the spread of infection.

Tourism has a multi-dimensional effect on any country's social and economic scenario. It enhances employment opportunities as well as foreign reserves. Now tourism is the most significant and fastest-growing industry in the world. According to WTTC (2019) latest annual report travel and tourism sector has secured 3.5% growth which is greater than the global economic growth of 2.5%, and most importantly, this is for the ninth time in consecutive that tourism sector growth has outrun the global economic growth. This sector contributes US$ 8.9 trillion to world GDP, which was 10.3% of global GDP, and around 330 million jobs were created (1 in every 10 jobs) in 2019.[2]

The tourism sector in India becomes the most important sector, and its share in GDP and total employment has risen significantly. The tourism sector has contributed a 6.8% share in India's GDP.[3] For the past few years, the Ministry of Tourism has been trying to diversify, develop, and promote the niche tourism products like cruise, sports, medical, MICE (marketing, incentives, conference, and events), eco-tourism, and sustainable tourism to make India as a 365-day destination.[4] Indian government initiated different marketing, such as "Incredible India!" and "*Atithi Devo Bhava*"[5] to enhance tourism growth. Easing visa regime, augmenting tourism infrastructure, boosting the hospitality sector by cutting GST, etc., are some steps taken by the government in the line.[6] India, with its vast cultural and natural diversities, has become a great attraction for tourists globally (Dhariwal 2003). After establishing the Indian Tourism Development Corporation and the initiation of the National Tourism Development Policy, the tourism sector is considered a significant engine of growth and start exercising planned tourism development (Chavan and Bhola 2014). According to WTTC (2020) annual report, the travel and tourism sector generated about 39 million jobs in India, constituting 8% of total employment in 2019.

The travel and tourism sector has been severely affected by the coronavirus outbreak across the globe. The WTTC has reported that, in 2020, the tourism sector suffered a loss of almost US$4.5 trillion in GDP resulting in a loss of 62 million jobs

[1] See the WHO website for the details which are available under the URL: https://covid19.who.int/.

[2] See the WTTC website for the details which are available under the URL: https://wttc.org/Research/Economic-Impact.

[3] See the WTTC website for the details which are available under the URL: https://wttc.org/Research/Economic-Impact.

[4] See the Annual Report 2020–2021, Ministry of Tourism, Government of India. Available at: https://tourism.gov.in/annual-reports/annual-report-2020-21.

[5] A phrase adopted from the Sanskrit literature which means "Guests are equivalent to God," prescribing the beauty of host–guest relationship, which embodies the traditional Indian Hindu–Buddhist philosophy of considering guests with the same respect as God.

[6] See the Annual Report 2019–2020, Ministry of Tourism, Government of India. Available at: https://tourism.gov.in/annual-reports/annual-report-2020-21.

globally.[7] India is no exception; tourism has experienced a noteworthy decline in 2020. Due to the COVID-19 crisis, travel restrictions and lockdown will lose domestic and international tourists and impact tourism and related industry (Balakrishnan and Christopher 2020).

The tourism sector is facing unprecedented challenges and a massive threat to livelihood from the current pandemic. UNWTO recognizes that tourism plays a supportive role in achieving the sustainable development goal.[8] It is a must that every government acknowledges this and guarantees that tourism can face the current threat to satisfy its vital role in triggering global economic recovery and growth. The present study aims to demonstrate the increasing tourism trend in India and examine the extent to which this sector is affected by coronavirus outbreaks. This paper is also intending to find the opportunities and challenges faced by the Indian tourism sector.

20.2 Literature Review

Large shreds of literature have discussed the impacts of tourism on the economic growth of a country. Ohlan (2017), Shahzad et al. (2017), Garay et al. (2017), Kumar (2014), Incera and Fernández (2015), Tang and Abosedra (2016) have done a comprehensive study on tourism-led growth.

McKinnon (1964), in his paper, asserted that tourism could be a great source of foreign exchange earnings and have great potential in fetching new technologies for production. Likewise, Blake et al. (2006) mentioned that tourism helps promote new infrastructural investment, enhancing human capital and raising competition, which in turn induces economic growth.

Kumar (2019), Endo (2006), Chen (2017), Arain et al. (2019), Dwyer and Forsyth (1993), Jalil et al. (2013), Pratt (2015) have shown the effect of inbound tourism on short-term and long-term economic growth. Furthermore, according to Cernat and Gourdon (2012), there is a spillover effect of inbound tourism, which triggers industrial development.

Ghosh (2011), Sharma (2018), Rout et al. (2018), Mishra et al. (2020), and Tang et al. (2016) found support for the legality of the tourism-led growth in India.

Mathieson and Wall (1982) analyzed in their work that the development of the tourism industry has a positive contribution to a country's balance of payment. This sector is more effective in terms of income and employment generation than the other ones (Shah et al. 2014; Lopez and Arreola 2019; Prasad and Kulshrestha 2015). The International Labour Organization (ILO) considered the tourism industry

[7] See the WTTC website for the details which are available under the URL: https://wttc.org/Research/Economic-Impact.

[8] See the UNWTO website for the details which are available under the URL: https://www.unwto.org/tourism-and-covid-19-unprecedented-economic-impacts.

a highly labor intensive and leading creator of jobs that require different skill sets and provide easy entry into the workforce by youth, women, and migrant workers.[9]

Poddar (2015) investigated a skill gap among the lower cadre of workers working in the tourism sector in India. Though the government had taken some schemes toward the development of such workers, it still needs more public and private initiatives.

Kakkar and Sapna (2012) described that tourism had played an auxiliary role in developing the industry in India. It can be the leading source of foreign exchange earnings and employment generating sector in India. Singh (2008) discussed sustainable tourism and economic development and, in that context, mentioned that the tourism sector is taking over the traditional agricultural sector in India.

In assessing the economic consequences of COVID-19 on tourism, UNCTED (2020), with the help of a computable general equilibrium model, has shown the direct and indirect effects of the decline in international tourism receipts on the global economy. Furthermore, the study showed incurred loss to GDP is much larger than the direct effects of the loss of tourism.

The tourism industry is presently considered a precursor of inclusive growth, which can generate employment both directly and indirectly (Prasad and Kulshrestha 2015). Therefore, many works asserted the growing importance of the travel and tourism sector in the economic growth of a country, including India. Many researchers have analyzed that tourism is the leading source of employment generation and foreign exchange earnings. The tourism industry is presently considered a precursor of inclusive growth, which can generate employment both directly and indirectly (Prasad and Kulshrestha 2015). The present study differs from the above research papers in that it systematically discusses the growing trend and tendency of the tourism sector in the Indian economy. In addition, the study intends to find how far the Indian economy can be affected by the unprecedented transmission tendency of the current pandemic.

20.3 Objectives and Methods

The tourism industry is one of the largest industries enhancing global GDP and employment. During the past few years, the Indian tourism sector has significantly developed and become a crucial sector in raising GDP and employment opportunities. However, due to high dependence on public transport and public communication, this sector is mainly affected by the coronavirus outbreak. This study aims to observe the importance and the growth trend of the Indian tourism industry and examines the effect of this pandemic on it.

---

[9] See the ILO website for the details which are available under the URL: https://www.ilo.org/global/publications/world-of-work-magazine/articles/WCMS_157893/lang%2D%2Den/index.htm.

This paper uses secondary research methodology to analyze the stated objectives. For this purpose, this paper maintains the standard procedure of performing secondary research:

- *firstly*, developing a research question, i.e., "*what is the effect of economic lockdown in the tourism sector of India*?"
- *secondly*, detecting the proper dataset (data taken from WTTC, UNWTO, Indian Tourism Statistics, Ministry of Tourism, different research reports, websites, different surveys, etc.),
- *thirdly*, analyzing the dataset (by constructing charts, graphs, diagrams, etc.), and,
- *finally*, deriving the conclusion.

20.4 The Direct and Indirect Effect of Tourism

Tourism has a multifaced impact on any nation's economy. The significant impact which can be readily noticed is on the government revenue and employment generation. In terms of contribution to government revenue and employment generation, the impact can be divided into three headings: direct, indirect, and induced (WTTC 2020). The direct contributions are revenue and employment generated by tourism industries directly related to tourists, including hotels, travel agents, airlines, and other transport services. It also includes restaurant and leisure activities that directly deal with the tourists. Alternatively, it is the total internal expenditure on travel and tourism industries less the purchase made by these industries. The indirect contributions of tourism involve the promotion of capital investment in these industries, collective government spending, supply-chain effect, etc. All the money spending by the individuals who are employed directly or indirectly by the tourism sector constitutes the induced contribution of the travel and tourism sector (Christian 2015).

20.4.1 Indian Tourism Scenario

Tourism is a part of the service sector, and its importance is growing day by day in India. According to WTTC (2019) global economic impact report, the tourism sector accounted for US$ 247.3 billion or 9.2% of India's GDP in 2018 and contributed 42.67 million jobs which are 8.1% of total employment. The WTTC anticipated that the tourism sector might grow at an annual of 6.8% and accounted for US$ 511.9 billion by 2028, which would be 9.9% of GDP.

Table 20.1 Foreign tourist arrival from different nations (Top 5)

Year	Bangladesh	USA	UK	Canada	Malaysia
2019	2,577,727	1,512,032	1,000,292	351,859	334,579
2018	2,256,675	1,456,678	1,029,758	351,040	319,172
2017	2,156,557	1,376,919	986,296	335,439	322,126
2016	1,380,409	1,296,939	941,883	317,239	301,961
2015	1,133,879	1,213,624	867,601	281,306	272,941

Source: India Tourism Statistics, Govt of India, (2015), (2017), (2018), (2019), (2020)

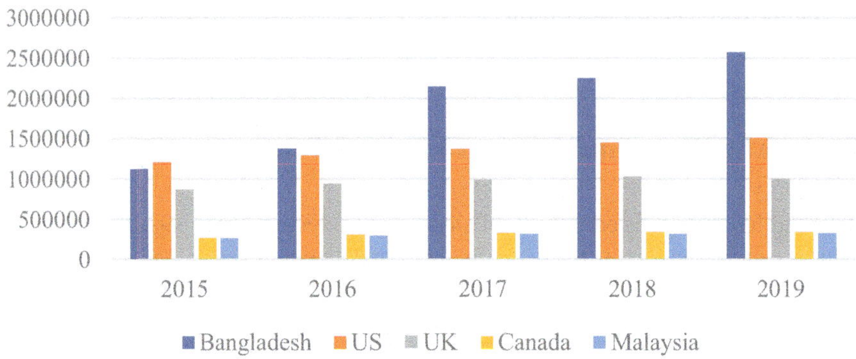

Fig. 20.1 Foreign tourist arrival in top 5 countries (Source: Prepared by the author)

Table 20.2 Foreign tourist arrival in India, 2013–2014 to 2019–2020

Year	2010	2011	2012	2013	2014	2015	2016	2017	2018	2019
FTAs (in million)	5.78	6.31	6.58	6.97	7.68	8.03	8.80	10.04	10.56	10.89
% change over previous year	11.8	9.2	4.3	5.9	10.2	4.5	9.7	14.0	5.2	3.1

Source: Annual Report, Ministry of Tourism, (2013–2014), (2014–2015), (2015–2016), (2016–2017), (2017–2018), (2018–2019), (2019–2020)

20.4.1.1 Foreign Tourist Arrival (FTA)

According to the India Tourism Statistics report (2016, 2017, 2018, 2019, 2020), the major source of tourism supply from 2015 to 2019 is given in Table 20.1 and illustrated in Fig. 20.1. From the figure, Bangladesh, the USA, and the UK are the leading countries constituting foreign tourist arrival in India. Tourist arrival from Bangladesh raised by more than double (127%) from 2015 to 2019, whereas for the same period, it was almost 25% for the USA, 16% for the UK, 25% for Canada, and 22% for Malaysia.

Table 20.2 shows the foreign tourist arrival from 2010 to 2019. This data shows tremendous growth in FTA arrival making India one of the most attractive tourist places. One of the reasons may be its price competitiveness, in which India ranks

13 (The Travel and Tourism Competitiveness Report 2019). Travel and tourism is a key sector of foreign exchange earning (Lee and Chang 2008), and it depends on the foreign tourist arrival. In 2019, according to Indian Tourism Statistics, foreign tourist arrivals were 10.89 million, which was almost 3.2% growth over the same period in 2018. India registered a growth of 23% over the previous year in foreign tourist arrival on e-Tourist Visa in 2019 (Annual Report, Ministry of Tourism, Govt of India 2019–2020).

20.4.1.2 Foreign Exchange Earnings (FEE)

India earned US$ 30 billion foreign exchange from travel and tourism in 2019, and this foreign exchange earnings (FEE) has risen exponentially from 2009. Considering the international tourism receipt, it is evident that India has shown remarkable growth between 2000 and 2019 compared to the Asia-Pacific region and the world. During this period, India's receipt is raised by almost ten times, which is very promising compared to the other. India's percentage share in the world has risen from 0.73 in 2000 to 2.03 in 2019, which pushed India up from 36th rank to 12th. In the Asia-Pacific region, the percentage share rises from 4.06 to 6.78 percent, making India holding the sixth rank during the same period (See Table 20.3, Figs. 20.2 and 20.3). The growth of FEEs, during the last decade, in the Asia-Pacific and in the world is 73% and 58%, respectively, whereas, during this period, India's FEEs have grown by 108%. All these figures exhibit tourism as the prime sector which can contribute significantly to the economic growth of this country.

20.4.1.3 Gross Domestic Product (GDP)

Tourism enhances economic growth (Kumar et al. 2015). As per the second Tourism Satellite Account for India (TSAI) for the reference year 2009–2010, the contribution of tourism in GDP remains steady and quite impressive during 2009–2010 to 2012–2013. In the third TSAI for the reference year, 2015–2016, the travel and tourism industry retains its percentage of contribution in the country's GDP (see

Table 20.3 Share of India in international tourism receipts in the world and the Asia-Pacific

	Int. Tourism Receipts (US $b)			% share and rank of India in world		% share and rank of India in Asia-Pacific region	
Year	World	Asia-Pacific	FEEs in India (in US $ b)	% share	Rank	% share	Rank
2000	475.3	85.3	3.460	0.73	36	4.06	10
2005	679.6	135.0	7.493	1.10	22	5.55	7
2010	931.0	255.3	14.490	1.56	17	5.68	7
2015	1217.0	355.6	21.013	1.73	14	5.91	7
2019	1478.0	443.1	30.058	2.03	12	6.78	6

Source: India Tourism Statistics 2020, Govt of India

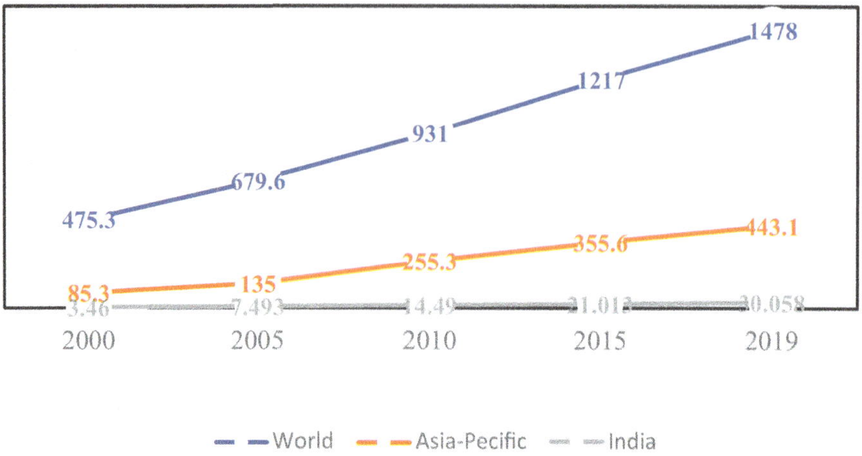

Fig. 20.2 International tourist earnings in US$b (Source: Prepared by the author)

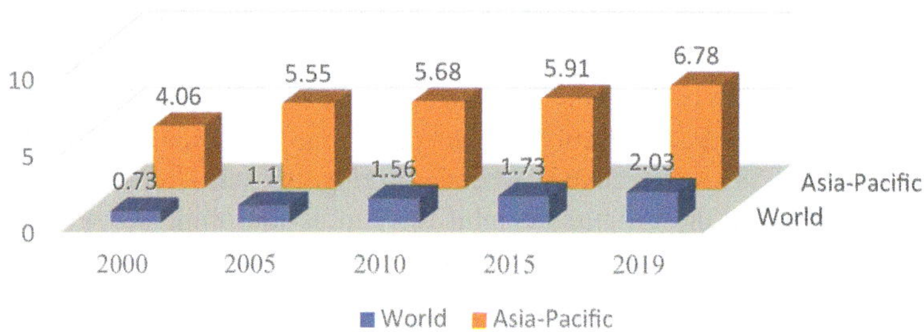

Fig. 20.3 India's percentage share in FEE (Source: Prepared by the author)

Table 20.4 Contribution of tourism in GDP (in percentage)

Year	Direct	Indirect	Total
2009–2010	3.68	3.09	6.77
2010–2011	3.67	3.09	6.76
2011–2012	3.67	3.09	6.76
2012–2013	3.74	3.14	6.88
2013–2014	3.06	2.62	5.68
2014–2015	3.14	2.67	5.81
2015–2016	2.70	2.50	5.20

Annual Report, Ministry of Tourism, Govt of India 2013–2014, 2019–2020, Ministry of Tourism

Table 20.4). Statista (2021) report explores the direct and total contribution of tourism in GDP has risen steadily from 2013 to 2019 (Fig. 20.4). It shows that travel and tourism have emerged as significant industries for India.

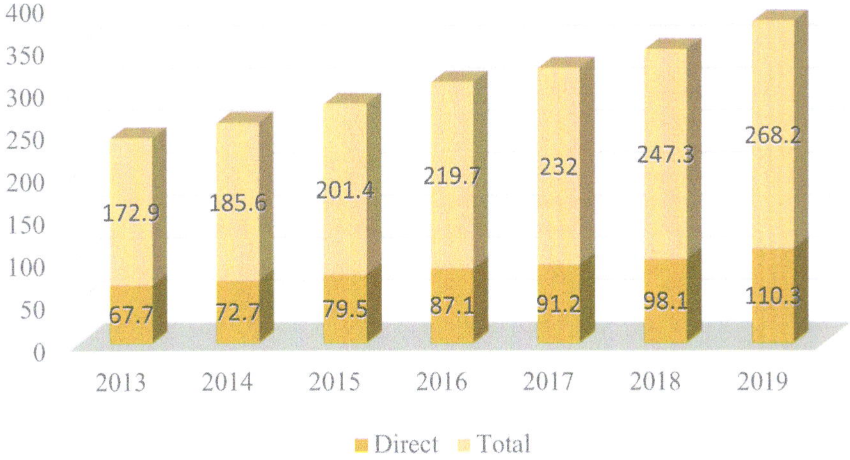

Fig. 20.4 Contribution of tourism in GDP in US$b (Source: Satista.com, 2021(See the Statista website for the details which are available under the URL: https://www.statista.com/statistics/313665/direct-contribution-of-travel-and-tourism-to-gdp-in-india/#:~:text=In%202018%2C%20the%20direct%20contribution%20of%20travel%20and,highest%20tourism%20GDP%20contribution%20in%20Asia-Pacific%20in%202,018.), (See the Statista website for the details which are available under the URL: https://www.statista.com/statistics/313724/total-contribution-of-travel-and-tourism-to-gdp-in-india-by-segment/.))

20.4.1.4 Employment

Travel and tourism have a considerable contribution to the employment generation of any country. Due to the price competitiveness (world rank 13) and cultural resources (world rank 9), India becomes a good destination for travelers in the Asia-Pacific region. During the last few decades, travel and tourism took a significant position in generating employment in India. According to Statista, India positioned second in terms of the direct contribution of travel and tourism to employment (number of job creation) in the world in 2019 (Fig. 20.5). During the period of 2014–2019, India created the highest number of jobs (6.36 million) in the tourism sector.[10]

As per the Tourism Satellite Account (TSA) for India prepared by the Ministry of Tourism, the direct, indirect, and total contributions of tourism in employment generation are shown in Table 20.5 for the year from 2013–2014 to 2018–2019. During these years, the direct, indirect, and total contribution of tourism in employment generation has increased significantly. In terms of total employment, it has increased by almost 30% during these years (see Fig. 20.6), which constituted the

[10] See the WTTC website for the details which are available under the URL: https://wttc.org/Portals/0/Documents/Reports/2020/Global%20Economic%20Impact%20Trends%202020.pdf?ver=2021-02-25-183118-360.

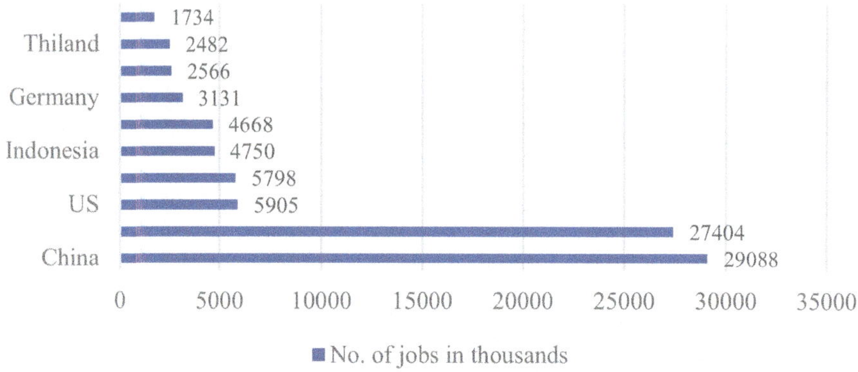

Fig. 20.5 Direct contribution of travel and tourism in the world in thousands (Source: Statista.com (See the Statista website for the details which are available under the URL: https://www.statista.com/statistics/292490/contribution-of-travel-and-tourism-to-employment-in-selected-countries/.))

Table 20.5 Contribution of tourism in employment

Year	Direct (in%)	Indirect (in %)	Total (in %)	Total employment (million)
2013–2014	4.96	6.41	11.37	67.21
2014–2015	5.34	6.91	12.25	69.75
2015–2016	5.40	6.98	12.38	72.26
2016–2017	5.32	6.88	12.20	75.34
2017–2018	5.29	6.84	12.13	80.63
2018–2019	5.56	7.19	12.75	87.50

Source: Annual Report 2019–2020, Ministry of Tourism

highest employment sector in any sector in India. With this increased number of employment and increase in tourist expenditure, the travel and tourism sector has made a significant contribution to India's GDP, and also this sector has great potential in raising income and employment.

20.5 COVID-19 and Indian Tourism

Tourism is the third-largest export industry followed by fuel and chemical industry and accounted for 7% of global GDP, and out of 10 jobs, one is provided by this sector. Therefore, probably this sector is the one that could be affected the most by COVID-19. According to the latest report of UNWTO (December 2020),[11] this pandemic lowers the export revenue by US$ 935 billion for the period January to December 2020 as compared to the same period in 2019. Based on this trend,

[11] See the UNWTO website for the details which are available under the URL: https://www.unwto.org/impact-assessment-of-the-covid-19-outbreak-on-international-tourism.

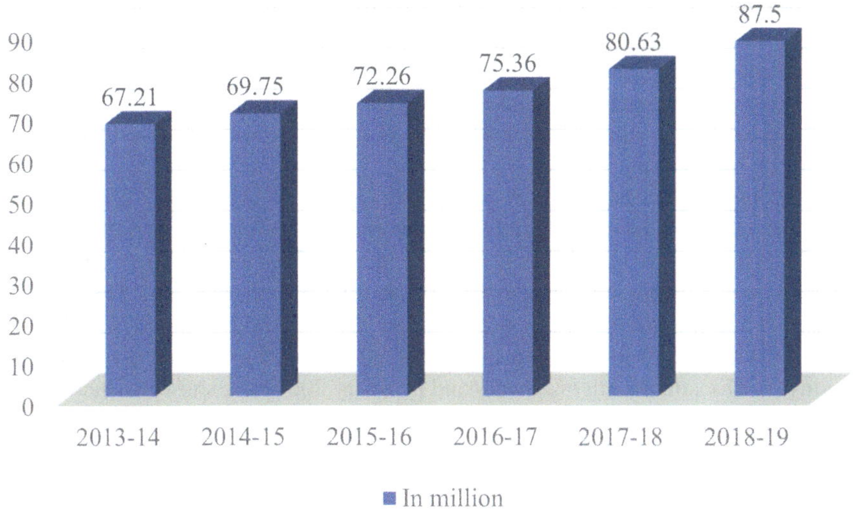

Fig. 20.6 Total contribution of tourism in employment generation in India in million (Source: Prepared by the author)

UNWTO predicted that it could result in an estimated loss of over US$ 2 trillion in global GDP.

The COVID-19 crisis is estimated to have a devastating impact on the Indian tourism sector, which involves hotels, tour operators, travel agencies, restaurants, different modes of transportation, and many more. The overall estimated loss of GDP could be US$69.4 billion, and a total of 8.3 million jobs were lost in 2020 as compared to 2019.[12] To provide some boost to this sector, the Ministry of Tourism has set up a national tourism task force headed by the Ministry of State for Tourism (I/C), Govt of India. This task force will include state tourism ministers, joint secretary-level officers of the concerned central ministries, and representatives from associations like FICCI, CII, ASSOCHEM, WTCII, as well as heads of tourism and hospitality associations.[13]

There has been a huge fall in both inbound and outbound tourism India is experiencing on account of the cancellation of travel plans by foreign as well as domestic tourists. The tourism minister reported in Lok Sabha that international tourist arrivals drop by about 97% from December 2019 to December 2020. There

[12] See the WTTC website for the details which are available under the URL: https://wttc.org/Research/Economic-Impact.

[13] For details see: https://travelobiz.com/ministry-of-tourism-sets-up-national-tourism-task-force/#:~:text=%E2%80%9CMinistry%20of%20Tourism%20has%20constituted%20the%20National%20Tourism,National%20Tourism%20Task%20Force%20shortly%2C%E2%80%9D%20reads%20the%20order.

were 470 international arrivals in April 2020 compared to 7.74 lakh in April 2019 and 79,910 in December 2020 compared to 12.26 lakh in December 2019.[14]

The Federation of Associations in Indian Tourism and Hospitality (FAITH), an apex body of the Indian tourism industry, forecasted that about 70% of the country's estimated workforce (direct and indirect) in this sector could be jobless, affecting near about 3.8 crore people.[15]

According to the annual report of the India Tourism Statistics (2020), foreign tourist arrival (FTA) has been found to be lowered by 53% from the period January–June 2019 to January–June 2020. During the month of February, March, and April, the FTA was downcast by 6.6%, 66.4%, and 100%, respectively. The foreign tourists availing e-tourist Visa also lowered from 907,528 to 837,721 during January–March 2019 to the same period of 2020. In terms of foreign exchange earnings (FEEs), the month of January and February 2020 have shown an increase of 9.9% and 0.9%, respectively, over the same period in 2019, but in March 2020, the figure drastically decreased by 66.3% over the same period in 2019.

According to a forecasting model developed by Jaypuria et al. (2020), if there was no COVID-19, then the predicted minimum foreign exchange earnings (FEE) from August 2020 to September 2021 could be US$ 31325.5 million. However, if the government could be able to manage tourism activities timely, then FEE would be UD$ 13351.07 million, and if not managed, then it could cost US$ 1790.53 million.

Due to the price competitiveness and cultural resources, India is one of the most preferred destinations for both domestic and international travelers. Because of the performance of the previous year, the tourism sector was expected to flourish strongly. But due to the running pandemic, India is facing huge revenue loss in 2020. According to a report by CARE Rating,[16] India could face a revenue loss of Rs. 1.25 trillion, which would be the 40% fall in revenue over 2019 in calendar 2020 owing to lockdown and travel restrictions imposed by the government. About 83% of total travel and tourism spending comes from domestic spending (WTTC 2020), and about 8% of total employment comes from this sector. Therefore, the coronavirus hits the job sector of the tourism industry of this country mostly. Amid the COVID-19 crisis, the travel and tourism sector lose around 50 million jobs.

India may lose a huge number of jobs as 93% of the total working force is engaged in the unorganized sector, which includes construction workers, people

---

[14]For details see: https://m.timesofindia.com/india/tourism-ministry-engages-ncaer-to-study-impact-of-pandemic-on-tourism-fhrai-writes-to-pm-modi-seeking-time-direct-relief-for-travel-sector/articleshow/80793051.cms.

[15]For details see: https://m.timesofindia.com/business/india-business/india-to-issue-free-5-lakh-tourist-visas-give-easy-loans-to-travel-agents-tour-guides/articleshow/83919347.cms.

[16]For details see: https://www.careratings.com/upload/NewsFiles/CareInPress/Tourism%20industry%20to%20incur%20Rs%201.25%20trn%20revenue%20loss%20in%202020%20-%20Business%20Standard.pdf.

associated with the travel and tourism sector, agriculture, etc.[17] These are the most vulnerable section of society and may be badly hit due to the current pandemic. In a press release on 10 April 2020, the Ministry of Tourism admits that the tourism sector is being affected severely in India and regularly discusses the current issue with its stakeholders.[18]

20.6 Opportunities and Challenges

India, with its vast geographical location and diverse landscapes, has excellent potential in the tourism sector. Having the long Himalayan range, a very long coastal area covering two seas and an ocean, forests at different altitudes, desert, rivers, and many more always attracted the tourists not in the present but from the past. The wide variety of landscapes and the cultural and religious diversity have in the favor of this country. Moreover, there is a huge potential in domestic tourism as India is the second-highest populated country. Tourism has been called a sunrise industry for India for its vast potential. The tourism sector emerges as an economic powerhouse, and its potential as a driver of economic growth (Ohlan 2017) is undeniable. It has the ability to create diverse employment opportunities and improves the quality of human life. The main objective of the Ministry of Tourism is to develop the tourism infrastructure and promote digital tourism. These continuous efforts make India 31 place up in the Travel and Tourism Competitiveness Index ranking over the period 2013 to 2019 (The Travel and Tourism Competitiveness Report 2019).

To increase inbound tourism Ministry of tourism extended the e-visa facility to 171 countries under 5 sub-categories, i.e., "e-Tourist visa," "e-Business visa," "e-medical visa," "e-Medical Attendant Visa," and "e-Conference Visa." Two major schemes have been launched by the Ministry of Tourism during 2014–2015, viz. "Swadesh Darshan-integrated Development of Theme-Based Tourist Circuits" and PRASHAD (Pilgrimage Rejuvenation and Spiritual, Heritage Augmentation Drive) for tourism development. An amount of Rs. 5686.67 crore has been sanctioned for 76 projects with a total release of Rs. 4278.88 crore till 31.12.2020 under Swadesh Darshan Scheme. In the PRASHAD scheme, 57 sites have been identified at present in 29 states for development and an amount of Rs. 1160.55 crore has been sanctioned for 36 projects in 24 states since the inception of the project. Till 12.01.2021, a total of Rs. 622.08 crore has been released in this project. A total of Rs. 39.77 crore has been released to various central agencies like Archeological Survey of India (ASI), Port Trust of India, India Tourism Development Corporation (ITDC), etc., for creating tourism-related infrastructure. The ministry has launched various

[17] Report of the working group on Labour Laws and other Labour Regulations, Planning Commission, Govt of India. Available at: http://planningcommission.nic.in/aboutus/committee/wrkgrp11/wg11_rplabr.pdf.

[18] For details see: https://www.indianbureaucracy.com/daily-bulletin-on-covid-19/.

schemes/programs for different purposes like RCS-UDAN for stimulating regional air connectivity, IITF certification program for online training and accreditation of Tourist Facilitators, "Incredible India" mobile application for accessing information, etc. All these programs facilitate and strengthen the tourism sector in India (Annual Report, 2020–2021).

An initiative called SAATHI (System for Assessment, Awareness and Training for Hospitality Industry) has been developed by the Ministry of Tourism to implement guidelines/SOPs effectively issued for COVID-19 and safe operations of hotels, restaurants, homestays and tourist service providers, and other units.[19]

Amidst of COVID-9 crisis, the GST council announced a cut in the tax rate on hotel room tariffs to boost the tourism sector. The council agreed to reduce the tax rate from 28% to 18% on tariffs of Rs. 7500 and above and for tariffs up to Rs. 7500 it is reduced to 12% from 18%. There will be no GST on room tariff of Rs. 1000 and below.[20]

Despite being its growing contribution to GDP and employment generation, this sector faces some challenges. Infrastructure is the intrinsic part of travel and tourism. Infrastructure, comprising all sorts of services like road, water, electricity, safety services, health services, communication, and public services, etc., is the player of tourism development which attracts national as well as international tourists (Khadaroo and Seetanah 2007; Currie and Falconer 2014; UKEssay 2018). Wells and Brandon (1992) have found a multiplier effect between infrastructure, tourism, and local economies. Only 16% of roads were four-lane in India, and the lane capacity is less than average.[21] Indian city roads are facing high congestion, and intra-city vehicle speed is among the lowest in the world.[22] India still lacks in terms of expressways (1583 km as of 2013)[23] that can connect major cities.

World Economic Forum constitutes Travel and Tourism Competitive Index for providing a strategic benchmark tool for business and government to develop the travel and tourism sector. This index helps the travel and tourism business and government take appropriate policies and actions to make this sector a more competitive one. This index covers four aspects of each respective country, enabling environment, T&T policy enabling condition, infrastructure, and natural and cultural resources. Though India ranked 34 among 140 nations and improves six places over 2017, it still faces significant challenges in terms of specific indicators. For example, India belongs to the bottom 20% of the nation regarding safety, security, and environmental sustainability. Along with that, health and hygiene, ICT readiness,

[19]For details see: https://pib.gov.in/PressReleaseIframePage.aspx?PRID=1654147.

[20]For details see: https://www.google.com/amp/s/www.hindustantimes.com/business-news/gst-council-hikes%2D%2Don-caffeinated-beverages-cuts-rates-on-hotel-tariffs/story-CQpzTvHez6X71Ump4EVDXI_amp.html.

[21]For details see: http://business.in.com/article/briefing/crossing-the-chasm/4202/1.

[22]For details see: https://www.thenewsminute.com/article/surprise-surprise-bengaluru-traffic-speed-slowest-says-study-ola-73943.

[23]https://www.rediff.com/business/slide-show/slide-show-1-indias-13-super-highways/20110705.htm.

prioritization of T&T, and tourist service information are indicators responsible for the below-average result (The Travel and Tourism Competitiveness Report 2019).

20.7 Conclusion and Policy Recommendations

COVID-19 crisis is the worst crisis that the travel and tourism sector has ever faced. On this note, UNWTO Secretary-General, Z. Pololikashvili stated, "This crisis is an opportunity to rethink the tourism sector and its contribution to the people and planet; an opportunity to build back better toward a more sustainable, inclusive and resilient tourism sector that ensure the benefits of tourism are enjoyed widely and fairly." The tourism sector needs to measure the impact of the current pandemic and prepare an effective policy involving both the government and the stakeholders to survive, revive, and thrive. This paper has discussed the extent to which COVID-19 can damage the Indian tourism sector. It is evident that no one can foretell how long this crisis will continue. However, this is probably the biggest challenge that the travel and tourism sector is facing and need to survive, revive, and thrive for the large section of people who directly or indirectly depend on this sector.

Tourism has a strong forward and backward linkage to other sectors of an economy. A sharp decline in inbound tourists (due to pandemic) affects income and employment, and this has manifold effects within the economy that cannot be easily calculated. The tourism sector employs largely unskilled labor and women, so these sections of the people are the most affected by this pandemic. According to the WTTC (2020), travel and tourism are that sectors that recover more slowly than any other economic sector. In the Indian tourism sector, 50 million jobs are at risk due to the COVID-19 crisis. FICCI (June, 2020) has recommended several measures to the government of India for the revival of the tourism sector, such as:

- 6-month to a 12-month moratorium on all the working capital, principal, interest payments, loans, and overdrafts;
- 12-month waiver of all statutory dues, including customs, excise, and license fees;
- stimulus package to provide salary support to businesses;
- short-term, low-interest, or interest-free loans for rebuilding business.

In the short run, establishing a healthy and safe tourism industry is very important. To prevent the worst effect of any pandemic appropriate social protection is needed. Government should protect casual and self-employed workers in the tourism industry. Not only workers but also tourism enterprises need special care. Government should help tourism enterprises that are affected mainly by giving low-interest loans or grants. The damage incurred in the tourism sector goes beyond cancelled flights and hotel bookings. Government and private enterprises should take all the necessary measures to bring back the confidence of the tourists to come to India. Public and private both the stakeholder should publicize the fact that India is one of the least affected regions by a coronavirus. There is a strong point to be made for governments to intervene and cooperate at an international level to protect the lives and livelihoods worldwide.

References

Annual Report, Ministry of Tourism, Govt of India (2013–2014). Available at: https://tourism.gov.in/sites/default/files/2019-10/021320150345924.pdf

Annual Report, Ministry of Tourism, Govt of India (2014–2015). Available at: https://tourism.gov.in/sites/default/files/2019-10/Report_0.pdf

Annual Report, Ministry of Tourism, Govt of India (2015–2016). Available at: https://tourism.gov.in/sites/default/files/2019-10/Annual%20Rreport%202015-16.pdf

Annual Report, Ministry of Tourism, Govt of India (2016–2017). Available at: https://tourism.gov.in/sites/default/files/2019-10/MoT%20Annual%20Report%202016-17_English.pdf

Annual Report, Ministry of Tourism, Govt of India (2017–2018). Available at: https://tourism.gov.in/sites/default/files/2019-10/Annual%20Report2017-18.pdf

Annual Report, Ministry of Tourism, Govt of India (2018–2019). Available at: https://tourism.gov.in/sites/default/files/2019-10/Ministry%20of%20Tourism%20Annual%20Report%20English%20for%20Web.pdf

Annual Report, Ministry of Tourism, Govt of India (2019–2020). Available at: https://tourism.gov.in/sites/default/files/2020-02/Annual%20Report%20Tourism%202019_20_Final.pdf

Arain H, Han L, Sharif A, Meo MS (2019) Investigating the effects of inbound tourism on FDI: the importance of quantile estimations. Tour Econ 26(4):682–703

Balakrishnan R, Christopher PB (2020) Impact of tourism in India due to coronavirus. IJARIIT 6(2):281–282

Blake A, Sinclair MT, Campos JA (2006) Tourism productivity: evidence from the United Kingdom. Ann Tour Res 33:1099–1120

Cernat L, Gourdon J (2012) Paths to success: benchmarking cross-country sustainable tourism. Tour Manag 33:1044–1056

Chavan RR, Bhola SS (2014) Indian tourism market: an overview of emerging trends and development. Global J Commerce Manag Perspect 3(4):113–122

Chen Y (2017) China's tourism-led foreign direct investment inflows: an empirical study. Mod Econ 8(01):39

Christian (2015) Economic contribution of travel and tourism. [online] *AALEP*. Available at: http://www.aalep.eu/economic-contribution-travel-and-tourism. Accessed 12 Dec 2020

Currie C, Falconer P (2014) Maintaining sustainable island destinations in Scotland: the role of the transport–tourism relationship. J Destin Mark Manag 3(3):162–117. https://doi.org/10.1016/j.jdmm.2013.10.005

Dhariwal, R. (2003). Tourist arrivals in India: how important are domestic disorders. Available at SSRN: https://ssrn.com/abstract=390666 or doi:https://doi.org/10.2139/ssrn.390666

Dwyer L, Forsyth P (1993) Assessing the benefits and costs of inbound tourism. Ann Tour Res 20(4):751–768

Endo K (2006) Foreign direct investment in tourism—flows and volumes. Tour Manag 27:600–614. Available at: https://www.scirp.org/journal/paperinformation.aspx?paperid=73447

FICCI (June, 2020) Travel and Tourism: survive, revive and thrive. Available at: https://www.grantthornton.in/globalassets/1.-member-firms/india/assets/pdfs/travel-and-tourism-in-times-of-covid-19.pdf

Garay L, Font X, Pereira-Moliner J (2017) Understanding sustainability behaviour: the relationship between information acquisition, proactivity and performance. Tour Manag 60:418–429. Available at: https://www.sciencedirect.com/science/article/pii/S0261517716302631

Incera AC, Fernández MF (2015) Tourism and income distribution: evidence from a developed regional economy. Tourism Manag 48:11–20

India Tourism Statistics, Govt of India (2015). Available at: https://tourism.gov.in/sites/default/files/2020-04/India%20Tourism%20Statistics_English_2015%20Final.pdf

India Tourism Statistics, Govt of India (2017). Available at: https://tourism.gov.in/sites/default/files/2020-04/INDIA%20TOURISM%20STATISTICS%202017.pdf

India Tourism Statistics, Govt of India (2018). Available at: https://tourism.gov.in/sites/default/files/2020-04/India%20Tourism%20Statistics%202018_0.pdf

India Tourism Statistics, Govt of India (2019). Available at: https://tourism.gov.in/sites/default/files/2020-04/India%20Tourism%20Statistics%20at%20a%20Glance%202019.pdf

India Tourism Statistics, Govt of India (2020). Available at: https://tourism.gov.in/sites/default/files/2021-05/INDIA%20TOURISM%20STATISTICS%202020.pdf

Jalil A, Mahmood T, Idrees M (2013) Tourism–growth nexus in Pakistan: evidence from ARDL bounds tests. Econ Model 35:185–191

Jaypuria S, Parida R, Ray P (2020) The impact of COVID-19 on tourism sector in India. Tour Recreat Res 12:28–32

Khadaroo J, Seetanah B (2007) Transport infrastructure and tourism development. Ann Tour Res 34(4):1021–1032. https://doi.org/10.1016/j.annals.2007.05.010

Kumar RA (2019) The impact of inbound tourism on economic growth of Nepal. Int J Acad Res Bus Soc Sci 9(3):183–195

Kumar RR (2014) Exploring the role of technology, tourism and financial development: an empirical study of Vietnam. Qual Quant 48(5):2881–2898

Kumar RR, Loganathan N, Patel A, Kumar RD (2015) Nexus between tourism earnings and economic growth: a study of Malaysia. Qual Quant 49(3):1101–1120

Lee CC, Chang CP (2008) Tourism development and economic growth: a closer look at panels. Tourism Manag 29:180–192. https://doi.org/10.1016/j.tourman.2007.02.013

Kakkar L, Sapna (2012) Impact of tourism on Indian economy. Int J Market Finan Serv Manag Res I(4):89–95

Lopez CSG, Arreola KSB (2019) Impacts of tourism and the generation of employment in Mexico. J Tour Anal 26(2):94–114. Available at: https://www.emerald.com/insight/content/doi/10.1108/JTA-10-2018-0029/full/pdf?title=impacts-of-tourism-and-the-generation-of-employment-in-mexico

Mathieson A, Wall G (1982) Tourism: economic, physical and social impacts. Longman, London. p. 14, 36, 87

McKinnon DRI (1964) Foreign exchange constraint in economic development and efficient aid allocation. Econ J 74:388–409

Mishra PK, Rout HB, Kestwal AK (2020) Tourism, foreign direct investment and economic growth in India. Afr J Hospitality Tourism Leisure 9(1) Available at: https://www.ajhtl.com/uploads/7/1/6/3/7163688/article_62_vol_9_1__2020_india.pdf

Ohlan R (2017) The relationship between tourism, financial development and economic growth in India. Future Business J 3:9–22

Page J et al (2021) In hunt for COVID-19 origin, patient zero points to second Wuhan market. Wall Street J. Available at: https://www.wsj.com/articles/in-hunt-for-covid-19-origin-patient-zero-points-to-second-wuhan-market-11614335404

Poddar S (2015) A study on role of tourism in Indian economy with reference to identification of skill gaps in tourism industry. Int J Manag Soc Sci 03(01):102–112

Prasad N, Kulshrestha M (2015) Employment generation in tourism industry: an input-output analysis. Ind J Labour Econ 58:563–575

Pratt S (2015) The economic impact of tourism in SIDS. Ann Tourism Res 52:148–160. https://doi.org/10.1016/j.annals.2015.03.005

Rout HB, Mishra PK, Pradhan BB (2018) Empirics of tourism-led growth in India, 1995 to 2016. J Environ Manag Tourism 9(6):1190–1201

Singh, P. (2008). Sustainable tourism and economic development: a case study of Agra and Varanasi published in conference proceedings on 'tourism in India - challenges ahead', *IIMK*

Shah AR, Beig MA, Ahmed M (2014) Role of tourism in income and employment in Kashmir valley. Int J Recent Sci Res 5(12):2301–2303. Available at: http://recentscientific.com/sites/default/files/1789.pdf

Shahzad SJH, Shahbaz M, Ferrer R, Kumar RR (2017) Tourism-led growth hypothesis in the top ten tourist destinations: new evidence using the quantile-on-quantile approach. Tourism Manag 60:223–232. Available at: https://www.sciencedirect.com/science/article/pii/S0261517716302552

Sharma N (2018) Tourism led growth hypothesis: empirical evidence from India. Afr J Hospitality Tourism Leisure 7(2) Available at: https://www.ajhtl.com/uploads/7/1/6/3/7163688/article_7_vol_7__2__2018.pdf

Tang CF, Abosedra S (2016) Tourism and growth in Lebanon: new evidence from bootstrap simulation and rolling causality approaches. Empir Econ 50(2):679–696

Tang CF, Tiwari AK, Muhammad S (2016) Dynamic inter-relationships among tourism, economic growth and energy consumption in India. MPRA paper no. 69848. Available at: https://mpra.ub.uni-muenchen.de/69848/1/MPRA_paper_69848.pdf

The Travel and Tourism Competitiveness Report (2019) World Economic Forum. Available at: http://www3.weforum.org/docs/WEF_TTCR_2019.pdf

UNCTED, (2020). COVID-19 and tourism: assessing the economic consequences. [online] *UNCTED*. Available at: https://unctad.org/system/files/official-document/ditcinf2020d3_en.pdf. Accessed 17 Dec 2020

UKEssays (November 2018) Tourism infrastructure and tourism development tourism essay. Retrieved from https://www.ukessays.com/essays/tourism/tourism-infrastructure-and-tourism-development-tourism-essay.php?vref=1

Wells M, Brandon K (1992) People and parks: linking protected area management with local communities. World Bank, Washington DC

WTTC (2019) Travel and tourism: global economic impact & trends. [online] WTTC. Available at: https://wttc.org/Research/Economic-Impac Accessed 25 Oct 2020

WTTC (2020) Travel and tourism: Global Economic Impact & Trends. [online] *WTTC*. Available at: https://wttc.org/Research/Economic-Impact. Accessed 30 Oct 2020

Arunava Kumar Choudhury is an assistant professor in Sripat Singh College, Jiaganj, Murshidabad, in India since 2016. He has been awarded M. Phil. in Economics in 2009 from University of Kalyani. Presently, he is pursuing his Ph.D. from Raiganj University, India. His areas of interest are agricultural economics and financial economics.

Subrata Saha was awarded with Ph.D. in Economics from the University of North Bengal. Dr. Saha specializes on issues of fiscal policies and effects of structural changes on fiscal relationships in Southeast Asian Countries and India. The areas of research interest are public finance and time series econometrics. He is a member of the editorial board of UGC CARE listed journal ENSEMBLE and a life member of the Indian Econometric Society and Indian Economic Association. Dr. Saha is presently working as an Associate Professor of Economics at Raiganj University, West Bengal, India. He has supervised four M. Phil. and one Ph.D. scholars and presently supervising four Ph.D. scholars at Raiganj University. He has contributed several research papers to national and international Journals and edited research volumes. Besides, he has authored two books to his credit. IRDP, India, has awarded him with prestigious Sarvepalli Radhakrishnan Lifetime Achievement National award in 2018 in recognition of his contribution to his activities toward promoting educational excellence.

Chapter 21
Invasion of the Pandemic in Indian Economy and the Government: A General Equilibrium Approach

Nilendu Chatterjee and Bappaditya Koley

Abstract Underdevelopment in the third world developing economies has been one of the essential features that have made a welfare state's role even more important. We need a strong government setup to break the curse of backwardness by using its various socio-economic policies. Again, within backward parts, few parts are 'advanced' in nature. Due to COVID-19, we have seen good health facilities being provided even in the so-called backward regions. So, even in the backward regions, we have seen a good healthcare system is provided. But, obviously, in a nation, there exists one strong urban manufacturing sector that plays an important, dominant role in the development of a country. All these sectors are connected, and they collectively enhance the process of development. In a welfare state, the government follows any of these two ways—the subsidies are provided either to the health and allied sectors or to the people directly to improve their standard of health and living. All these will not only help to pass through the COVID period but will also improve the human development indicators. Unfortunately, few governments have been following both these policies together. Both these forms are supposed to improve the condition of the people of the backward parts. In this paper, with the help of a three-sector general equilibrium framework, we have considered both these forms of government intervention and seen what effect they have on the economy.

Keywords Health sector · Subsidy · Human development · General equilibrium · COVID-19

N. Chatterjee
Department of Economics, Bankim Sardar College, South 24 Parganas, West Bengal, India

B. Koley (✉)
Department of Geography, Bankim Sardar College, South 24 Parganas, West Bengal, India

S. Saha et al. (eds.), *Economic and Societal Transformation in Pandemic-Trapped India*, New Frontiers in Regional Science: Asian Perspectives 55,
https://doi.org/10.1007/978-981-16-5755-9_21

21.1 Introduction

There is no doubt that COVID-19 has hit the world hard, especially in three nations, namely the USA, India, and Brazil, regarding the number of cases and number of deaths. The world has witnessed governments step up to protect the economies as expected during a pandemic. Being developing in nature, the welfarist attitude of governments becomes even more critical in cases of Brazil and India as these economies suffer from proper infrastructures. Although vaccination is going on, we all know it will take at least 2–3 years, if not more, to vaccinate all, and even after vaccination, people are getting attacked by the virus and even cases of deaths, in spite of completion of vaccination doses, are being reported.

Hence, we all know that the governments, both at the regional level and at the national level, have to come up to save the nation. The spread of the virus depends on people's consciousness, how far they can follow the health norms. But, both the first wave and second wave have destroyed the base of economies, especially the developing ones.

In India, a huge proportion of labour forces are migrant workers who go to the big cities searching for work. Research by Ajeevika Bureau has reported that almost 140 million workers in India migrate seasonally to the nearby towns or cities or even to other states in search of work. (Ajeevika Bureau 2014). Another report by United Nations (2021) reveals that almost one-third of India's population are migrants. They have estimated that around 60% of urban migrants have migrated from rural areas searching for work, and around 36% of them are engaged in the construction sector, followed by agriculture, manufacturing, trade, and transport. But, most of them are casual labours who have not completed primary education and belong to the youth age group of below 30 (United Nations). COVID's second wave has seen lockdown in several states, [1] including all the big states. Consequently, it is quite natural that the economy of these states and India has been hit very badly. In India, a large proportion of the working population is engaged in the informal sector, and they are associated with small industries. In the year, the size of employment in the informal sector was 81% of the total labour force. In 2005, this figure was around 86%. However, if we consider the contractual and casual workers who work within the formal sector or organized sector, [2] this figure goes up to 92% or even more (Punia 2020). Hence it is quite natural that their economic condition is very brittle, and it is quite possible for them to suffer due to the sudden economic shock of the second wave and, thus, face its negative consequences. During the first wave, many migrant labours lost their works and got back to their home states. The economic and political consequences of such return are known to the whole world. The endless affliction of millions of migrant workers and other informal sector workers who lost their works received a wage cut cannot be described in words.

[1] Although there is no national lockdown that occurred during the first wave in 2020.

[2] But the nature and conditions of their work are very much similar to that of informal sector.

As the number of unemployed kept increasing, it increased the pressure on the governments at all levels to provide employment opportunities and basic amenities for survival. Being a welfare economy, [3] the government could not deny its responsibilities, and in a locked-down economy, we all know how tough it is for any government to continue its welfare programmes due to a lack of revenues. [4] As the first wave gradually disappeared, workers began to get back to their working places, which reduced the unemployment a bit, but we all know the economy never got back to the smooth path of growth or functioning which it used to in the pre-COVID era.

Now, the second wave, its effects have been even deadlier than the first one. It has been hitting almost all parts of India at a greater pace; numbers of cases have seen an upsurge, so has the number of deaths. Consequently, the economic sufferings of people, especially those at the backward parts of the nation, rural economy, financially weaker section have also gone up. During the first wave, governments gave their best efforts to continue their welfarist approach, such as announcing special economic packages, providing relief from paying the instalments for few months, supplying free ration to every family, etc. The government had announced Rs. 20 lakh crore package, including 1.7 lakh crores for free food grain supply to the poor and cash transfer to the bank account of poor and elderly people (The Economic Times 2020) by continuing their subsidized programmes. [5] We know that such approaches will continue during the second wave, but since the socio-economic effects of this ongoing second wave are even worse, the question arises—how far and what sort of help from the government would help overcome this situation? States are in locked-down mode; economic sectors have been functioning poorly. Consequently, suffering is rising, and no one can guarantee when such sufferings would end because even if this second wave goes away, it will take time for the economy to get back to normalcy and again, there is no guarantee that all the sectors of the economy would start functioning smoothly or efficiently. Hence, the welfarist approach by the government would be necessary and would continue to get an important place, as far as the livelihood of millions of Indians belonging to the financially weaker section along with control of COVID is concerned.

In the post-second war era, we have witnessed the formation of several developing economies featured with a combination of developing and developed regions. These economies are featured with multiple types of dualism, starting from financial

---

[3] Although in the post-liberalization era, the nature of government's activities has changed a lot but the very basic feature of governments in India is still welfarist because in such a huge nation with so many socio-economic problems governments cannot deny their developmental role. Government's function has shifted from core economic activities to improvement of social sectors and infrastructure development. During this worldwide pandemic, this role has become even more important.

[4] Governments also understand that it is very tough to continue such activities for a long time, especially when the revenues are not getting generated but they have been trying more than their level best to continue such programmes to help the poor and perhaps that is why no nationwide lockdown was announced during the second wave and even the state governments allowed economic activities but in a restricted manner, like in West Bengal, Maharashtra, etc.

[5] We all know that these helps were far lower than what was expected as well as needed.

dualism to developmental and regional dualism (Kathuria et al. 2013; Kesar and Bhattacharya 2020). These economies have big developed cities with all the modern facilities, and however, these economies have regions with poverty where people are in hunger. These backward regions even suffer from the lack of basic amenities. India also belongs to this category of the dual economy.

On the one hand, we have seen backward regions suffering from COVID without proper infrastructural facilities such as hospitals and beds, and on the other hand, we have seen cities with all facilities. [6] So, it is much tougher for the backward regions to fight COVID, its second wave, and get back to normalcy. But, the COVID period has witnessed a considerable amount of investment by the governments in rural areas' health infrastructure. For example, one can say that in 2016 there were 13 medical college hospitals in West Bengal with 144,798 sanctioned beds, but in 2018 there were 18 medical colleges with 153,804 beds, and most of these new hospitals were set up in rural or semi-urban areas to help the needy people (SBHIDHS 2018). Along with the waves of COVID, the government started to take up private hospitals, increased the number of beds in all hospitals, set up testing labs even in the remotest part, converted non-COVID hospitals into COVID hospitals with facilities. All these have helped to increase the investment in health in backward areas, which definitely has a multiplier effect [7, 8] that has been helping these regions to fight this second wave.

Backward regions, [9] as we all know, are generally defined based on economic and social indicators, for example, HDI. Such indicators help to understand the degree or intensity of backwardness and the necessary steps governments need to follow to break the jinx of a vicious circle of poverty. During COVID, we have been witnessing such steps being followed. Backward areas, its population were given priorities not only to fight COVID but also to continue their livelihood.

During the first wave of COVID, India did suffer. Still, the extent of suffering has increased immensely during the second wave as the number of cases has been increasing leaps and bounds along with the number of deaths, and these numbers

[6] It does not mean that the big cities did not suffer from COVID. They also suffer, but they are well equipped with infrastructural facilities to fight it.

[7] One should not ignore the increased investment in health equipment such as sanitizer, mask, PPE kits, etc.

[8] Regarding hospital facilities, testing facilities, availability of doctors, bed facilities in hospitals, oxygen, medicines, etc. are concerned.

[9] As identified by various reports of NITI Aayog (2018) (on the basis of Health & Nutrition, Education, Agriculture & Water Resource, Financial Inclusion, Skill Development and Basic Infrastructure). For example, one can name Bokaro of Jharkhand, Muzaffarpur of Bihar, etc. there is one more list of Backward districts prepared by Income Tax Department (1979) of India in which West Bengal have 13 districts.

are almost more than double than that occurred during the first wave. [10] So, it is a worry for all.

Coming back to the economic scenario, we can see that certain sectors are a bit advanced in nature in the backward area of an economy. For example, in Rajasthan, the sector of Rajasthani dresses is advanced in nature, in Purulia of West Bengal, the sector of 'Chhau' dance and its associated industries are advanced. These industries have survived for a long time and have earned worldwide fame, which has resulted in an increasing demand for their products and services. So, they have constantly been upgrading themselves in order to increase their demand further. Hence, economic activities and investments in these industries are of better quality than other sectors of those districts. Hence, such sectors may be referred as 'high linkage sectors' by which growth can be promoted in the backward areas (Bhattacharrya and Rajeev 2014).

Similarly, in the coastal part of West Bengal, fisheries sectors are advanced. So, within the backward part, there are few significant sectors or economic activities that enjoy an advantageous position. COVID has forced governments to invest in the healthcare sector of each and every part of a nation, with special emphasis being given to the backward regions. People also have begun to increase their out-of-pocket expenditures for healthcare purposes during COVID. Be it the purchase of masks, sanitizers, or even day-to-day maintenance of high immunity level for which we need a good amount of money to be invested on a daily basis. So, health and its allied sectors, due to COVID and consequential measures taken by the government, are in an advanced position compared to other economic sectors even within the backward parts. So, investment by the government in such advanced sectors would definitely have positive multiplier effects on the economy. As per the government's data, the Indian economy has witnessed a negative growth rate of −7.3% in the 2020–2021 financial year, which is the worst this nation ever had since independence (Business Standard 2021). The World Economic Outlook of IMF confirmed this fact and reported a contraction of GDP by 7.965% in 2020 (IMF 2021). We cannot deny the shock given to the economy by the pandemic in 2020 is the sole reason for this, and it may further worsen due to its second wave in 2021. [11] The way predictions are made from the doctors regarding upcoming waves of COVID; it looks like these negative impacts of COVID would persist. [12]

Under such circumstances, it becomes even more important for governments to invest more in the backward areas in both ways. First, to invest in the little bit advanced parts or sectors, including health being emphasized, its benefits can be enjoyed by the rest of the economy within the backward parts. Second, governments

[10] In the first wave the highest number of positive cases recorded in a day was over one lakh but this second wave has witnessed this number to be over four lakhs. The number of deaths reported also has increased significantly (As per the Reports of Ministry of Health).

[11] Although no nationwide lockdown is announced in 2021, like 2020. But, the negative impacts of state-wise prolonged lockdowns cannot be ignored.

[12] Since it is not easy to get back to normalcy immediately even if COVID disappears.

are expected to increase and continue socio-developmental programmes, such as providing ration and unemployment benefits to the poor people that will help them survive during this period. [13]

Here, in this paper, all these aspects will be considered and discussed with the help of a general equilibrium model based on the Heckscher–Ohlin–Samuelson structure. The rural economy is primarily informal, and the urban or developed economy is formal. It is pretty reasonable to assume these things here as well.

In the field of literature, there are many works on the general equilibrium framework. We have seen works on the informal sector by Banerjee (1983), Fields (1990), and Chaudhuri and Mukhopadhyay (2009) on developing economies. Marjit and Kar (2011) have compiled several works regarding the informal sector's existence and unemployment. Chaudhuri et al. (2017) have inquired whether a subsidy by the government in education would help reduce wage inequality. Regarding COVID and its economic impact, few works are there. In their study, Yuen et al. (2020) have discussed nine most essential questions regarding COVID, beginning from its cause of occurrence to vaccination. Mustafa (2021) has discussed and explained facts and figures on various aspects of COVID, starting from analysing the daily reports of COVID positive cases to analysing the rise and fall of its curve along with the death rate. Gupta et al. (2020) have detailed the COVID-19 outbreak in India in 2020. Using polynomial and exponential regression analysis, they have predicted the number of COVID cases and deaths in India in April 2020. Agarwal and Singh (2020) have analysed the impact of COVID-19 on the Indian economy. They have outlined the possible government policies and highlighted those sectors that are supposed to get a boost during COVID. Roy (2020) has also analysed the impact of COVID 19 impact on few important sectors of India, such as tourism, financial sector, healthcare, oil, etc. She has shown how these sectors are going to incur losses due to this pandemic. Mou (2020) has analysed how far both developing and developed economies are going to be affected due to COVID 19 as far as fall in GDP is concerned and international business with the rest of the world is concerned in an era of globalization.

Hence, from the above discussion, it becomes clear that almost all the works on COVID are either analysing or econometric in nature. As far as the impact of COVID and corresponding necessary policy measures by governments are concerned, there is almost no paper involving general equilibrium analysis. Again, there is perhaps no work that has discussed the possible policies in the backward part of a nation by dividing the economy into two parts. We shall look to fill up these lacunae by this paper, and these are our prime motivations behind this study.

The rest of the paper is organized as follows: Sect. 21.2 discusses the general equilibrium model of our study built on the H-O-S framework. This section has two parts. First, it will discuss the impact of a subsidy on social capital given by the

[13] One can mention the names of successful programmes, such as "Duare Sarkar," "Free Ration," "Kanyasree" initiated by West Bengal Government, programmes like "Ujjwala," "Direct Benefit Transfer," etc. by Central Government.

government. Here we shall discuss the impact of a subsidy on the advanced sector within the backward part of an economy. This advanced part contains the health sector. The second part will show the impact of government intervention in externality-augmented social indicators. This section also includes the comparative static analysis in the presence of externality. Finally, concluding remarks are made in Sect. 21.3.

21.2 The Model

Here we have considered three sectors. Sector *B* is the traditional rural sector which is backward in nature. Sector *H* is the advanced rural sector that incorporates health [14] as well, and Sector *Z* is the modern, most advanced urban sector. Sector *B* uses labour (*L*) and land (*T*) as inputs. The product of sector *B* is regarded as numeraire, and its price has been set equal to unity. Sector *H* uses labour (*L*) and capital (*K*) and enjoys subsidy (*S*) from the government. So, its price is inclusive of subsidy. Sector *Z* uses labour (*L*), advanced capital (*N*), and intermediate inputs or goods from advanced rural sector *H*. [15] Production function in each sector exhibits constant returns to scale (CRS), variable co-efficient technology, and diminishing marginal productivity to the variable factors. The urban sector uses labour at a higher wage ($\overline{w}$) as it needs more efficient labour compared to the other two sectors.

The following notations are used in the formal presentation of the model:

a_{Li} = Labour-output ratio in the *i*th sector, $i = B, H, Z$
a_{Ki} = Capital-output ratio in the *i*th sector, $i = H$
a_{Ni} = Capital-output ratio in the *i*th sector, $i = Z$
P_i = Price of the product of the *i*th sector, $i = B,H,Z$
w = Competitive wage rate of sector *b* and sector *h*
$\overline{w}$ = Fixed or higher wage rate of sector *z*
r = Rate of return on traditional capital
R = Rate of return on advanced capital
L = Total endowment of labour in the economy
K = Total endowment of traditional capital
N = Total stock of advanced capital
λ_{ji} = Use of *j*th factor by *i*th sector with respect to total factor endowment ($j = L, K$; $i = B, H, Z$)
θ_{ji} = Distributive share of the *j*th input in the *i*th sector

14 Here health has been considered as a part of advanced sector within the backward parts or areas of the economy because of high investment in health sector even in the backward areas, since the beginning of COVID, which has resulted in an improvement in overall health facilities within backward areas.

15 Intermediate input may come from health sector or from any other advanced sector within rural backward economy.

X_i = Output of sector *I* ($i = B, I, Z$)
$\mu = K_F/K$

21.2.1 Equational Structure of the Model

The competitive equilibrium conditions of the three sectors are given as

$$a_{LB}w + a_{TB}P_T = P_B = P_B^* = 1 \tag{21.1}$$

$$a_{LH}w + a_{KH}r = P_H^*(1 - S) = P_H \tag{21.2}$$

$$a_{LZ}\overline{w} + a_{HZ}P_H + a_{NZ}R = P_Z^* \tag{21.3}$$

Input market equilibrium conditions are

$$a_{LB}X_B + a_{LZ}X_Z + a_{LH}X_H = LE(\text{HDI}) \tag{21.4}$$

Equation (21.4) gives us the full employment condition of the labour market along with a meaningful insight. For instance, it is quite reasonable to assume that the average efficiency of the workers, *E*, generally depends on their human development conditions. Therefore, an improvement in any of the human development parameters is expected to raise workers' efficiency and the following expression illustrates that

$$E = E(\text{HDI}); \quad E' > 0 \tag{21.5}$$

$$a_{TB}X_B = T \tag{21.6}$$

$$a_{KH}X_H = H \tag{21.7}$$

$$a_{NZ}X_Z = N \tag{21.8}$$

$$a_{HZ}X_Z = X_H \tag{21.9}$$

21.2.2 Government Intervention Regarding Investment in Industries in the Advanced Part of the Rural Sector

Here, it has been assumed that the government provides a subsidy to the advanced rural sector of which health is also a part or invests in the social capital.

Using '^' mathematics and following Jones (1965, 1971), from Eqs. (21.1), (21.2), and (21.3) we get the following specifications:

$$\widehat{w} = -\frac{\theta_{TB}}{\theta_{LB}}\widehat{P}_T \tag{21.10}$$

$$\widehat{w} = \frac{1}{\theta_{LH}}\widehat{P}_H \tag{21.11}$$

$$\widehat{P}_H = \frac{\theta_{NZ}}{\theta_{HZ}}\widehat{R} \tag{21.12}$$

Using Eqs. (21.6), (21.24), and (21.21) we get (see Appendix for details)

$$\widehat{X}_B = \left(\frac{\sigma_B \cdot \theta_{LB}}{\theta_{TB}}\right)\widehat{w} \tag{21.13}$$

Similarly, using Eqs. (21.7), (21.27), and (21.26) we get (see Appendix for detailed derivation)

$$\widehat{X}_H = \left(\frac{\mu_K}{\lambda_{KH}}\right)\widehat{K} - (\theta_{LT} \cdot \sigma_H)\widehat{w} \tag{21.14}$$

From Eq. (21.8) we get

$$\begin{aligned} \theta_{NZ}\widehat{X}_Z &= \widehat{N} \\ \widehat{X}_Z &= \left(\frac{1}{\theta_{NZ}}\right)\widehat{N} \end{aligned} \tag{21.15}$$

From Eq. (21.9) we get

$$\begin{aligned} \theta_{HZ}\widehat{X}_Z &= \widehat{X}_H \\ \widehat{X}_Z &= \left(\frac{\mu_K}{\theta_{HZ} \cdot \lambda_{KH}}\right)\widehat{K} - \left(\frac{\theta_{LT} \cdot \sigma_H}{\theta_{HZ}}\right)\widehat{w} \end{aligned} \tag{21.16}$$

From expressions (21.10) to (21.12) we get to see an increase in S which implies a fall in $P_H^*(1 - S)$. This again leads to a fall in the P_H, which again leads to a fall in the w and an increase in P_T, a fall in R. From Eq. (21.13), we get a fall in w which leads to a fall in X_B. From Eq. (21.14) we get a fall in w which leads to an increase in X_H, and following Eq. (21.16) we get a fall in w which leads to an increase in X_Z.

The above-described procedure leads us to the following proposition.

Proposition 1 When a subsidy is given to the advanced rural sector, it leads to two outcomes. First, an increase in output of that sector, that is, advanced rural sector along with the output of urban sector, but it leads to a fall in the output of traditional rural sector which is backward in nature. Second, it leads to a fall in the competitive wage in the rural economy, the price of advanced capital, and a fall in price of advanced rural sector also, but an increase in the price of land.

21.2.3 Government Intervention in Terms of Externality-Augmented Social Indicators

During COVID and its devastating second wave, people have suffered immensely just to acquire their basic needs. Be it health-related goods, be it food, almost everything. The severity of the prolonged lockdown and its economic curse has been felt by all, especially the middle class and underprivileged sections. Hence, the role of the government in a welfare state becomes even more critical to help people during this time and even after COVID to get out of this problem.[16] So, we can say, if the governments decide to invest directly in people, by providing them subsidy directly through various developmental schemes which are expected to bring a desired change in the values of social indicators like health, education, etc., then the governments are actually looking to achieve a larger goal which is to improve the value of HDI. Improvement in such indicators of backward areas is expected to be reflected through an improvement in the efficiency of labour force of such areas. The next sub-section reveals the economic feasibility of such programmes.

21.2.3.1 Macroeconomic Foundation of Externality-Augmented Efficiency Function

Following Shapiro and Stiglitz (1984), the profit of a firm is expressed as

$$\Pi = F(LE(\text{HDI})) - wL \tag{21.17}$$

Total derivative of Eq. (21.17) gives us

$$\begin{aligned} d\Pi &= \frac{\partial F}{\partial (LE(\text{HDI}))} LdE - Ldw \\ &\text{Or} \\ \frac{d\Pi}{dw} &= \frac{\partial F}{\partial (LE(\text{HDI}))} L \cdot \frac{dE}{dw} - L \end{aligned} \tag{21.18}$$

In Eq. (21.18) $\frac{\partial F}{\partial (LE(\text{HDI}))} > 0$ and $\frac{dE}{dw} > 0$ implies higher w which leads to higher E which again leads to higher profit.

Again from Eq. (21.17) we get

[16] Because COVID has long lasting effects as far as economic angle is concerned. It has pushed back millions of Indians below the poverty line again and it is not at all an easy task to pull them out of it again, at least it would not happen overnight, certainly.

$$\frac{d\Pi}{dE} = \frac{\partial F}{\partial E} - \frac{Lw}{E}\left(\frac{dw}{dE}\cdot\frac{E}{w}\right)$$
$$\frac{d\Pi}{dE} = \frac{\partial F}{\partial E} - \frac{Lw}{E}\xi_{W,E} \tag{21.19}$$

As $\frac{\partial F}{\partial E} > 0$ and $\xi_{W,\,E} < 0$, it implies $\frac{d\Pi}{dE} > 0$, which again implies higher E which leads to higher profit. Expressions (21.18) and (21.19) claim that direct subsidy will help workers to earn more and which in turn gives scope to improve productivity in terms of efficiency. Apart from this both expressions show possibilities of increase in profit as a result of improved efficiency. Following this, next we move to comparative static analysis with respect to subsidies in the presence of efficiency augmented externalities.

21.2.4 Comparative Static with Externality

From Eq. (21.4) we get

$$\lambda_{LB}\widehat{X}_B + \lambda_{LZ}\widehat{X}_Z + \lambda_{LH}\widehat{X}_H = \xi_L\widehat{\text{HDI}} \tag{21.20}$$

An increase in HDI regarding improvement in human development indicators implies labour supply increases in terms of efficiency, that is, (LE (HDI)) increases. This will ensure an increase in the bargaining power of the labour, that is, competitive wage (w) may or should increase. This leads to the following proposition.

Proposition 2 An improvement in the value of any or all the human development indicators is expected to increase the efficiency, bargaining power, and wage of the labour.

Therefore, an increase in HDI leads to an increase in w, a fall in P_T, an increase in P_H. Again, an increase in HDI leads to an increase in w, an increase in a_{KH}, a fall in X_H. Again, a fall in P_T implies an increase in a_{TB}, a fall in X_B, this again leads to an increase in the availability of labour to X_H and X_Z.

An increase in P_H leads to a fall in R, an increase in a_{NZ}. So, using more efficient labour, along with more N, sector Z will expand. We can refer to this effect as the absorption effect due to the efficient use of factor (AEEF). Again as P_H rises, it leads to an increase in W, a fall in $\frac{\bar{r}}{w}$, an increase in a_{KH}, a fall in X_H, from Eq. (21.16) we get to see a fall in X_Z. This effect can be referred as factor price effect of intermediate input (FPII). If AEEF is higher than or if its effect is more than that of FPII, then it leads to an increase in X_Z and a fall in X_B and X_H. But, if the opposite happens, that is, if effect of FPII is higher than that of AEEF, then we will get an increase in X_B and a fall in X_Z and X_H. All these effects give us the following proposition.

Proposition 3 An improvement in the value of any or all the human development indicators generates two significant effects. First, if the absorption effect due to efficient use of factor (AEEF) dominates over factor price effect of intermediate input (FPII), this leads to an increase in the output of the urban sector but a fall in that of both the rural sectors. Second, if factor price effect of intermediate input (FPII) dominates over absorption effect due to efficient use of factor (AEEF), this will increase the output of traditional rural sector but a fall in both advanced rural sector and urban sector.

21.3 Concluding Remarks

The second wave of COVID-19 has hit India worst most compared to any nation in the world, and it obviously is far more severe than that of the first wave. The health sector got a shock because it spread at a much faster rate than it did before. The nature of the negative shock it has given to the economy is also very severe. Several sectors have been suffering badly, millions of people have been becoming workless, and these effects are surely very tough to overcome in the near future. Hence, it is quite expected that the backward parts are even more affected due to the lack of various infrastructural and financial facilities. Under such circumstances, government intervention has become a necessity in the modern-day welfare state to fight this battle of COVID-19 and continue the livelihood of people residing in the backward parts and overcome the backwardness in various parts of a nation. Development economists, ever since the inception of COVID-19, have always advocated in favour of the government's intervention.

Intervention may be in the form of investing in industries of the backward region, with emphasis given on health and thus increasing employment opportunities and connecting the formal, advanced urban sector with those backward industries by means of input supply. But, the form of a government initiative that has been immensely popular as well as effective is a direct help to the poor in the form of various programmes aimed to improve the values of several socio-economic indicators and thus increase the efficiency of labour and help in their development. Such help may be meant for any or all of the social sectors. In reality, both forms of government help exist, and the magnitude or degree of both forms of help need to be intensified if we want to fight and win the battle against the second wave of COVID from all aspects and stop the spread of a possible third wave.

Here we have looked to capture both these forms by means of a three-sector general equilibrium framework, with two rural sectors, one traditional and another one is rural-advanced, which incorporates health as well, but they both represent the backward region of the society. The third sector is the urban developed sector. We have seen when a subsidy is given to the advanced rural industries, it increases its output along with the urban sector's output but causes a fall in the output of the traditional rural sector. Again, when investment is made in human capital and social capital, it may or may not improve the condition of the rural sectors.

Government, therefore, needs to look after the different sectors of the rural economy or backward economy with more importance so that overall development of the economy can be achieved.

Appendix

Derivation of Useful Mathematical Expressions

From Eq. (21.6) we get

$$\lambda_{TB}\widehat{X}_B + \lambda_{TB}\widehat{a}_{TB} = 0 \tag{21.21}$$

Using the elasticity of substitution from sector B we get

$$\sigma_B = \frac{\widehat{a}_{TB} - \widehat{a}_{LB}}{\widehat{w} - \widehat{P}_T}$$
Or
$$\widehat{w} - \widehat{P}_T = \frac{1}{\sigma_B}(\widehat{a}_{TB} - \widehat{a}_{LB}) \tag{21.22}$$

Using envelop condition we get

$$\theta_{LB}\widehat{a}_{LB} + \theta_{TB}\widehat{a}_{TB} = 0$$
Or
$$\widehat{a}_{LB} = -\left(\frac{\theta_{TB}}{\theta_{LB}}\right)\widehat{a}_{TB} \tag{21.23}$$

Using Eq. (21.23) in Eq. (21.22), we get

$$\widehat{a}_{TB} = \left(\frac{\sigma_B.\theta_{LB}}{\theta_{TB}}\right)\widehat{w} \tag{21.24}$$

Using Eq. (21.24) in Eq. (21.21), we get

$$\widehat{X}_B = \left(\frac{\sigma_B \cdot \theta_{LB}}{\theta_{TB}}\right)\widehat{w} \tag{21.13}$$

From Eq. (21.7) we get

$$\lambda_{KH}\widehat{a}_{KH} + \lambda_{KH}\widehat{X}_H = \mu_K\widehat{K} \tag{21.25}$$

Again from the elasticity of substitution from Sector H we get

$$\begin{aligned} \sigma_H &= \frac{\widehat{a}_{KH} - \widehat{a}_{LH}}{\widehat{w} - \widehat{\bar{r}}} \\ \text{Or} \\ \widehat{w}\sigma_H &= \widehat{a}_{KH} - \widehat{a}_{LH} \end{aligned} \tag{21.26}$$

Using envelop condition we get

$$\begin{aligned} \theta_{LH}\widehat{a}_{LH} + \theta_{KH}\widehat{a}_{KH} = 0 \\ \widehat{a}_{LH} = -\frac{\theta_{KH}}{\theta_{LH}}\widehat{a}_{KH} \end{aligned} \tag{21.27}$$

Using Eq. (21.27) in Eq. (21.26), we get

$$\widehat{a}_{KH} = (\theta_{LT} \cdot \sigma_H) \cdot \widehat{w} \tag{21.28}$$

Using Eq. (21.28) in Eq. (21.25), we get

$$\widehat{X}_H = \left(\frac{\mu_K}{\lambda_{KH}}\right)\widehat{K} - (\theta_{LT} \cdot \sigma_H)\widehat{w} \tag{21.14}$$

From Eq. (21.8), we get

$$\begin{aligned} \theta_{NZ}\widehat{X}_Z &= \widehat{N} \\ \widehat{X}_Z &= \left(\frac{1}{\theta_{NZ}}\right)\widehat{N} \end{aligned} \tag{21.15}$$

From Eq. (21.9), we get

$$\begin{aligned} \theta_{HZ}\widehat{X}_Z &= \widehat{X}_H \\ \widehat{X}_Z &= \left(\frac{\mu_K}{\theta_{HZ} \cdot \lambda_{KH}}\right)\widehat{K} - \left(\frac{\theta_{LT} \cdot \sigma_H}{\theta_{HZ}}\right)\widehat{w} \end{aligned} \tag{21.16}$$

References

Agarwal S, Singh A (2020) COVID 19 and it's impact on Indian economy. Int J Trade Commerce 9 (1):72–79

Ajeevika Bureau (2014) Labour and migration in India. https://www.aajeevika.org/labour-and-migration.php. Accessed 19 June 2021

Banerjee B (1983) The Role of the Informal Sector in the Migration Process: A Test of Probabilistic Migration Models and Labour Market Segmentation for India. Oxford Economic Papers 35 (3):399–422

Bhattacharrya T, Rajeev M (2014) Identifying the high linked sectors for India: an application of import-adjusted domestic input-output matrix, working paper 329. The Institute for Social and Economic Change, Bangalore

Business Standard (2021) Indian economy to contract by 8% in 2020–21, show govt. estimates. https://www.business-standard.com/article/economy-policy/india-s-gdp-may-shrink-8-in-fy21-on-COVID-blow-second-advance-estimate-121022600857_1.html#:~:text=GDP%20at%20Current%20Prices%20or,per%20cent%20in%202020%2D21. Accessed 19 June 2021

Chaudhuri S, Mukhopadhyay U (2009) Revisiting the informal sector: a general equilibrium approach. Springer, New York

Chaudhuri S, Ghosh A, Banerjee D (2017) Can public subsidy on education necessarily improve inequality. Int Rev Econ Financ 54:165–177

Fields GS (1990) Labour market modeling and the urban informal sector: theory and evidence. In: Turnham D, Salome B, Schwarz A (eds) The informal sector revisited. Organisation for Economic Co-operation and Development, Paris, 49–69

Gupta R, Pal SK, Pandey G (2020) A comprehensive analysis of COVID-19 outbreak situation in India. MedRxiv Preprint. https://doi.org/10.1101/2020.04.08.20058347

IMF (2021) World economic outlook, April 2021. International Monetary Fund

Income Tax Department (1979) List of backward areas [section 80HH, as amended by the Finance Act, 1979] https://www.incometaxindia.gov.in/Acts/Income-tax%20Act,%201961/1979/102120000002043087.htm. Accessed 19 June 2021

Jones RW (1965) The structure of simple general equilibrium models. J Polit Econ 73:557–572. https://doi.org/10.1086/259084

Jones RW (1971) A three-factor model in theory, trade and history. In: Bhagwati J, Jones RW, Mundell RA, Vanek J (eds) Trade, balance of payments and growth. North-Holland, Amsterdam, 3–21

Kathuria V, Rajesh Raj SN, Sen K (2013) The effects of economic reforms on manufacturing dualism: evidence from India. J Competit Econ 41(4):1240–1262. https://doi.org/10.1016/j.jce.2012.10.003

Kesar S, Bhattacharya S (2020) Dualism and structural transformation: the informal manufacturing sector in India. Eur J Dev Res 32:560–586

Marjit S, Kar S (2011) The outsiders: economic reform and informal labor in a developing economy. Oxford University Press, Oxford

Mou J (2020) Research on the impact of COVID19 on global economy. IOP Conf Ser Earth Environ Sci 546:032043. https://doi.org/10.1088/1755-1315/546/3/032043

Mustafa N (2021) Research and statistics: coronavirus disease (COVID-19). Int J Syst Dynam Appl 10(3):67–83. https://doi.org/10.4018/IJSDA.20210701.oa1

Niti Aayog (2018) Deep dive: insights from champions of change: the aspirational districts dashboard. Niti Aayog, Government of India

Punia K (2020) Future of unemployment and the informal sector of India, observer research foundation. https://www.orfonline.org/expert-speak/future-of-unemployment-and-the-informal-sector-of-india-63190/. Accessed 19 June 2021

Roy S (2020) Economic impact of COVID-19 pandemic, preprint version. https://www.researchgate.net/profile/Shohini-Roy/publication/343222400_ECONOMIC_IMPACT_OF_COVID-19_PANDEMIC/links/5fa1e11e92851c14bc036d68/ECONOMIC-IMPACT-OF-COVID-19-PANDEMIC.pdf

SBHIDHS (2018) Health on the March 2016–2017 & 2017–2018 (combined). State Bureau of Health Intelligence Directorate of Health Services, Government Of West Bengal, West Bengal

Shapiro C, Stiglitz JE (1984) Equilibrium unemployment as a worker discipline device. Am Econ Rev 74(3):433–444

The Economic Times (2020) https://economictimes.indiatimes.com/news/economy/finance/latest-stimulus-package-among-largest-in-the-world/articleshow/75701976.cms. Accessed 19 June 2021

United Nations in India (2021) Title of decent work for migrant workers in India. https://in.one.un.org/page/decent-work-for-migrant-workers-in-india/. Accessed 19 June 2021

Yuen KS, Ye ZW, Fung SY, Chan CP, Jin DY (2020) Sars-Cov-2 and COVID-19: the most important research questions. Cell Biosci 10:40. https://doi.org/10.1186/s13578-020-00404-4

Nilendu Chatterjee is an Assistant Professor in the Department of Economics, Bankim Sardar College, West Bengal, India. He has research interest in Resource Economics, General Equilibrium, Development Economics, etc. He has published research papers in various national and internationally renowned journals including International Journal of Sustainable Economies Management, Economic Affairs, and Foreign Trade Review.

Bappaditya Koley is presently the Head of the Department in Geography, Bankim Sardar College, South 24 Parganas, West Bengal, India. He is investigating in the field of Applied Geomorphology and Landslide Disaster Mitigation Studies. His areas of research interest cover Landslide Hazard, Vulnerability and Risk Assessment; Landslide Early Warning System; Coastal Risk Assessment, Coastal Planning and Management; Climate Change; and Environmental issues. He has published research papers in various national and international reputed journals.

Part VI
Silver Lining in the Cloud of Uncertainty

Chapter 22
Rural Livelihood Options During the Pandemic in India: Finding Avenues for Revival

Govinda Choudhury and Debjani Choudhury

Abstract The rural population has a higher proportion of poor people, and during the pandemic, their vulnerability has increased. The main objective of this chapter is to study the challenge in rural livelihood and to understand how economic lockdown and the pandemic have aggravated the rural livelihood problem. Migration as a coping strategy is not an option for the rural workforce during the pandemic. In the absence of recent data on rural livelihood, the study is based on secondary sources and publications. The pandemic has increased vulnerability of the rural population, which will require measures that have an immediate impact on livelihood. Medium- and long-term measures are also required to reduce rural vulnerability. The role of livelihood programmes is immense for short- and long-term outcomes. The role of agriculture and allied sector in reviving the rural economy is highly important. Agriculture is the only sector that has shown resilience during the crisis and growth of this sector will not only provide income but through its linkages help to start other sectors of the economy.

Keywords Farm · Rural non-farm economy · Reverse migration · COVID-19 · Economic recovery

22.1 Introduction

Indian economy is under recession and most of the major economies in the world are struggling to keep their economy buoyant. Generally, in recession the real gross domestic product (GDP) of a country declines and along with it there is an increase

G. Choudhury (✉)
Department of Economics, University of North Bengal, Siliguri, West Bengal, INDIA
e-mail: govinda.c2006@nbu.ac.in

D. Choudhury
B.S.F. Senior Secondary Residential School, Kadamtala (Siliguri), West Bengal, India

S. Saha et al. (eds.), *Economic and Societal Transformation in Pandemic-Trapped India*, New Frontiers in Regional Science: Asian Perspectives 55,
https://doi.org/10.1007/978-981-16-5755-9_22

in unemployment in all the sectors of the economy. The poorest section of the population is the worst affected group in an economic downturn. The global multidimensional poverty index (MPI) 2020 report indicated that the rural and urban incidence of poverty is 36.8% and 9.2%, respectively, in India (Oxford Poverty and Human Development Initiative 2020). The rural population will be the worst sufferers during the recession, as most of the indicators of MPI are likely to deteriorate further. The mass reverse migration of workers from urban centres to their native villages during the first few months of lockdown has added more pressure on the rural economy. Most of the migrants had earlier gone to cities and towns because there were not enough livelihood opportunities for them in the villages. The scope to increase employment in the agriculture activities is limited, as it is not possible to bring more land under cultivation and the employment elasticity in the agriculture sector has remained low. The objective of this study is to identify livelihood activities that may employ rural workers during the period of the present crisis.

Estimates of employment elasticity[1] using compound annual growth rate (CAGR) method show that it was 0.57 in 1972–1973 to 1977–1978 and it has declined to 0.18 in 1993–1994 to 2011–2012 (Misra and Suresh 2014: 10). The sectoral employment elasticity of the Indian economy shows negative employment elasticity for agriculture, which explains why a large rural population had migrated to urban centres in search of livelihood (ibid: 11). However, not all rural workers were able to get employment in the urban manufacturing sector. The main reason is that the urban manufacturing sector is capital-intensive and does not increase the employment opportunity substantially. In other words, the growth of employment in manufacturing lagged behind the growth of the sector.

The rural and urban usual status[2] unemployment rate was 5.0% and 7.7%, respectively, in 2018–2019. The average usual status unemployment rate was 5.8% (PLFS[3] 2020). However, immediately after the country was placed under lockdown the unemployment rate increased from 8.75% in March to 23.52% in April and 21.73% in May 2020. As the lockdown restrictions were gradually lifted from May onwards, the unemployment rate began to fall and it was 6.54% in November. The urban and rural unemployment for the same month was 7% and 6.3%, respectively (CMIE[4] 2020a).

Two broad features of the rural economy in India has been (a) decline in the share of agriculture in gross value added over the years and its share was 16.5% in 2019–2020 (Economic Survey 2020: 15) and (b) decline in the share of agriculture workers[5] in the rural workforce. In India, out of 481.7 million workers, as many as 348.6 million workers reside in rural areas. The cultivators and agriculture labourers

[1] Employment elasticity is ratio of percentage change in employment to 1% change in economic growth.

[2] It refers to the activity status of a person of last 365 days preceding the date of survey.

[3] Periodic Labour Force Survey.

[4] Centre for Monitoring Indian Economy.

[5] Cultivators and agriculture labourers combined is considered as agriculture workers.

were 118.7 million and 144.3 million, respectively, and they collectively account for 54.6% of the total workers in the country (Census 2011). Agriculture workers as a proportion of total workers in the country have declined from 58.4% in 2001 to 54.6% in 2011 (Census 2001 & 2011). Agriculture activities are no longer able to provide a livelihood to increasing labour force and hence more and more workers are joining non-farm activities. The share of agriculture workers in rural employment has declined from 76.3% in 1999–2000 to 64.1% in 2011–2012 (Chand et al. 2017). The non-farm sector comprising of manufacturing, construction and services provides rest of the rural employment in the country.

The rural labour market in India suffers from segmentation and adversely affects the vulnerable sections of the rural workforce. Segmentation of the labour market usually manifests into problems like inequality in income distribution, unemployment and discrimination. Duality does not need to be limited to two groups, the favoured and the ones who are discriminated. The formal and informal labourers constitute two broad groups in the segmented labour market. The rural informal workers are the floating population who migrate to urban centres for livelihood. This group of migrants has been increasing during the last two decades because of increasing distress in the rural economy. At an average, 9 million people migrated annually for livelihood during 2011–2016 (Economic Survey 2017). These figures are much higher 3.3 million reported in the census of 2011. Short duration migration is usually part of the coping strategy by poor people. This category of migrants is informal workers either self-employed like street vendors or casual labourers.

Both urban and rural informal labour markets are inter-connected by the circulatory migrants (Breman 1996 as *cited* in Srivastava 2019). As a result, a similar type of segmentation of the labour market exists in both places, which gives the employers a higher ability to control wages. Migrant workers are relatively cheaper than the local labour and easier to exploit. Besides, the rural migrant workers lack social security, health benefits and minimum safety measures. The living condition of the migration labourers lacks basic amenities like adequate space for living, sufficient water, drainage, cleanliness, which affects their health. Also due to lack of portability of ration card, the rural migrant workers do not get access to the public distribution system (PDS) in the place of their work. The Inter-State Migrant Workmen Act 1979 has provisions for regulation of migrant workers and protects their interest. The law had provision for registration of the workers and employers were responsible for free medical facilities, protective clothing and safety gears. However, the implementation of the law is not only poor but also the awareness about the law among the migrant workers is low. During this recent crisis, it was amply clear that there was no reliable official data of the rural migrant population. The proper record would have helped in the orderly distribution of assistance and avoided the chaos in the transportation of the rural migrants to their native places. No doubt that the hardship faced by the reverse migrants could have been avoided, but we need to ask if distress migration was avoidable in the first instance and livelihood could be provided to them in the village itself. It is therefore important to know whether agriculture growth can provide enough livelihood opportunities to poor, or will it be the non-farm sector, that will generate employment and thereby eliminate poverty. If it is the latter then not only it would mean agriculture workers will move out of agriculture, but it also means workers would out-migrate to urban centres where non-farm livelihood are more abundantly available.

The role of agriculture in economic development is a much-debated issue. In the 1950s and 1960s, the dominant opinion inspired by Lewis (1954) considered agriculture as backward and unproductive. The general prescription was to transfer labour from the agriculture sector to industry, which promoted urbanisation as a strategy for development. Indian experience in the 1960s showed that neglect of agriculture has consequences for food and livelihood security of the people. The Green Revolution in the 1960s and 1970s amply demonstrated that it is possible to transform the agriculture sector into a modern vibrant production sector with high industrial linkages. Even when the agriculture sector grows at a moderate rate, it helps the economy to grow faster (Schultz 1964). The argument put forward was that growth in agriculture reduces the price of food and raw materials and thereby cost. Growth in agriculture is a necessary condition to address the two important problems of food security and malnutrition (Gulati et al. 2012). However, during the 1980s and 1990s experience from various regions of the world showed that it was possible to achieve a high level of development through the growth of the non-agriculture sector. The East-Asia region was an example of the reduction of poverty achieved through high growth of the manufacturing sector. By the end of the 1990s, the focus was on poverty reduction and it eventually became the target of Millennium Development Goals (MDGs).

The whole purpose of MDGs was to make development more inclusive and participatory. Reduction of poverty is possible only when the sectors in which poor people are mostly employed have a high rate of growth. In developing countries, a large proportion of the poor depend on agriculture for livelihood and the reduction of poverty is possible only when the agriculture sector grows at a high rate. Several studies substantiated that the agriculture sector plays an important role in reducing poverty (Ravallion and Datt 1996; Christiaensen et al. 2011). The extent to which agriculture sector can reduce poverty in a country depends on both its direct and indirect growth component, the proportion of workforce engaged in the agriculture activity and its share in the economy (Ivanic and Martin 2018). This implies that even when the growth rate in agriculture is lower as compared to the other sectors of the economy, a higher proportion of work participation in the farm sector enables to increase the income level and thereby reduce poverty. Unlike the non-agriculture sector, the growth of agriculture benefits mostly the people with little or no skill who are the poorest among the poor (ibid). This naturally raises a question whether the benefits of economic growth in the non-farm sector bypass the poorest section of the population.

For many years, the dominant view about the rural non-farm economy (RNFE) was that it was less remunerative and only provided supplementary earning to rural households. The perception about RNFE began to change when studies were published from across the world, which showed that in a situation of declining farm income the participation of rural workers in non-farm sector increases (Lanjouw and Lanjoouw 2001; Haggblade et al. 2010; Himanshu et al. 2013). The

RNFE plays a pivot role in enhancing livelihood opportunities thereby reducing poverty, reduces distress migration and vulnerability of the rural poor. However, there is little consensus on the RNFE impact on rural inequality. RNFE is a mix of heterogeneous activities and each requires a different level of skill and resource endowment. People with low skill and little resource endowment did not benefit much from the non-farm economy. The non-farm activities are neither equally remunerative nor people have equal access to jobs in this sector. Therefore, the net impact of the RNFE on rural income distribution in a region depended on the mix of activities and skill of the rural workforce (Haggblade et al. 2010: 1432). Even when people in different income group receives higher income from the non-farm sector, but if a large share of the non-farm income goes to the wealthier section, then it will increase the existing inequality (Lanjouw and Lanjoouw 2001: 8). However, studies show that the inequality initially increases and then decline which is even valid at the village level (Himanshu et al. 2013: 462).

Agriculture growth plays an important role in the expansion of RNFE. Increase in rural income increases demands for non-agriculture products, which in turn creates an opportunity for capital investment in the non-farm manufacturing sector. Regions which were first to experience higher agriculture growth are also regions where at present the non-farm sector has a larger share of the rural economy. The growth of the non-farm sector not only provides enhanced livelihood opportunity for the rural workforce but the withdrawal of excess agriculture workers pushes agriculture wage rate (Thirupathy and Ravisankar 2015: 37). Besides, factors like rural employment programme, irrigation, literacy rate push rural wage rates. Rising wage rate reduces the rural poverty level, but it also increases the cost of cultivation, thereby reducing farm income. Besides, the small size of the agriculture landholdings makes it difficult to generate a sizeable marketable surplus. Thus, agriculture is no longer a preferred livelihood option for rural youth.

The non-farm comprises three types of livelihoods: first, those who have a fixed salary and a secure tenure of employment, second those who work for daily wage rate or piece rate and third who are self-employed. The fixed salary category of non-farm livelihood is not more than one-fifth of the total non-farm employment. This type of employment requires higher skill, has a greater impact in reducing rural poverty and is the most preferred activity. The low skilled non-farm livelihoods also provide earning opportunities to the rural poor who otherwise would not find any work.

It is in this background that the impact of COVID-19 pandemic on India's rural economy must be interpreted. The paper enquires the extent of impact the pandemic may have on rural economy and identify the livelihood options to address this challenge. The following four sections of the paper are as follows: Section 22.2 of the paper looks into the impact of the pandemic on rural economy of India. Section 22.3 of the paper examines the changing rural livelihood scenario in post-1991 period. The fourth section elaborates the policy approach of Government of India for economic revival during global recession. The last section is the concluding section, which elaborates the policy options for revival of the rural economy of India.

Since there is a dearth of data at the country level, the analysis is based on both official publications and information published in the public domain by researchers and organisations. Even the official data released by the government are based on substitutes and proxies, as collection of data is severely restricted due to pandemic.

22.2 Impact of COVID-19 Pandemic on the Rural Economy of India

The International Monetary Fund (IMF) projected global economic growth at −4.4% in 2020, which is the lowest projection ever announced by the organisation (IMF 2020). The world economy is experiencing the worst economic downturn since the second quarter of 2020 due to the lockdown imposed by most countries of the world. Never before in recorded human history had a pandemic spread across the globe at such a rapid pace. The world did not have any contingency plan to address a challenge of such a large dimension. Almost all nations considered the best recourse to stop the spread of the pandemic were to isolate themselves, without any clue to what the likely economic consequences of such a decision. In every country, the priority of the governments was to stop the spread of COVID-19 pandemic and save lives. However, the consequence of lockdown has resulted in the large retrenchment of the workforce, reduced salary, business shutdown and reverse migration of the rural population. The two factors that seem to have prompted the people to return to their villages was the loss of income with little hope of any support. The second was exemption given to agriculture work in the lockdown, which gave them hope of getting work and food.

The migrant people were not wrong, as agriculture was the only sector that has shown some ray of hope during this economic crisis. In the first quarter of the financial year 2020–2021, India's GDP at constant (2011–2012) prices has contracted 23.9% and it declined 7.5% in the second quarter (see Table 22.1). Among all the sectors of the economy, agriculture was the only sector that grew 3.4% in both the first and second quarter of the financial year. All the sectors except agriculture have contracted, as it was the only livelihood activity permitted during the 70 days lockdown.

The decline of gross value added (GVA) of India in the first quarter was 22.8% and it contracted 7% in the second quarter of the financial year 2020–2021 (see Table 22.2). Other than agriculture, forestry and fisheries sector, it was electricity, gas, water supply and other utility services and manufacturing sector that grew at 4.4% and 0.6%, respectively, in the second quarter. It would be too early to conclude that the small increase in the manufacturing sector is due to enhanced production. A more likely reason can be a reduction in cost due to low input prices and lower-wage payment, which may have pushed the growth rate in the second quarter. The agriculture sector has grown in both the quarters of the financial year, but the production of rice in the first and second quarter has declined by 0.3% and 2.9%,

Table 22.1 Quarterly estimate of expenditure on GDP for the first and second quarter at constant (2011–2012) prices

	Expenditure (in ₹ billion)				Percentage change over the previous year	
	2019–2020		2020–2021			
Industry	April–June	July–September	April–June	July–September	April–June	July–September
1. Private final consumption expenditure	19,929.67	20,254.88	14,611.64	17,962.90	–[1]26.7	−11.3
2. Government final consumption expenditure	4182.49	4656.43	4866.36	3623.68	16.4	−22.2
3. Gross fixed capital formation	11,321.95	10,357.36	5991.92	9596.28	−47.2	−7.4
4. Change in stocks	673.28	669.99	533.36	712.08	−20.8	6.3
5. Valuables	513.47	517.61	46.45	209.95	−90.9	−59.4
7. Exports	7085.46	7032.82	5679.61	6925.68	−19.8	−1.5
8. Imports	8257.88	7796.37	4922.86	6458.51	−40.4	−17.7
9. Discrepancies	−95.76	150.62	89.08	569.62	193.0	278.9
GDP	**3,535,267**	**3,584,335**	**2,689,556**	**3,314,167**	**−23.9**	**−7.6**
Percentage change over same quarter for previous year	–	–	**−23.9**	**−7.5**	–	–

Source: Press note on estimates of GDP for the second quarter (July–September) 2020–2021

respectively, in 2020–2021 (NSO[6] 2020: 1–2). This is a matter of serious concern, as rice is the main food grain for most of India's population. The figure for the second quarter is an advance estimate; hence, the extent of change in rice production will be known once we have the full estimate.

The decline in construction (−50.3%), trade, hotel, transport, communication and service related to broadcasting (−47.0%) and manufacturing (−39.3%) in the first quarter dealt a severe blow to the workers (see Table 22.2). Due to the complete shutdown of activities, even government sector spending on public administration has declined by 10.3% in the first quarter and it further declined by 12.2% in the second quarter. In the time of economic crisis, the government sector spending keeps the economy of the nation afloat. In the first quarter of 2020, the government final consumption expenditure (GFCE) has increased ₹683.87 billion over the same period of the previous year. In the second quarter, the government final consumption expenditure (GFCE) declined by ₹1242.68 billion compared to the second quarter of 2019. The GFCE for the second quarter is 22.18% less than the second quarter of the previous year. The private final consumption expenditure (PFCE), which accounts four-fifth of the total consumption had declined 26.7% in the first quarter and then

[6] National Statistical Office, Ministry of Statistics & Programme Implementation, Government of India.

Table 22.2 Quarterly estimate of GVA at constant (2011–2012) prices

Industry	GVA at constant price (in ₹ billion) 2019–2020 April–June	July–September	2020–2021 April–June	July–September	Percentage change over the previous year April–June	July–September
Agriculture, forestry and fishing	439,843	367,758	454,658	380,239	3.4	3.4
Mining and quarrying	92,807	68,978	71,209	62,674	−23.3	−9.1
Manufacturing	578,936	576,112	351,396	579,683	−39.3	0.6
Electricity, gas, water supply and other utility services	81,628	79,525	75,877	83,051	−7.0	4.4
Construction	262,828	244,092	130,750	223,121	−50.3	−8.6
Trade, hotel, transport, communication and service related to broadcasting	630,860	611,609	334,284	516,500	−47.0	−15.6
Financial, real estate and professional services	803,322	864,974	760,491	794,995	−5.3	−8.1
Public administration, defence and other services	417,483	465,096	374,656	408,326	−10.3	−12.2
GVA at constant price	**3,307,707**	**3,278,144**	**2,553,321**	**3,048,589**	**−22.8**	**−7.0**

Source: Press note on estimates of GDP for the second quarter (July–September) 2020–2021

contracted 11.3% in the second quarter of 2020–2021 as compared to the previous year. It is a general tendency of people to postpone their consumption in the time crisis. However, this decline in private consumption expenditure is not entirely because of increased thriftiness of the people. Loss of income due to layoff, salary reduction and losses in business during the lockdown have reduced PFCE. The gradual opening of the economy has helped to revive the private consumption expenditure and as a result, the decline is less in the second quarter.

The share of the fisheries and livestock sector in GVA was 1% and 4.2%, respectively, in 2018–2019 (NAS 2020). The impact of lockdown and COVID-19 pandemic on dairy, fisheries, livestock and poultry activities was immense. The growth rate in production of fisheries for the first and second quarter was 5.3% and 12.4%, respectively, looks impressive, but growth figures in fisheries may not reveal the true impact on the fishing communities. As disaggregated data on fisheries is not available it is difficult to know the production of freshwater fisheries and marine fisheries. Marine fisheries produce almost one-third of the output in India. India exported ₹450 billion worth of marine products in 2019–2020 (Balasubramanian 2020). Due to lockdown and restrictions in the harbour and landing centres, the fishing community have suffered a lot. Both freshwater and marine fisheries have suffered due to restriction in transport.

Export of fish had come to a halt and all this has affected the entire value chain linked with fisheries. Ice-factory, cold storage, processing units and the markets were affected due to closure and loss of workers and it will take time before they can function in full capacity. Most of these production units had sizeable migrant labour force, a large number of these workers have migrated back to their native villages. The economic crisis was not limited to fisheries, but it has also affected dairy, livestock and poultry sector. All these products are highly perishable products and without storage facilities especially in summer, the shelf lives of the products are highly reduced.

The marginal and small agriculture landholders and landless labourers mostly do livestock production in India. The livestock owners suffered not only due to the lockdown but also because of the rumour that spread through the media. Consumption of chicken and eggs had declined as rumour spread that poultry birds can be the carrier of the virus. Consumption demand for dairy, livestock and poultry products also declined due to closure of eateries and restaurants. Restriction on all social and religious ceremonies and celebration had resulted in a huge fall in demand for dairy products, livestock and poultry products. Dairy products like ice cream, milkshakes and *lassi* (yoghurt drink) witnessed a fall in demand by the consumer and supply was interrupted due to restriction in transport and operation of storage facilities. Instead, consumption demands of cottage cheese, *ghee* (clarified butter) and butter has increased during the lockdown. The increase in household consumption demand was not sufficient to compensate for the loss of demand from sweet shops, eateries, hotels and caterers. Some of the large milk cooperatives continued to procure milk at pre-lockdown level, but most of the smaller milk cooperatives were unable to sustain the procurement level (Biswal et al. 2020: 1929). However, with the gradual opening of the economic activities, the demand for dairy products has increased and the producers have a better return now.

The impact of COVID-19 pandemic on RNFE was stronger than its effect on the agriculture sector. The closure of manufacturing sector, eateries and trade, a complete halt to construction and public works increased the overall rural unemployment rate from 7.34% in February to 8.44% in March and then it increased sharply to 22.89% in April 2020 during the first 4 weeks of lockdown. The rural employment fell to 21.11% in May and thereafter as the lockdown was gradually withdrawn, the unemployment declined to one-digit figure at 9.49% in June. The rural unemployment rate declined to 5.86% in September, but in the following month, it increased to 6.9% in October (CMIE 2020b). The increase in employment rate in October is not usual, as Kharif crops are harvested in this month and it is a festival month. It will take some time to understand why the rural unemployment rate has increased in October.

It is during this time of economic crisis that rural livelihood programmes employ poor people. The MGNREGA[7] 2005 guaranteed 100 days of wage workers to the poor and vulnerable rural population, especially during lean periods. The

[7] Mahatma Gandhi National Rural Employment Guarantee Act.

Table 22.3 Work provided under MGNREGA in April–November 2020 (in million)

Month (2020)	Rural unemployment rate[a] (%)	Person-days	Change of person-days from the previous month	No of the household employed	Change in number of households employed from previous months
March	8.44	17.64	–	1.53	–
April	22.89	14.13	(−) 3.51	1.10	(−) 0.43
May	21.11	56.45	(+) 42.32	3.25	(+) 2.15
June	9.49	63.04	(+) 6.59	3.81	(+) 0.56
July	6.51	38.21	(−) 24.83	2.67	(−) 1.14
August	7.65	25.42	(−) 12.79	1.94	(−) 0.73
September	5.86	25.49	(+) 0.07	1.92	(−) 0.02
October	6.90	25.17	(−) 0.32	1.90	(−) 0.02
November	6.26	22.26	(−) 2.91	1.73	(−) 0.17

[a] Source for Rural Unemployment Rate is CMIE (2020a, b)
The source of the rest of the data in the table is: https://nrega.nic.in/netnrega/home.aspx. Accessed on 12 Dec 2020

programme is implemented through the Gram Panchayats (rural local government). The centre and state governments share the funding for the scheme in 9:1 ratio. The central allocation for the programme was increased by ₹400 billion which is in addition to the budgetary allocation of ₹615 billion during 2020–2021. In the first five months of 2020–21 the Government of Uttar Pradesh issued 2.1 million new job cards, followed by the state of Bihar where 1.1 million job cards were issued. The other major states were West Bengal and Rajasthan both issued 0.7 million new job cards. All these states are also a major source of internal rural migrants.

During April to November 2020, under the MGNREGA programme wage employment provided was 270.17 million person-days to 18.23 million households. The number of person-days in May was 56.45 million, which then increased to 63.04 million in June. In July when the rules were relaxed, the person-days fell sharply to 38.21 million, which was a decline of 24.83 million days within a month (see Table 22.3). During this period when the unemployment rate was at its peak in April and May the wage employment provided under the MGNREGA was much-needed income to the rural poor. This was the period when thousands of migrants were returning to their villages. In May and June MGNREGA work was provided to 3.25 and 3.81 million households. In August 12.79 million fewer jobs were offered under the programme, but in September a marginal increase of 0.07 million person-days was noted. Livelihood programme undertaken by the government is required to provide the much-needed income to the poor when there is little opportunity to get work in any economic activity.

22.3 Changing Rural Livelihood Opportunity in Post-1991 Period

Agriculture and allied sector contribute 16.87% of the GVA in 2019–2020 (Economic Survey 2020: A7). Agriculture livelihood is 72.28% of total livelihood in rural areas of India (Census 2011). Almost 86% of the agriculture landholdings are less than 2 ha (Agriculture Census 2015–16). Rural livelihood data shows that while the number of cultivators has declined the number of agriculture labourers has increased during 2001–2011. Within this decade, the number of cultivators declined by 8.5 million and the number of agriculture labourers increased 37.55 million. Cultivator makes for 24.65% of total workers in 2011, which is 7% lower than in 2001. If we disaggregate the data, we find that main category[8] cultivators have declined 7.68 million and marginal cultivators declined by 0.82 million between the inter-census years. The decline in the rural cultivator population is because agriculture is no longer a remunerative livelihood option; especially in case of food grains, the increasing cost of inputs make it difficult to recover the cost of production. Though the production of food grains has increased from 244.5 in 2010–2011 to 285 in 2018–2019 at a 1.91% per annum, but the area under it has decreased by 0.9 million hectares during this period (Economic Survey 2020: A34 & A35). The high cost of inputs and increase in demand for horticulture crops, dairy and livestock product has encouraged the expansion of these sectors. It also follows the Engels law, which states that with an increase in income people's consumption will shift away from staple food to more nutritious food items. Increase in income in India during the last two decades and export opportunity has encouraged the production of vegetables, fruits, fish, milk, poultry and livestock products.

Between 2010–2011 and 2017–2018, the production of plantation crops increased from 12.0 to 18.1 million metric tons, respectively, which is equal to 6% growth per annum. During the same period, the production of vegetables increased from 146.6 to 184.4 million metric tons[9]. The overall growth of crop production at constant (2011–2012) price was 1.02% per annum during 2011–2012 to 2018–2019. During the same period, the growth of GVA by livestock and fisheries and aquaculture was 7.3% and 9.45%, respectively. As a result, the share of livestock and fisheries and aquaculture increased from 21.8 and 4.5% in 2011–2012 to 28.6 and 6.8 in 2018–2019 in the agriculture sector GVA (Ministry of Agriculture 2020:23). The impact of diversification in the agriculture sector on labour absorption depends on the extent of technological change and crop productivity. A distinct pattern of labour employment was observed in the first phase (1975–1976 to 1995–1996) and the second phase (1995–1996 onwards) of technological intervention in the agriculture sector. In the first phase, the technological intervention was largely labour augment that helped to increase labour absorption in the agriculture sector. The

[8] Workers are categorized as main category workers if they have worked more than 180 days during a year prior to the date of enumeration.

[9] See the Horticulture Statistics at a Glance 2018.

second phase of the intervention was mainly mechanisation, higher integration with the market resulting in reduced absorption of labour in agriculture (Raju et al. 2015). This creates a condition for releasing farm labour to be employed in non-farm work. However, the expansion in non-farm sector in India has provided security of employment to only one-fifth of the labour who are in salaried work, the rest of workers are either wage earners or self-employed and each has a share of two-fifth of the non-farm work. For rural distressed agriculture workers, non-farm employment is a coping strategy. RNFE not only provides an opportunity for the rural population to earn a living but also helps to reduce their debt burden.

A comparison of rural indebtedness between the farm and non-farm workers shows that while the incidence of indebtedness for agriculture households was 52.5%, it was 42.8% in case of the non-agriculture household (NABARD 2018: 25). One of the primary reasons that drive rural out-migration is the indebtedness of the households. Studies have found households who have access to microfinance are more likely to out-migrate (Shonchoy 2015). This is in contrary to the standard economic theory of migration, which considers a lack of access to credit to be an immediate cause for rural out-migration. The amount of credit provided through the microfinance system is too little to give a big push to the rural economy. Most of the loan is used for consumption and the little amount that is invested are usually in small livestock, poultry or in petty trade. This usually lowers the return from such investments as the scale of operation is low and the product market is highly competitive because large numbers of creditors compete to sell similar products. The second factor is, once a person takes a loan, they are under immense pressure to repay the loan regularly. In most cases, the repayments start immediately after taking the loan and usually repaid on weekly basis, which gives little time to a debtor to generate income from their investment. It is not surprising that microfinance loans are used to smooth household consumption, repay previous loans, to pay health bills and to spend on ceremonies and celebrations (Taylor 2011). If repayment starts after a period of moratorium, then it helps people to invest in productive activities, though in such cases repayment default rate has increased (Field et al. 2013). Distress migration of the rural population will continue unless poverty is reduced through enhanced livelihood opportunities in the villages. The reverse migration since the outbreak of the pandemic in India has put additional pressure on the rural economy.

The outbreak of zoonotic[10] diseases is not new, rather there occurrence has increased during the last couple of decades. Diseases like influenza A (H1N1), severe acute respiratory syndrome (SARS), Ebola were reported from various parts of the world. COVID-19 pandemic is different from the earlier zoonotic diseases as it has posed not only a health shock but has also severely disrupted the global economy and social fabric of the nations. Countries have imposed lockdown, which not only declined output and income but also increased unemployment to a high level. Activities that were considered non-essential have remained suspended

[10] Zoonotic diseases refer to those diseases caused by microbes that have jumped from non-human animals to humans.

ever since the lockdown was imposed. Inter-state travel was restricted and railways only provided limited transport for migrants and stranded people to return to their native places. It is a catch 22 situation, as any macroeconomic stimulus to increase aggregate demand especially consumption demand would also require an increase in production as well. Production will require the mobilisation of the labour force, which is difficult due to restrictions under the present pandemic situation. In absence of any successful treatment for COVID-19, the general health prescription is to maintain a safe distance between people to help stop the spread of the pandemic. This has forced a reduction in the number of workers in economic activities to ensure the safety of the workers. It means that in the immediate future not all the workers who have migrated back to their villages can expect to get back their work in the cities. There is an immediate need to increase livelihood opportunities in rural areas or else a large rural population will suffer and be impoverished.

Rural livelihood data shows that while the number of cultivators has declined the number of agriculture labourers has increased during 2001–2011. Within this decade, the number of cultivators declined by 8.5 million and the number of agriculture labourers increased 37.55 million. Cultivator makes for 24.65% of total workers in 2011, which is 7% lower than in 2001. If we disaggregate the data, we find that main category[11] cultivators have declined 7.68 million and marginal cultivators declined by 0.82 million in 10 years. The decline in the number of cultivators was due to low returns from agriculture and small and marginal landholders are selling or leasing out their lands (Gupta 2016: 195). Neglect of agriculture and three droughts in 2002, 2004 and 2009 have pushed the cultivators to distress. The increase in household industry workers and other workers to the extent of 8.1% and 32.6%, respectively, in between 2001 and 2011 shows a higher reliance on non-farm livelihood by the rural workforce. Agriculture growth can reduce poverty if it raises farm income and labour wages. To increase farm income capital investment in agriculture plays a crucial role, especially public investment. Increase in public investment encourages private investment in agriculture (Bathla 2017).

Post-1991 New Economic Policy the agriculture sector did not receive enough attention from the policymakers. Many of the policy decisions taken by the government for price rationalisation had adversely affected the income of the cultivators that has reduced the net return in crop production. The rise in the price of diesel has gradually increased irrigation cost and has substantially reduced the profitability of water-intensive crops like paddy and sugarcane. The pattern of precipitation has changed over the years in different agro-climatic regions of the country. The high temperature not only reduces the yield of crops (Mukherjee et al. 2019) but also increases the loss of surface water due to evaporation. Any delay in rainfall or shortage of water for irrigation can damage standing crops in the field. There is an urgent need to increase the efficiency of irrigation in the country. This will require the promotion of appropriate irrigation technology through price incentives.

[11] Workers are categorised as main category workers if they have worked more than 180 days during a year prior to the date of enumeration.

In 2010, the Nutrient Based Subsidy (NBS) scheme for fertilisers was launched to correct the soil nutrient balance and achieve the optimum balance of nitrogen: potassium:phosphorus at 4:2:1. While the deregulation of potassium and phosphorus-based fertilisers has increased almost three times, but the price of urea a nitrogen-based fertiliser has increased 10% in the last decade. During 2010–2011 and 2018–2019 due to distortion in price, the consumption of phosphate and phosphorus fertilisers declined from 80.50 and 35.14 million tonnes in 2010–2011 to 69.68 and 27.79 million tonnes in 2018–2019. On the contrary, the consumption of nitrogen fertilisers increased from 165.58 to 176.28 million tonnes in this period. The overall consumption of fertilisers fell from 281.22 to 273.75 million tonnes. From 2000–2001 to 2010–2011 the consumption of fertilisers in the country has increased from 197.02 to 281.22 million tonnes (Economic Survey 2020: A41). The consumption of fertiliser in India fell from 179 kg in 2010 to 165 kg per hectare in 2016. The consumption of fertiliser was always less as compared to other agriculturally developed countries any further decline in its consumption will adversely affect agriculture growth and cultivators income. Besides, unbalanced consumption of fertiliser lowers crop yield and reduces the efficacy of fertilisers (Wen et al. 2016 as cited in Motesharezadeh et al. 2017).

Agriculture growth has a profound impact on rural poverty and a decline in agriculture growth increases agrarian distress. In addition to this, the growth of agriculture is a key to food and nutritional security of a growing population. The forward and backward linkage of the agriculture sector plays an important role in the industrial development of a nation. Another argument for growth in agriculture is that it helps to reduce the dependence of poor on natural resources and promotes conservation. Higher growth in agriculture will help to reduce inequality by increasing the income of agriculture workers and by providing the capital for RNFE. The role of RNFE is crucial in the eradication of rural poverty. In India, the growth in income of non-agriculture workers far exceeded the growth in income of agriculture workers during 1983–1984 and 2003–2004. Between 2003–2004 and 2011–2012, the gap between agriculture and non-agriculture income has narrowed down as income of cultivators grew at 7.3% per annum (Chand et al. 2015: 145). From the early 2000s, it was realised that neglect of agriculture has resulted in agrarian distress and higher capital investment in agriculture will be required to push agriculture growth. The flow of institutional credit to agriculture was increased and along with it, the minimum support price (MSP) for major crops was increased. The increase in public expenditure also encouraged private investment in agriculture to grow at a rate of 9% per annum (at 2004–2005 prices). Within a short period, the irrigation intensity increased from 30 to 50% and 3.8% agriculture growth was achieved annually between 2000 and 2013. The impact of public expenditure on technological intervention and irrigation is high (Bathla 2017). This calls for increased public expenditure on agriculture.

However, agriculture growth was sustained after 2012–2013 and the growth in farm income declined 1% per annum, which increased distress of rural population (Chand et al. 2015). A reason for this is that the agriculture production function in India is not shifting outwards as technological progress is lagging (Raju et al. 2015).

The growth of agriculture will require higher public investment in R&D and increase in irrigation intensity. Recurrence of droughts and floods along with very hot summer will require technological intervention to develop better drought resistance crops, increasing efficiency of irrigation through drip and sprinklers. However, the argument for a focus on agriculture does not mean neglect of the non-farm sector. Both the sectors are complementary to each other and a balance between the two sectors in rural development policy will be required.

The present challenge that the RNFE faces in this pandemic situation is a lack of overall demand and slowing down of the construction sector. In the present pandemic crisis, rural final consumption demand can increase if the agriculture sector attains a high growth rate. For this public investment in agriculture and allied activities will be essential for increasing production and along with price support for output will reduce risk and incentivise farmers to produce. The growth rate of 3.4% during the first two quarters of 2020–2021 should give enough confidence to the policymakers to understand that the recovery of the economy will be led by the agriculture sector.

22.4 India's Policy Approach to Economic Recovery During the Global Recession

The relief package under plan I was announced by the Government of India (GoI) on 25 March 2020 and its amount was ₹1.7 trillion to take care of poor, workers and those who needed immediate help during the lockdown. Under the Pradhan Mantri Garib Kalyan Yojana (PMGKY), 800 million poor people were included in a cash transfer and food security programme. Each person was to be provided 5 kg of rice or wheat every month freely for 3 months and along with it, a kilogram of pulse per household was also to be given for 3 months. The targeted group for direct cash transfer benefits were the farmers, job card holders of MGNREGA, widow pensioners, physically challenged persons (PCPs), women with Jan Dhan accounts, women self-help groups (SHGs) and construction workers.

The wage rate under MGNREGA was increased by ₹20, which meant an additional ₹1000 for 100 person-days of work. Almost 30 million poor, widows, PCPs were to receive a one-time ex gratia of ₹1000 in two instalments. An ex gratia of ₹500 for 3 months was to be given to 200 million women Jan Dhan account holders. Also, 83 million women below the poverty line were to be provided free cooking gas under the Ujjwala scheme. The amount for the collateral-free loan for women SHGs was doubled to ₹2 million, which was to benefit 70 million households (The Hindu 2020). Also, ₹310 billion under the Building and Other Construction Workers Fund was to be spent to support 35 million construction workers.

In this time of crisis, the generous donations from corporate and individuals were deposited in PM care relief fund and to state government relief funds. Philanthropic organisations like non-government organisations (NGOs), missionaries, clubs, associations and individuals provided relief materials to the needy people. The work done by civil society is highly commendable, but it is difficult to sustain such work for a long period. In due course, the quantum of assistance began to fade away and poor people were left to fetch for themselves. The role of government is crucial in the time of crisis. The immediate task of the government is to increase aggregate demand and take measures to remove supply bottlenecks.

In May 2020, the GoI announced a ₹20 trillion economic package to boost the fragile economy. This was inclusive of the ₹1.7 trillion packages announced in March. No doubt, this is the largest stimulus package declared by a developing country (The Economic Times 2020). The focus of the package is not only to stimulate the Indian economy but also to make the country self-reliant. Two things that shaped the idea of self-reliant were a realisation that too much dependence on the international value chain even for basic and life-saving products can be disastrous in the time of a pandemic. The other was a growing public sentiment against China for its belligerent action along the LAC. People were in no mood to accept that business can be as usual with a country that threatens the country's security. Slogans like "*Atmanirbhar Bharat*" and "vocal for local" were meant to blend the idea of nationalism with economic revival.

The idea of self-reliant gave a boost to the production of life-saving machines like ventilators and products like facemasks, sanitisers and personal protection equipment (PPE). Production of life-saving drugs and vaccines has increased and exported in large quantities to other nations. A large share of the facemasks, sanitisers and PPE is produced by the SHGs and micro, small and medium enterprises (MSME). The importance of MSME in India's economy is very high. Almost 30% of the GDP comes from them and it provides 110 million jobs (Ray and Subramanian 2020). The MSMEs will play an important part in increasing the domestic supply and hence adequate capital must be made available to this sector. However, the economic crisis is not just a supply-side problem; the lack of aggregate demand has been a problem for quite some time even before the lockdown. The stimulus package to provide cash in the hands of the people will help to push the demand. This will help to increase demand in the economy. The various liquidity measures announced by Reserve Bank of India (RBI) from February to April amounts to ₹8.02 trillion, which is a part of the ₹20 trillion economic packages. Under this scheme, not all MSMEs are eligible and only 7% of the MSMEs who already have an outstanding loan are eligible for the concessional rate of interest (Ghosh 2020: 12). However, the government realised that the MSMEs will play an important role in the economic revival and employment generation and have expedited the payment of the outstanding amount. Between May and November 2020, the government has paid ₹210 billion of the outstanding bill. During this period the procurement by GoI has increased 2.5 times and measures taken to expedite the payment at the earliest (The Times of India 2020).

The government has announced stimulus packages even after the marathon 5 days announcement in May. In June an additional stimulus of ₹900 billion for PMGKY was announced by the prime minister. Further announcements were made in November. A notable shift is that the government now recognises that rural economy in general and agriculture sector, in particular, will be the sectors for economic revival. The urban centres are struggling with the pandemic and it will be some time before the urban economy can start the economic recovery. The MGNREGA will stimulate the rural economy and ₹605.99 billion was already released until 15 September 2020[12]. All the cash transfer to the rural population mentioned above adds to ₹1.3–1.4 trillion. If we also add the value of food that was distributed, then the total amount will be much higher. The financial transfer is almost 1% of the GDP and this will have an impact on the rural economy. In this time of economic crisis, the deficit level can be increased and the allocation for livelihood programmes enhanced. A plausible reason for the slowing down of MGNREGA work is the exhaustion of the fund. Additional allocation will be required to avoid any humanitarian crisis.

22.5 Conclusion

The pandemic has eroded livelihood of millions of workers, who need some work immediately. It is evident from the discussions in the previous sections that to increase rural livelihood the choice is not between farm and non-farm. We cannot rely on any one of them for reviving the economy. Allied sectors like a plantation, dairy, livestock, fisheries each on its own cannot pull itself on its own. Nor can the farm and non-farm sector perform separately. Close coordination between all the sectors is required to come out successfully from the crisis. The rural household income is pooled from agriculture work, dairy, livestock, forestry, non-farm. The average income of a farmer in 2015–2016 was ₹96,703 of which the farm and non-farm income were ₹ 54,246 and ₹38,457, respectively (Dalwai 2018: 9). Not all states had a higher share of farm income in farmer's income. States like Goa, Himachal Pradesh, Kerala, Nagaland, Sikkim, Tamil Nadu and West Bengal had a higher share of non-farm income. It means depending upon various factors like capital investment, irrigation intensity, agro-climatic conditions, infrastructure, level of market integration the share of the farm and non-farm income varies across states. It is also true that with a share of 15% of GDP the agriculture sector cannot pull the economy out from the recession. However, this is the only sector that has shown resilience during the pandemic and provides livelihood to millions of rural people.

The diversification of the farm sector has helped to increase income but lack of storage and other infrastructure has shown how vulnerable the agrarian community can be during the pandemic. A similar crisis may hurt the agrarian community when a disaster of a global proportion due to climatic factors or epidemic occurs. The present pandemic is a challenge, as well as a learning experience for the global community. We need to find ways to find out how we can best address this challenge. Our experience during the lockdown showed that too much dependence

---

[12]As per the statement of the Ministry of Rural Development, Govt. of India.

on a source in the value chain could be a problem during any crisis. Self-reliance in food, health and other basic needs will give a strong foundation to manage better during any future crisis. Decentralisation of storage of food and other essentials must be an integral part of the disaster management policy of a district. The construction of the warehouse to store food and essentials can be done through a convergence of MGNREGA and National Nutrition Mission and Public Distribution System.

The MGNREGA will not only provide the much-needed livelihood to the rural poor but also should be used to develop rural asset base. The recurrence of flood and drought has adversely affected the agriculture sector. At the same time, there is a need to increase irrigation intensity in the country. Since MGNREGA fund can be used for rainwater harvesting and water conservation, such work should be given high priority. Afforestation programme, renovation of water bodies like ponds and tanks should be a high priority in drought-prone areas. Flood control programme like the construction of drainage in waterlogged areas and renovation of wetlands will help to damage from floods. Expansion of micro-irrigation especially in water-scarce regions will help to increase irrigation efficiency. Soil and water management must be made an integral part of agriculture practices. Optimal use of resources should be encouraged and to do that information and communication technology (ICT) application in farming requires promotion. Farmers should be linked with the information highway that will increase awareness level and keep them updated. Promotion of Electronic National Agriculture Market (e-NAM) will require better promotion among the farmers. Farmers should be directly linked with food programmes like mid-day meal scheme, community kitchens, hostels, armed forces canteen and mess. This will help them to get better price and guarantee of payment.

Marginal and small cultivators cannot reap any economics of scale, so collective farming especially under the present COVID-19 crisis must be promoted in the villages. An incentive can be given to form farmer's cooperative, similar to SHGs and WUGs. Subsidy for irrigation can be linked with the formation of cooperatives. This will help to reduce competition in water use and achieve an optimal water use level. Contract farming between a buyer and farmer's cooperative can be more equal than an individual farmer entering into a contract. Insurance premium can be subsidised for a farmer cooperative. Farmer's cooperative should be given priority under mid-day meal, government hostels and canteens. Such incentives will help to increase cooperative farming, which will help the farmers during the time of crisis.

The share of the dairy industry is almost 25% of the GVA in the agriculture and allied sector. India's export of dairy products was 51.4 thousand metric tons and its value was ₹13.4 billion in the year 2019–2020 (APEDA[13] 2020). To increase the export of dairy products India on a war footing the eradication of the foot and mouth disease (FMD) must be given high priority. Eradication of FMD will help to expand the dairy sector and increase the earning from this activity. Decentralisation in dairy infrastructure will help to procure, process and store dairy products like skimmed milk, butter and ghee and will ensure a more stable price of the milk. This will

[13] Agricultural & Processed Food Products Export Development Authority.

encourage the expansion of dairy activities and employment in this sector. All fresh food, fishery, dairy and livestock products require high hygiene and safety standards. SHGs have risen to the occasion and have supplied a large share of the facemasks, sanitisers and PPEs. Such activities certainly increase health and safety awareness among rural workers. It will help to attain a higher hygiene and safety standards in processing, production and distribution of the food products.

The non-farm sector at present is faced with a cost rationalisation problem. Salary reduction and wage cuts, lower recruitment of workers and reduced capital formation, all these have led to a drastic fall in non-farm work. Expanding non-farm activities in rural areas can be focused on value addition of agriculture products. Skill enhancement programme should be strengthened. However, to avoid any future humanitarian crisis migrant's data bank at *GramPanchayat* level must be maintained. During this crisis, it has become evident that none of the law related to migrants was adhered to by the state administrations. A reason for this is each state is in a race to project a business-friendly environment, which is not wrong, but to flout rules and law at the cost of human life is not the right thing to do. At present, the livelihood programme like MGNREGA must be increased and not just kept at 100 days only. Extraordinary circumstances require extraordinary measures.

References

APEDA (2020) Ministry of Commerce & Industry, Government of India. http://www.apeda.gov.in/apedawebsite/SubHead_Products/Dairy_Products.htm#:~:text=India%20Facts%20and%20Figures%20%3A,Egypt%20A%20Rp%2C%20U%20S%20A. Accessed 16 Dec 2020

Balasubramanian S (2020) Climate change, covid-19 burden India's fisherfolk: sustaining livelihood in lockdown. The Energy and Resource Institute. https://www.teriin.org/article/climate-change-covid-19-burden-indias-fisherfolk-sustaining-livelihood-lockdown. Accessed 15 Dec 2020

Bathla S (2017) Public investment in agriculture and growth: an analysis of relationship in the Indian context. In: Bathla S, Dubey A (eds) Changing contours of Indian agriculture. Springer, Singapore. https://doi.org/10.1007/978-981-10-6014-4_2. Accessed 14 Nov 2020

Biswal J, Vijayalakshmy K, Rahman H (2020) Impact of COVID-19 and associated lockdown on livestock and poultry sectors in India. Vet World 13(9):1928–1933. https://doi.org/10.14202/vetworld.2020.1928-1933. Accessed 15 Dec 2020

Breman J (1996) Footloose labour: working in India's informal sector. Cambridge University Press, Cambridge

Chand R, Saxena R, Rana S (2015) Estimates and analysis of farm income in India, 1983–84 to 2011–12. Econ Polit Wkly 50(22):139–145. https://www.im4change.org/siteadmin/tinymce/uploaded/Estimates_and_Analysis_of_Farm_Income_in_India_198384_to_201112.pdf. Accessed 23 Nov 2020

Chand R, Srivastava SK, Singh J (2017) Changing structure of the rural economy of India—implications for employment and growth. NITI Aayog. https://niti.gov.in/writereaddata/files/document_publication/Rural_Economy_DP.pdf. Accessed 31 Oct 2020

Christiaensen L, Demery L, Kuhl J (2011) The (evolving) role of agriculture in poverty reduction—an empirical perspective. J Dev Econ 96(2):239–254. https://doi.org/10.1016/j.jdeveco.2010.10.006. Accessed 11 Dec 2020

CMIE (2020a) Unemployment rate in India. https://unemploymentinindia.cmie.com/. Accessed 7 Dec 2020

CMIE (2020b) Unemployment in India: a statistical profile May-Aug 2020. Centre for Monitoring Indian Economy Pvt. Ltd. https://unemploymentinindia.cmie.com/kommon/bin/sr.php?kall=wstatmore. Accessed 7 Dec 2020

Dalwai A (2018) Summation of recommended reforms and implementation items covering topics discussed in all volumes of the DFI report. Report of the Committee on Doubling Farmers' Income, vol XIV. http://agricoop.nic.in/doubling-farmers. Accessed 19 Nov 2020

Economic Survey (2017) Government of India, vol I. https://www.indiabudget.gov.in/budget2017-2018/es2016-17/echapter.pdf. Accessed 2 Nov 2020

Economic Survey (2020) Government of India, vol II. https://www.indiabudget.gov.in/economicsurvey/. Accessed 31 Oct 2020

Field E, Pande R, Papp J, Rigol N (2013) Does the classic microfinance model discourage entrepreneurship among the poor? Experimental evidence from India. Am Econ Rev 103(6): 2196–2226. https://doi.org/10.1257/aer.103.6.2196. Accessed 16 Dec 2020

Ghosh S (2020) Examining the COVID-19 relief package for MSMEs. Econ Polit Wkly 55(22): 10–12. https://www.academia.edu/43220662/Examining_the_COVID_19_Relief_Package_for_MSMEs. Accessed 20 Dec 2020

Gulati A, Ganesh-Kumar A, Shreedhar G, Nandakumar T (2012) Agriculture and malnutrition in India. Food Nutr Bull 33(1):74–86. https://doi.org/10.1177/156482651203300108. Accessed 11 Dec 2020

Gupta N (2016) Decline of cultivators and growth of agricultural labourers in India from 2001 to 2011. Int J Rural Manag 12(2):179–198. https://doi.org/10.1177/0973005216665939. Accessed 12 Nov 2020

Haggblade S, Hazell P, Reardon T (2010) The rural non-farm economy: prospects for growth and poverty reduction. World Dev 38(10):1429–1441. https://doi.org/10.1016/j.worlddev.2009.06.008. Accessed 12 Nov 2020

Himanshu LP, Murgai R, Stern N (2013) Nonfarm diversification, poverty, economic mobility, and income inequality: a case study in village India. Agric Econ 44(4–5):461–473. https://doi.org/10.1111/agec.12029. Accessed 12 Nov 2020

International Monetary Fund (2020) World economic outlook: a long and difficult ascent. Washington, DC. https://www.imf.org/en/Publications/WEO/Issues/2020/09/30/world-economic-outlook-october-2020. Accessed 3 Dec 2020

Ivanic M, Martin W (2018) Sectoral productivity growth and poverty reduction: national and global impacts. World Dev 109:429–439. https://doi.org/10.1016/j.worlddev.2017.07.004. Accessed 7 Dec 2020

Lanjouw JO, Lanjoouw P (2001) The rural non-farm sector: issues and evidence from developing countries. Agric Econ 26(1):1–23. https://doi.org/10.1016/S0169-5150(00)00104-3. Accessed 12 Nov 2020

Lewis WA (1954) Economic development with unlimited supply of labor. Manchester Sch Econ Soc Stud 22(2):139–191. https://doi.org/10.1111/j.1467-9957.1954.tb00021.x. Accessed 12 Dec 2020

Ministry of Agriculture (2020) Agriculture statistics at a glance 2019. Ministry of Agriculture. https://eands.dacnet.nic.in/PDF/At%20a%20Glance%202019%20Eng.pdf. Accessed 24 Nov 2020

Misra S, Suresh AK (2014) Estimating employment elasticity of growth for the Indian economy. RBI working paper series. WPS (DEPR) 06/2014. https://rbidocs.rbi.org.in/rdocs/Publications/PDFs/06WPSN240614.PDF. Accessed 24 Nov 2020

Moteshar ezadeh B, Etesami H, Bagheri-Novair S, Amirmokri H (2017) Fertilizer consumption trend in developing countries vs. developed countries. Environ Monit Assess 189:103. https://doi.org/10.1007/s10661-017-5812-y. Accessed 11 Dec 2020

Mukherjee A, Wang S-YS, Promchote P (2019) Examination of the climate factors that reduced wheat yield in Northwest India during the 2000s. Water 11:343. https://doi.org/10.3390/w11020343. Accessed 10 Dec 2020

NABARD (2018) All India rural financial inclusion survey 2016–17. https://www.nabard.org/auth/writereaddata/tender/1608180417NABARD-Repo-16_Web_P.pdf. Accessed 4 Dec 2020

National Accounts Statistics (2020) Ministry of Statistics and Program Implementation (MOSPI). Government of India. http://mospi.nic.in/national-accounts-statistics-0

National Statistical Office (2020) Press note on estimates of gross domestic product for the second quarter (July–September) 2020–2021. Ministry of Statistics & Programme Implementation, Government of India. https://www.mospi.gov.in/documents/213904/416359//PRESS_NOTE-Q2_2020-211606480008567.pdf/f2b98a11-a06d-8b6f-6f37-621f33ca8f25. Accessed 6 Dec 2020

Oxford Poverty and Human Development Initiative (2020) "India country briefing", multidimensional poverty index data bank. Oxford Poverty and Human Development Initiative, University of Oxford. www.ophi.org.uk/multidimensional-poverty-index/mpi-country-briefings/. Accessed 4 Dec 2020

PLFS (2020) Periodic labour force survey—annual report [July 2018–June 2019]. https://pib.gov.in/PressReleaseIframePage.aspx?PRID=1629366. Accessed 23 Nov 2020

Raju SS, Suresh S, Chand R, Chauhan S (2015) Pattern and trend in labour use in Indian agriculture: an analysis across major crops and states. Econ Affairs 60(1):99–108. https://doi.org/10.5958/0976-4666.2015.00014.5. Accessed 16 Dec 2020

Ravallion M, Datt G (1996) How important to India's poor is the sectoral composition of economic growth? World Bank Econ Rev 10(1):1–25. https://doi.org/10.1093/wber/10.1.1. Accessed 7 Nov 2020

Ray D, Subramanian S (2020) India's lockdown: an interim report. Indian Econ Rev 55:81. https://doi.org/10.1007/s41775-020-00098-y. Accessed 16 Dec 2020

Schultz TW (1964) Transforming Traditional Agriculture. Yale University Press, New Haven

Shonchoy AS (2015) Seasonal migration and microcredit during agricultural lean seasons: evidence from northwest Bangladesh. Dev Econ 53(1):1–26. https://doi.org/10.1111/deve.12063. Accessed 16 Dec 2020

Srivastava R (2019) Emerging dynamics of labour market inequality in India: migration, informality, segmentation and social discrimination. Indian J Labour Econ 62:147–171. https://doi.org/10.1007/s41027-019-00178-5. Accessed 2 Nov 2020

Taylor M (2011) 'Freedom from poverty is not for free': rural development and the microfinance crisis in Andhra Pradesh, India. J Agrar Chang 11(4):484–504. https://doi.org/10.1111/j.1471-0366.2011.00330.x. Accessed on 15 Dec 2020

The Economic Times (2020) https://economictimes.indiatimes.com/news/economy/finance/latest-stimulus-package-among-largest-in-the-world/articleshow/75701976.cms. Accessed 16 Dec 2020

The Hindu (2020) Coronavirus: centre rolls out ₹ 1.7-lakh-crore lockdown package. https://www.thehindu.com/business/Economy/17-lakh-cr-package-with-doubled-food-rations-cash-transfers-for-poor/article31172100.ece. Accessed 16 Dec 2020

The Times of India (2020) Finance minister takes stock of MSME dues. https://timesofindia.indiatimes.com/business/india-business/finance-minister-takes-stock-of-msme-dues/articleshow/79664909.cms#:~:text=As%20a%20result%2C%20over%20Rs,of%20over%20Rs%204%2C100%20crore. Accessed 20 Dec 2020

Thirupathy P, Ravisankar T (2015) Agricultural labourers and wages recent trends in rural employment in India. Shanlax Int J Econ 3(4):34–38. http://www.shanlaxjournals.in/pdf/ECO/V3N4/ECO_V3_N4_006.pdf. Accessed 14 Dec 2020

Wen Z, Shen J, Blackwell M, Li H, Zhao B, Yuan H (2016) Combined applications of nitrogen and phosphorus fertilizers with manure increase maize yield and nutrient uptake via stimulating root growth in a long-term experiment. Pedosphere 26(1):62–73. https://doi.org/10.1016/S1002-0160(15)60023-6. Accessed 10 Dec 2020

Govinda Choudhury teaches economics in the Department of Economics, University of North Bengal. He teaches economic theory, resource economics and research methodology. His primary interest is in regional economic issues and has focused extensively on Eastern Himalayan region. He has worked on forest economics, community rights, ecosystem services, tribal livelihood and regional agricultural issues. The focus of his current projects is on the management of wetland and water issues in the context of climate change. He is also working in a project on sustainable rural livelihood of Eastern Himalayan community. At present, he is a consultant for the District Human Development Report of Dakshin Dinajpur district of West Bengal.

Debjani Choudhury teaches economics in B.S.F. Residential School, Kadamtala. Her primary interest is in economics of environment, gender and livelihood. She has worked in a project entitled 'Socioeconomic Perspective Plan of North Bengal'. She has presented papers on national and regional issues. At present she is working on local community and environment conservation.

Chapter 23
COVID-19 and Lockdown: Key Constraints and Surviving Strategies for the Micro, Small, and Medium Enterprises (MSMEs)

Arindam Metia

Abstract Micro, small, and medium enterprises (MSMEs) are at the heart of our economic lives. The COVID-19 pandemic has given rise to most of the devastating economic crises of our lifetime. MSMEs are at the center of it. *Social distancing*, *safe home*, *lockdown* are new words in our life, but they have changed our way of life. MSMEs have seen declining demand for goods and services, resulting in a lack of working capital, unsold goods accumulation, delayed payment and cash shortage, layoffs, and eventually business. The purpose of the research paper is to study key factors that affect MSMEs during the lockdown. The paper analyzed the constraint factors concerning financing, market, technology, infrastructure, and technology. Atmanirbhar Bharat (self-reliant) will be built based on the significant contribution of MSMEs in all fields. It has been observed that the epidemic situation has changed partially or entirely in business activities, and we will have this change with the help of the fourth industrial revolution. This chapter aims to develop some strategic issues which are most relevant in this current scenario.

Keywords Lockdown · Indian economy · Finance · Market · Infrastructure · Technology

23.1 Introduction

The micro, small, and medium enterprises (MSMEs) sector is the most vital constituent of the industrial sector. The SMEs cover a broad spectrum of industries, and they are established in almost all the major sectors in the Indian industry. MSME is the largest employment generator after agriculture in India, as per the 73rd round National Sample Survey (2015–2016).

A. Metia (✉)
Department of Management, Raiganj University, Raiganj, Uttar Dinajpur, West Bengal, India

S. Saha et al. (eds.), *Economic and Societal Transformation in Pandemic-Trapped India*, New Frontiers in Regional Science: Asian Perspectives 55,
https://doi.org/10.1007/978-981-16-5755-9_23

According to Prime Minister Narendra Modi's vision, a self-reliant (Atmanirbhar Bharat) India needs five pillars, namely economy, infrastructure, system, vibrant population, and demand. Indian MSMEs will be supporting Atmanirbhar Bharat[1] Abhiyan as it is totally made in India's concept. MSMEs have been recognized as the engine of economic growth and for promoting equitable development by providing larger employment opportunities through the industrialization of rural and backward areas (MSME 2021). Due to low investment and operational flexibility, MSMEs can operate all over India. MSME sectors have been contributing significantly to the country's GDP. The MSME sector shared 30.2% of India's total GDP in 2018–2019 (MSME 2019).

The Swadeshi Movement of India (1905) was symbolized by Khadi and led by M.K. Gandhi. The financial independence of rural India was promoted by M.K. Gandhi. According to Gandhi's philosophy, self-reliance in the economy will be the pillar of a self-sufficient society. India is the second-largest textile operator in the world (Textiles & Apparel 2020). The government has plans to provide employment and promotion for Khadi brands nationally and internationally. Great opportunities and hopes in the Khadi industry will help a lot in production, sales, and employment.

The Khadi sector in India is a potential segment for creating employment for rural people with minimal investment. Indian Khadi is the symbol of Atmanirbhar (self-reliance). Khadi and village MSMEs have been significantly increasing in terms of production, sales, and employment. Total production, sales, and employment were Rs. 45,725.69 cr, Rs. 55,750.40 cr, and 137.79 lakh, respectively, in 2017–2018, which are satisfactorily higher than previous financial years (MSME 2018). Village industries which include different sectors like mineral based-industry, forest-based industry, handmade paper and fiber industry generated the total production of Rs. 41,110.26 cr in 2016–2017 (MSME 2018). The Ministry of MSME provides different facilities regarding credit and financial assistance, skill development training, infrastructure development, marketing assistance, technological and quality upgradation.

Figure 23.1 shows that while MSMEs have been contributing more enormous employment opportunities worldwide, India has been capable to generate comparatively lesser employment opportunities in MSMEs compared to other countries. The Indian economy has been trying to enrich its development through the creation of employment by MSMEs. India needs to create 10–15 million job opportunities per year over the next decade to provide gainful employment to its population. Under the provision of the Micro, Small and Medium Enterprises Act (01st July 2020), the micro, small, and medium enterprises (MSME) are classified into three (Table 23.1):

[1] Atmanirbhar Bharat implies "self-reliant" by providing a 20-lakh crore economic package to help the country and its citizen. Economy, infrastructure, system, vibrant demography, and demand are identified as the key pillars of Atmanirbhar Bharat. For more details, see https://aatmanirbharbharat.mygov.in/.

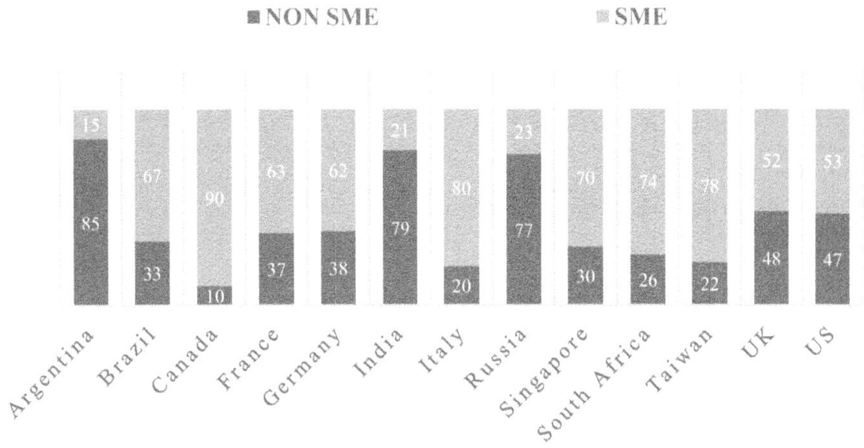

Fig. 23.1 Employment generation by MSME as a percentage of overall employment globally. (Source: Country-specific MSME Reports, KPMG data & estimates, 2015)

Table 23.1 Classification of manufacturing enterprises and enterprises rendering services

Classification	Investment in plant and machinery or equipment	Annual turnover
Micro	Not more than Rs. 1 crore and annual turnover	Not more than Rs. 5 crore
Small	Not more than Rs. 10 crore	Not more than Rs. 50 crore
Medium	Not more than Rs. 50 crore	–

Source: (Compiled by the author based on The Gazette of India, No-1532, Dated 01st June 2020)

MSMEs account for over 48% of India's exports. Thus, the Indian MSME plays a significant role in the national economic structure, and it helps India stands in the world economy. With around 63.4 million units throughout the geographical expanses of the country, MSMEs contribute 6.11% of the manufacturing GDP and 24.63% of the GDP from service activities, as well as 33.4% of India's manufacturing output.[2] India stands as the second-largest country in respect to MSMEs after China.

Table 23.2 shows that 6.34 crores MSME are in the country and approximately 51.12% are operating from rural India. Table 23.3 reveals that about 11 crores of employment are generated from MSME. Out of total employment in MSME, 55% of the employees were found in the urban area. On average, less than two people are employed per MSME. MSMEs are categorized into three segments: micro, small, and medium, respectively (Table 23.1). However, 99.5% of all MSME fall in the micro category.[3] From Table 23.2, out of a total of 6.34 crores MSMEs, 6.30 crores

[2]Micro, medium, and small-scale industry in the official website under the URL https://www.cii.in/Sectors.aspx?enc=prvePUj2bdMtgTmvPwvisYH+5EnGjyGXO9hLECvTuNuXK6QP3tp4gPGuPr/xpT2f.

[3]Indian weekly magazine Economic and Political Weekly reported on 30th May 2020, which is available on their archive under the URL https://www.epw.in/journal/2020/22/commentary/examining-covid-19-relief-package-msmes.html.

Table 23.2 Estimated number of MSMEs (in Rs. lakh)

Activity category	Rural	Urban	Total	Share (%)
Manufacturing	114.14	82.5	196.64	31.02
Trade	108.71	121.64	230.35	36.34
Other services	102	104.85	206.85	32.63
Electricity (non-captive electricity generation and transmission)	0.03	0.01	0.03	0.00
All	324.88	309	633.87	100.00

Source: (Annual Report, 2018–2019, Ministry of MSME, GOI)

Table 23.3 Estimated employment in MSMEs (in lakh)

Activity category	Rural	Urban	Total	Share (%)
Manufacturing	186.56	173.86	360.41	32.00
Trade	160.64	226.54	387.18	35.00
Other services	150.53	211.69	362.22	33.00
Electricity (non-captive electricity generation and transmission)	0.06	0.02	0.08	
All	497.79	612.11	1109.89	100.00

Source: Annual Report, 2018–2019 Ministry of MSME, GOI

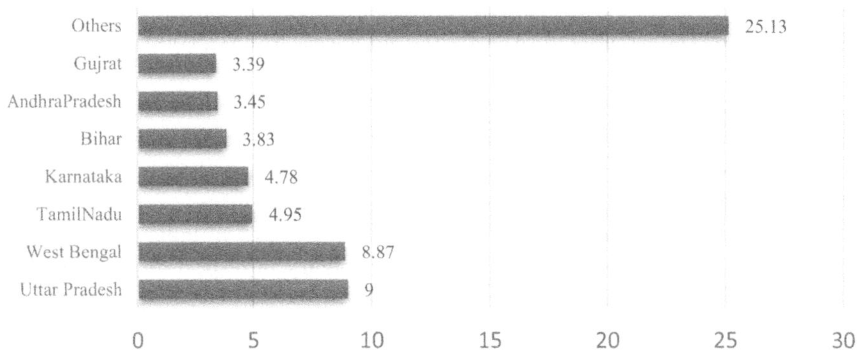

Fig. 23.2 Number of estimated MSMEs in millions. (Source: MSME Annual Report, 2018–2019, Ministry of MSME, GOI)

micro-enterprises significantly contribute to GDP in the Indian economy. In India, micro-enterprises essentially refer to a single man or a woman working independently from their home. Out of a total, 6.34 crores, 98.88% of MSMEs are proprietary enterprises. MSMEs are, however, predominated by males (79.63%) compared to females in terms of ownership (MSME 2019). MSMEs encourage inclusive growth by providing employment (44.85%) in rural areas, especially people belonging to the weaker section of the society. In terms of geographical distribution, seven Indian states alone account for 50% of all MSMEs. Figure 23.2 exhibits that Uttar

Table 23.4 Distribution of enterprises by categories

Sector	Micro	Small	Medium	Total	Share (%)
Rural	324.09	0.78	0.01	324.88	51
Urban	306.43	2.53	0.04	309	49
All	630.52	3.31	0.05	633.88	100

Source: Annual Report, 2018–2019 Ministry of MSME, GOI

Pradesh is having maximum numbers of MSMEs (14.20%), followed by West Bengal (13.9%), Tamil Nadu (7.80%), Karnataka (7.55%), and Bihar (6.14%).

Indian MSMEs have been badly affected by the COVID-19 pandemic situation. On 24th March 2020, India announced a nationwide lockdown to control the spread of the coronavirus for 21 days initially. After that, the lockdown was extended in the subsequent three phases up to 31st May 2020.[4] This total of 68 days of lockdown across India impacted MSMEs terribly. IMF has projected a GDP growth of 1.9%. COVID-19 influences a profound impact on the Indian economy; however, at the same time, The New Wave Indian MSME-An Action Agenda for Growth (2015), in its report, expressed an optimistic view toward the economic recovery of India. As the MSMEs were shut down, millions of migrant workers left the workplace and searching for new destinations for surviving. Some of the workers have perished in transit. As no data is maintained by the government, it is impossible to ascertain the number of deaths. All the MSMEs suffer from low labor-power as most of the labors returned to their homeland to survive.

In addition to that the global supply chain is being disturbed, and the majority of the MSMEs are dependent on China for its raw materials like pharmaceutical, cotton yarn, iron ore, organic chemicals, plastic items, electrical machinery, gems, and jewelry. The prolonged lockdown has damaged the supply chain of raw materials from China. The lockdown results in the shrinkage of exports, increased production cost, weakening workforce, loss of consumer confidence, and cash crunch among MSMEs. According to a study commissioned by the All India Manufacturers' Organisation (AIMO 2020),[5] India is the home to over 75 million MSMEs, and close to 25% of these firms will face closure. Table 23.4 reveals that about 51% of MSMEs are located in rural areas and the rest in urban areas. This micro-scale industry is confined to most of the villages. Therefore, the rural economy is highly dependent on it. This large number of closures of MSME could slow down the Indian economy in every sense. If the local industry is not surviving, it will affect the extensive industry and result in a higher cost of production and fewer jobs in MSMEs.

[4]Indian daily newspaper, The Hindu reported on 24th March 2020, which is available in their archive under the URL https://www.thehindu.com/news/national/pm-announces-21-day-lockdown-as-covid-19-toll-touches-10/article31156691.ece (24.03.2020).

[5]Indian daily newspaper, Financial Express reported on 2nd June 2020, which is available in their archive under the URL https://economictimes.indiatimes.com/small-biz/sme-sector/over-one-third-msmes-start-shutting-shop-as-recovery-amid-covid-19-looks-unlikely-aimo-survey/articleshow/76141969.cms?from=mdr.

In this respect, the present study was based on the following objectives:

- To analyze key constraint factors on MSMEs during lockdown period;
- The study aims at developing strategies for the survival of MSMEs.

23.2 Lockdown and MSMEs: Key Constraints

The first case of COVID-19 was detected in Kerala, India, on 27th January 2020 (Andrews et al. 2020). The nationwide lockdown was declared on 24th March 2020 to break the coronavirus transmission cycle. It was assumed that lockdown would prevent community transmission and helps the health system to prepare itself to fight against COVID-19. The sudden outbreak of the coronavirus has given the shortage of a large number of migrant workers of India, and their absence in the workplace has affected all MSME units in respect of production. Countries with greater dependence on the service sector, higher levels of informality, and weak safeguards against the termination of employment have experienced much higher initial job losses (ILO 2020). MSMEs have long been playing a vital role in handicrafts in terms of manufacturing and job creation. However, people have been buying only necessary product items, and handicrafts and fashionable items are losing demand to the cash crunch.[6] Thus, the labor shortage and the fall in demand, these double crises, have put the entire MSME sector in terrible shape. As prime minister declared the five pillars of "Atmanirbhar Bharat" but in this pandemic situation, MSMEs are mostly suffering from six constraints:

- Access to finance
- Access to markets
- Access to technology and environment
- Access to infrastructure
- Access to people

23.2.1 Access to Finance

Access to finance always plays a pivotal for the expansion and growth of the entrepreneur. While large entrepreneurs do not encounter problems, small enterprises have been vulnerable to the pandemic-induced shock. A survey by BCG and Omidia (2018) found that 40% of MSMEs' lending was accessible through informal

---

[6]Indian daily newspaper, Financial Express reported on 9th July June 2021, which is available in their archive under the URL https://www.businesstoday.in/opinion/columns/story/covid19-impact-on-handicrafts-sector-handloom-artisans-suffering-due-to-coronavirus-lockdown-270592-2020-08-20.

sources. Several factors were identified that discouraged MSMEs from accessing formal sources. The reasons are long processing time, lack of transparency in a timeline, amount of loan, term of the loan, and high-interest rate.

A 59-min loan portal was launched by Prime Minister Narendra Modi in Nov 2018 to ease credit access for MSMEs. The Economic Survey (2019–2020) has disclosed that a total of 159,422 loans has been sanctioned involving Rs. 49,330 crore within 59 min. This digital platform helped MSMEs to get eligibility letters for approval of loans within 59 min. Due to too much formality, paperwork, and stringent norms, MSMEs are relatively dependent on informal sources for finance (PM Launches Historic Support and Outreach Initiative for MSME Sector 2018). Delay in loan processing time, higher interest rate, loan tenure, lack of transparency, insufficient loan size discourage MSMEs from availing credit from formal sources (Making India MSMEs Globally Competitive 2019). The post lockdown period will be more stressful for MSMEs to get finance as cash flows will be lower or nil. A 5% contraction in the Indian economy is expected, as per CRISIL (2020–2021). The consulting company Dun and Bradstreet revealed that MSMEs managed 67% of the total funds from alternative sources, including friends and family, trade credit, etc. MSME will access only 30–35% of total funds through capital market and FIIs.[7] Informal sources of financers are not interested in providing a good source of finance like earlier because a shortage of cash flows in MSMEs may not be able to repay the dues.

23.2.2 Access to Market

Most of the MSMEs provide services to local or regional markets by direct sales. According to All India Motor Transport Congress, the daily movement of trucks has collapsed to less than 10% during the lockdown.[8] Disruption of the supply chain has severely affected the supply of raw materials and sales to MSMEs. Inventories have piled up at factories in the form of raw materials, work-in-progress, and finished goods. MSMEs had been suffering from profitability and cash earning capacity due to low inventory turnover. The scarcity of the labors is adding up the problems of the distribution of products in different markets. The concept of "work from home" is not applicable as most of the MSMEs lack the facility of internet service, or the nature of work does not permit them to do so. The survival of the food and beverage

[7] Indian daily newspaper, Financial Express reported on 26th November 2019, which is available in their archive under the URL www.financialexpress.com/industry/sme/msme-fin-informal-channel-financing-formal-financing-bank-loan-government-scheme-for-msmes-small-business-finance-nitin-gadkari/1775597.

[8] Indian daily newspaper, Financial Express reported on 27th April 2020, which is available in their archive under the URL https://www.financialexpress.com/industry/msme-logi-logistics-industry-to-suffer-rs-50000-crore-losses-due-to-lockdown-msmes-impacted-most-by-coronavirus/1940936/
.

industry is the biggest challenge. These industries suffer huge losses due to a drastic drop in sales, delay in the movement of perishables, and deficiency of human resources, respectively. It could result in 1.2 crore job losses in the hospitality industry at the end of the crisis.[9]

23.2.3 Access to Technology and Environment

In the open market economy era, the MSMEs sector must have the facility of modern technology to compete with the local, regional, national, and international levels. Enabled with IT not only makes the MSME firms competitive, but it will also be affordable and meet the quality desire and needs of the customer.

The maximum requirement is to adopt information technology to effectively connect with other enterprises during the lockdown. This will create business opportunities and make it easier to get loan applications or payments, payroll, cash flow management, product delivery, and the creation of potential customers. Jimenez and Lim (2018) list the benefits of digitization such as improved profit margins, productivity, and customer loyalty and retention as well as cost reduction and the ability to deliver new products and services and increase revenue. Karr et al. (2020) reports that digital solutions can help MSMEs to manage long-distance transactions, efficiently deliver goods, and facilitate access to financial services. MSMEs suffer from three types of crises during the lockdown: adoption, access, and lack of innovation concerning technology. IT enabled services like cloud computing, online booking applications, and other advanced digitization systems could be a survival strategy during the post-COVID period. Indian MSMEs are lagging in knowledge and market research, and they are not finding a proper marketplace to showcase their product.[10] Availability and adoption of technology could help fix the marketing strategy to compete with prominent players in the lockdown market. A physical customer has now been converted to a digitized customer. MSMEs are constantly losing a broader customer base due to the non-availability of digital marketing. An effective management information system will be able to link between users (customers) and manufacturers.

23.2.4 Access to Infrastructure

Infrastructural development is also closely associated with railways, waterways, roadways, and airways. Due to the non-availability of the materials, revenues have

[9]Indian daily newspaper, The Economic Times reported on 24th June 2021, which is available in their archive under the URL https://economictimes.indiatimes.com/small-biz/sme-sector.
[10]Ibid.

dropped by 50–60% overall. Most of the micro industries are very small in size; they are not registered. Due to non-registration, sometimes it is challenging to identify and access them.

The current coronavirus pandemic situation made the business environment challenging due to the non-availability of physical labors, raw materials, and transport infrastructure. Cost and complexity are the major issues in respect of infrastructure during the lockdown. The facility of power, water supply, and waste management is inadequate, especially in rural areas where maximum MSMEs are found. Sometimes MSMEs carry water from distant places for the manufacturing process. As a result, it costs higher in production (Vision 2020).

23.2.5 *Access to People*

A study conducted by IIM Lucknow & CMEE (2020) stated that 79% of the people are worried and surrounded by the feeling of fear (40%) and sadness (22%) even though they are confident of India curbing the pandemic. The feeling of fear and worry has influenced people to purchase all categories of products online. The majority of the consumers prefer online services because of safety and reduction of the risk of infection. They also prefer it since it is more convenient, saves time, and provides online solutions at all times in an online environment. The availability of skilled workers and consumers is seen to be very tough during a pandemic.

23.3 Survival Package: Let Us Discover and Recover

Govt. of India declared an economic relief package of 20 lakh crores to help the MSMEs during this pandemic situation in 2020. All MSMEs with a turnover of up to Rs. 100 crore and outstanding credit of up to Rs. 25 crore will be eligible to borrow up to percent of their total outstanding credit as of 29th February 2020. These loans will have 4-year tenure, and the scheme will be open until 31st October. A total of Rs. 3-lakh crore has been allocated for this.[11] This arrangement shall provide oxygen to all above mentioned all MSMEs during this crisis moment and helps to collect the required amount of working capital. Eligible MSMEs will get a collateral-free Rs. 3 lakh crore scheme under The Guaranteed Emergency Credit Line (GECL), and it will benefit 45 lakh units to resume their business activities. Around Rs. 50,000 crore as equity funds will be provided to MSMEs for the purpose of

[11] Indian daily newspaper, The Hindu reported on 17th May 2020,which is available in their archive under the URL https://www.thehindu.com/business/Economy/coronavirus-package-how-will-the-COVID-19-relief-for-msmes-help/article31603575.ece.

market listing and capacity expansion. Another package is that MSMEs can participate in government tender, as up to Rs. 200 crore, no global tender will be called.

In a move to provide emergency credit support to MSMEs impacted by coronavirus lockdown, the government has operationalized Rs. 20,000 crore stressed fund, which is likely to benefit around 2 lakh medium and small entrepreneurs. In addition, the government has issued guidelines for the Credit Guarantee Scheme for Subordinate Debt (CGSSD), a part of the Rs. 20.97 lakh crore Atmanirbhar Bharat Abhiyan package announced by Prime Minister Narendra Modi.[12]

These initiatives stand on the following five topics, which are discussed below. Every unit of MSME should care about the following strategies:

23.3.1 Analyzing the Financial Statement

Understanding the financial statement, preparing the budget for the short and long term is the most crucial job for entrepreneurs. The financial statement discloses the expected deficit, cash flows, and future expenses and revenues. Without authentic and reliable information, MSMEs will be facing difficulties in getting financial assistance from the government. MSMEs should plan for strong financial analysis with experts. MSMEs should analyze the risks and returns before applying for loans to banks and other financial institutions. MSMEs must look into cash-to-cash conversion cycle. The entire cash cycle should be careful re-designed from cash collection from customers to pay to vendors. In this current scenario, cash collection is a more critical function. A sound plan is required to maintain a balance between supply chain management and the minimization of working capital requirements.

23.3.2 Redesigning the Existing Business Plan

On the financial evaluation of future profitability and solvency, it is essential to redesign or redefine the business goal or mission in this situation. COVID pandemic situation teaches us to plan for an unforeseeable crisis. Every business unit should plan to access the customer 24 × 7 without interruption and maintain a good customer network. A new plan should be formulated which will benefit all the stakeholders in the post-COVID situation. New business opportunities may come up in the computer-related service due to the massive demand for virtual meetings and communication. Telecommunication (manufacturing of routers, switches,

---

[12]Indian daily newspaper, Business Today reported on 23rd June 2021,which is available in their archive under the URL https://www.businesstoday.in/current/economy-politics/govt-operationalises-rs-20000-crore-stressed-fund-for-msmes-issues-guidelines-avail-loan-from-banks/story/413442.html.

low-cost mobile, other value-added services), electronics (low-cost consumer electronics and consumer durables), IT/ITES (cloud computing, E-governance, mobile apps, and software development), media (manufacturing of set-top boxes, digital screen), pharmaceuticals (generic manufacturing), biotechnology (agri-products), automotive, food and agriculture, tourism and hospitality sectors may provide multiple employment and surviving opportunities in the post-COVID era.

Intensive communication with the customer during this epidemic crisis will help in formulating business plans. In addition, textile MSMEs can restart their business by producing PPEs, masks, sanitizers, and other essentials.

23.3.3 *Digital System*

Connection and togetherness with internal and external stakeholders are solely dependent on the robust digital system. The digital system enables business enterprises to create a positive brand image in the minds of the customer. As a result, CHAMPIONS,[13] a technology platform, has been launched to empower MSME. MSMEs should be planned for affordable robotic support. Bravo Robot is made in India and has been developed indigenously for Indian MSMEs.[14] We are now at the beginning of the fourth Industrial revolution (4.O). This period is the era of revolutionary digitization, and it has evolved based on the digital integration value chain in cyber-physical models and systems. It is possible to convert all activities, from raw material collection to digitization, to provide goods or services. In the aftermath of the COVID, MSMEs must achieve operational efficiency, more significant productivity gains, security, cost savings, and desired profits. To achieve this, every MSME must adopt a digital transformation system leading to the fourth industrialization revolution (4.O).

23.3.4 *Crisis Recovery Strategy*

In this time of crisis, there is an urgent need to maintain close contact with customers, owners, and stakeholders, requiring a crisis-free communication plan.

---

[13]Champion Stands for Creation and Harmonious Application of Modern Process for Increasing Output and the Nation Strength. The basic effort has been taken by the government to make MSMEs national and international champions. For further details, please see https://www.champions.gov.in/Government-of-India/Ministry-of-MSME-Portal-for-handholding/welcome.html.

[14]Tata Group of Companies, TL Manufacturing Solutions, has unveiled India's first-ever robot called "BRABO" for micro, small, and medium enterprises. It can be used for manufacturing process and it will cost much less than foreign robots. For further details, please see https://www.tataelxsi.com/insights/tata-brabo.

Barton (2008) identified six characteristics of a crisis: surprise, lack of information, increased incidents, loss of control, panic, and rapid absence of basic situations.

We identified all of the features mentioned above during this epidemic. Izz al-Din (1990) and Maher (2006) defined three stages of crisis management: pre-crisis stage, crisis response stage, and post-crisis stage. Currently, Indian MSMEs are undergoing the second and waiting for the third stage. All MSME procedures should be followed to overcome the crisis and coordinate with the behavioral, psychological, organizational, and financial aspects. Thus, crisis management requires a strong financial plan with adequate reserves and a digitally enabled system. A robust strategy has to be developed and innovated to overcome the crisis. With the help of digitization, modern equipment, and training, it is possible to reactivate old businesses. SWOT analysis will help to identify positive, negative, internal, and external factors related to business performance.

23.4 Conclusions

The COVID-19 situation affects all the countries in the world. This global economic downturn for this epidemic began in this century. The paths to economic recovery are ultimately uncertain and fragile without cent percent vaccination to the citizens. Though the Indian economy has felt the tremor of the pandemic shock, it is still performing better than other G20 countries. The FITCH (Organisation for Economic Co-operation and Development) estimates FY (2021–2022) Indian GDP growth at 6.7%. According to RBI's latest position, India's GDP could grow 7.2%.

Similarly, various agencies have predicted a GDP growth of 8% (Asian Development Bank, 2020–2021), 12.5% (International Monetary Fund, 2020–2021), 10.3% (Goldman Sachs, 2020–2021), and 8% (Federation of Indian Chambers of Commerce and Industry, 2020–2021). The global economies will slow down due to economic downturn in first world countries like the USA, Europe, and other countries. India has many potential strengths such as a growing young workforce, the middle-class consumer, a large workforce, and consequently, MSMEs can increase productivity by relying on these strengths. Winston Churchill (1940) famously said after World War II, "Never let a good crisis go to waste." In the current pandemic, we are trying to educate and make ourselves aware of the virus and how to prevent it.

COVID-19 has provided an opportunity for confidence and reconstruction. The B2C (business to consumer) segment will suffer due to low demand and low purchasing power. However, we have to buy daily necessities. For this self-sufficiency of both MSMEs and us, we must buy indigenous products. When demand increases, so do the production. MSMEs will be able to recover from past losses slowly. The government will have to take many initiatives to bring migrant workers back to work safely for the new restart.

Thus, the attitude in future MSMEs will be based on the development of entrepreneurial skills and strategy. This whole attitude and collaborative effort will help establish MSMEs and make India self-reliant. Thus, with the help of MSMEs, India will appear as Atmanirbhar (i.e., self-sufficient) in the post-pandemic world.

References

Andrews MA, Areekal B, Rajesh KR et al (2020) First confirmed case of COVID-19 infection in India: a case report. Indian J Med Res 151(5):490–492. https://doi.org/10.4103/ijmr.IJMR_2131_20

Barton L (2007) Crisis leadership now: a real-world guide to preparing for threats, disaster, sabotage, and scandal. McGraw-Hill, New York, NY

Barton L (2008) Crisis leadership now: a real-world guide to preparing for threats, disaster, sabotage, and scandal. McGraw-Hill, New York

Boston Consulting Group & Omidyar Network (2018) Credit disrupted digital MSME lending in India. https://omidyar.com/wp-content/uploads/2020/09/18-11-29_Report_Credit_Disrupted_Digital_FINAL.pdf

Not the virus but economic crisis is the biggest source of worry under lockdown (2020): Centre for Marketing in Emerging Economies (CMEE) & Indian Institute of Management (IIM), Lucknow. http://www.iiml.ac.in/sites/default/files/upload/news/1858590256CMEE%20Study-Option%201%20(1).pdf. Accessed 15 Dec 2020

GDP Growth Forecast (2021) Asian Development Bank. https://www.adb.org/news/india-growth-rebound-11-fy2021-moderate-7-fy2022. Accessed 10 Jul 2021

ILO (2020) ILO policy brief on COVID-19: a policy framework for responding to the COVID-19 crisis. https://www.ilo.org/global/topics/coronavirus. Accessed 14 Dec 2020

Izz al-Din A (1990) Crisis management in the terrorist event. Naïf Arab University for Security Sciences, Riyadh

Jimenez D-Z et al (2018) Unlocking the economic impact of digital transformation in Asia Pacific, IDC Singapore. https://news.microsoft.com/wp-content/uploads/prod/sites/43/2018/11/Unlocking-the-economic-impact-ofdigital-transformation.pdf. Accessed 14 Dec 2020

Karr J, Loh K, Wirjo A (2020) Supporting MSMEs' digitalization Amid COVID-19. APEC (Asia-pacific Economic Cooperation) Policy Support Unit Policy Brief No. 35. https://www.apec.org/-/media/APEC/Publications/2020/7/Supporting-MSMEs-Digitalization-Amid-COVID-19/220_PSU_Supporting-MSMEs-Digitalization-Amid-COVID-19.pdf. Accessed 30 Nov

Maher A (2006) General rules for dealing with crises (crisis management). Alexandria University House, Alexandria

Making India MSMEs Globally Competitive (2019) Ministry of Micro, Small & Medium Enterprises & Confederation of Indian Industry (CII), Govt. of India. https://static.pib.gov.in/MSME/Making India MSMEs Globally Competitive. Accessed 11 Dec 2020

MSME (2018) Annual report 2017-18. Ministry of Micro, Small and Medium Enterprises. Govt. Of India. https://msme.gov.in/sites/ Annual Report (2017-18). Accessed 30 Nov 2020

MSME (2019) Annual report 2018-19. Ministry of Micro, Small and Medium Enterprises. Govt. Of India. https://msme.gov.in/sites/ Annual Report (2018-19). Accessed 30 Nov 2020

MSME (2021) Annual report 2020–21. Ministry of Micro, Small and Medium Enterprises. Govt. of India. https://msme.gov.in/sites/default/files/MSME-Ammual-Report-English%202020-21.pdf. Accessed 30 Nov 2020

CRISIL. MSMEs Face Existential Crisis, Revenue to Fall a Fifth (2020, June 6) https://www.crisil.com/en/home/newsroom/press-releases/2020/06/msmes-face-existential-crisis-revenue-to-fall-a-fifth.html. Accessed 14 Dec 2020

PM Launches Historic Support & Outreach Initiative for MSME Sector (2018) Press Information Bureau, Govt. Of India, Prime Minister's Office. https://pib.gov.in/Pressreleaseshare.aspx?PRID=1551771. Accessed 14 Dec,2020

Textiles & Apparel (2020) India Brand Equity Foundation (IBEF). https://www.ibef.org/download/Textiles-and-

The New Wave Indian MSME-An Action Agenda for Growth (2015) KPMG & CII. https://assets.kpmg/content/dam/kpmg/pdf/2016/03/The-new-wave-Indian-MSME.pdf. Accessed 11 Dec 2020

Vision 2020: Implication for MSMEs (2011) http://ficci.in/spdocument/20143/grant-thornton-ficci%20msme.pdf. Accessed 21 Dec 2020

Arindam Metia is presently working as an Assistant Professor in the department of Management in Raiganj University. He has pursued Masters in Commerce (Accounting) from University of Calcutta. He has received his Ph.D. degree from the University of North Bengal. He has total teaching experience of more than 12 years.

Chapter 24
Where Classical Ends, Keynes Proceeds: Arguments Under the Perspective of Reviving India from the Impact of COVID-19

Indrani Basu

Abstract India has been opting for social containment measures with poor health infrastructural facilities since the last week of March 2020 to protect its citizens from the pandemic COVID-19. This timely and strict measure may help the country lessen the spread of community infection until proper medicine and vaccine reach its hand. However, such steps have miserable impacts on the economic health of the country. A trade-off has arisen between the economic health and human health of the country. We can assure one of them at the cost of others. Nevertheless, one can apprehend that this transitory shock of lockdown immediately sets an alteration in the employment profile of the country. Keynes successfully solved the great controversy in 1930 by favoring the relevance of fiscal and monetary measures to stimulate aggregate demand that could be a significant weapon to mitigate the vast unemployment that classicists had failed to alleviate. This chapter aims to identify the changes of macroeconomic variables that have been occurring during the period of COVID-19 and to revisit Keynesian prescription to see how the economy will breathe life into by following Keynesian suggestions.

Keywords Macroeconomic variables · Unemployment · Demographic dividend · Interventionist policy · Poverty alleviation program · Migration crisis

24.1 Introduction

The outbreak of a new coronavirus disease, known as COVID-19, has started its journey from Wuhan in the People's Republic of China (PRC) since January 2020. Within a short span, it has been spread to several regions of our earth. Significantly, American, European, and Asian countries were severely affected by the outbreak of

I. Basu (✉)
Berhampore College, Berhampore, Murshidabad, West Bengal, India

S. Saha et al. (eds.), *Economic and Societal Transformation in Pandemic-Trapped India*, New Frontiers in Regional Science: Asian Perspectives 55,
https://doi.org/10.1007/978-981-16-5755-9_24

Table 24.1 Fatality rate and infection rate of COVID-19 and other epidemics

Disease	Fatality rate (death/case) (%)	Infection rate (per infected person)
Ebola	50	1.5–2.5
MERC	34.30	0.42–0.92
SARS	10	3
COVID-19	1–3.4	1.5–3.5
Seasonal flu	0.05	1.3

Source: World Health Organization (2020)

COVID-19. However, the mortality rate of COVID-19 is relatively lesser than of its predecessors like Ebola, severe acute respiratory syndrome (SARS) outbreak in 2003, the Middle East respiratory syndrome (MERS) outbreak in 2012, etc. However, it has outweighed its predecessors in spreading the infection. From Table 24.1, we can observe that its infection rate is higher than the seasonal flu with whom we always stay, and substantially the mortality rate of COVID-19 is much higher than that of seasonal flu, which is less than 0.1%. So, the effects of COVID-19 are definitely much higher than the regular impact of seasonal flu. Additionally, even though it emerged from animal hosts, it now spreads through human-to-human contact and continuously mutating itself.

Affected developed countries are rushing to protect their citizens from this virus by utilizing their existing developed health care system. At the same time, developing countries are shivering by thinking about how they can protect their poor citizens and poor economy from this fatality where prominent tycoon of Worlds become befuddled? WHO has anticipated in its situation report 43, published on 3rd March 2020, that COVID-19 outbreak will have severe global, regional, and economic, and sector-specific adverse economic impacts. Irrespective of the economic strength of the countries, the immediate impact will be on the level of employment and level of output.

A country like India with poor health infrastructural facilities that means inadequate health centers, non-availability of respiratory support system which is a prerequisite to get cure from infection of COVID-19, insufficient number of intensive care units, less number of health workers has opted social containment measures like lockdown within the entire economy from 24th March 2020. This timely and tuff measure may help the country lessen the spread of community infection until proper medicine and vaccine reach its hand. However, such a step has a miserable impact on the economic health of the country. So there arises a trade-off between the economic health and human health of the country. We can assure one of them at the cost of others. It is expected that the majority of the real and nominal sectors have been affected but which cluster of the economy will be affected most; it could not be predicted accurately at this initial stage. In many instances, epidemic causes short-term costs, while the long-term costs are unclear. However, one can apprehend that this transitory shock of lockdown immediately set alteration in the employment profile of the country as the working population of the country being trapped within their working place without work.

Union Government is making a rough estimate that there are at present 80 million internal migrant workers, or more than 50% of the internal migrant workers in India, affected by the crisis (Dasgupta 2020). Being jobless, lacking food, and shelter, a vulnerable section of the working population, i.e., migrant workers, traveled more than 1000 km in 40 °C temperature to their native land. The miserable condition of unemployed people could easily be realized if we recall the great words of Alfred Marshall (1922):

> Those whose livelihood is secure gain physical and mental health from happy and well-spent holidays. But want of work, with long-continued anxiety, consumes a man's best strength without any return. His wife becomes thin, and his children get, as it were, a nasty notch in their lives, which is perhaps never outgrown.

This lays a profound impact on their productivity. All such matters restrain the economy from resurging. Unemployment is such a macroeconomic variable, the solution of which constitutes forever debates among economists and policymakers over the causes and the proper policy response. History told us about the experiences of the world at the early phase of the 1900s when the Great Depression sank the entire world into distress. The unemployment rate rose from 3.2% of the labor force in 1929 to 25.2% in 1933 (Froyen 2013). The debate was raised on the policy measures to curb unemployment. The British Economist John Maynard came with his book "*The General Theory of Employment, Interest, and Money.*" During the course, he developed his revolutionary macroeconomics theory and recommended interventionist policy conclusions that did not appear before Keynesian revolutionary policy prescription.

India has enjoyed a distinct advantage in the labor market compared to developed and less developed countries due to the fast-changing age distribution of the population. After four decades since independence, when the population growth rate was maintained between 2.1 and 2.2% per annum, it has come down to 1.57% only as per the Registrar General (2008). Consequently, the percentage of the population in the age group 15–59 years is likely to go up in the coming decades. The country would, thus, enjoy significant demographic dividends during the next few decades, unavailable to most other countries. The country's vision should be maintaining its growth performance for the next few years—at least not slowing down dramatically and emerging as an economic power gets further support from this process of demographic transition (Kundu and Sarangi 2007). Migration for employment from rural to urban areas emerges as an effective tool of poverty alleviation. Migrants fuel the Indian economy by carrying human capital to regions where needed and enabling the acquisition of new skills and a better standard of living (Korra 2011).

These unemployment and poverty alleviation processes have got struck when we have opted for social containment measures to protect our citizens. We seem to move towards condensing again. Economists are prescribing that governments should intervene into the market with expansionary monetary and fiscal measures so that economy will regain its capacity of enjoying demographic dividend which other countries do not have.

Here we need again to pursue the philosophy of Keynes (1883–1946). The world has to get shelter frequently to this personality who first favored fiscal policy and monetary measures, especially primarily government spending on public works projects, to stimulate demand. To understand the implication of this revolutionary nature of this theory at present state in the Indian economy, we consider the state of unemployment that occurs due to the adaptation of social containment measures to protect the country's physical health. This chapter will attempt to explore the alterations in macroeconomic variables that have already taken place or going to take place within the Indian economy due to COVID-19. It also attempts to revive the Indian economy following Keynesian prescription. A brief conclusion will be made to evaluate the need to follow Keynesian prescription by the Government of India to mitigate this transitional shock arising out of the outbreak of COVID-19. This paper proceeds with four sections. Section 24.2 discusses the relevance of the Keynesian approach for reviving the economy. Section 24.3 presents India's recent experience at the backdrop of the outbreak of COVID and probable effect on the significant macroeconomic variable, which is followed by Sect. 24.4 suggesting necessary steps keeping Keynesian prescription, those should be followed to outweigh the effect of COVID-19 in the course of reviving Indian economy. Finally, Sect. 24.5 concludes the paper with the hope that India will win the battle.

24.2 Why Keynes Matter All-Times

The turn of the World Depression 1930s has popularized the term "*Macroeconomics*." The forces that determine real and nominal variables like income, employment, prices, rate of interest had been receiving greater attention since then. Since then it has become important to study and solve the macro-economic questions. The products of this research originated the uses of policy prescriptions for stabilizing economic activities. The book containing this theory was "*The General Theory of Employment, Interest, and Money* by John Maynard Keynes, and the process of change in economic thinking that resulted from this work has been called the *Keynesian revolution*. But revolution against what? *Keynes termed it "classical economics*" (Froyen 2013).

By applying the policies prescribed by John Maynard Keynes, the world had successfully survived from the major meltdown that persisted 100 years ago. Keynes successfully solved the great controversy that emerged during the Great Depression in 1930 by favoring the relevance of fiscal and monetary measures as well as negating the separation between real and monetary sectors to stimulate aggregate demand that could be the major weapon to mitigate the vast of unemployment that the classical system could not explain and for which classical economics provided no remedy.

This was not the end. Again in 1950–1960, when the USA sank into a severe recession, the unemployment rate rose to 6.8%. Again, Keynesian stabilization measures successfully ensured rapid growth in output and employment and arrested unemployment 23.8% in 1966 via fiscal expansion. In the recent past, during the deep recession of 2007–2009, the monetary policy failed to carry out the whole burden, and the government again took shelter under the fiscal initiative. The American Recovery and Reinvestment Act (ARRA) in February 2009 incorporates several fiscal measures like aid to the states, expanded unemployment benefits, the funding of construction and other public works projects, the tax cut for individuals and businesses, etc. The ARRA provided another potential test of Keynesian fiscal stabilization's effectiveness, and this fiscal initiative successfully raised GDP from 1.1 to 3.5% during the fourth quarter of 2010 and created between 1.8 and 3.5 million jobs by the end of 2010 (Froyen 2013).

24.3 India's Recent Experiences and Effect of the Outbreak of COVID-19 on Major Macroeconomic Indicators

The structure of the Indian economy does not match with those of developed tycoons. Its occupational structure differs from those of developed countries. It has a flawed health care system that strongly influences the age composition of its demographic profile. Now when India is forced to receive the external shock via outbreak COVID-19 in the early days of March 2020, it seems to clear that the economic health of the country would be dismantled as the Indian economy was already under an extended period of stagnation before the outbreak of the COVID-19 pandemic. Before the outbreak, India's economic performance in its GDP growth rate was unstable. It is an essential indicator of the economic potency of a country, as GDP refers to the total market value of all goods and services produced within a country per year (Statista 2020). Table 24.2 represents the growth of the real gross domestic product (GDP) in India from 2010 to 2019; one may observe that the growth rate has been immensely fluctuating over the last 10 years. In 2019, India's

Table 24.2 Real gross domestic product (GDP) growth rate of India from 2009 to 2021

Year	GDP growth rate (%)
2010	10.26
2011	6.64
2012	5.46
2013	6.39
2014	7.41
2015	8.00
2016	8.26
2017	7.04
2018	6.12
2019	4.23
2020	1.87
2021	7.43

Source: Statista (2020)

real gross domestic product growth was about 4.23% compared to the previous year. However, due to the pandemic outbreak, the value of the real GDP growth rate is estimated at around 1.87%.

However, we still hope for a resurgence of the Indian economy in the coming year. How could it be possible for such an economy to sustain its economic health from stagnation? How its significant macroeconomic variables reciprocate during its tuff situation? Firstly, we need to look at the effect of the outbreak of COVID-19 on India's demographic profile.

24.3.1 COVID-19 and Demographic Profile of India

The demographic age profile of an economy illustrates the profile of supply of human capital, i.e., labor. The age structure indicates the extent to which a country's population is productive from the economic point of view. Population in the age group of 15–60 years is known as the working population or non-dependent population, where the population in the age group of 0–14 years and above 60 years is the non-working or dependent population. Therefore, for a country with a higher dependency ratio (the population below and above working age divided by the working-age population), lesser will be productivity growth. A country having a higher proportion of the working population refers to an opportunity before it to boost economic growth. Demographic structural changes can be expected to affect the real income, the real rate of interest, and rate of inflation as they have different consumption and saving behavior, different productivities, contribute differently to the innovation process in the long and short term, either directly or via some indirect effect on expectations on the future course of key variables.

Gordon (2012, 2014) and Fernald and Jones (2014) observed that changes in demographic age profile in favor of the aging population slowdown macroeconomic performance in advanced countries. This is because the aging population has a more substantial income effect that outweighs the substitution effect in labor-leisure choice, and the labor supply curve becomes backward bending. So they furnish less labor in the labor market, and as a result amount of real income in the product market will be reduced. Following the economic model, we can treat alteration of the structure of demography in favor of the aging population as an exogenous factor that determines output and employment by altering the positions of the labor supply and demand curves in the labor market and the position of the aggregate production function in the product market. This is classical thinking. Both dependent cohorts (both young and old) tend to impact all real macroeconomic variables, including real returns negatively, and add positive inflationary pressures in the long run (Basso et al. 2015). He found that the decrease in working-age population and fertility in most OECD countries and the increase in the proportion of retirees expected for the next 20 years would result in a strong decrease in trend output growth.

Table 24.3 The recent trend of age structure of India during 2014–2018

Year	0–14 years (%)	15–64 years (%)	Above 65 years (%)
2014	28.93	65.60	5.485
2015	28.44	65.94	5.61
2016	27.93	66.27	5.79
2017	27.48	66.54	5.98
2018	27.05	66.77	6.18

Source: Statista (2020)

Conversely, despite having few negative consequences like social unrest, crime, high divorce rates, etc., many East Asian countries could achieve high economic growth rates by utilizing their demographic dividend. However, demographic dividend can only be effective if it is accompanied by supportive national policies that improve literacy, provide employment, and extend health care support (Doran 2007).

If we consider India's age structure trend, one can observe a prolonged trend during 1881–2001 (Nangia and Kumar 2005). It is seen that the proportion of the productive age group remains more or less stable at 55%. After the share of the working population has been rising. Table 24.3 gives us the recent trend of the age structure of India during 2014–2018.

But, how this composition of age structure is relevant in determining the effect of COVID-19 on macroeconomic variables? Medical research has stated that while the new coronavirus can infect people of all ages, the aging population with pre-existing medical conditions (such as asthma, diabetes, heart disease) appears to be more vulnerable to becoming severely ill with the coronavirus and even dying from it (India today 2020). In addition, *co-morbid conditions* like cancer, diabetes, heart ailments, and respiratory difficulties, which are more dominant in older populations, make the problem more acute. A study done by the Chinese Center for Disease Control found that the case fatality rate was 22.8% for those 70 years of age and above. The case mortality rate was also elevated among those with pre-existing co-morbid conditions: 10.5% for cardiovascular disease, 7.3% for diabetes, 6.3% for chronic respiratory disease, 6.0% for hypertension, and 5.6% for cancer (Guan et al. 2020).

Italy has the second oldest population globally, with 23% of people aged 65 years or more (UNDP 2019). WHO Situation Report 116 (15 May 2020) stated that 223,096 people out of nearly 60 lakhs were infected there, and Italy became the second country, after the USA, in terms of mortality, with 26,977 deaths. A deeper analysis was made by the Italian National Institute of Health (Istituto Superiore di Sanità [ISS]) at the outset of the COVID-19 outbreak (de Leo and Trabucchi 2020). They found that the fatality rate, based on the data up to 17 March 2020, was 7.2%. However, considering the age-specific stratified sample of a subsample of 355 patients with COVID-19 revealed that among these patients, the mean age was 79.5 years. 117 patients, which was 30% of this observation, had ischemic heart disease, 35.5% had diabetes, 20.3% had active cancer, 24.5% had atrial fibrillation, 6.8% had dementia, and 9.6%had a history of stroke. Overall, only

0.8% had no diseases, 25.1% had a single disease, 25.6% had two diseases, and 48.5% had three or more underlying diseases. Medical experts have stated that these comorbidities might have raised the risk of mortality among Italians (Rezza et al. 2020).

The latest data from the Union Health Ministry India on COVID-19 deaths corroborates with that of China and Italy. The Union Health Ministry said that those above 60 years of age account for 60% of COVID-19 deaths in India. However, the age group analysis of COVID-19 patients in India showed that the maximum 42% are of 21–40 years, 33% of 41–60 years, 17% above 60 years, and 9% of 0–20 years (India Today 2020). India's position in the health care system is weak relative to other developing countries. Over the past decades, India's decaying healthcare system repeatedly has disgraced its people (D'Cunha 2017). According to the Global Burden of Disease study, published in *The Lancet* (02 June 2018), India ranked 154 out of 195 countries regarding healthcare access, far behind nations like Bangladesh, Nepal, Ghana, and Liberia(Lozano 2018). Despite improving its infant mortality rate from 57 per 1000 live births in 2005–2006 to 41 per 1000 live births in 2015–2016 (OECD 2015), it is still higher than 150 middle- and low-income countries, many poorer than India, including Colombia, Costa Rica, and Slovenia. The actual scenario is that, on average, one government doctor for every 10,189 people, one hospital bed for every 2046, and one government-run hospital are available for every 90,343 in India. Death incidents in government hospitals cases in Uttar Pradesh, Jharkhand in the recent past,[1] which are in tatters, are not rare occurrences (D'Cunha 2017). Government spending has been abysmally low (near about 4.7% of GDP) in recent years. As Jean Drèze speciulates:

> The Gorakhpur tragedy is just another example of the general lack of attention being given in India to basic needs, especially those of poor or marginalized people, (D'Cunha 2017)

From our above analysis, one may expect how India's age composition demographic structure gets affected due to the outbreak of COVID-19 having such a poor health care system. Table 24.4 presents the relative position of India among South-East Asian countries where it is situated and among such countries which are developed countries having a prominent share of aging people in their population regarding fatality and death ratio of COVID-19.

In Table 24.4, one can surprisingly observe that in comparison with neighboring and also with some developed countries, India with a vast population, the burden of high population density and poor health care system does better to protect its citizens from the fatality of COVID-19. Its fatality and death rate are well lesser than those of those developed countries that are placed in a much better position than India in the

[1] In the month of August 2017, in a hospital in Gorakhpur in the northern state of Uttar Pradesh, the largest province of India, more than 400 children died due to oxygen shortages (among the variety of reasons) after a private supplier cut the supply over unpaid bills. In another government-run hospital in the same province 49 children died allegedly due to oxygen and medicine shortages in the same month. In the state of Jharkhand also, more than 800 children died in two state-run hospitals in the same year (source: www.forbes.com).

Table 24.4 Relative position of India

Location	Name of the country	0–14 years	15–64 years	Above 65 years	Population	Population density (per km^2)	Confirmed COVID-19 infected case	Fatality ratio	Total death due to COVID-19	Death rate	Type of infection
(1)	(2)	(3)	(4)	(5)	(6)	(7)	(8)	(9)	(10)	(11)	(12)
Southeast Asia	India	27.8	66.2	6	1,362,227,886	417	81,970	0.000060	2649	0.032317	Cluster of case
Southeast Asia	Bangladesh	28.9	66.5	5.1	168,608,212	1171	18,863	0.000112	283	0.015003	Cluster of case
Southeast Asia	Indonesia	27.4	67.3	5.3	268,074,600	141	16,006	0.000060	1043	0.065163	Community transfer
Southeast Asia	Thailand				66,504,880	130	3025	0.000045	56	0.018512	Cluster of case
Southeast Asia	Maldives				374,775		982	0.002620	4	0.004073	Cluster of case
Southeast Asia	Sri Lanka	24	65.9	10.1	21,881,913	332	925	0.000042	9	0.00973	Cluster of case
Southeast Asia	Nepal	30.9	63.3	5.8	29,609,623	201	258	0.000009	0	0	Cluster of case
Southeast Asia	Myanmar	26	67.4	5.7	54,339,766	80	181	0.000003	6	0.033149	Cluster of case
Southeast Asia	Bhutan	26.6	68.6	4.9	826,200	22	20	0.000024	0	0	Cluster of case
Western Pacific	Australia	19	65.5	15.5	25,708,537	3	6989	0.000272	98	0.014022	Community transfer
EU	Austria	14.1	66.7	19.2	8,902,600	106	16,005	0.001798	626	0.039113	Community transfer

(continued)

Table 24.4 (continued)

Location	Name of the country	0–14 years	15–64 years	Above 65 years	Population	Population density (per km^2)	Confirmed COVID-19 infected case	Fatality ratio	Total death due to COVID-19	Death rate	Type of infection
(1)	(2)	(3)	(4)	(5)	(6)	(7)	(8)	(9)	(10)	(11)	(12)
EU	Canada	16	67	17	38,028	4	72,536	1.907437	5337	0.073577	Community transfer
Western Pacific	China	17.7	71.7	10.6	1,402,640,000	145	84,469	0.000060	4644	0.054979	Community transfer
EU	France	18.1	62.2	19.7	67,060,000	123	139,152	0.002075	27,378	0.196749	Community transfer
EU	Germany	13.1	65.5	21.5	83,149,300	233	173,152	0.002082	7824	0.045186	Community transfer
EU	Greece	14.2	65.4	20.4	10,724,599	81	2770	0.000258	156	0.056318	Community transfer
EU	Italy	13.5	63.5	23	60,252,824	200	223,096	0.003703	31,368	0.140603	Community transfer
Western Pacific	Japan	12.9	60.1	27	126,010,000	333	16,193	0.000129	710	0.043846	Community transfer
Western Pacific	New Zealand	19.8	64.9	15.3	4,985,605	18	1148	0.000230	21	0.018293	Community transfer
EU	Russia	17.6	68.2	14.2	146,877,088	9	262,843	0.001790	2418	0.009199	Community transfer
EU	Spain	14.7	65.9	19.4	46,934,632	93	229,540	0.004891	27,321	0.119025	Community transfer
EU	Sweden	17.5	62.5	19.9	10,382,805	23	28,582	0.002753	3529	0.123469	Community transfer

EU	Switzerland	14.9	66.7	18.4	8,586,550	208	30,380	0.003538	1588	0.052271	Community transfer
	UK	17.7	63.8	18.5	67,886,004	280	233,155	0.003435	33,614	0.14417	Community transfer
	USA	18.9	65.7	15.4	329,685,899	34	1,361,522	0.004130	82,119	0.060314	Community transfer

Source: WHO. Coronavirus disease 2019 (COVID-19) situation report—116, 15th May, 2020-05-15; https://www.en.wikiepedia.org (accessed on 16.05.2020)
Note: Column (2), column (3), and column (4) represent the age composition of the demographic profile. Column (2) represents the age group 0–14 years, i.e., children and adolescents; column(3) is for the age group of 15–64 years, i.e., working population; and column (4) is for the age group of >65 years, i.e., retirees and elderly people

human developed index (HDI) table. One reason behind this performance is its relatively lesser share of aging people who are more susceptible to get infected by COVID-19. Another reason for which India has received worldwide praise is for its quick and timely implementation of quarantine social measures at the earlier stage of the outbreak of COVID-19 throughout the country.

Our above study assures us that outbreak of COVOD-19 may not be to reduce our labor force in a mass. So at present, having almost 47.3 (The Periodic Labor Force Survey, 2018–2019) working population ratio, India can emerge as a growing economy. However, how far would it be possible if containment measures extend for future dates also? Demographic dividends cannot be enjoyed with having poor governance. Having migrated working population who became jobless, shelterless have started to go back to their native land to protect themselves from the severity of virus, hunger, and distress. The pandemic has caused a humanitarian crisis that threatens the life of a large mass of people in a country or in a region in terms of their health, safety, and well-being and like (Dasgupta 2020) Is it possible for such an economy that is already under economic meltdown to sustain its economic health from stagnation? How its significant macroeconomic variables reciprocate during its challenging situation? What measures should be opted for? Either self-adjusting free-market mechanism or government-led stabilization policy or a combination of both? What would be the next step?

24.3.2 *Effect of COVID on Labor Market and Level of Unemployment*

In the classical economics, level of output or real income is determined in the labor market. A central relationship in the classical model is the aggregate production function. The production function, which is based on the technology of individual firms, is a relationship between the level of output and the level of factor inputs, i.e., especially the level of employment. The hallmark of classical labor market analysis is the assumption that the market works well. Firms and individual workers optimize. They have perfect information about relevant prices. There are no barriers to the adjustment of money wages; the market *clears*. The labor demand curve is downward sloping due to the law of diminishing returns. The higher is the real wage, for example, the lower is the level of labor input that will equate the real wage to the marginal productivity of labor. Conversely, regarding the supply of labor, as the *real wage increases* (or decreases), leisure decreases (or increases), and *hours of work increase* (or decrease). *Demand for labor* (Nd) and *supply of labor* (Ns) determine equilibrium conditions for the labor market, i.e., "$Ns = Nd$." This condition determines output, employment, and the real wage (Fig. 24.1).

Before the outbreak of COVID-19, the economy was at a low employment level N_1 (as reported in Table 24.2), which further fell to N_2 due to the strict implementation of containment measures. Persisted lockdown shifts both the curves in the

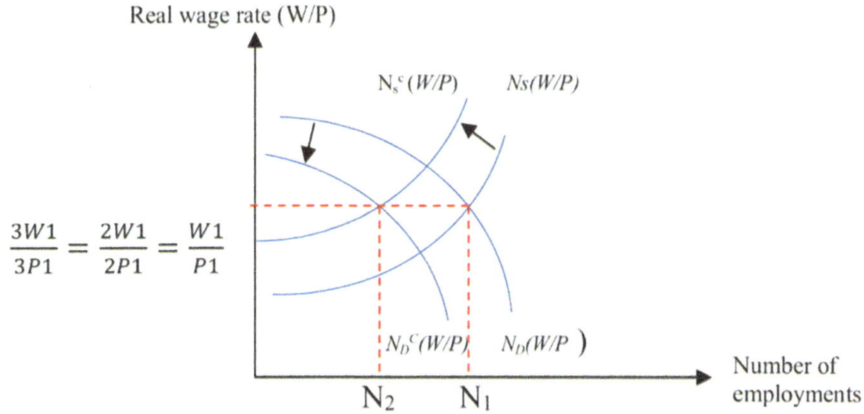

Fig. 24.1 Equilibrium in labor market: $N_s^c(W/P)$ is the labor supply curve under containment and $N_D^C(W/P)$ is the labor demand curve under containment. (Source: Author's perception)

downward direction. Initially, at the first round lockdown, the workers seem it was a transitory shock. Some kind of cash and food aid might protect the labor from hunger and retain them within the adjacent area of their working place. However, as the lockdown has been extended for a longer time, the labor market started to fall, making the situation more vulnerable. Those vulnerabilities, keeping the classical prescription, stress the level of employment, which determines the level of production. Thus, this persisted lockdown really reaches the economy to distress situation with the low level of production and rising level of unemployment with fixed real wage rate.

It need not end here. To take a broader view, we have analyzed the sectoral decomposition of GDP and occupation profile because as sector-wise relevancies vary, magnitudes of adverse impact ought to be different. To earn a livelihood, people pursue different activities based on their education, skill, and family traditions. We usually classify them into three different sectors of the economy, such as *primary sector*, *secondary sector*, and *tertiary sector*. Here at first, we consider the relative contribution of the sector to real GDP over the last few years as it gives us some additional information about the economic health of our economy (Table 24.5).

The sectoral decomposition of GDP in the Indian economy asserts that the lion-share (i.e., almost half) of India's GDP has been coming from the service sector over the last 10 years. Among the leading services industries in the country are telecommunications, IT, and software. The agriculture sector in India is still a global power, producing more wheat or tea than anyone in the world except for China (Plecher 2020a, b).

The next important area in which the performance of the Indian economy should be judged closely is the sectoral decomposition of the labor market that is prevailing

Table 24.5 Distribution of gross domestic product (GDP) across economic sectors from 2008 to 2018

Year	Agricultural sector	Industrial sector	Service sector
2010	17.52	30.8	45.18
2011	17.19	30.16	45.44
2012	16.85	29.4	46.3
2013	17.15	28.4	46.7
2014	16.79	27.66	47.82
2015	16.17	27.35	47.18
2016	16.25	26.64	47.82
2017	15.62	26.5	48.45
2018	14.6	26.75	49.13

Source: Statista (2020)

within the economy. Table 24.6 presents the occupational structure of the Indian economy over the last few years (Table 24.6).

While the service sector generates the lion-share of India's real GDP, agriculture still plays a pivotal role in generating employment. The majority of Indians still do not live in cities where IT jobs are generated. Even workers who work in the service and industrial sectors are generally migrant workers from rural India. Despite having underreported nature of the estimation of migrating workers of different occupations due to inefficiency in capturing the actual extent of short-term circular migration, there have been definitive trends of an increasing number of internal migrants moving from rural to urban areas for work between the decadal census findings of 2001 and 2011. Studies using railway data found that between 2011 and 2016, an average of 9 million workers undertook interstate travel (Table 24.7).

From Table 24.7, it is clear that the share of migrant workers is so prevalent that the economy could not sustain neglecting them. The majority of the migrant workers reside in urban sectors and work in the secondary and service sectors. Though it is argued that migration for employment from rural to urban areas emerges as a major tool of poverty alleviation (Kundu and Sarangi 2007), but people are migrated for joining the labor market from their local place out of the two reasons: *search for better opportunities* and *lack of local options*. Through the process of *pull migration,* a group of migrants left the local place to seek better opportunities. They are stimulated growth process and engaged in the non-agricultural jobs in specific cities and areas. They usually diversified their source of income and risk across the farm and non-farm income, making them more resilient to sectoral shocks. However, there is also the *push migration*, i.e., people leaving due to lack of local options. A large part of them has been absorbed in various low-paid services. Among this group's share of women, labor is very high. Formal industries and businesses owe their growth and profitability partly to employing such workers on an informal basis without being covered under the social security system (GOI 2017).

Under the containment measures, conditions of such push migrants become more vulnerable in the urban labor market. Without an alternative income source, they are

Table 24.6 Distribution of the workforce across economic sectors in India from 2009 to 2019

Sector	2009	2010	2011	2012	2013	2014	2015	2016	2017	2018	2019
Agricultural sector	51.12	51.06	48.96	47	46.6	46.07	45.56	45.12	44.52	43.86	43.21
Industry sector	21.61	22.38	23.52	24.36	24.36	24.38	24.34	24.29	24.47	24.69	24.89
Service sector	26.27	26.57	27.52	28.64	29.04	29.55	30.1	30.59	31.01	31.45	31.9

Source: Statista (2020)

Table 24.7 Share of migrant workers in total workers by major sectors

Sector	Rural (%)	Urban (%)
Primary	79	85
Manufacturing	72	89
Public services	85	96
Construction	81	99
Traditional services	75	84
Modern services	82	92
Total	79	89

Source: Based on NSS 2007–2008 as cited in GOI (2017)* Using the National Industrial Classification codes of 2004 (NIC) primary includes agriculture, hunting, forestry, fishing, mining, and quarrying (NIC 01–14), manufacturing is NIC 15–37, public services are NIC 40–41, transport via railways (NIC 6010), national postal activities (NIC 64110), and public administration (NIC 751, 752, and 753), construction is NIC 45, traditional services include wholesale and retail trade, hotels and restaurants, transport, storage and communications (NIC 50–52, 55,60–64, except 6010 and 64,110), and modern services include financial intermediation, real estate, renting and business, education, health, social work, other community, social and personal services (NIC65–74, 80, 85, 90–99, excluding 751, 752, 753)

compelled to return to the local region, which they had left due to the lack of options (GOI 2017). They are not the lone sufferer, however, the receiving end of this story of distress. Understanding how these forced returns will affect the labor market needs a sector-wise analysis of the labor market.

24.3.3 Effect of COVID-19 on Sector-wise Occupational Structure

If we decompose our entire labor market in the agrarian-based rural sector (say ARS herein forth) and industrial and service-based urban sector (say IUS herein forth), we can explore the impact of COVID-19 on real variables like level of employment, real wage rate, amount of production, and many more. At the initial stage, when the economy restarts, we have an excess supply of labor in the ARS, where the IUS will suffer from a lack of labor supply. The IUS will demand labor but would fail to collect as the labor (mainly migrant laborers) had left the place. How the macroeconomic variables reciprocate? We represent two figures (Fig. 24.2a, b). Figure 24.2a will state the condition of the ARS, and Fig. 24.2b will be that of the IUS.

From the above figures, it is clear that there will be downward pressure on the real wage rate {as $(W/P)_0$ ARS falls to $(W/P)^C$ ARS} in the ARS and upward pressure on real wage rate in the IUS {$(W/P)_0$ IUS rise to $(W/P)^C$ IUS}. Due to the containment measures that create an obstacle to the smooth mobility of migrant workers from rural to urban, this wage differential will persist among the sectors. Despite higher

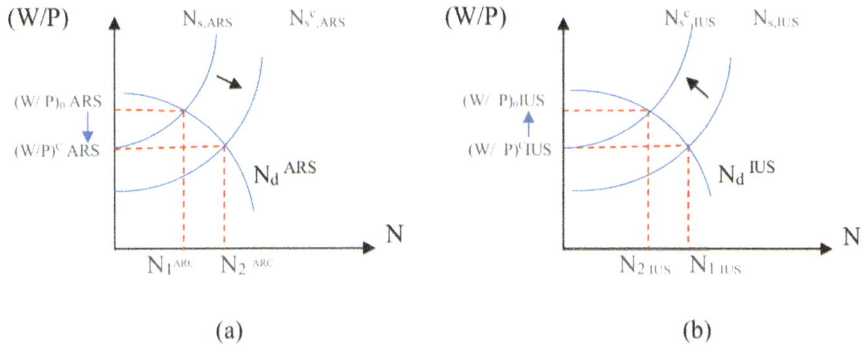

Fig. 24.2 (**a**) Determination of equilibrium real wage rate and level of employment in agrarian-based rural sector labor market; (**b**) Determination of equilibrium real wage rate and level of employment in industrial and service-based urban sector labor market. (Source: Author's perception)

wages in the industrial and service sector, labor may not be willing to return to their workplace due to their anxiety of loss of life and painful experiences that they gathered at their previous workplace. Production in industry and service sectors will be curtailed due to the shortage of labor supply and the agriculture sector becoming overburdened. However, as the agricultural sector is already experiencing the law of diminishing returns, additional gathering in this sector may reduce the volume of production. Finally, real GDP will fall.

24.4 Keynesian Prescriptions for Reviving the Economy

What economy does? The supply-driven mechanism is unable to exterminate these shortfalls. It can explain the situation, but supply-driven automatic mechanisms fail to solve these problems. Though high wages in the non-agricultural sector may beckon the labor to be migrated to this sector again but will not get success unless and until safety health measures appear within the economy and laborers are able to get out themselves from recent painful memories of the death of fellow laborers, experiences of anxiety and uncertainty, emotional setback, fear of to be infected from this urban disease(!) those they experienced in recent past at their working place. For the time being, laborers are unwilling to return. So lately, no self-adjustment policy will have appeared. A meager amount of cash transfer and food aid may not uproot the vulnerability of these poor migrant workers, especially those who pushed migrants. We should remember that the Indian economy was already under an extended period of stagnation before the outbreak of the COVID-19 pandemic. Lack of consumption demand has already prevailed. The slump in consumption was most evident in rural India in the months leading up to the pandemic, showing no signs of recovery. This demand crisis has been attributed prominently to the fall in days of employment available to manual workers along with a slowing down of the wage rates. The classical solution that recommends *noninterventionist* policy conclusions, such as supply-determined nature of output and employment, would not gear up employment and output for the time being.

The economy should revisit to Keynesian prescription for its revival. Fall in production would be geared up by stimulating aggregate demand that consists of consumption expenditure, investment expenditure, government expenditure, and net export. But at that time, relying on export-led growth does little as we are shutting down foreign trade in almost all goods and services. The foreign transaction may raise more chances to face adverse external shock. Among the three components, the nature of investment expenditure is quite unstable. One of the significant determinants of investment is the expected yield on investment projects. Business managers' expectations about the future profitability of investment projects are the central element in Keynes's analysis. Keynes emphasized the "uncertain knowledge" which is required to predict the profitability of a project that will produce output over 20 or 30 years. So a manager should have a great deal of knowledge about the future consumer tastes and the state of aggregate demand, knowledge about future costs, including money wages, interest rates, tax rates, etc. So the lack of government stabilizes policies income will be unstable because of the instability of investment. Due to this uncertain nature of the investment, we could not rely first on this wing. What next?

Following the prescription of Keynes, the government, through its fiscal and monetary measures, enhances consumers' purchasing capacities. An additional economic package should be flown to those low-income strata that enhance the purchasing power in plentiful amounts and reduce inequality for having a higher marginal propensity to consume. Implementation of several social security measures (like cash transfer, food aid via PDS, health package, etc.) helps reduce inequality.

Fiscal policy must focus on the domestic sphere and yoke the country's immense potential of the working population for the resurgence of the economy. More initiatives should be taken to enhance agricultural yield; more avenues have to be opened to get agricultural credit at a cheaper rate of interest and more straightforward terms and conditions (some loan exemption may attract farmers to take the loan for investment in farm activities). It is urgent to create suitable infrastructure for the marketability of agricultural products; to enhance the capacity of storage of farm products; to ensure more intensive farming that turns agriculture into profitable employment opportunities that could protect the pull-migrants who already have sources of farm income.

We must recognize the role of employment guarantee schemes such as Mahatma Gandhi National Rural Employment Guarantee Scheme (MGNREGS),[2] rural

---

[2]**Mahatma Gandhi National Rural Employment Guarantee Act (MGNREGA)**, also known as Mahatma Gandhi National Rural Employment Guarantee Scheme (MNREGS) is Indian legislation enacted on August 25, 2005. The MGNREGA provides a legal guarantee for 100 days of employment in every financial year to adult members of any rural household willing to do public work-related unskilled manual work at the statutory minimum wage with an aim of improving the

housing schemes such as Prime Minister Awas Yojana-Rural (PMAY-Rural)[3] in mitigating pressures for push migration. Government should generate alternative employment opportunities, especially in the rural sector, that help absorb those laborers (i.e., push migrants, who are now trying to be crowded on the agricultural sector despite its nature of diminishing returns) and make them into skilled labor through some skill generation policies. The outcome would be that these poor can arrange basic sustenance for the time being when they are not willing to go out of their local place as they previously did due to a lack of opportunities. To solve the emerging wage differential problem, the government should enhance the payment structure and create more man-days for participants under such government projects.

A large share of migrant workers belonged to self-employed categories, and they are unaware and unskilled for generating agricultural production. However, if they get money, then they could start up their new venture. Through its monetary measures relaxing the terms and conditions of getting the institutional loan and disbursed additional funds to its disbursing authorities so that they extend more liquidity to this stratum of the lower-income group, this will stimulate future investment. Along with this, small entrepreneurs should be encouraged by providing a stimulus package.

Government should centrally opt for cash transfer and food aid programs on a large scale for at least a sixth month. Though the central government has already adopted such measures, they should keep in mind that people are now trying to hold money as they are out of jobs and the future is uncertain. This repercussion in the money market causes a rise in demand for money that will negatively impact investment expenditure in the product market. To overcome such a crowding-out effect, the government should inject more liquidity by enhancing the supply of real balances by such an amount that instigates the people to spend more into the circular flow of income. Such monetary expansion reduces the rate of interest that will further encourage the private investor. The monetary authority has announced a lower repo rate. However, there is no such actual demand for credit as the future is uncertain. Thus expansionary monetary policy should be combined with a sizeable fiscal stimulus for generating employment and raise productivity.

Reviving the package needs to incorporate some non-economic factors which have indirectly damaged the potentiality of the laborer. These non-economic factors (e.g., the emotional stress; anxiety; painful experiences in returning their home; uncertainty about the availability of alternative occupation, ambiguity about cop-up with the recent skill that will be required for alternative jobs; unaccustomed

---

purchasing power of the rural people, primarily semi- or unskilled work to people living below poverty line in rural India (source: https://vikaspedia.in/agriculture/policies-and-schemes/rural-employment-related-1/mgnrega/rural-employment-related).

[3]The **Pradhan Mantri Awas Yojana-Gramin** (**PMAY**-G) has been devised in line with government's commitment to provide "Housing for All" by 2022 in the **rural** areas. The scheme aims at providing a pucca house with basic amenities to all houseless householder living in kutcha and dilapidated houses by 2022 (source: https://transformingindia.mygov.in/scheme/pradhan-mantri-awas-yojana-gramin/).

experiences with those safety measures those are required to be protected from COVID-19; lack of symmetrical information; prevailing misconceptions about the vulnerability of pandemic like COVID-19 is an urban disease) not only arrest them within their local village, but it hinders their potentiality. All such matters restrain the economy from resurgence. So government should boost up its investment in social infrastructures like health, education, vocational training, and information technology which have a robust forward linkage for human capital development.

Based on the Union Budget 2019S, Singh (2020) documents the Government of India's budgeted spend on healthcare plus education is around Rs 150,000 core or $22 billion, i.e., less than 1% of GDP. COVID-19 has bared our performance of the health care system. We do not have the option but to restrict our citizens within the home to protect them from COVID-19. Time has come to invest a bulk amount of resources in the upgradation of the health care system. The Indian health sector has a *mix* of both public and private providers of health services. States like Kerala, Andhra Pradesh, Goa are reaping the benefit in the current time from their previous investment in the health sector. However, few states have taken such initiatives. All the state governments should build up the small pocket-health center at each tiny electoral zone under the shade of the SHG scheme both in rural and urban areas.[4] It may train and involve local educated unemployed youth to such an extent that they become able to provide some basic health services like measuring blood-pressure, stitch a small wound, collect the blood sample, sputum, urine for pathological test, aware the local people about a balanced diet, provide information of public health policies. Such steps uphold the health care system within society and generate skills and employment with higher potential capacities.

The government should build up technical know-how hubs in the rural sector that generate technical skills among the laborers who are now staying in rural areas and engaging in farm or non-farm activities. Such knowledge on information technology would help them continue their working activities with maintaining social distance measures and generate additional skills to join the IT industry through the knowledge on software management, online services, business process management (BPM) after firing of lockdown.

24.5 Conclusion

Therefore, we need a comprehensive interventionist strategy that includes both fiscal and monetary measures to protect the working population from hunger and vulnerability during this transition period and strengthen effective demand. All such steps

[4]**Indian SHGs** consist of 5–15 members. Usually, participants are women of similar social and economic backgrounds. These groups convene on regular basis, at which time members make routine deposits into a joint savings account. As savings accumulate, **SHG** members can borrow from the poor (source: https://aif.org/self-help-groups-an-overview/).

mentioned above focus on reviving demand following Keynesian prescription. The government has already announced an economic package aimed at expanding liquidity within the economy. However, its efficacy will depend on the government's potentiality to implement such projects and, of course, on its will. Controversy may arise that this enormous amount of government expenses in the context of the meager growth rate of GDP will generate a larger fiscal deficit. However, such fiscal deficit could be wiped out if the government makes more investment in social infrastructure. These have both backward and forward linkages. At present, when the economy is not functioning well, such a rise in government expenditure can be turned to have a "ripple effect" of a stone dropped in a pond. There is an initial effect as the stone disturbs the water. This initial disturbance would affect the rest of the surface as the water displaced by the stone spreads out to the adjoining water, with the intensity that diminishes with the distance from the initial point of impact. In the present context, the government should create stimulus packages that will initially create waves at the primary sector where a larger population has crowded. Eventually, waves spread out to adjoining the urban industrial sector. The resurgence of this sector with additional opportunities beckons the worker in the primary sector to come back to their previous working place, and the economy could succeed to resurge.

References

Basso H, Aksoy Y, Grasl T, Smith R (2015) Demographic structure and macroeconomic trends. Birkbeck working papers in economics and finance no. 1501

D'Cunha S (2017) Despite a booming economy, India's public health system is still failing its poor. www.forbes.com. Accessed Sept 2017

Dasgupta B (2020) Corona pandemic and migrant workers' crisis. http://www.macroscan.org/fet/jul20/fet06072020Migrant_Workers.htm. Accessed 6 Jul 2020

de Leo D, Trabucchi M (2020) COVID-19 and the fears of Italian senior citizens. Int J Environ Res Public Health 17(10):3572. https://doi.org/10.3390/ijerph17103572

Doran J (2007) The effects of age structure on economic growth: an application of probabilistic forecasting to India. Int J Forecast 23(4):587–602. https://doi.org/10.1016/j.ijforecast.2007.08.001

Fernald JG, Jones CI (2014) The future of U.S. economic growth NBER working paper no. 19830

Froyen RT (2013) Macroeconomics—theories and policies, 10th edn. Global edition University of North Carolina—Chapel Hill, Pearson

GOI (2017) Report of the working group on migration. Ministry of Housing and Urban Poverty Alleviation, Government of India, New Delhi

Gordon RJ (2012) Is U.S. economic growth over? Faltering innovation confronts the six headwinds NBER working paper no. 18315

Gordon RJ (2014) The demise of U.S. economic growth: restatement, rebuttal, and reflections NBER working paper 19895

Guan WJ, Ni ZY, Hu Y et al (2020) Clinical characteristics of coronavirus disease 2019 in China. N Engl J Med 382(18):1708–1720. https://doi.org/10.1056/NEJMoa2002032

India Today (2020). https://www.indiatoday.in/india/story/63-of-coronavirus-deaths-in-india-in-60-age-group-health-ministry-1663951-2020-04-06. Accessed 6 Apr 2020

Korra V (2011) Short-duration migration in India. In: Irudaya Rajan S (ed) Migration, identity and conflict: India migration report 2011. Routledge, New Delhi, pp 52–71
Kundu A, Sarangi N (2007) Migration, employment status and poverty: an analysis across urban centres in India. Econ Polit Wkly 42(4):299–306
Lozano R (2018) Measuring performance on the Healthcare Access and Quality Index for 195 countries and territories and selected subnational location: a systematic analysis from the Global Burden of Disease study 2016. Lancet 391(10136):2236. https://doi.org/10.1016/S0140-6736(18)30994-2
Marshall A (1922) Money, credit and commerce. Macmillan, London, p 260
Nangia S, Kumar A (2005) Change in the age structure of India's population (1881–2001), dialogue January–March, 2005. 6(3)
OECD (2015) https://www.oecd.org/officialdocuments/publicdisplaydocumentpdf/?cote=ECO/WKP(2015)2&docLanguage=En. Accessed 29 June 2021
Plecher H (2020a) Distribution of the workforce across economic sectors in India 2019. https://www.statista.com. Accessed 15 May 2020
Plecher H (2020b) Distribution of gross domestic product (GDP) across economic sectors from 2008 to 2018. https://www.statista.com. Accessed 15 May 2020
Registrar General (2008) Sample registration system. SRS Bull 43(1). Registrar General, Government of India, New Delhi
Rezza G, Onder G, Brusaferro S (2020) Case-fatality rate and characteristics of patients dying in relation to COVID-19 in Italy. JAMA 323:1775–1776. https://doi.org/10.1001/jama.2020.4683
Singh R (2020) Indian economy. McGraw Hill Education (India) Private Limited
Statista (2020) Distribution of gross domestic product (GDP) across economic sectors from 2008 to 2018. https://www.statista.com/statistics/263771/gross-domestic-product-gdp-in-india/. Accessed 19 May 2020
UNDP (2019) World population prospects 2019. https://www.prb.org/resources/countries-with-the-oldest-populations-in-the-world/. Accessed 29 June 2021
World Health Organization (WHO) (2020) Coronavirus disease (COVID-19) situation reports 43. https://www.who.int/emergencies/diseases/novel-coronavirus-2019/situation-reportsWHO. Accessed 5 Mar 2020

Indrani Basu is an Assistant Professor in the Department of Economics at Berhampore College, under the University of Kalyani, India. She has 18 years of experience in the teaching and research playing field. She has completed her post-graduate studies in economics at the University of Kalyani and was awarded M.Phil. and Ph.D. degree by the University of Kalyani. She has published several research papers in peer-reviewed journals and books, especially on women studies. Her areas of interest are to do something for the women, who, despite serving the world a lot, get very meager recognition from society. She acted as an "external monitoring officer" in a project seeking the evaluation of the "Pulse-Polio Immunization program" financed by the Government of India and WHO during 2003–2006 in West Bengal. She likes to compassionate services to others by fostering social justice and build strong human relationships.

The manufacturer's authorised representative in the EU is Springer Nature Customer Service Centre GmbH, Europaplatz 3, 69115 Heidelberg, Germany. If you have any concerns regarding our products, please contact ProductSafety@springernature.com

Printed and bound by CPI Group (UK) Ltd, Croydon, CR0 4YY
30/09/2026
02225332-0016